Progress in Optical Science and Photonics

Volume 30

Series Editors

Indexed by Scopus

The purpose of the series Progress in Optical Science and Photonics is to provide a forum to disseminate the latest research findings in various areas of Optics and its applications. The intended audience are physicists, electrical and electronic engineers, applied mathematicians, biomedical engineers, and advanced graduate students.

Vijay Kumar · Irfan Ayoub ·
Yogendra Kumar Mishra · Hendrik C. Swart
Editors

Persistent Luminescence

Fundamentals, Mechanisms and Applications

Springer

Editors
Vijay Kumar
Department of Physics
National Institute of Technology, Srinagar
Srinagar, Jammu and Kashmir, India

Irfan Ayoub
Department of Physics
National Institute of Technology, Srinagar
Srinagar, Jammu and Kashmir, India

Yogendra Kumar Mishra
Mads Clausen Institute
University of Southern Denmark
Sønderborg, Denmark

Hendrik C. Swart
Department of Physics
University of the Free State
Bloemfontein, South Africa

ISSN 2363-5096 ISSN 2363-510X (electronic)
Progress in Optical Science and Photonics
ISBN 978-981-97-4942-3 ISBN 978-981-97-4943-0 (eBook)
https://doi.org/10.1007/978-981-97-4943-0

This Springer imprint is published by the registered company Springer Nature Singapore Pte Ltd.
The registered company address is: 152 Beach Road, #21-01/04 Gateway East, Singapore 189721, Singapore

Preface

We are delighted to present this edited volume on the exciting topic of persistent luminescence: fundamentals, mechanisms and applications. This book combines in-depth contributions from well-known experts in the field, providing a broad overview of innovative advancements, research findings and emerging trends in persistent luminescence. Although the concept of luminescence has been explored for many decades, the different types of luminescence have always led to versatile breakthroughs in various fields of science and technology. Among the same is one more well-established phenomenon known as "persistent luminescence," which is characterized by the maintenance of luminescence for periods ranging from minutes to hours without an external excitation source. Because of their remarkable properties, many persistent luminescent materials have been developed and are now widely used in various fields, particularly bioimaging, biosensing, therapeutics, etc.

This book attempts to look at the fundamental and applied properties of different phosphor materials that show the phenomenon of persistent luminescence. The book aims to offer a basic understanding of persistent luminescence, other mechanisms, models, related spectroscopic techniques and a few applications essential for anyone interested in learning about persistent luminescent materials. As we dig into the pages of this book, we uncover the underlying mechanisms that govern persistent luminescence, unraveling the intricate interplay of electrons, defects and energy states within luminescent materials. Through insightful discussions and in-depth analyses, we strive to provide a comprehensive understanding of the fundamental principles driving this extraordinary phenomenon. This is partly because the property of long-lasting luminescence is closely associated with other properties and functionalities (e.g., electronic, magnetic and thermal), which, besides being essential for many technological applications, such as bioimaging and biosensing, involve in-depth physics and chemistry.

This book covers many aspects of persistent luminescence in different materials, structure-property relationships, structural fabrications and their applications. The fundamental science and mechanism behind this phenomenon are described in detail to help readers understand the structure-property relationship in nanostructures showing the phenomenon. This book will present a set of state-of-the-art review chapters written by well-known specialists in the field. Each chapter will be self-contained, with adequate cross-references, and written clearly and concisely.

As we embark on this exploration of persistent luminescence, we sincerely thank the contributors, collaborators and mentors who have enriched this endeavor with their expertise and insights. We hope this book will serve as a valuable resource for scholars, students and enthusiasts seeking to deepen their understanding of persistent luminescence and its transformative impact on science and society.

Together, let us embark on a journey of discovery and enlightenment guided by the enduring glow of persistent luminescence.

Srinagar, India — Vijay Kumar
Srinagar, India — Irfan Ayoub
Sønderborg, Denmark — Yogendra Kumar Mishra
Bloemfontein, South Africa — Hendrik C. Swart

Contents

Persistent Luminescence in Comparison to Phosphorescence

Sitender Singh, Pawan Kumar, Isha Gupta, Preeti Siwach, and Devender Singh

Abstract Persistent luminescent (PersLumin) materials are extremely attractive for energy-efficient applications due to their exceptional ability for temporary storage and gradual release of light. Persistent luminescence is a fascinating optical phenomenon in which light can persist at room temperature after the UV–vis irradiation source has been removed. Thermally driven recombination of charge carriers trapped at host intrinsic defect locations produces such phosphorescence decay. It is nevertheless essential to throw some light on the basic terminology such as persistent luminescence and phosphorescence. For the economic development and rapid rise in global energy demand, the investigation of unconventional renewable energy sources is critical. Over the past few years, mechanistic studies appear to have matured due to the great efforts of researchers. This has aided in the development of novel materials for applications in areas like in vivo bioimaging, security signs, emergency lighting, interior decorating, AC-LEDs and optical storage media. The innovative use of LED lighting powered by renewable energy can have a positive impact on the environment by reducing greenhouse gas emissions and less deforestation, as well as providing people with a more affordable source of lighting. This chapter highlights the key differences between phosphorescence and PersLumin materials including their history. The chapter also includes a summary of the present-day status, trends in research of PersLumin materials and their practical applications nowadays. In the subsequent sections, the earlier as well as advanced PersLumin mechanisms are divided into various models, mainly the electron trapping–detrapping and hole trapping-detrapping models. Furthermore, the advancement made in this field through valuable research inputs such as bandgap engineering is discussed in detail. Moreover, we cover phosphorescence and PersLumin in both inorganic and organic phosphor materials. Besides, we highlight the upcoming potential, targets

S. Singh · P. Kumar · I. Gupta · D. Singh (✉)
Department of Chemistry, Maharshi Dayanand University, Rohtak, Haryana 124001, India
e-mail: devjakhar@gmail.com

S. Singh
Department of Chemistry, Baba Mastnath University, Asthal Bohar, Rohtak 124021, India

P. Siwach
Department of Chemistry, Kurukshetra University, Kurukshetra 136139, India

V. Kumar et al. (eds.), *Persistent Luminescence*, Progress in Optical Science and Photonics 30, https://doi.org/10.1007/978-981-97-4943-0_1

and challenges (like under trial bioimaging technique) and probable solutions to these challenges.

Keywords Persistent luminescence · Phosphorescence · Phosphors · Trap depth · Bandgap engineering

1 Introduction

Persistent luminescence (PersLumin) commonly called glow-in-dark is the scientific term for the phenomenon in which light can remain visible at room temperature (RT) for hours after terminating the excitation (Heggen et al. [1]; Xu et al. [2]; Xie et al. [3]). The persistent luminescent materials possess light-storing and releasing properties and are extremely attractive for energy-efficient applications due to their exceptional ability for temporary storage and gradual release of light (Xie et al. [4]). The outstanding phenomenon of PersLumin has been known to human civilization for long and is still extremely prevalent today. The advantages of these materials ensure their practical applications in everyday operations and in an emergency including AC-LEDs, security signs, emergency lighting, solar cells, optical storage media, interior decorating, night displays, in vivo bioimaging, detectors of high energy radiation, night-vision surveillance, etc. (Song et al. [5]; Peng et al. [6]; Lozano et al. [7]; Poelman et al. [8]). Technically, PersLumin is also defined as a type of thermoluminescence, i.e., thermal-stimulated luminescence at a specified temperature. Nowadays, materials exhibiting high temperatures (> 450 K) PersLumin have been explored for optical storage. Other persistent phosphors working around RT or physiological temperature are used for ultrasensitive biodetection and bioimaging (Lozano et al. [7]; Zhang et al. [9]; Gao et al. [10]; Yan et al. [11]).

PersLumin phosphors consist of a luminescence (recombination) center and a trap center. The function of the emission center is to emit radiation of a particular frequency upon excitation by an external source. On the other hand, the function of the trapping center is to store the energy which thereafter is released to the emitting center via thermal stimulation (Lin et al. [12]; Qiu et al. [13]; Singh et al. [14]). Furthermore, the trapping process involves either electron or hole trapping. PersLumin process involves the spontaneous release of the trapped holes/electrons at RT followed by recombination at the emission center (Swart et al. [15]; Qiu et al. [16]; Lyu et al. [17]). In PersLumin phosphors, the luminescence properties like emission wavelength, PersLumin intensity and lifetime can be tuned depending on the choice of recombination and trap centers (Liu et al. [18]; Zhang et al. [19]). For instance, the emission center decides the emission wavelength (λ_{em}), while a trap affects the intensity and duration of PersLumin (Xu et al. [20]; Głuchowski et al. [21]). In general, the emissive and trapping states both are associated with the structure of the host lattice and their composition (Zhang et al. [22]). Here, it is worth mentioning that obtaining a recombination center with a desirable emission wavelength and trap state with a proper depth within a particular host is difficult (Wang et al. [23]). PersLumin phosphor materials usually consist of inorganic host doped with activator ions, generally, including either lanthanides Ce(III), Pr(III), Nd(III), Sm(III), Eu(II), Eu(III), Tb(III), Dy(III), Er(III) and Tm(III), or transition/main group metal ions Ti(IV), V(III), Cr(III), Mn(II), Mn(IV), Co(II), Cu(II), Bi(III), Sn(II) and Pb(II) (Qiu et al. [16]; Singh et al. [24, 25]; Qiu et al. [26]; Qiu et al. [27]; Singh et al. [28]).

PersLumin has a well-established history starting at the beginning of the seventeenth century. The popular Bologna stone, also known as barite, $BaSO_4$, was first discovered by V. Casciarolo (Italian shoemaker) in 1602. However, the substance responsible for PersLumin was a reduced form of $BaSO_4$, i.e., BaS (Virk et al. [29]; Singh et al. [30]). In addition, the impurity present in the barite as well as the charcoal used for the reduction makes us predict that the change in emission color (orange or red) might be due to the presence of different impurities. PersLumin was first discovered, studied and investigated for many uses among the various forms of luminescence in the twentieth century despite no explanation responsible for luminescence being found. However, the root cause of PersLumin was mysterious until the twenty-first century (Smet et al. [31]). Luminous paints have been used based on PersLumin from ZnS:Cu, but luminescence was feeble and lasted for few hours (Lehmann et al. [32]). Doping ZnS with a radioactive element (e.g., Pm) was used to get prolonged PersLumin upon external excitation (Aitasalo et al. [33]); however, it was an unethical technique due to the radioactive nature of Pm (Panigrahi et al. [34]). These techniques were simply postponing the inevitable rejection of the ZnS-based persistent materials, which were also prone to humidity instability (Singh et al. [35]). Around the 1960s, with the emergence of rare-earth (RE)-activated phosphor materials, the dominant position of PersLumin as the emerging luminescence subject was eventually lost, and other sectors of luminescence started growing like cathode ray tubes, lighting and scintillator materials. The rapid introduction of new phosphor materials caused the negative side of PersLumin, i.e., the afterglow. This caused the emission to be retarded and weakened, although different solutions were developed to resolve the problem. Until the mid-1990s, nothing was done to understand the phenomenon or develop new materials (Hölsä et al. [36]).

In 1995, first generation modern PersLumin phosphors, i.e., Eu^{2+} activated and trivalent rare-earth (RE^{3+}) co-activated aluminates ($SrAl_2O_4:Eu^{2+},RE^{3+}$ and $CaAl_2O_4:Eu^{2+},RE^{3+}$) became commercial and laboratory materials (Singh et al. [14]; Duan et al. [37]; Richard et al. [38]). Owing to the increased interest of scientists and researchers, significant growth in this research field was observed as a large number of reports were published; however, the standard of research was not the best (Lü et al. [39]). There are now PersLumin materials of prime colors fulfilling the commercial requirements of such as stability of the material and strong as well as long PersLumin. For instance, $Y_2O_2S:Eu^{3+},Mg^{2+},Ti^{4+}$, $SrAl_2O_4:Eu^{2+},Dy^{3+}$ and $Sr_2MgSi_2O_7:Eu^{2+},Dy^{3+}$ persistent materials exhibit red, green and blue emission in dark upon UV excitation (Liu et al. [40]; Mao et al. [41]). The intensity and lifetime of red-emitting materials, on the other hand, are significantly lower than those of blue and green PersLumin materials. It is worth mentioning that aluminate and silicate-based host materials are more stable, eco-friendly and economical than sulfide–based materials (Lin et al. [42]; Singh et al. [43, 44]). Table 1 depicts the PersLumin phosphors based on different types of hosts with their wavelength of excitation as well as emission, color/spectral range and persistent time (Singh et al. [45]). In addition, $M_2MgSi_2O_7:Eu^{2+},RE^{3+}$ (M = Ca and Sr) PersLumin materials are more advantageous compared to the corresponding aluminates (MAl_2O_4) in terms of stability, efficiency and synthesis process (no flux required) (Singh et al. [46]). Among the emission centers, Eu^{2+} ions are dominating over the others owing to the suitable match of its ground state with the electronic structure of host materials (Meijerink et al. [47]; Dorenbos et al. [48]). The rising price of Eu^{2+} activator ions is a concern; still, the requirement of its very small concentration makes this useful for various luminous applications including security indications (e.g., Euro bank notes).

It is theoretically feasible to combine three separate blue-, green- and red-emitting phosphors to produce a white-emitting persistent light source (Singh et al. [49, 50]), although this application could be suspended or impossible due to the imbalanced decay behavior of these prime colors, particularly, weakly intense and short red persistent emission (Wang et al. [51]). The development of novel materials and knowledge of the mechanism underlying the phenomenon have been the two main goals of persistent luminescence research. However, the most crucial component of the mechanism, i.e., determining the nature of charge carriers, their energy level positions or the defects that store the excitation energy is further used to get PersLumin. In various compounds with the same energy scale, the energies at the top and bottom of the valence band (VB) and conduction band (CB), respectively, are compared using a diagram called vacuum referred binding energy (VRBE) (Sontakke et al. [52]; Ueda et al. [53]). It is worth mentioning that the phenomenon of trap depth engineering (TDE) is undoubtedly quite helpful in designing the PersLumin phosphors for specific applications (Xu et al. [54]). Along with the knowledge of energy locations of VB and CB, an appropriate combination of dopant (emission or recombination center) and co-dopant (trapping center) should be used to get the desired PersLumin properties (Stanek et al. [55]; Singh et al. [56]). For example, the phosphors with long persistent time and optical storage applications can be designed using "medium" + "deep" trap depths while the "shallow" + "deep" combination provides the materials with intense PersLumin suitable for optical information storage. The possible dopant (emission center) and co-dopant (trap center) combination used for the persistent phosphors involves the lanthanide–lanthanide combination, lanthanide–transition metal as well as transition metal–lanthanide (Yen et al. [57]). The strategy used for making desired PersLumin phosphor materials has been discussed in detail under the section "*Advanced mechanisms and engineering of persistent luminescence*".

Based on the VRBE diagram for the electron-trapping mechanism, the emission center as a deep hole trap (e.g., Eu^{2+}) and trap center with a small trap depth (e.g., Dy^{3+}, Nd^{3+}, Ho^{3+} or Er^{3+}) are preferred (Yu et al. [58]; Singh et al. [59]). On the other hand, the hole-trapping mechanism involves hole-trapping center with a narrow trap width and recombination centers having deeper electron traps. In this mechanism, the desired combination should include the easy detrapping of the trapped hole from the trapping center to VB compared with the movement of a stimulated electron into CB. Moreover, in $Y_3Al_2Ga_3O_{12}:Cr^{3+}$ co-doped with transition metal (TM) ions (V^{3+}, Fe^{3+} and Ti^{3+}), the energy gap between the bottom of conduction and ground states of divalent TM ions changes which is mainly affected by the number of d-electrons, nuclear charge and crystal field. Based on the knowledge of the VRBE diagram, CB engineering can be used to design innovative PersLumin materials doped with lanthanides and TM ions (Dorenbos et al. [60]). Based on the electron trapping–detrapping mechanism, to design the PersLumin phosphors with desirable luminescence performance, more attention should be paid to determining the trap depth and the energy level position of the top of VB and the ground state of the emission center (Zhang et al. [61]).

2 Current Status and Disagreements

Due to the sharp increase in the world's energy consumption, the study of unconventional renewable energy sources is essential (Rojas-Hernandez et al. [105]). The design and development of energy-efficient and ecologically benign phosphors has become the foundation of modern lighting technologies, enabling the expansion of social and commercial activities. The way the world lights up has evolved significantly in a short time, especially with the wide acceptance of light-emitting diodes (LEDs). There is a great deal of interest in developing materials with unusual emission properties, with the majority of efforts focused on the fabrication of LEDs having enhanced efficiency and color-tunable phosphors used for color tuning in illumination (Singh et al. [106]). Engineered structural defects are crucial in PersLumin materials, contrary to normal phosphors, where defects are harmful to luminescence efficiency (Singh et al. [107, 108]). PersLumin materials enable energy storage along with controlled release through charge trapping. The defect engineering enables us to achieve an acceptable combination having ample storage capability, PersLumin decay time and luminous efficacy (Stanek et al. [55]).

Since the discovery of $SrAl_2O_4:Eu^{2+},Dy^{3+}$ green persistent phosphors, various aluminates and silicates doped with Eu^{2+} and co-doped with different RE^{3+} ions have been designed and analyzed (Singh et al. [109, 110]). In addition, some garnets ($Y_3Al_2Ga_3O_{12}$ or $Gd_3Al_2Ga_3O_{12}$) doped with Ce^{3+} (Liu et al. [111]) and co-doped with Cr^{3+} as well as some perovskite host ($NaNbO_3$, $CaTiO_3$) activated with Pr^{3+} have been explored as an alternate to Eu^{2+} activated phosphors (Lozano et al. [7]; Viana et al. [112]; Marí et al. [113]). The methods used to obtain the PersLumin with improved luminescence properties are normally complex and costly, which include doping different RE ions or changing the synthesis procedure normally requiring high temperatures (Singh et al. [114]). Designing a simple, economical method to enhance the PersLumin properties is still a challenge (Song et al. [5]; Singh et al. [115]). At present, to enhance electron–hole recombination in p–n junctions, commercially used LEDs are facilitated by direct current (DC). Thus, to operate the LED module, the input alternating current (AC) is transformed into DC. This AC-to-DC conversion is what causes the uneconomical use of electric power. By focusing more on AC-LEDs,

Table 1 PersLumin phosphors involving various hosts with their wavelength of excitation/emission, color and persistent time

Host type	Phosphor	λ_{ex} (nm)	λ_{em} (nm) Color/Spectral range	Persistent time	Year	References
Bologna Stone	BaS	–	–	–	1603	Lastusaari et al. [62]
Sulfide	$ZnS:Cu^{+}$	–	525 Green	40 min	1936	Smet et al. [31]
	$ZnS:Cu^{+},Co^{2+}$	–	525 Green	~ 80 min	–	Chen et al. [63]
	$Y_2O_2S:Eu^{3+},Mg^{2+},Ti^{4+}$	UV	617 Red	> 5 h	1999	Zhang et al. [64]
Aluminates	$SrAl_2O_4:Eu^{2+},Dy^{3+}$	UV	520 Green	> 30 h	1996	Matsuzawa et al. [65]
	$SrAl_4O_7:Eu^{2+},Dy^{3+}$	UV	480 Greenish-blue	–	1998	Katsumata et al. [66]
	$SrAl_{12}O_{19}:Eu^{2+},Dy^{3+}$	UV	400 Green, UV	–	1998	Katsumata et al. [66]
	$CaAl_2O_4:Eu^{2+},Nd^{3+}$	UV	440 Blue	19 h	2003	Lin et al. [67]
	$Y_3Al_5O_{12}:Ce^{3+}$	460	Yellow 580 nm	65 ns	2009	Meijerink et al. [68]
	$Y_3Sc_2Al_{3-x}Al_xO_{12}:Ce^{3+}$ (x = 0–3)	365	495 Green	> 8 min	2014	Xia et al. [69]
	$Sr_4Al_{14}O_{25}:Ce^{3+}$	γ-ray	472, 511 Blue–green	> 10 min	2009	Sharma et al. [70]
	$Y_3Al_2Ga_3O_{12}:Ce^{3+}$	460	505 Green	~ 1 min	2014	Ueda et al. [71]

(continued)

Table 1 (continued)

Host type	Phosphor	λ_{ex} (nm)	λ_{em} (nm) Color/Spectral range	Persistent time	Year	References
	$Y_3Al_2Ga_3O_{12}$:Ce^{3+},Cr^{3+}	460	505, 700 (Cr^{3+}) Green	> 5 h	2014	Ueda et al. [71]
	$Y_3Al_2Ga_3O_{12}$:Ce^{3+},Pr^{3+}	430	520, 607 (Pr^{3+}) Green	~ 1 h	2014	Lin et al. [72]
	$Y_3A_{12}Ga_3O_{12}$:Ce^{3+},Yb^{3+}	365	520 Green	138.8 h	2018	Ueda et al. [53]
	$CaAl_2O_4$:Ce^{3+}	280	400 Blue–violet	> 10 h	2002	Jia et al. [73]
	$SrAl_{12}O_{19}$:Eu^{2+}	300	400 Blue	–	2006	Zhang et al. [74]
	$SrAl_{12}O_{19}$:Eu^{2+},Cr^{3+}	300	400, 693 (Cr^{3+}) Blue	–	2006	Zhang et al. [74]
	$LaAlO_3$:Mn^{4+}	336	731 NIR	> 24 h	2014	Qiu et al. [27]
	$GdAlO_3$:Mn^{4+}	326	719 NIR	> 24 h	2014	Qiu et al. [27]
	$GdAlO_3$:Mn^{4+},Ge^{4+}	324	690, 713 NIR	> 24 h	2016	Wang et al. [75]
	$Sr_4Al_{14}O_{25}$:Cr^{3+},Eu^{2+},Dy^{3+}	330	490(Eu^{2+}), 693(Cr^{3+}) Blue-NIR	> 10 h	2006	Zhang et al. [74]
	$SrAl_2O_4$:Cr^{3+},Eu^{2+},Dy^{3+}	370	515(Eu^{2+}), 695(Cr^{3+}) Green-NIR	–	2014	Qiu et al. [76]
	$CaAl_2O_4$:Mn^{2+},Ce^{3+}	UV	525 Green	> 10 h	2003	Zhang et al. [64]

(continued)

Table 1 (continued)

Host type	Phosphor	λ_{ex} (nm)	λ_{em} (nm) Color/Spectral range	Persistent time	Year	References
	$MgAl_2O_4:Mn^{2+}$	209	520 Green	–	2010	Lin et al. [77]
	$SrAl_2O_4:Ce^{3+},Mn^{2+}$	273	374(Ce^{3+}), 515 Green	> 2.5 h	2011	Wang et al. [78]
Silicates	$Sr_2MgSi_2O_7:Eu^{2+},Dy^{3+}$	250	476 Blue	–	2005	Dorenbos et al. [79]
	$Ca_3MgS_{i2}O_8:Eu^{2+},Dy^{3+}$	UV	476 Blue	> 5 h	2001	Lin et al. [80]
	$Sr_2SiO_4:Eu^{2+}$	313	479, 493, 560 -	–	2008	Lakshminarasimhan et al. [81]
	$Sr_2SiO_4:Eu^{2+},Dy^{3+}$	313	479, 493, 560 Yellow	> 5 min	2008	Lakshminarasimhan et al. [81]
	$Sr_2SiO_4:Eu^{2+},Ce^{3+}$	313	406, 479, 493, 560 -	–	2008	Lakshminarasimhan et al. [81]
	$Ba_4(Si_3O_8)_2:Eu^{2+}$	365	506 Green	~ 2 min	2011	Wang et al. [82]
	$Ba_4(Si_3O_8)_2:Eu^{2+},Dy^{3+}$	367	506 Green	> 24 h	2011	Wang et al. [82]
	$CdSiO_3:Tb^{3+}$	247	544–552 Green	–	2012	Brito et al. [83]
	$Sr_2SiO_4:Pr^{3+}$	254	~ 390, 534, 604 White	> 40 min	2008	Zhou et al. [84]
	$CdSiO_3:Mn^{2+}$	233	575 Orange	~ 1 h	2004	Shi et al. [85]

(continued)

Table 1 (continued)

Host type	Phosphor	λ_{ex} (nm)	λ_{em} (nm) Color/Spectral range	Persistent time	Year	References
	$Zn_2SiO_4:Mn^{2+}$	248	520 Green	–	2005	Patra et al. [86]
	$Mg_2SiO_4:Mn^{2+}$	254	656 Red	~ 6 min	2003	Zhang et al. [64]
	$Mg_2SiO_4:Dy^{3+},Mn^{2+}$	254	656 Red	~ 7 min	2003	Zhang et al. [64]
	$MgSiO_3:Mn^{2+},Eu^{2+},Dy^{3+}$	UV	660 Red	~ 4 h	2003	Wang et al. [87]
	$CdSiO_3:Bi^{3+}$	254	344 UV	> 30 min	2016	Ci et al. [88]
	$CdSiO_3:Pb^{2+}$	254	498 Green	> 2 h	2006	Liu et al. [89]
	$Ca_3Si_2O_7:Eu^{2+},Sm^{3+}$	UV	350 – 550 –	20 min	2017	Ueda et al. [90]
	$Ca_3Si_2O_7:Eu^{2+},Tm^{3+}$	UV	620 –	50 min	2017	Ueda et al. [90]
	$CaMgSi_2O_6:Mn^{2+},Pr^{3+}$	254	685 Red	–	2011	Richard et al. [91]
Nitrides and oxy-nitrides	$Ca_2Si_5N_8:Eu^{2+}$	Xe arc	610 Orange	> 41 min	2011	Eeckhout et al. [92]
	$Ca_2Si_5N_8:Eu^{2+},Nd^{3+}$	254	582 Orange	> 200 min	2013	Liu et al. [93]
	$Sr_2Si_5N_8:Eu^{2+}$	Xe arc	620 Red	> 1 min	2011	Eeckhout et al. [92]

(continued)

Table 1 (continued)

Host type	Phosphor	λ_{ex} (nm)	λ_{em} (nm) Color/Spectral range	Persistent time	Year	References
	$Ba_2Si_5N_8:Eu^{2+}$	Xe arc	580 Yellow	> 6 min	2011	Eeckhout et al. [92]
	$CaSi_2O_2N_2:Eu^{2+}$	395	560 Yellow	6.8 min	2005	Hintzen et al. [94]
	$SrSi_2O_2N_2:Eu^{2+}$	387	539 Green Yellow	100 min	2005	Hintzen et al. [94]
	$BaSi_2O_2N_2:Eu^{2+}$	406	498 Blue Green	46 min	2005	Hintzen et al. [94]
	$BaAlSi_5O_2N_7:Eu^{2+}$	363	516 Green	40 min	2015	Dhoble et al. [95]
Phosphates	$Ca_6BaP_4O_{17}:Eu^{2+}$	365	553 Yellow	1 h	2015	Wang et al. [96]
	$Ca_6BaP_4O_{17}:Eu^{2+},Dy^{3+}$	365	553 Yellow	1.2 h	2015	Wang et al. [96]
	$Ca_6BaP_4O_{17}:Eu^{2+},Tb^{3+}$	365	553 Yellow	4.3 h	2015	Wang et al. [96]
	$Ca_6BaP_4O_{17}:Eu^{2+},Ce^{3+}$	365	553 Yellow	9.9 h	2015	Wang et al. [96]
	$Ca_6BaP_4O_{17}:Eu^{2+},Gd^{3+}$	365	553 Yellow	18.8 h	2015	Wang et al. [96]
	$Ca_6BaP_4O_{17}:Eu^{2+},Nd^{3+}$	365	553 Yellow	37.9 h	2015	Wang et al. [96]
	$Sr_2P_2O_7:Eu^{2+},Lu^{3+}$	365	420 Blue	–	2013	Hu et al. [97]

(continued)

Table 1 (continued)

Host type	Phosphor	λ_{ex} (nm)	λ_{em} (nm) Color/Spectral range	Persistent time	Year	References
	$ZrP_2O_7:Mn^{2+}$	UV	467 -	> 10 min	2019	Zheng et al. [98]
Borates	$CaB_2O_4:Ce^{3+}$	254	365, 460 -	15 h	2019	Sharma et al. [99]
	$SrB_4O_7:Eu^{2+}$	306	367 UV	–	2007	Brito et al. [100]
	$Al_4B_2O_9:Eu^{2+}$	330	434 Blue	–	2010	Chen et al. [101]
	$Ba_2Mg(BO_3)_2:Eu^{2+}$	UV	564 Yellow	–	2021	Qiu et al. [102]
Halides	$CaF_2:Eu^{2+}$	UV	425 Blue	3 s	1991	Silfsten et al. [103]
	$BaLiF_3:Eu^{2+}$	254	408 Blue	–	2004	Fujihara et al. [104]

the drawbacks of DC-LEDs may be overcome due to advantageous characteristics of AC-LEDs like low price, compact volume, longer life and improved efficiency (Wang et al. [116]).

In addition, there are certain doubts about the practical feasibility of some mechanisms. For example, in the hole trapping–detrapping model, the liberated hole to VB from the excited Eu^{2+} ion is later trapped by the co-activator Dy^{3+} ion resulting in a Dy^{4+} ion, which results in the creation of Eu^{+} ion from the excited Eu^{2+} ion. It is exceedingly improbable, though, considering the large amount of energy required to produce Eu^{+} in ambient conditions. No experimental method has also been able to confirm the generation of Eu^{+} (Hölsä et al. [117]). On the other hand, in the electron trapping–detrapping model, the experimental confirmation of the reduction in Dy from 3+ to 2+ by EPR or XPS, however, has not yet been confirmed. Instead of playing a redox function, it's plausible that the co-dopant Dy^{3+} ion enables $SrAl_2O_4$ to generate new defects or stabilize/increase lattice or inherent defects like oxygen vacancies.

3 Earlier Persistent Luminescence Mechanisms

Before the development of the persistent aluminate phosphors after 1990, only a little was understood about the processes underlying persistent luminescence. This is not surprising as for almost a century ZnS:Cu phosphors were commercially accessible. In actuality, there is still debate on the process underlying the typical luminescence from Cu. In $SrAl_2O_4$:Eu^{2+}, Dy^{3+}, it was assumed that electrons and holes behave as charge carriers. Initially, it was suggested that the electron from VB is trapped by the emission center Eu^{2+} and the hole is trapped by co-dopant Dy^{3+} traps resulting in the formation of Eu^{+} and Dy^{4+} ions. However, the formation of the Eu^{+} is extremely implausible requiring enormous quantities of energy (Dorenbos et al. [48]). Simply said, this energy cannot exist in the surrounding environment. Soon after, more logical researchers proposed a reverse mechanism, where Eu^{2+} would be oxidized and Dy^{3+} would be reduced and again the charge carriers used were the same. The issue with "redox" mechanisms, despite being far more plausible, is that $SrAl_2O_4$:Eu^{2+}, Dy^{3+} and $Sr_2MgSi_2O_7$:Eu^{2+},Dy^{3+} exhibit more intense PersLumin even without the Dy^{3+} (co-dopant) although with short duration. Furthermore, in experimental studies (EPR, XPS or XANES), only Eu^{2+} to Eu^{3+} oxidation occurs while the valence of the co-dopants remains unchanged (Jobic et al. [118]; Lastusaari et al. [119]). Based on this, it can be concluded that it may not always be required to design a dopant–co-dopant redox pair. For instance, in $CdSiO_3$:Tb^{3+} material, Tb^{3+} ions behave as both emission centers and co-dopants. Thus, the co-dopant's function can be entirely different. Based on the above facts, it was realized that the defects generated by co-dopants may be used to trap the electrons and holes (Fernández-Rodríguez et al. [120]). The presence of defects is more or less evident due to charge compensation, especially in situations having different valence states of the co-dopant and host cation. However, it is important to mention that the valence state of host ion and emission centers are often the same (Singh et al. [121–123]). This is further justified by the fact that the emission center should be efficient, and there should not be luminescence quenching due to uncontrollable lattice defects (Heggen et al. [1]; Qiu et al. [16]).

It is worth mentioning that normally used efficient emission in various PersLumin phosphors exists as chemically as well as physically significant redox couples such as Eu(II)-Eu(III), Ce(III)-Ce(IV) Pr(III)-Pr(IV), Eu(III)-Eu(II), Tb(III)-Tb(IV), Mn(II)-Mn(III), Ti(III)-Ti(IV), Cr(III)-Cr(II)/Cr(IV), Bi(III)-Bi(V), Pb(II)-Pb(IV) and Cu(I)-Cu(II) (Hölsä et al. [124]). The other important point is that effective PersLumin phosphors mainly involve the oxidation of the emission center; however, the reduction is completely ignored, and it occurs much less commonly. This observation leads us to conclude that the charge carrier should be electrons, and the mechanism should involve CB, wherein the mobility of electrons is significantly greater compared to holes in VB. Therefore, there are higher probabilities of having an electron trapping center just below the CB. In addition, the redox behavior of the emission center and co-dopant could simply be hypothetical; perhaps RE^{n+}–h^+ and RE^{n+}–e^- pairs are generated instead of the actual oxidized and reduced entities. For only the Eu^{2+}–Eu^{3+} pair, the oxidized species have been detected. RE^{3+} ions, having a small size, high charge as well as electron density, present ion lattice containing M^{2+} ions, and have a higher tendency to attract the electron making the pair formation possible (Smet et al. [125]; Qi et al. [126]). Based on the early mechanisms proposed for PersLumin, it can be concluded that at least a fundamental understanding of the chemical nature of different ions in the phosphor studied could have helped to reject the most unlikely mechanisms.

4 Advanced Mechanisms and Engineering of Persistent Luminescence

Enhancing the performance of discovered PersLumin materials or fabrication of novel phosphors remains a question under research. In recent times, it has become conceivable to explain PersLumin and trapping–detrapping process of electrons/holes using the energy level strategy of VB and CB of the host matrix, as well as the position of ground and excited levels of dopant/co-dopant ions (Ueda et al. [127]).

4.1 Electron Trapping–detrapping Model

J. Ueda presented an electron trapping–detrapping model in 1996, which is today the most well-known and popular mechanism for describing the intriguing phenomenon of PersLumin (Ueda et al. [53]). Schematic diagrams of the electron trapping-detrapping approach-based phenomenon of persistent luminescence are shown in

Fig. 1. Two different active centers, comprising emission and trap centers, are involved in the persistence luminescence process in the host lattice. Such active centers are located between the conduction and valence bands (CB and VB). For example, in $SrAl_2O_4:Eu^{2+},RE^{3+}$, Eu^{2+} and RE^{3+} act as emission and trap centers respectively (Heggen et al. [1]). The electron trapping–detrapping mechanism has been explained based on two approaches including **(a)** VB → CB excitation and **(b)** excitation among the energy level of emission ions. The former requires the higher energy excitation source involving a UV lamp (254 nm), electron beams and X- or γ-rays radiation sources while the latter is lower energy excitation using UV or visible light.

In band-to-band (VB → CB) excitation (Fig. 1), the excitation of electrons from VB to CB generates a hole in the former **(Step 1)**. Consequently, the free hole moves through VB and electrons move through CB. **Step 2** involves the trapping of the hole by ground state (GS) of the luminescent center and free electron by trap state. After the excitation terminates, thermally stimulated electrons are consequently released or detrapped. The probability of detrapping strongly depends on the trap depth (ΔE = gap across CB edge and trapping state). If the detrapping of the trapped electron is thermally accessible at room temperature, then the electron in the trap state return to the excited state (ES) of the luminescent center via CB **(Step 3)** and ultimately recombines with the hole present in its GS, producing PersLumin **(Step 4)**. It is noteworthy to mention that change in trap depth considerably affects the luminous intensity along with its duration. The trap depth varies from one trivalent rare earth ion to another, and this phenomenon of changing trap depth via changing the co-activator ion in the Eu^{2+} host is called trap depth engineering.

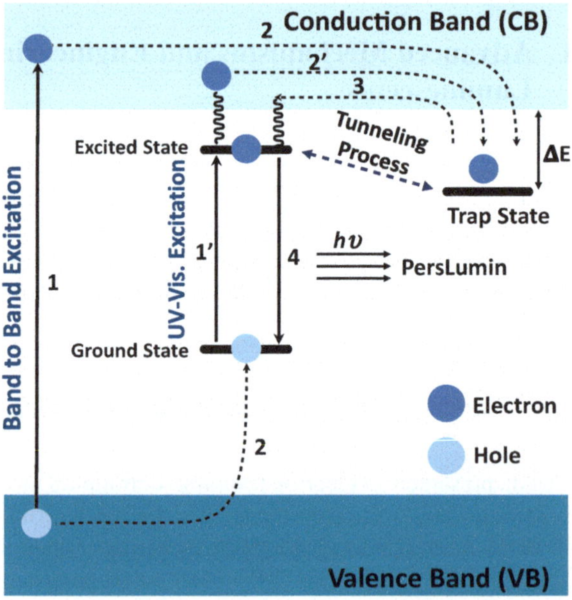

Fig. 1 The pictorial representations of electron trapping–detrapping mechanism involving high-energy excitation (band-to-band) and low-energy UV or visible light excitation (pumping the emission ions)

Low energy excitation using UV or visible light explaining the trapping and detrapping phenomenon has also been presented in Fig. 1. This mechanism differs from the former in **Step 1** only. Herein, the hole is already located at GS of the luminescent ion instead of getting captured via VB, whereas the electron gets excited to ES of the luminescent ion. **Step 1** also involves the jumping of an excited electron to the conduction band which is conceivable in case the gap between ES and CB edge is sufficiently tiny or when these levels overlap, e.g., the jumping of an electron from the 5d excited state of Ce^{3+} or Eu^{2+} due to thermal-ionization process. **Steps 2, 3** and **4** are almost the same as that described in the previous process. Because their vibrational parabolas overlap, the tunneling process between the excited state of the emission center and the trap state may take place. To provide solid proof of the tunneling phenomenon, it is advised to monitor the decay curves at extremely low temperatures and pump the lower excited states via lower energy stimulation. This will stop electrons in an excited state from jumping into CB during the thermal-ionization process. It is worth mentioning that electrons trapped in a shallow trapping state are released immediately at RT, allowing for short persistent duration but strong initial luminescence (Ueda et al. [53]; Wang et al. [96]). However, in phosphors with comparatively deeper trap depth (ΔE), electrons are released slowly at RT providing long persistent duration but weak luminescence (Xu et al. [54]).

Generally, the dopant (emission center) and co-dopant (trap center) combination used for the persistent phosphors involves the lanthanide-lanthanide combination (Luo et al. [128]). Instead, lanthanide–transition metal as well as transition metal-lanthanide doping strategies have been designed and applied. Using lanthanides as a trap center is more advantageous owing to the negligible interaction of the crystal field with well-shielded 4f electrons (Singh et al. [129, 130]). Therefore, knowing the energy position of any lanthanide ion concerning VB and CB, vacuum-referred binding energy (VRBE) diagrams can be constructed for a particular host to get the materials with desired PersLumin properties. For instance, $Gd_3Sc_2Ga_3O_{12}:Cr^{3+},Ln^{3+}$ (Ln = Sm, Eu, Tm, Yb) exhibit intense emission at 770 nm owing to Cr^{3+} ion where Ln^{3+} ions function as electron trap as GS of their Ln^{2+} ions lies below the CB edge. VRBE diagram of $Gd_3Sc_2Ga_3O_{12}$ host has been depicted in Fig. 2. In the case of Sm^{3+} and Tm^{3+}, the gap between CB edge and GS of Sm^{2+} and Tm^{2+}, i.e., trap depth is very small. However, in the case of Eu^{3+} and Yb^{3+}, the trap depth is deeper which leads to shifting of thermo-luminescence (TL) glow peak at higher temperature. Therefore, the shift in TL glow temperature in the case of Eu^{3+} is greater than Yb^{3+} (Xu et al. [54]).

In addition, $SrSi_2O_2N_2:Eu^{2+},RE^{3+}$ and $SrSi_2O_2N_2:Yb^{2+},RE^{3+}$ (where RE = Dy/Ho/Er) exhibit green and orange luminescence, respectively. Herein, these RE^{3+} ions act as trap centers or sensitizers with varying energy gaps between the CB edge and ground state of their divalent ions, i.e., $Dy^{2+}/Ho^{2+}/Er^{2+}$. Among these, Er^{3+} possesses the smallest trap depth while Dy^{3+} act as the deepest trap center. Therefore, based on their trap depth, the temperature required to release the trapped electrons should follow the order: $Dy^{2+} > Ho^{2+} > Er^{2+}$. Conclusively, $SrSi_2O_2N_2:Eu^{2+},Er^{3+}$ and

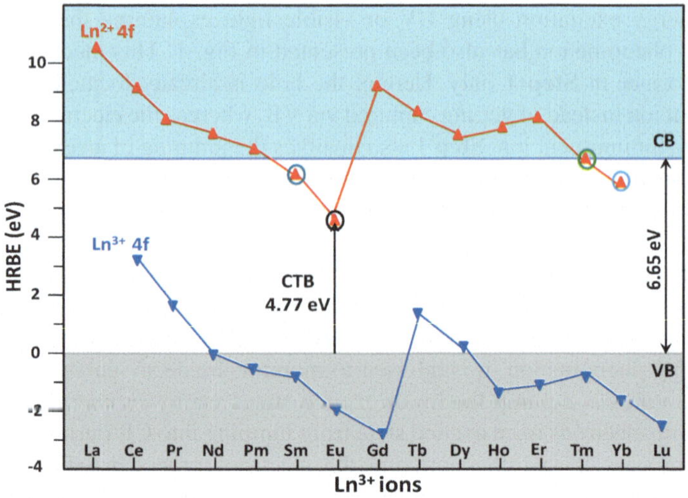

Fig. 2 HRBE diagram for $Gd_3Sc_2Ga_3O_{12}$ host

$SrSi_2O_2N_2$:Yb^{2+},Er^{3+} materials give intense initial PersLumin with a short duration, while Dy^{3+} or Ho^{3+} co-doped material exhibits less intense luminescence with relatively longer persistent time (Xie et al. [3]).

4.2 Hole Trapping–detrapping Model

Hole trapping-detrapping process includes the release of a hole by a hole trap followed by its recombination at the emission center. Hole trapping–detrapping phenomenon (Fig. 3) is a mirror image of the former represented in Fig. 2. In **Step 1,** holes and electrons get produced due to stimulation involving **(a)** VB → CB excitation and **(b)** excitation among the energy level of emission ions. VB is an electronic state free of holes making the movement of stimulated holes free. **Step 2** in the mechanism involves trapping of excited hole at the hole trap and an electron at ES of luminescent ion. A trapped hole releases back when thermal energy available at RT compensates ΔE, i.e., the gap between VB edge and trap (**Step 3**). Lastly in **Step 4**, the hole after returning to GS recombines with the electron present at ES giving PersLumin. This model requires understanding the relationship of energy levels between GS of emissive ion and hole trap. With proper understanding of the energy level position of luminescence and hole trap center, this model can be used to analyze the PersLumin of various phosphors, despite its failure to elucidate the formation of Eu^{2+}/Eu^{+} and Dy^{3+}/Dy^{4+} redox pairs in case of $SrAl_2O_4$:Eu^{2+},Dy^{3+} (Xu et al. [54]).

To date, the infrequent reports on this model have been mostly due to the failure to understand the detection of a hole trap and, in particular, trap depth. Meanwhile, the ground states (4f) of the lanthanides (Ln^{2+} and Ln^{3+}) are practically not affected

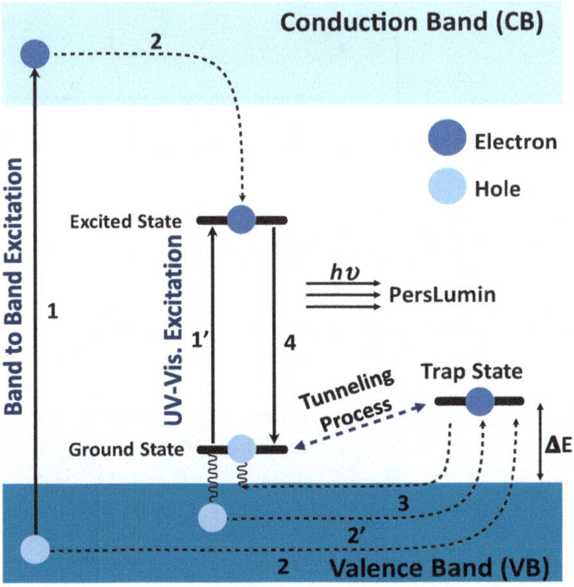

Fig. 3 The pictorial representations of hole trapping–detrapping mechanism involving high energy excitation (band-to-band) and low energy UV or visible light excitation (pumping the emission ions)

by the type of compounds. VRBE diagram of YPO_4 including lanthanide ions (Lyu et al. [131]) has been shown in Fig. 4. It is noteworthy to mention that the hole and electron trap widths are tailored by shifting VRBE at VB and CB edge, respectively. In this model, for easier moving of the trapped hole to VB, the desired combo is the emission center having a deeper electron trap and co-dopant having shallow hole traps. As a result, the gap between the depths of the electron and hole trap is critical when assessing the mechanism of PersLumin. For instance, in YPO_4:Eu^{3+}, RE^{3+} (RE = Tb and Pr), Eu^{3+} acts as an emission center while Tb and Pr act as hole-trapping centers. TL emission spectra of these compounds exhibited peaks corresponding to Eu^{3+} only. The absence of emission peaks for Tb^{3+} and Pr^{3+} confirms that these ions do not act as luminescence centers but instead behave as hole traps. In addition, Eu^{3+} cannot act as a hole-trapping center as it has a deeper electron-trapping state compared to Pr^{3+} and Tb^{3+} (Lyu et al. [131]).

4.3 Self-Electron/hole Trapping–detrapping Model

The understanding of energy levels of Ln within the bandgap of considered hosts enables us to alter trap width (ΔE) via CB or VB modifications in the case of electron trap or hole trap, respectively. In addition to that, phosphor materials doped even with a single ion (main group or transition metal ion) can show strong PersLumin, despite the lattice defects or aliovalent substitution including the formation of electron or hole traps. For instance, phosphors having Cr^{3+} or Mn^{2+} or Bi^{3+} as activation

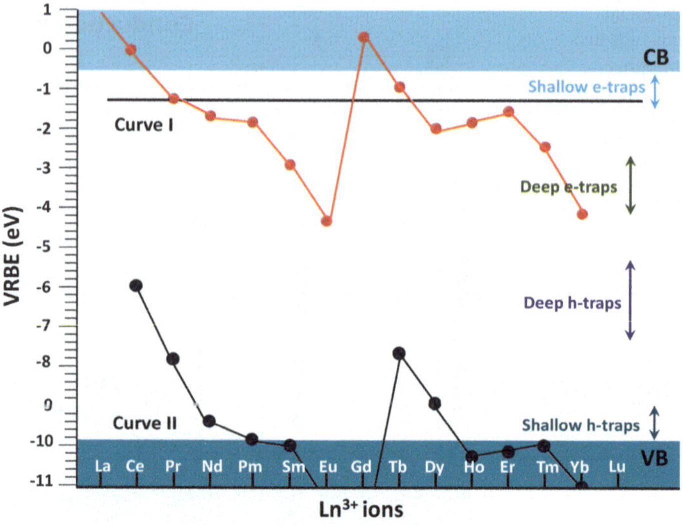

Fig. 4 VRBE diagram of YPO_4 including lanthanide ions

centers can act as emission as well as trap centers due to their self-redox behavior. This model including the self-electron or hole trapping-detrapping process can be understood by knowing the energy level positions of Cr(III)/Cr(II) and Bi(III)/Bi(II) using their VRBE diagram with different host materials. The important condition for PersLumin to take place via electron trapping is that the hole trap (recombination center) should be deeper than the electron trap (co-dopant ion). In perovskite $LaAlO_3$ materials, the energy of Cr(III) and Cr(II) can be predicated using VRBE diagram (Fig. 5a). For $LaAlO_3$, Cr^{3+} ion functions as recombination center (hole trap) having a deeper trap width (~ 1.5 eV), while its other redox state, i.e., Cr^{2+} having shallow trap depth (~1.3 eV) can act as an electron trap state. The local field near TM ions typically affects the location of their energy levels. However, the VRBE diagrams of $LaAlO_3$ and $GdAlO_3$ perovskites and $YAlO_3$ and $LuAlO_3$ compounds show similar GS positions of Cr(II) and Cr(III). Also, the VRBE diagrams of the garnet family ($Y_3Al_{5-x}Ga_xO_{12}$) possess quite similar energy level positions of ground and excited state of Cr^{2+} and Cr^{3+} ions. The similar energy positions of GS of Cr(III) and Cr(II) in VRBE diagrams particularly for hosts having similar crystal lattices like garnets or perovskites leads us to predict the self-electron or hole trapping–detrapping phenomenon for chromium ions (Xu et al. [54]).

Like Cr(III), the self-redox behavior of Bi(III) ions has been observed in the YPO_4 host lattice. With the knowledge of the VRBE diagram (Fig. 5b) including the energy level position of Bi(III) and Bi(II) and lanthanide ions in a particular, different types of charge trapping and detrapping process can be designed. For instance, in YPO_4:Bi^{3+},Ln^{3+} (Ln = Pr and Tb), Bi(III) having a deeper electron trap level acts as the emission center while Pr^{3+} and Tb^{3+} with shallow hole trap depth act as the trap

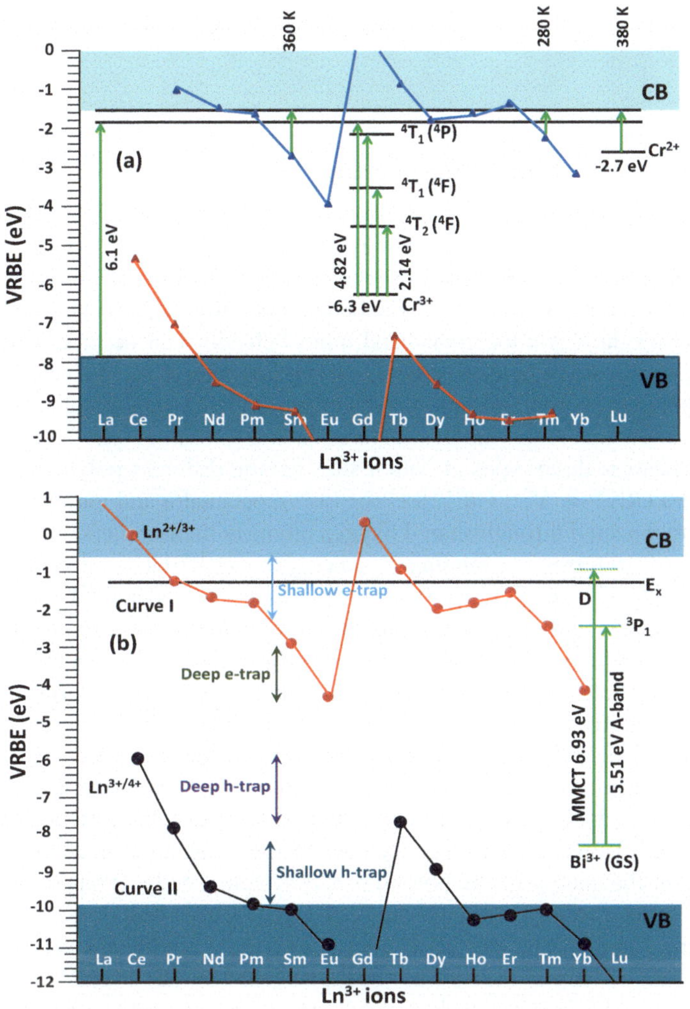

Fig. 5 VRBE illustration for **a** Ln and Cr in $LaAlO_3$ and **b** Ln and Bi in YPO_4

center. At a lower temperature, the holes released from Ln(IV) further recombine with Bi(II) through VB to yield Bi(III) emission (Xu et al. [54]).

However, in YPO_4:Ln^{3+},Bi^{3+}, where Ln = Sm, Eu, Tm and Yb act as recombination/emission centers due to deep electron trapping, and Bi^{3+} with shallow hole trap depth acts as trap state. Holes are detrapped from Bi(IV) at a lesser temperature compared to the electrons from Ln(II) and the former recombine via VB with Ln^{2+} to generate emission in Ln^{3+} ions. Lanthanides do not possess more than two valency normally, thus, do not show self-redox behavior like transition or main group metal. By using bandgap engineering, the band structures of the host lattice can be directly modified—these promising discoveries open up new opportunities for the construction of novel PersLumin materials having TM or main group ions as activators.

4.4 Conduction Band Engineering

Knowing the positions of various energy levels including VB, CB, emission and trap centers, the PersLumin behavior (intensity and duration of luminescence) of the phosphors can be controlled. As predicted, when trap width is changed by carefully changing the CB's location, PersLumin behavior can be carefully manipulated using the termed conduction band (CB) engineering. CB engineering has been successful in improving scintillation performance and the development of innovative long PersLumin materials. Garnet structures having formulae $A_3B_2C_3O_{12}$ and cubic symmetry possess three types of cation sites having different polyhedra including dodecahedra ($RE^{3+} = Y^{3+}$, Gd^{3+}, La^{3+} or Lu^{3+}), octahedra and tetrahedra (Al^{3+} or Ga^{3+}) interconnected through shared oxygen atoms at their corners.

Vacuum referred binding energy (VRBE) illustration of $Y_3Al_{5-x}Ga_xO_{12}$:Ce^{3+},Cr^{3+} (x = 0 – 5) enables us to understand the change in energy of CB edge with the replacement of Al by Ga in the host (Fig. 6) (Ueda et al. [127]). Such conduction band engineering is responsible for the tuneable trap depth which is a gap between CB edge and GS of Cr^{2+} (Fig. 7). It is worth mentioning that this systematic modification of trap depth is due to the substitution of Al^{3+} by Ga^{3+} ions. The d^{10} configuration of Ga^{3+} ions makes them more favorable for the substitution of Al^{3+} at tetrahedral rather than octahedral sites in garnets. On moving from $Y_3Al_5O_{12}$ to $Y_3Al_4GaO_{12}$, the shifting of CB to higher energy has been observed. However, a further increase in the concentration of Ga^{3+} leads to a lowering in the energy of the bottom of CB systematically. It has been observed that the electronic band structure of the $A_3B_2C_3O_{12}$ garnet host does not vary much by changing the [A] site, i.e., RE^{3+} ions; however, a significant variation in the band structure has been observed by the replacement of [B] site. For instance, the substitution of Al^{3+} by Ga^{3+} or In^{3+} results in a lowering in the energy of the CB edge while substitution by As^{3+} or Sb^{3+} raises the energy of the VB edge, leading to a significant decrease in the bandgap in both the cases. Conclusively, based on the knowledge of the VRBE diagram and CB engineering, one can design novel persistent phosphors doped with various rare-earth as well as transition metal ions (Ueda et al. [127]).

In $Gd_3Sc_2Ga_3O_{12}$:Cr^{3+},Ln^{3+} (Ln = Eu, Yb), the electron trap for Eu^{3+} is much deeper than that of Yb^{3+}. Therefore, the initial PersLumin intensity for Eu^{3+} co-doped material is lesser than the case of Yb^{3+} at RT. Using CB engineering, the energy of the bottom of CB can be lowered to decrease the gap between the CB edge and GS of Eu^{2+}. In such a modified host, the tendency of Eu^{3+} to act as an efficient trapping center and give enhanced PersLumin increases as in the case of $Gd_3Ga_5O_{12}$:Cr^{3+},Eu^{3+}.

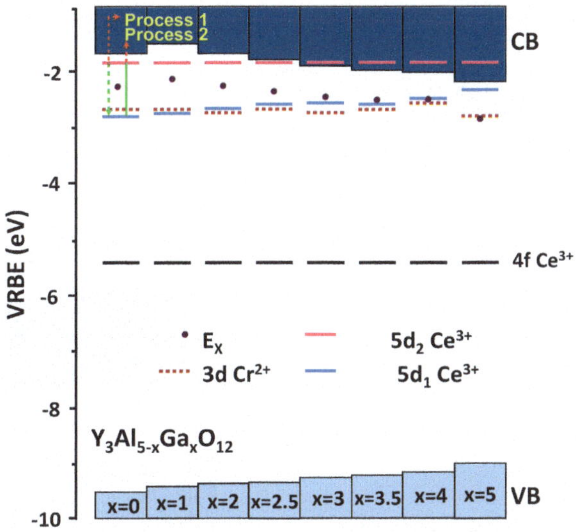

Fig. 6 Combined VRBE diagram of $Y_3A_{5-x}Ga_xO_{12}$:Ce^{3+},Cr^{3+} (x = 0 – 5)

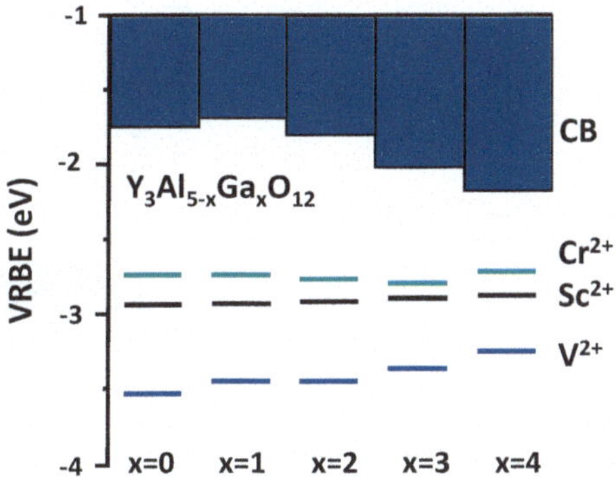

Fig. 7 Combined VRBE diagram of $Y_3Al_{5-x}Ga_xO_{12}$ for TM^{2+} (TM = Sc, V, Cr)

4.5 Valence Band Engineering

Considering the VRBE diagrams of MPO_4 (M = La, Gd, Y and Lu) (Fig. 8a), the PersLumin phosphors can be designed by carefully regulating the detrapping of holes in the hole-trapping level (Lyu et al. [131]). Additionally, on substitution of Y(III) ions in YPO_4 by La(III), Gd(III) or Lu(III) ions, the position of the VB edge is changed systematically to alter the energy gap between the top of the VB and hole-trapping center. This phenomenon of trap width alteration by modifying the position of the VB edge is called valence band (VB) engineering (Lyu et al. [131]; Dorenbos et al. [132]]. In $La_xGd_{1-x}AlO_3$ perovskite materials, bandgap engineering was employed by substitution of Gd(III) sites by La(III) sites. Figure 8b depicts the combined VRBE illustrations of **(b)** $La_xGd_{1-x}AlO_3$ (x = 0 – 1) including GS binding

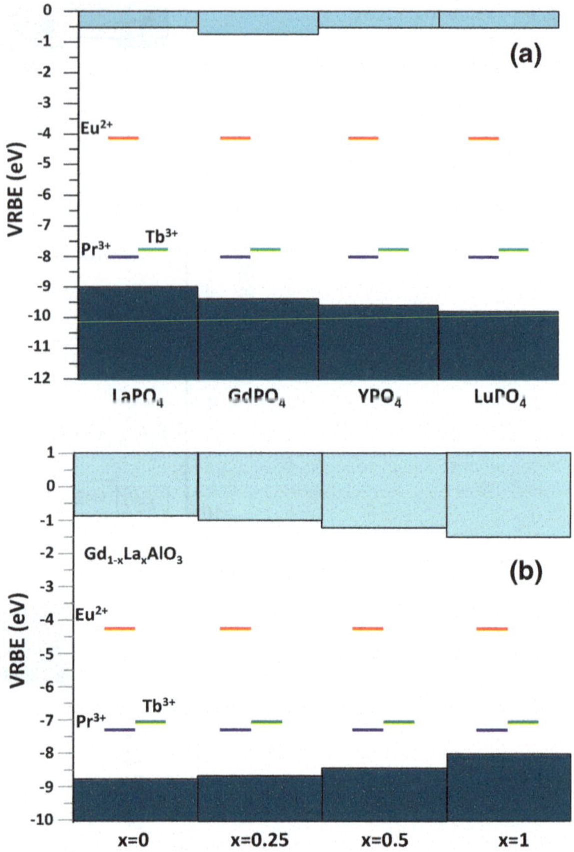

Fig. 8 Combined VRBE diagram of **a** MPO_4 (M = La, Gd, Y and Lu) and **b** $La_xGd_{1-x}AlO_3$ (x = 0 – 1) respectively including GS binding energy of Eu(II), Pr(III) and Tb(III) ions

energy of Eu(II), Pr(III) and Tb(III) ions (Dorenbos et al. [133]). Replacement of Gd^{3+} by La(III) results in a regular increase in the energy of the CB edge as the x value varies from 0 to 1. Therefore, in Eu(III) doped and Tb(III) or Pr(III) co-doped $Gd_{1-x}La_xAlO_3$, the trap depths decrease continuously with an increase in La content which is further responsible for the shifting of glow peak to lower temperature in the TL emission spectrum of Eu^{3+} (Dorenbos et al. [133]).

4.6 Bandgap Engineering

The origin of PersLumin is affected by two fundamental processes including electron charging and detrapping and to meet the desired application of the materials, careful control of these processes is required. Electron charging involves the promotion of an electron from the excited state of RE ions to the conduction band (CB) while electron detrapping involves the combining of electrons released from the traps with

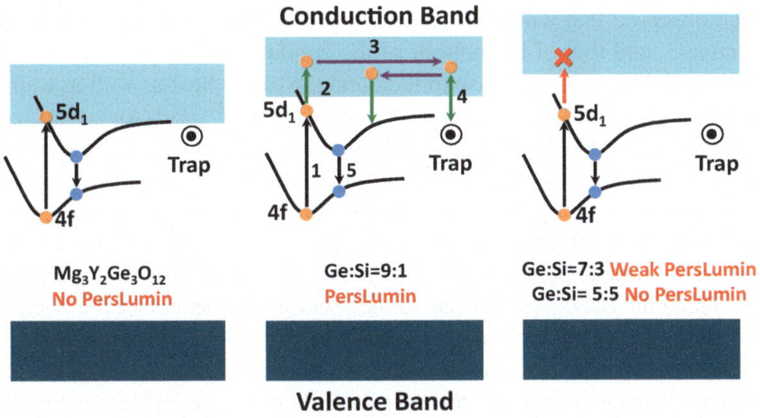

Fig. 9 The pictorial representation of bandgap engineering in $Mg_3Y_{1.99}(Ge_{1-x}Si_x)_3O_{12}$:$0.01Ce^{3+}$ ($x = 0 - 0.5$) by changing the Ge/Si ratio

the emission center. These phosphor materials are used with the commercially used InGaN chips, therefore, the charging wavelength of the persistent phosphor needs to be tailored to the blue region. Furthermore, to get a fast-detrapping step, the trap depth should be shallow (Lozano et al. [7]).

For example, a plausible PersLumin mechanism for $Mg_3Y_{1.99}(Si_xGe_{1-x})_3O_{12}$:$0.01Ce^{3+}$ ($x = 0 - 0.5$) has been outlined in Fig. 9 which includes the five major steps: **(a)** external excitation, **(b)** electron jumping from excited state ($5d^1$) of Ce^{3+}, **(c)** movement of electron through CB, **(d)** trapping-detrapping and **(e)** recombination of electron. Herein, the gap between ES of luminescent ion is tailored in a way to photo – ionize the electron to CB without changing the trap depth, which severely affects the PersLumin intensity. Instead, the decrease in the trap depth leads to a shorter PersLumin lifetime and enhanced PersLumin intensity (Dorenbos et al. [134]; Liu et al. [135]).

$Mg_3Y_2Ge_3O_{12}$ host possesses a narrow bandgap which can be tailored using Si doping and both the electron charging and detrapping in the persistent luminescence process can be optimized. In the undoped host, the excited state of Ce^{3+} almost overlaps with the bottom of CB. As the concentration of Si increases in the host, the bandgap (energy gap between VB and CB) increases. Figure 9 also includes the effect of increased Si content on the afterglow lifetime and PersLumin intensity (Wang et al. [116]). In the absence of Si content, low brightness is observed as some of the thermal-stimulated electrons from the trap do not combine with the emission center and retrap (i.e., return to the trap). With an increase in Si content, trap depth gets deeper which further gives PersLumin with prolonged lifetime (Wang et al. [116]).

In addition, in the case of $Y_3Al_{5-x}Ga_xO_{12}$:Cr^{3+} also the shifting of peak proposes the trap depth modification with the change in Ga^{3+} content which is further used for tuning the PersLumin properties like initial intensity and duration of luminescence.

It has been observed that with the increase in the Ga content, the energy of the top of VB increases and that of the bottom of the conduction band decreases which is further responsible for the decrease in the bandgap of the host as well as trap depth (Tanabe et al. [136]). Similarly, in $Y_3Al_{5-x}Ga_xO_{12}$:Ce^{3+}, $Gd_3Al_{5-x}Ga_xO_{12}$:Ce^{3+} and $Lu_3Al_{5-x}Ga_xO_{12}$:Ce^{3+}, the trap depth and bandgap can also be engineered by changing the Ga content (Stanek et al. [55]; Dorenbos et al. [137]; Zhu et al. [138]).

Also, in $Gd_3Al_xGa_{5-x}O_{12}$:Tb^{3+}, the bandgap can be modified by changing the ratio of Ga:Al. Using the bandgap engineering, the spectroscopic properties of PersLumin materials can be tuned. Higher Al^{3+} content stabilizes the metastable garnet structure and widens the bandgap while increased Ga content decreases the energy gap. Conclusively, with the increase in the Ga fraction, the valence band moves upwards and the conduction band moves downwards. In garnets, PersLumin can be efficient when 5d band is located near the bottom of CB. In case, the 5d band is too deep in the CB or below its bottom, the electron escapes directly to the excited states of Tb^{3+} and no PersLumin is detected. Moreover, the gap between the excited state of Tb^{3+} (i.e., 5d) and CB with the increase in Ga content in the host structure suppresses the energy transfer between the 5d to 4f level of Tb^{3+} ion. Therefore, the trap depth and $4f^7 5d^1$ transitions are important for tuning the PersLumin properties (Głuchowski et al. [21]).

5 Persistent Luminescence in Inorganic Phosphors

Two important properties of PersLumin are (i) initial intensity of luminescence due to electronic transition and (ii) extended luminescence even after ceasing the excitation (Wang et al. [139, 140]). PersLumin phosphors exhibit the same electronic transition in luminescent ion when excitation is turned on and off. For instance, green PersLumin in $SrAl_2O_4$:Eu^{2+},Dy^{3+} is due to Eu^{2+} emission center (Jung et al. [141]). Blue PersLumin in $Y_3Al_2Ga_3O_{12}$:Ce^{3+},Cr^{3+}, where Ce^{3+} and Cr^{3+} behave as recombination and electron trap centers respectively, is obtained due to emission by Ce^{3+} and Cr^{3+} in the green and red range respectively and does not change when excitation source blue LED is turned on or off (Ueda et al. [53]; Liu et al. [135]). In addition, owing to the overlapping of the emission spectrum of Ce^{3+} and the absorption spectrum of Cr^{3+}, Ce^{3+} functions as an energy donor while Cr^{3+} behaves as an energy acceptor. PersLumin of $Y_3Al_2Ga_3O_{12}$:Ce^{3+},Cr^{3+} is composed of two processes: (a) electron detrapping from trap center (Cr^{3+}) followed by PersLumin due to Ce^{3+} and (b) persistent energy transfer process from Ce^{3+} to Cr^{3+} (Ueda et al. [127]; Tanabe et al. [142]). In addition to typical situations, there are also certain "exceptional" ones, such as the difference in the shape of the spectrum between PersLumin and PL. A prominent case is dominating N_2 line of $ZnGa_2O_4$:Cr^{3+} in PersLumin, which is caused by anti-site defects (GaZn or ZnGa) (Srivastava et al. [143]), but in PL, the Cr^{3+} $^2E(^2G) \rightarrow {}^4T_2(^4F)$ R-line transition is considerably more noticeable (Jing et al. [144]; Richard et al. [145]). Additionally, because the efficacy of energy transfer changes due to changes in pumping wavelengths for PL, the spectral ratio between PL and PersLumin can be altered in energy transfer processes across various activators. In inorganic phosphors, the difference of PL and PersLumin from a particular emission center is quite small, allowing for a clear distinction between PersLumin and phosphorescence. Fluorescence and phosphorescence, on the other hand, emerged from quite distinct optical transitions in organic molecules (Itoh et al. [146]). The varied decay duration with the removal of the sources of excitation is the other distinguishing characteristic between PersLumin and phosphorescence. The exceptionally long duration in inorganic PersLumin materials can be explained by the period that is required for a trapped electron or hole to be completely liberated from the trap state to luminescent ion via thermal stimulus at RT, a process that typically takes minutes to even days and has a completely distinct duration (in seconds) for phosphorescence (Monteiro et al. [147]). When comparing distinct PersLumin decay-profiles of different PersLumin phosphors, particular care must be taken because the often-employed arbitrary unit is essentially useless in this context. Real brightness of PersLumin materials is instead typically measured using various physical parameters from optical physics, most notably luminance (mcd/m^2). The International Maritime Organization (IMO), for instance, specifies that PersLumin materials appropriate for night-vision usage must give brightness ~ 15 and > 2 mcd/m^2 at 10 and 60 min respectively (Zhuang et al. [148]) when the excitation source is turned off. The period during which brightness decays to a specific value (often 0.32 mcd/m^2) or around a hundred times the visual threshold of dark adaptation of human eye can be used to determine the lifetime of PersLumin materials (Wang et al. [23]; Liu et al. [40]; Lin et al. [80]). More specifically, brightness cannot be physically measured simply rather an ideal measure taking into account the sensitivities of human vision, as highlighted by Y. Zhuang et al. The power of light is estimated using its luminous flux which is distinct from the radiant flux, a measurement of the general intensity of the emitted light. On the other hand, radiant flux accounts for the human eye's variable sensitivity to various wavelengths of light. Simply, it is radiant flux (power) that has been photometrically weighted. Lumens (lm) are used as the SI unit of luminous flux where a source of light-emitting 1 cd of light intensity across a 1 sr (steradian) solid angle is said to create one lumen of light (Zhang et al. [149]). Thus, an overall measure of luminescence is very helpful in evaluating radiation that can be sensed by the human eye from smooth surfaces, especially flat lighting equipment or night-vision emergency signatures, undoubtedly, within the visible light field (Heggen et al. [1]).

Following the lanthanide-lanthanide co-doping approach, new PersLumin phosphor Ba_2SiO_4:Eu^{2+};Ho^{3+} (bright green) was synthesized in which Eu^{2+} and Ho^{3+} behave as emission and trap centers respectively Ba_2SiO_4:Eu^{2+},Ho^{3+} (termed as super-long-persistent material) exhibited improved PersLumin behavior (~ 5 times) in comparison to commercial green emitting $SrAl_2O_4$:Eu^{2+},Dy^{3+} where the true brightness luminance value, i.e., 2.97 mcd/m^2 for the former is observed at 24 h while the latter exhibit absolute luminance at ~ 5 h upon removal of the excitation source (Hu et al. [150]). However, the wavelength and flux of the stimulation source are too significant in comparing criteria in addition to the duration of the luminescence. Knowing the absolute luminance and wavelength of emission, the

PersLumin materials can be compared. However, the use of these physical quantities is only sometimes explained, and measurements are much less commonly performed. Using night-vision monoculars, high-sensitivity CCD cameras or commercial digital cameras, people often take images at various points after stopping external stimulation to demonstrate the performance of persistent phosphors (Joos et al. [151])]. In $Y_3Al_2Ga_3O_{12}$:Ce^{3+},Yb^{3+} phosphor, Yb^{3+} behaves as a deeper trap center which is responsible for the prolonged PersLumin of Ce^{3+}. PersLumin behavior of this phosphor owing to $5d^1 \rightarrow 4f$ transition was observed for about 110 h after shutting off the excitation source, i.e., blue LED (Zhang et al. [152]).

6 Persistent Luminescence in Organic Molecules

The key point for the generation of long-term persistent luminescence in inorganic phosphors is the segregation of holes and electrons and subsequent trapping by charge carriers. However, the production of charge-separated states in organic molecules is very challenging. Extreme conditions such as an intense source of excitation and low temperature (liq. He) are typically needed for photon ionization. R. Kabe and C. Adachi have demonstrated that PersLumin can be achieved in organic molecules by mixing two different types of organic moieties. They give persistent luminescence of 1 h on excitation under a weak source (white LED) at room temperature or above 100 °C. Among the mixing components, one should be a strong electron-donating molecule and the other will be a strong electron-accepting molecule. Electron donating N,N,N′,N′-tetramethylbenzidine (TMB) has a very stable radical cation while the electron acceptor such as (2,8-bis(diphenylphosphoryl)dibenzo[b,d]thiophene) [PPT] has high triplet state energy, which offers rigid amorphous atmosphere to reduce the relaxation by non-radiative processes (Chen et al. [101]). The acceptor molecule (PPT) by absorbing light fulfills the energy requirement for the removal of electrons from TMB. The additional electron on the acceptor species can move to another acceptor molecule responsible for the charge separation. In comparison to the fluorescence spectra of TMB and PPT, their phosphorescence spectra exhibit bathochromic shifted emission bands measured at 77 K. The phosphorescence (triplet to singlet transition) lifetime of these donor and acceptor moieties measured in toluene is 2.20 and 0.63 s, respectively.

TMB/PPT blend, having 1 mol % optimal concentration of TMB, displayed similar spectral position under and after excitation from exciplexes between the two. This is the exact characteristic of persistent luminescence noticed in inorganic phosphors (Xu et al. [54]; Pan et al. [153]). The lifetime of persistent luminescence in the above-mentioned mixture could continue for more than one h at 300 K and be captured for at least 30 min by employing a CCD camera. The long phosphorescence lifetime of the TMB/PPT blend does not correspond to the $T \rightarrow S$ transition, although it is exhibited due to classical trapping and detrapping processes as in inorganic PersLumin phosphors. Using the process of recombination, the photon-generated radical species must remain in the organic mixture in spite of losing energy, whereas OLEDs or solar cells use a charge separation technique in which the charges are swept out of the device. TMB/PPT film on photo-excitation gives the TMB radical cation corresponding to their absorption and the formation of this cation was confirmed by the broad band ranging 600–1400 nm in the absorption spectra. Greenish-blue, red as well warm white PersLumin using Föster energy transfer from TMB to emissive moieties has already been attained previously by R. Kabe et al. This method resembles the persistent energy transfer mechanism found in inorganic persistent phosphors, where energy is transferred from donor ions to acceptor ions. This new approach of using an organic mixture having a donor–acceptor combination opens up exciting opportunities for advanced applications in biomarkers, fabrics, windows and paints (Kabe et al. [154]).

7 Phosphorescence

Phosphorescence is the emission of light by a material in response to stimulation by an external source that continues to emanate as an afterglow long after the excitation source is turned off. Due to the quantum mechanical spin-flipping across states having different spin multiplicity, phosphorescence is a spin-forbidden radiative transition, from triplet ES (T_1) to singlet GS (S_0). Raising the heating temperature can prolong the duration of phosphorescence, which is longer than that of fluorescence. The gradual emission of photons occurs when a material absorbs energy. Based on the phenomenon of phosphorescence, glow-in-dark materials get charged on absorption of photons from an excitation source (Fig. 10). The absorbed light is instantaneously emitted after excitation due to a transition between states having the same spin called fluorescence having a lifetime of about 10^{-8} – 10^{-9} s, whereas phosphorescence requires additional excitation and might continue from 10^{-3} s to days or years, based on the conditions (Valeur et al. [155]). However, the criterion of a lifetime is insufficient to distinguish the two phenomena as the durations of the two may also be comparable. For instance, there may be long-lived fluorescence and short-lived phosphorescence as shown by Eu^{2+} salts and zinc sulfide, respectively. Firstly, Francis Perrin (1929) differentiated the phosphorescence from the fluorescence and stated that for the former before emission the excited species should pass through an intermediate state. Specifically, spin multiplicity retained in fluorescence changed in phosphorescence. The triplet-to-singlet phosphorescence transition mainly depends on the kind of molecular aggregates, temperature conditions and availability of oxygen. Due to their weaker oscillator strength, triplet-to-singlet transitions possess a considerably higher probability for non-radiative compared to radiative transitions (Tang et al. [156]).

Nowadays, the number of studies on RTP is rising on purely organic compounds which are metal-free, and exhibit promise as materials for advanced applications such as sensors, solid-state lighting, and display technologies because of their favorable characteristics including broad design windows, simple processing and low material costs (Kim et al. [157]). This rise has been observed due to molecular design efforts

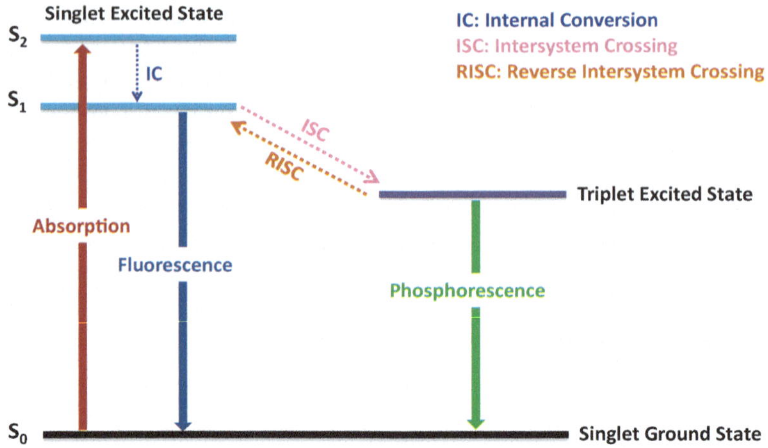

Fig. 10 Jablonski diagram showing various radiative and non-radiative transitions

for the formation of triplet-excitons which require rigorous conditions like careful selection of host, rigid structure of crystalline solid, inert atmosphere, etc. With the advancements in excited states manipulation, organic RTP having a luminous lifetime of the order of seconds has been used for afterglow organic light-emitting diodes (OLEDs), optical storage, bioimaging, biosensing and oxygen/pH-sensors.

The term "phosphorescence" has a phenomenological definition that distinguishes it from fluorescence in terms of time duration, and it has been correctly applied in organic fields up until this point. It is now broadly used to denote the delayed luminescence phenomenon of both inorganic and organic materials, despite having entirely dissimilar mechanistic origins. Such a misunderstanding is extensively reported in the literature, notably about the "glow-in-dark" inorganic PersLumin phosphors' luminous qualities. However, only a small number of studies have raised serious concerns regarding this terminological misunderstanding. We'll give a quick overview of the phosphorescence mechanism and three characteristics that differentiate it from PersLumin. To further emphasize the evident distinction between phosphorescence and PersLumin, recent substantial advancements in the phosphorescence of inorganic phosphors and PersLumin of organic compounds are also presented in separate sections in this chapter.

8 Phosphorescence in Organic Molecules

Numerous characteristics of phosphorescence (P) such as bathochromic shifted emission band in comparison to fluorescence (F) and increase in lifetime with decreasing temperature having range in microseconds to seconds make it different from the PersLumin. The lifetime feature of phosphorescence is dissimilar to PersLumin in inorganic phosphors as the separation of charges from compactly bound excitons to free holes and electrons is quite challenging in organic moieties. This is because the absorption energy retained by traps and charge carriers are gradually released from traps with thermally assisted energy. Hence, the condition of low temperature becomes a promising feature to stabilize the triplet excited state which increases the radiative transition probability and leads to the long lifetime of phosphorescence. Amorphous inorganic materials generated through wet chemical synthesis exhibit ultra-long room temperature phosphorescence (RTP). In these materials, the organic moieties encapsulated in the glassy matrix displayed distinct temperature dependency of a lifetime in comparison to conventional inorganic persistent phosphors. Another way to differentiate phosphorescence from PersLumin is the period after the excitation stops, which may range from a second (RTP) to three orders of magnitude greater in PersLumin. The explanation for this significant difference appears to be very simple as the phosphorescence's duration has a direct relation with the triplet → singlet transition ($\Delta S \neq 0$) may be lengthened by special molecular design. High RTP efficiency can be obtained by fulfilling three requirements: effective intersystem crossing from singlet → triplet to have sufficient triplet exciton, reduction or suppression of non-radiative decay as well as suppression of quenching of triplet (Tang et al. [156]).

Organic luminophores usually show non-efficient ISC which can be amplified by spin–orbit coupling (SOC). A number of factors are responsible increasing this spin–orbit coupling. For instance, the SOC of aromatic compounds is ~ 10^{-1} cm^{-1} which leads to slow intersystem crossing (Nau et al. [158]). This problem can be resolved by introducing groups containing heteroatom (e.g., O, N, S, P) (Chi et al. [159]; Yang et al. [160]; Voskuhl et al. [161]; Takeda et al. [162]) which have abundant lone pair. Another strategy to strengthen SOC is using the Heavy Atom effect. SOC is directly related to Z^4, wherever Z represents the nuclear charge. So, a rapid phosphorescent decay rate and large SOC can be produced using heavy atoms like Cl, Br, I, S, Se and Te (Yang et al. [163]). Intense RTP is observed when heavy atoms are closely packed with phosphorescent centers (Tang et al. [164]). Molecular aggregates show excitonic orbital overlap due to intermolecular interactions (H-aggregation, π–π and n–π stacking) (Li et al. [165]; Bryce et al. [166]) which leads to energy splitting, generating more ISC channels hence boosting the efficiency of RTP. The intersystem crossing can also be increased by narrowing ΔE_S by introducing intermediate states like intramolecular charge-transfer states with the help of donor–acceptor type molecules thus increasing k_{ISC} (Zhang et al. [167]). Another way to achieve efficient RTP in small molecules is structural rigidification which can be achieved by crystallization (Tang et al. [168]). Crystallization-driven RTP can be categorized into various pure organic crystals on the basis of their modes of packing (Kim et al. [169]; Feng et al. [170]). Restraining the phosphorescent cores with the help of crosslinked hydrogen-bonding or halogen-bonding networks leads to a reduction of k_{nr} and k_q which further leads to efficient RTP. This can be achieved by the help of polymerization (Li et al. [171]). Apart from the above-mentioned factors, extremely long RTP lifetimes can be achieved by deuterium substitution, $^3(\pi, \pi^*)$ characteristic enrichment. Deuterium substitution in polycyclic aromatics results in the suppression of intrinsic non-radiative decay. The dominant $^3(\pi, \pi^*)$ characteristics of the triplet state reduce the rate of phosphorescence (Shuai et al. [172]). S. Xu et al. also reported a similar phenomenon for single-component organic molecules (1.35 s), polymers (0.38 s), host–guest systems (5.58 s), CDs (0.66 s) and MOFs (22.4 s). Such RTP lifetimes are estimated utilizing time-resolved spectrophotometers as these lie in their range. The radiative emission among the energy levels of a specific luminescent ion in PL has a substantially shorter lifetime than PersLumin in inorganic phosphors. For example, PersLumin in $Y_3Al_2Ga_3O_{12}$:Ce^{3+},Cr^{3+}, due to Ce^{3+} may persist for almost 14 h at room temperature; however, their fluorescence lifetime is 50 ns. It should be noted that, similar to PersLumin, it is always advised to compare different RTP organic materials by using the luminance parameter instead of duration claimed by different people by observing through the bare eye/digital camera. Furthermore, research on the comparative study of phosphorescence (organic compounds) and PersLumin (inorganic phosphors) has not been published, despite the fact that this comparison should be fascinating and a great method for researchers in two different fields to engage with one another.

The movement of the emission band toward a longer wavelength in comparison to the fluorescence is the distinguishing characteristic of phosphorescence. The photoluminescence (PL) spectrum is associated with various transitions occurring from ES to GS, and the emitted photon's energy is equivalent or proportional to their energy difference. Fluorescence and phosphorescence could be recognized from one another by comparing their PL spectra under stimulation with phosphorescence spectra upon ceasing the excitation. An et al. found that stabilizing triplet excited states for RTP could be used as an intriguing example of this problem. 4,6-diphenyl-2-carbazolyl-1,3,5-triazine (DPCT) molecule quickly stabilizes triplet excitons due to their strong H-aggregation coupling, which promotes the growth of stabilized triplet ES that act as extra traps to reduce radiationless decay. Trapping of holes/electrons in inorganic PersLumin materials promoted this process. Furthermore, a group of organic moieties capable of promoting ISC, through $n \rightarrow \pi^*$ transition comprising –O, –N and –P atoms, were successfully developed. These molecules have shown a comparatively longer lifetime of the triplet state ranging from 0.21 – 1.35 s which caused a significant variation in the luminous background color. Green to red phosphorescence emission bands may be seen after excitation, which can be distinctly separated from fluorescence in the UV region. The long-lived duration and color distinction features of RTP compounds demonstrate their potential application in anticounterfeiting. DPCT-based flexible organic materials possessing long-lasting phosphorescence or excellent RTP characteristics could advance various horizons including molecular imaging, data security, storage encoding, organic optoelectronics, etc.

9 Phosphorescence in Inorganic Phosphors

In inorganic phosphors, PersLumin is the more frequently observed in comparison to phosphorescence and it was supposed that the latter is limited to only organic compounds. Recently, yttrium–aluminum–borate (a–YAB) amorphous inorganic phosphors, produced by wet chemical techniques, challenged this conventional behavior and showed ultra-long RTP. A. D. Sontakke et al. observed the afterglow in a–YAB in comparison to persistent phosphors and phosphorescent dyes to confirm the presence of RTP instead of PersLumin. This RTP in a-YAB was observed due to the formation of a stable triplet-state of trapped luminous organic moiety in a glassy matrix (Sontakke et al. [173]). The intense RTP in the sample was observed for about 10 s even with naked eyes and well-analyzed by digital cameras. Meanwhile, the detailed mechanism for this luminous behavior is still unclear and is supposed to be due to the trapping of carbogenic impurities during their synthesis. However, the interpretation of results for a-YAB clarifies that the intense afterglow luminescence is owing to phosphorescence comprising triplet-to-singlet transitions. The phosphorescent behavior of a-YAB was also confirmed by the comparison of red-shifted bands obtained in the emission spectra of a-YAB (recorded at 365 nm excitation wavelength) with steady-state fluorescent under excitation (Lin et al. [174]). The intense luminescence, high quantum yield, suitable color rendering property and greater physical and chemical stability have projected dopant-free a-YAB as promising materials for solid-state illumination (Sontakke et al. [175]). The pioneering research on phosphorescence behavior in inorganic materials opens up novel opportunities for maintaining phosphorescence performance at high temperatures, which could be particularly attractive for several desirable applications, such as temperature-gradient imaging, time-resolved bioimaging and nano-thermometry.

10 Mechanism of Phosphorescence

Organic luminescence, a kind of cold-body emission induced upon stimulation, includes mainly fluorescence and phosphorescence (Itoh et al. [146]; (Singh et al. [176, 177]). Phosphorescence includes light emission between levels of distinct spin multiplicities, while fluorescence involves light emission across energy levels having the same spin multiplicities (Ågren et al. [178]). Generally, the Jablonski diagram is used to explain the basic mechanism of light emission from organic materials, however, due to abundant flexible excited states, numerous luminescence phenomena including fluorescence and phosphorescence become possible making it more complicated (Singh et al. [179]; Liu et al. [180]; Singh et al. [181]). Fluorescence occurs on excitation of an electron by incident radiation in singlet GS (S_0) to singlet ES (S_1) which thereafter undergoes radiative relaxation to the ground state in a very little period (ns order) by releasing a photon (Reineke et al. [182]). Phosphorescence involves the transfer of electrons from the S_1 state to the triplet excited state

(T_1) by non-radiative intersystem crossing (ISC) followed by the radiative relaxation from T_1 to S_0 showing delayed luminescence (Shuai et al. [172]; Marder et al. [183]; Singh et al. [184]). Nevertheless, the phosphorescence is not always practical at RT because mostly T_1 is unstable and can covert to S_0 in radiationless mode by (1) oxygen-quenching, (2) triplet–triplet annihilation and (3) non-radiative relaxation [185]:

$$T_1 + 3O_2 \rightarrow S_0 + 1O_2 \tag{1}$$

$$T_1 + T_1 \rightarrow S_0 + S_1 \tag{2}$$

$$T_1 \rightarrow S_0 + heat \tag{3}$$

In organometallic complexes, due to the heavy atom effect of noble heavy metals, T_1 gets stabilized which enables ISC across singlet (S_1) and triplet (T_1) ES with varying multiplicity (Singh et al. [186]). This concept of the heavy-atom effect has been applied in designing organic phosphorescent materials (He et al. [187]; Huang et al. [188]). In addition, molecular design has enabled substantial improvements in the lifetime modification of ES. For example, some pure organic compounds without heavy atom effect have exhibited an afterglow duration of > 1.35 s, and multi-component materials like poly(methyl methacrylate) (PMMA) and MOFs also demonstrated long-lasting luminescence having a duration as long as 5.58 and 22.4 s respectively (Reineke et al. [182]).

A phosphorescence with long-lasting spin-forbidden radiative transitions is referred to as room temperature phosphorescence (RTP), sometimes recognized as organic ultra-long RTP (OURTP). It is important to differentiate between two kinds of delayed luminescence, i.e., room temperature phosphorescence (RTP) (Strassert et al. [189]) and thermally activated delayed fluorescence (TADF) that are common in organic molecules [183]. TADF, a singlet-to-singlet relaxation, occurs when reverse intersystem crossing (RISC), i.e., the thermal activation of electrons present at T_1 to S_1 state at RT which is feasible for a small energy gap ($\Delta E \sim 0.5 - 1.0$ eV) between T_1 and S_1 state (George et al. [190]). A highly active field of organic materials with electronic applications is the designing strategy, development and investigations of TADF samples, which is mainly driven by third-generation OLED displays having about 100% internal quantum efficiency (IQE) (Singh et al. [191]). Spin statistics predict that during electrical stimulation, carrier recombination will generate singlet and triplet excitons in a 1:3 ratio (Ågren et al. [178]). The efficiency of electroluminescence (EL) as a whole is increased by the frequent usage of phosphorescent metal–organic complexes, which take advantage of the non-radiative triplet excitons (Singh et al. [192]). However, the novel optical and electrical features of metal-free TADF compounds are due to the effective transition and interconversion between the S_1 and T_1 excited states. Their capacity to use facilitated RISC to collect triplet excitons for fluorescence may have an immediate effect on the characteristics and functionality of the devices (Su et al. [193]). As a result, light may be produced using up to 100% of all excitons, which makes a variety of affordable optoelectronic devices (Singh et al. [194]).

11 Conclusion and Future Perspective

Nowadays, the field of PersLumin phosphors has attracted significant attention owing to their use in biomedical imaging. Although very effective materials have been found, along with vital information on their mechanism of storage, it is still difficult to create phosphors with a suitable particle size for a given application while keeping a high storage capacity. The bioimaging technique which is under trial should become a regularly used technique for bioimaging purposes (Kim et al. [195]). Additionally, some social demands may be met by the creation of certain persistent materials for surveillance and security reasons (Lin et al. [12]; Kim et al. [196]). There is still more work to be done to understand the mechanism for the trapping and detrapping process. The knowledge of the nature of trapping centers and their interaction with recombination centers is critical to understanding the mechanism of PersLumin. The chemical nature of the traps and how they interact with luminescence centers are crucial issues. At times, even identifying the emission center needs additional consideration since the significance of unknowingly inserted contaminants must be thoroughly understood (Liu et al. [18]). Even though we were able to produce identical ultra-small nanophosphors, there are still a lot of difficulties to overcome before we can use persistent luminescence for technological purposes. Utilizing tiny persistent luminous nanophosphors is not required for in vitro biosensing (Hu et al. [197]). However, for the accurate labeling of cellular structures in imaging, smaller particles (~10 nm) are frequently used (Mattheakis et al. [198]). For storing excitation energy and attaining unity production of light having little radiationless loss, persistent luminous nanophosphors with a lot of traps are also extremely desirable to make. There is also a requirement to investigate new methods for preserving the shape of nanomaterials after annealing as well as looking for novel synthetic techniques not involving post-annealing.

Another frequently occurring problem is the poor visible and NIR light charging of all PersLumin nanophosphors. Heterojunction materials could address this issue along with producing materials having tiny bandgaps. By sensitizing PersLumin phosphors, using Vis–NIR sensitizer agents such as quantum dots, it is feasible to efficiently harvest deep-red and NIR light (Qiu et al. [199]). This is made possible by the interfacial stimulated transfers of charge from surface-bound sensitizing agents to luminescent nanomaterials. Bioimaging usages of up-converted PersLumin nanophosphors and quantum dots are now confined to tiny animals. Because of substantial light attenuation, human imaging using persistent luminous nanoparticles is confined to extremely small depths under the skin. These materials are very desirable for photoreactions in deep layers of tissue in which X-ray imaging performs well due to their ability to emit continuously after activation. Although nanomaterials are typically larger than 10 nm, the kidneys cannot easily discharge them through urine, but must instead undergo exceedingly inefficient hepatobiliary clearance. The ultimate goal of the community is to produce PersLumin nanoparticles with an essential afterglow and an ultra-small diameter (5 nm).

References

1. D. Van der Heggen, J.J. Joos, A. Feng, V. Fritz, T. Delgado, N. Gartmann, B. Walfort, D. Rytz, H. Hagemann, D. Poelman, B. Viana, P.F. Smet, Persistent luminescence in strontium aluminate: a roadmap to a brighter future. Adv. Funct. Mater. **32**, 2208809 (2022). https://doi.org/10.1002/adfm.202208809
2. X. Fan, Z. Liu, X. Yang, W. Chen, W. Zeng, S. Tian, X. Yu, J. Qiu, X. Xu, Recent developments and progress of inorganic photo stimulated phosphors. J. Rare Earths **37**, 679–690 (2019). https://doi.org/10.1016/j.jre.2018.12.003
3. Y. Zhuang, Y. Lv, L. Wang, W. Chen, T.L. Zhou, T. Takeda, N. Hirosaki, R.J. Xie, Trap depth engineering of $SrSi_2O_2N_2$:Ln^{2+}, L^{n3}+ (L^{n2}+=Yb, Eu; L^{n3}+=Dy, Ho, Er) persistent luminescence materials for information storage applications. ACS Appl. Mater. Interfaces **10**, 1854–1864 (2018). https://doi.org/10.1021/acsami.7b17271
4. Y. Zhuang, Y. Lv, Y. Li, T. Zhou, J. Xu, J. Ueda, S. Tanabe, R.J. Xie, Study on trap levels in $SrSi_2AlO_2N_3$:Eu^{2+}, L^{n3}+ persistent phosphors based on host-referred binding energy scheme and thermoluminescence analysis. Inorg. Chem. **55**, 11890–11897 (2016). https://doi.org/10.1021/acs.inorgchem.6b01971
5. X. Shi, L. Shi, M. Li, J. Hou, L. Chen, C. Ye, W. Shen, L. Jiang, Y. Song, Efficient luminescence of long persistent phosphor combined with photonic crystal. ACS Appl. Mater. Interf. **6**, 6317–6321 (2014). https://doi.org/10.1021/am501420w
6. Z. Zhou, Y. Li, M. Peng, Near-infrared persistent phosphors: synthesis, design, and applications. Chem. Eng. J. **399**, 125688 (2020). https://doi.org/10.1016/j.cej.2020.125688
7. V. Castaing, E. Arroyo, A.I. Becerro, M. Ocaña, G. Lozano, H. Míguez, Persistent luminescent nanoparticles: challenges and opportunities for a shimmering future. J. Appl. Phys. **130**, 080902 (2021). https://doi.org/10.1063/5.0053283
8. J. Du, A. Feng, D. Poelman, Temperature dependency of trap-controlled persistent luminescence. Laser Phot. Rev. **14**, 2000060 (2020). https://doi.org/10.1002/lpor.202000060
9. S. Wu, Y. Li, W. Ding, L. Xu, Y. Ma, L. Zhang, Recent advances of persistent luminescence nanoparticles in bioapplications. Nano-Micro Lett. **12**, 70–95 (2020). https://doi.org/10.1007/s40820-020-0404-8
10. X. Wei, X. Huang, Y. Zeng, L. Jing, W. Tang, X. Li, H. Ning, X. Sun, Y. Yi, M. Gao, Longer and stronger: improving persistent luminescence in size-tuned zinc gallate nanoparticles by alcohol-mediated chromium doping. ACS Nano **14**, 12113–12124 (2020). https://doi.org/10.1021/acsnano.0c05655
11. A. Abdukayum, J.T. Chen, Q. Zhao, X.P. Yan, Functional near infrared-emitting Cr^{3+}/Pr^{3+} co-doped zinc gallogermanate persistent luminescent nanoparticles with superlong afterglow for *in vivo* targeted bioimaging. J. Am. Chem. Soc. **135**, 14125–14133 (2013). https://doi.org/10.1021/ja404243v
12. J. Du, S. Lyu, K. Jiang, D. Huang, J. Li, R. Van Deun, D. Poelman, H. Lin, Deep-level trap formation in Si-substituted Sr_2SnO_4:Sm^{3+} for rewritable optical information storage. Mater. Today Chem. **24**, 100906 (2022). https://doi.org/10.1016/j.mtchem.2022.100906
13. Y. Wu, Y. Li, X. Qin, R. Chen, D. Wu, S. Liu, J. Qiu, Near-infrared long-persistent phosphor of $Zn_3Ga_2Ge_2O_{10}$:Cr^{3+} sintered in different atmosphere. Spectrochim. Acta A. Mol. Biomol. Spectrosc. **151**, 385–389 (2015). https://doi.org/10.1016/j.saa.2015.06.117
14. D. Singh, V. Tanwar, A.P. Samantilleke, B. Mari, S. Bhagwan, P.S. Kadyan, I. Singh, Preparation and photoluminescence properties of $SrAl_2O_4$:Eu^{2+}, R^{E3}+ green nanophosphors for

display device applications. J. Electron. Mater. **45**, 2718–2724 (2016). https://doi.org/10.1007/s11664-015-4318-z
15. R. Krishnan, H.C. Swart, Recent progress in phosphor technology. Phosphor. Handb. Process. Prop. Appl. **99–115**(2023). https://doi.org/10.1016/B978-0-323-90539-8.00015-2
16. Y. Li, M. Gecevicius, J. Qiu, Long persistent phosphors - from fundamentals to applications. Chem. Soc. Rev. **45**, 2090–2136 (2016). https://doi.org/10.1039/c5cs00582e
17. T. Lyu, P. Dorenbos, Bi^{3+} acting both as an electron and as a hole trap in La–, Y–, and $LuPO_4$. J. Mater. Chem. C. **6**, 6240–6249 (2018). https://doi.org/10.1039/c8tc01020j
18. L. Liang, J. Chen, K. Shao, X. Qin, Z. Pan, X. Liu, Controlling persistent luminescence in nanocrystalline phosphors. Nat. Mater. **22**, 289–304 (2023). https://doi.org/10.1038/s41563-022-01468-y
19. Y. Liu, X. Zhang, Z. Hao, X. Wang, J. Zhang, Tunable full-color-emitting $Ca_3Sc_2Si_3O_{12}:Ce^{3+}$, Mn^{2+} phosphor *via* charge compensation and energy transfer. Chem. Commun. **47**, 10677–10679 (2011). https://doi.org/10.1039/c1cc14324g
20. J. Xu, D. Murata, Y. Katayama, J. Ueda, S. Tanabe, Cr^{3+}/Er^{3+} co-doped $LaAlO_3$ perovskite phosphor: a near-infrared persistent luminescence probe covering the first and third biological windows. J Mater Chem B **5**, 6385–6393 (2017). https://doi.org/10.1039/c7tb01332a
21. R. Tomala, K. Korkus, V. Boiko, D. Hreniak, P. Głuchowski, Electronic structure engineering of $Gd_{2.97}Tb_{0.03}Ga_{5-x}Al_xO_{12}$ persistent luminescence phosphors. J. Alloys Compd. **889**, 161745 (2022). https://doi.org/10.1016/j.jallcom.2021.161745
22. S. Zheng, J. Shi, X. Fu, C. Wang, X. Sun, C. Chen, Y. Zhuang, X. Zou, Y. Li, H. Zhang, X-ray recharged long afterglow luminescent nanoparticles $MgGeO_3:Mn^{2+}$, Y^{b3}+, Li+in the first and second biological windows for long-term bioimaging. Nanoscale **12**, 14037–14046 (2020). https://doi.org/10.1039/c9nr10622g
23. W. Chen, Y. Wang, W. Zeng, S. Han, G. Li, H. Guo, Y. Li, Q. Qiang, Long persistent composite phosphor $CaAl_2O_4:Eu^{2+}$, N^{d3}+/$_Y$3A_l5$_O$$_1$2:$C^{e3}$+: a novel strategy to tune the colors of persistent luminescence. New J. Chem. **40**, 485–491 (2016). https://doi.org/10.1039/c5nj02438b
24. S. Kadyan, S. Singh, S. Sheoran, A. Samantilleke, B. Mari, D. Singh, Synthesis and optoelectronic characteristics of $MGdAl_3O_7:Eu^{3+}$ nanophosphors for current display devices. Trans. Indian Ceram. Soc. **78**, 219–226 (2019). https://doi.org/10.1080/0371750X.2019.1690583
25. S. Singh, V. Tanwar, A.P. Simantilleke, H. Kumar, D. Singh, Structural and spectroscopic properties of $CaMgSi_2O_6:RE^{3+}$ (Eu^{3+} and Tb^{3+}) nanophosphors under UV-illumination. Optik **221**, 165364 (2020). https://doi.org/10.1016/j.ijleo.2020.165364
26. X. Qin, Y. Li, D. Wu, Y. Wu, R. Chen, Z. Ma, S. Liu, J. Qiu, A novel NIR long phosphorescent phosphor:$SrSnO_3:Bi^{2+}$. RSC Adv. **5**, 101347–101352 (2015). https://doi.org/10.1039/c5ra22375j
27. Y. Li, Y.Y. Li, K. Sharafudeen, G.P. Dong, S.F. Zhou, Z.J. Ma, M.Y. Peng, J.R. Qiu, A strategy for developing near infrared long-persistent phosphors: taking $MAlO_3:Mn^{4+}$, G^{e4}+ (M = La, Gd) as an example. J. Mater. Chem. C. **2**, 2019–2027 (2014). https://doi.org/10.1039/c3tc32075h
28. I. Gupta, D. Singh, S. Singh, P. Kumar, S. Bhagwan, V. Kumar, Structural and photophysical measurements of Er^{3+} doped $Gd_4Al_2O_9$ nanophosphors for NUV excitable solid-state lighting applications. Chem. Phys. Lett. **814**, 140350 (2023). https://doi.org/10.1016/j.cplett.2023.140350
29. H.S. Virk, History of luminescence from ancient to modern times. Defect. Diffus. Forum **361**, 1–13 (2015). https://doi.org/10.4028/www.scientific.net/ddf.361.1
30. I. Gupta, S. Singh, S. Bhagwan, D. Singh, Rare earth (RE) doped phosphors and their emerging applications: a review. Ceram. Int. **47**, 19282–19303 (2021). https://doi.org/10.1016/j.ceramint.2021.03.308
31. P.F. Smet, I. Moreels, Z. Hens, D. Poelman, Luminescence in sulfides: a rich history and a bright future. Materials **3**, 2834–2883 (2010). https://doi.org/10.3390/ma3042834
32. W. Lehmann, Alkaline earth sulfide phosphors activated by copper, silver, and gold. J. Electrochem. Soc. **117**, 1389 (1970). https://doi.org/10.1149/1.2407328

33. T. Aitasalo, J. Hölsä, H. Jungner, M. Lastusaari, J. Niittykoski, Thermoluminescence study of persistent luminescence materials: Eu^{2+}– and R^{3+}–doped calcium aluminates, $CaAl_2O_4:Eu^{2+}$, $^{R3}+$. J. Phys. Chem. B **110**, 4589–4598 (2006). https://doi.org/10.1021/jp057185m
34. N. Suriyamurthy, B.S. Panigrahi, Effects of non-stoichiometry and substitution on photoluminescence and afterglow luminescence of $Sr_4Al_{14}O_{25}:Eu^{2+}$, Dy^{3+} phosphor. J. Lumin. **128**, 1809–1814 (2008). https://doi.org/10.1016/j.jlumin.2008.05.001
35. S.D. Han, I. Singh, D. Singh, Y.H. Lee, G. Sharma, C.H. Han, Crystal growth of electroluminescent ZnS:Cu, Cl phosphor and its TiO_2 coating by sol-gel method for thick-film EL device. J. Lumin. **115**, 97–103 (2005). https://doi.org/10.1016/j.jlumin.2005.02.020
36. J. Hölsä, Persistent luminescence beats the afterglow: 400 years of persistent luminescence. Electrochem. Soc. Interface **18**, 42–45 (2009). https://doi.org/10.1149/2.f06094if
37. H. Duan, Y.Z. Dong, Y. Huang, Y.H. Hu, X.S. Chen, The important role of oxygen vacancies in $Sr_2MgSi_2O_7$ phosphor. Phys. Lett. Sect. A. Gen. At. Solid. State Phys. **380**, 1056–1062 (2016). https://doi.org/10.1016/j.physleta.2016.01.001
38. T. Lécuyer, E. Teston, G. Ramirez-Garçia, T. Maldiney, B. Viana, J. Seguin, N. Mignet, D. Scherman, C. Richard, Chemically engineered persistent luminescence nanoprobes for bioimaging. Theranostics **6**, 2488–2524 (2016). https://doi.org/10.7150/thno.16589
39. X. Lü, W. Shu, Q. Fang, Q. Yu, X. Xiong, Roles of doping ions in persistent luminescence of $SrAl_2O_4$:Eu2$^{+,}$ RE $^{3+}$ phosphors. J. Mater. Sci. **42**, 6240–6245 (2007). https://doi.org/10.1007/s10853-006-1097-1
40. Z. Wang, Z. Song, L. Ning, Z. Xia, Q. Liu, Enhanced yellow persistent luminescence in $Sr_3SiO_5:Eu^{2+}$ through Ge incorporation. Inorg. Chem. **58**, 8694–8701 (2019). https://doi.org/10.1021/acs.inorgchem.9b01020
41. X. Wang, Y. Mao, Recent advances in Pr^{3+}-activated persistent phosphors. J. Mater. Chem. C. **10**, 3626–3646 (2022). https://doi.org/10.1039/d2tc00208f
42. X. Li, Y. Zhang, D. Geng, J. Lian, G. Zhang, Z. Hou, J. Lin, $CaGdAlO_4:Tb^{3+}/Eu^{3+}$ as promising phosphors for full-color field emission displays. J. Mater. Chem. C. **2**, 9924–9933 (2014). https://doi.org/10.1039/c4tc01909a
43. S. Sheoran, K. Singh, V. Tanwar, S. Singh, A. Samantilleke, D. Singh, Synthesis and spectroscopic investigations of trivalent europium-doped $Z_2Si_3O_8$ ($Z =$ Mg, Ca and Sr) nanophosphors for display applications. Rare Met. **40**, 2610–2617 (2020). https://doi.org/10.1007/s12598-020-01391-8
44. S. Singh, V. Tanwar, A.P. Simantilleke, H. Kumar, D. Singh, Synthesis and photoluminescence behavior of of $SrMg_2Al_{16}O_{27}:Eu^{2+}$ nanocrystalline phosphor. Optik **225**, 165873 (2021). https://doi.org/10.1016/j.ijleo.2020.165873
45. S. Kadyan, S. Singh, A.P. Simantilleke, D. Singh, Synthesis and optical studies of nanocrystalline Eu^{2+}-doped and RE^{3+}(Nd^{3+}, Dy^{3+})–codoped $Ba_4Al_{14}O_{25}$ materials for UV-LEDs. Optik **212**, 164671 (2020). https://doi.org/10.1016/j.ijleo.2020.164671
46. D. Singh, P.S. Kadyan, I. Singh, V. Tanwar, A.P. Simantilke, B. Mari, Rapid synthesis and photoluminescent characterization of $MAl_2O_4:Eu^{2+}$, Dy^{3+} (M = Ca/Ca + Ba/Ca + Mg) blue nanophosphors for white lighting display applications. Adv. Mater. Lett. **7**, 47–53 (2016). https://doi.org/10.5185/amlett.2016.6103
47. D. Dutczak, C. Ronda, T. Jüstel, A. Meijerink, Anomalous trapped exciton and d-f emission in $Sr_4Al\ _{14}O_{25}:Eu^{2+}$. J. Phys. Chem. A **118**, 1617–1621 (2014). https://doi.org/10.1021/jp500947q
48. P. Dorenbos, Mechanism of persistent luminescence in Eu^{2+} and Dy^{3+} codoped aluminate and silicate compounds. J. Electrochem. Soc. **152**, H107 (2005). https://doi.org/10.1149/1.1926652
49. S. Singh, I. Gupta, D. Singh, Sm^{3+}–activated YAG nanocrystals: Synthesis, structural and spectroscopic analysis for orange-red emitting LEDs. Optik **238**, 166482 (2021). https://doi.org/10.1016/j.ijleo.2021.166482
50. S. Singh, D. Singh, Structural and optical properties of green emitting $Y_2SiO_5:Tb^{3+}$ and $Gd_2SiO_5:Tb^{3+}$ nanoparticles for modern lighting applications. Rare Met. **40**, 3289–3298 (2021). https://doi.org/10.1007/s12598-020-01585-0

51. B. Wang, H. Lin, J. Xu, H. Chen, Z. Lin, F. Huang, Y. Wang, Design, preparation, and characterization of a novel red long-persistent perovskite phosphor: $Ca_3Ti_2O_7:Pr^{3+}$. Inorg. Chem. **54**, 11299–11306 (2015). https://doi.org/10.1021/acs.inorgchem.5b01894
52. A.D. Sontakke, J. Ueda, Y. Katayama, Y. Zhuang, P. Dorenbos, S. Tanabe, Role of electron transfer in Ce^{3+} sensitized Yb^{3+} luminescence in borate glass. J. Appl. Phys. **117**, 013105 (2015). https://doi.org/10.1063/1.4905317
53. J. Ueda, S. Miyano, S. Tanabe, Formation of deep electron traps by Yb^{3+} codoping leads to super-long persistent luminescence in Ce^{3+}-doped yttrium aluminum gallium garnet phosphors. ACS Appl. Mater. Interf. **10**, 20652–20660 (2018). https://doi.org/10.1021/acsami.8b02758
54. J. Xu, S. Tanabe, Persistent luminescence instead of phosphorescence: history, mechanism, and perspective. J. Lumin. **205**, 581–620 (2019). https://doi.org/10.1016/j.jlumin.2018.09.047
55. M. Fasoli, A. Vedda, M. Nikl, C. Jiang, B.P. Uberuaga, D.A. Andersson, K.J. McClellan, C.R. Stanek, Band-gap engineering for removing shallow traps in rare-earth $Lu_3Al_5O_{12}$ garnet scintillators using Ga^{3+} doping. Phys. Rev. B. Condens. Matter. Mater. Phys. **84**, 081102 (2011). https://doi.org/10.1103/PhysRevB.84.081102
56. D. Singh, V. Tanwar, A.P. Samantilleke, B. Mari, S. Bhagwan, K.C. Singh, P.S. Kadyan, I. Singh, Synthesis of $Sr_{(1-x-y)}Al_4O_7:Eu_x{}^{2+}$, $L_n{}^{y3}+$ (Ln = Dy, Y, Pr) nanophosphors using rapid gel combustion process and their down conversion characteristics. Electron. Mater. Lett. **13**, 222–229 (2017). https://doi.org/10.1007/s13391-017-6038-4
57. W.M. Yen, D. Jai, W. Jia, X.J. Wanf, United States Patent US 6,953,536 B2 (2007)
58. N. Yu, F. Liu, X. Li, Z. Pan, Near infrared long-persistent phosphorescence in $SrAl_2O_4$: Eu^{2+}, Dy^{3+}, Er^{3+} phosphors based on persistent energy transfer. Appl. Phys. Lett. **95**, 231110 (2009). https://doi.org/10.1063/1.3272672
59. S. Kadyan, S. Singh, S. Sheoran, A. Samantilleke, B. Mari, D. Singh, Synthesis, luminescent and structural characteristics of $Sr_4Al_{14}O_{25}:Eu^{2+}$ and $Sr_4Al_{14}O_{25}:Eu^{2+}$, RE^{3+} (RE = Nd, Dy) long persistent nanophosphors for solid state lighting. Optik **204**, 164159 (2020). https://doi.org/10.1016/j.ijleo.2019.164159
60. P. Dorenbos, Modeling the chemical shift of lanthanide 4f electron binding energies. Phys. Rev. B. Condens. Matter. Mater. Phys. **85**, 1–10 (2012). https://doi.org/10.1103/PhysRevB.85.165107
61. J. Li, C. Wang, J. Shi, P. Li, Z. Yu, H. Zhang, Porous $GdAlO_3$: Cr^{3+}, Sm^{3+} drug carrier for real-time long afterglow and magnetic resonance dual-mode imaging. J. Lumin. **199**, 363–371 (2018). https://doi.org/10.1016/j.jlumin.2018.03.071
62. M. Lastusaari, T. Laamanen, M. Malkamäki, K.O. Eskola, A. Kotlov, S. Carlson, E. Welter, H.F. Brito, M. Bettinelli, H. Jungner, J. Hölsä, The Bologna stone: history's first persistent luminescent material. Eur. J. Mineral. **24**, 885–890 (2012). https://doi.org/10.1127/0935-1221/2012/0024-2224
63. L. Ma, W. Chen, ZnS: Cu, Co water-soluble afterglow nanoparticles: synthesis, luminescence and potential applications. Nanotechnology **21**, 385604 (2010). https://doi.org/10.1088/0957-4484/21/38/385604
64. L. Lin, M. Yin, C. Shi, W. Zhang, Luminescence properties of a new red long-lasting phosphor: $Mg_2SiO_4:Dy^{3+}$, Mn^{2+}. J. Alloys Compd. **455**, 327–330 (2008). https://doi.org/10.1016/j.jallcom.2007.01.059
65. T. Matsuzawa, Y. Aoki, N. Takeuchi, Y. Murayama, A new long phosphorescent phosphor with high brightness, SrAl2O4: Eu2+, Dy3+. J. Electrochem. Soc. **143**, 2670–2673 (1996). https://doi.org/10.1149/1.1837067
66. T. Katsumata, K. Sasajima, T. Nabae, S. Komuro, T. Morikawa, Characteristics of strontium aluminate crystals used for long-duration phosphors. J. Am. Ceram. Soc. **81**, 413–416 (1998). https://doi.org/10.1111/j.1151-2916.1998.tb02349.x
67. Y. Lin, Z. Tang, Z. Zhang, C. Nan, Influence of co-doping different rare earth ions on the luminescence of $CaAl_2O_4$-based phosphors. J. Eur. Ceram. Soc. **23**, 175–178 (2003). https://doi.org/10.1016/S0955-2219(02)00080-8

68. V. Bachmann, C. Ronda, A. Meijerink, Temperature quenching of yellow Ce^{3+} luminescence in YAG:Ce. Chem. Mater. **21**, 2077–2084 (2009). https://doi.org/10.1021/cm8030768
69. Y. Luo, Z. Xia, Effect of Al/Ga substitution on photoluminescence and phosphorescence properties of garnet-type $Y_3Sc_2Ga_{3-x}Al_xO_{12}:Ce^{3+}$ phosphor. J. Phys. Chem. C **118**, 23297–23305 (2014). https://doi.org/10.1021/jp507695n
70. S.K. Sharma, S.S. Pitale, M.M. Malik, R.N. Dubey, M.S. Qureshi, Luminescence studies on the blue-green emitting $Sr_4Al_{14}O_{25}:Ce^{3+}$ phosphor synthesized through solution combustion route. J. Lumin. **129**, 140–147 (2009). https://doi.org/10.1016/j.jlumin.2008.09.002
71. J. Ueda, K. Kuroishi, S. Tanabe, Bright persistent ceramic phosphors of Ce^{3+}-Cr^{3+}-codoped garnet able to store by blue light. Appl. Phys. Lett. **104**, 101904 (2014). https://doi.org/10.1063/1.4868138
72. B. Wang, H. Lin, Y. Yu, D. Chen, R. Zhang, J. Xu, Y. Wang, Ce^{3+}/ Pr^{3+}: YAGG: a long persistent phosphor activated by blue-light. J. Am. Ceram. Soc. **97**, 2539–2545 (2014). https://doi.org/10.1111/jace.12967
73. D. Jia, R.S. Meltzer, W.M. Yen, W. Jia, X. Wang, Green phosphorescence of $CaAl_2O_4:Tb^{3+}$, C^{e3}+ through persistence energy transfer. Appl. Phys. Lett. **80**, 1535–1537 (2002). https://doi.org/10.1063/1.1456955
74. R. Zhong, J. Zhang, X. Zhang, S. Lu, X.J. Wang, Energy transfer and red phosphorescence in strontium aluminates co-doped with Cr^{3+}, Eu^{2+} and Dy^{3+}. J. Lumin. **119–120**, 327–331 (2006). https://doi.org/10.1016/j.jlumin.2006.01.001
75. J.M. Liu, Y.Y. Liu, D.D. Zhang, G.Z. Fang, S. Wang, Synthesis of $GdAlO_3:Mn^{4+}$, G^{e4}+@Au core-shell nanoprobes with plasmon-enhanced near-infrared persistent luminescence for *in Vivo* Trimodality Bioimaging. ACS Appl. Mater. Interf. **8**, 29939–29949 (2016). https://doi.org/10.1021/acsami.6b09580
76. Y. Teng, J. Zhou, S.N. Khisro, S. Zhou, J. Qiu, Persistent luminescence of $SrAl_2O_4$: Eu^{2+}, Dy^{3+}, Cr^{3+} phosphors in the tissue transparency window. Mater. Chem. Phys. **147**, 772–776 (2014). https://doi.org/10.1016/j.matchemphys.2014.06.018
77. J. Lin, Y. Huang, J. Zhang, F. Shi, S. Wei, J. Gao, X. Ding, C. Tang, Synthesis and photoluminescence properties of $MgAl_2O_4:Mn^{2+}$ hexagonal nanoplates. Mater. Res. Bull. **44**, 106–109 (2009). https://doi.org/10.1016/j.materresbull.2008.03.031
78. X. Xu, Y. Wang, X. Yu, Y. Li, Y. Gong, Investigation of Ce–Mn energy transfer in $SrAl_2O_4:Ce^{3+}$, M^{n2}+. J. Am. Ceram. Soc. **94**, 160–163 (2011). https://doi.org/10.1111/j.1551-2916.2010.04061.x
79. P. Dorenbos, Mechanism of persistent luminescence in $Sr_2MgSi_2O_7:Eu^{2+}$; Dy^{3+}. Phys. Status Solidi. Basic Res. **242**, R7–R9 (2005). https://doi.org/10.1002/pssb.200409080
80. Y. Lin, Z. Zhang, Z. Tang, X. Wang, J. Zhang, Z. Zheng, Luminiscent properties of a new long afterglow Eu^{2+} and Dy^{3+} activated $Ca_3MgSi_2O_8$ phosphor. J. Eur. Ceram. Soc. **21**, 683–685 (2001). https://doi.org/10.1016/S0955-2219(00)00252-1
81. N. Lakshminarasimhan, U.V. Varadaraju, Luminescence and afterglow in $Sr_2SiO_4:Eu^{2+}$, RE^{3+} [RE = Ce, Nd, Sm and Dy] phosphors–Role of co-dopants in search for afterglow. Mater. Res. Bull. **43**, 2946–2953 (2008). https://doi.org/10.1016/j.materresbull.2007.12.005
82. Y. Gong, Y. Wang, Y. Li, X. Xu, W. Zeng, Fluorescence and phosphorescence properties of new long-lasting phosphor $Ba_4(Si_3O_8)_2$: Eu^{2+}, Dy^{3+}. Opt. Express **19**, 4310–4315 (2011). https://doi.org/10.1364/oe.19.004310
83. L.C. Rodrigues, H.F. Brito, J. Hölsä, R. Stefani, M.C. Felinto, M. Lastusaari, T. Laamanen, L.A. Nunes, Discovery of the persistent luminescence mechanism of $CdSiO_3:Tb^{3+}$. J. Phys. Chem. C **116**, 11232–11240 (2012). https://doi.org/10.1021/jp212021k
84. L. Zhang, X. Zhou, H. Zeng, H. Chen, X. Dong, White-light long-lasting phosphor $Sr_2SiO_4:Pr^{3+}$. Mater. Lett. **62**, 2539–2541 (2008). https://doi.org/10.1016/j.matlet.2007.12.043
85. B. Lei, Y. Liu, Z. Ye, C. Shi, Luminescence properties of $CdSiO_3:Mn^{2+}$ phosphor. J. Lumin. **109**, 215–219 (2004). https://doi.org/10.1016/j.jlumin.2004.02.010
86. A. Patra, G.A. Baker, S.N. Baker, Effects of dopant concentration and annealing temperature on the phosphorescence from Zn_2SiO_4: Mn^{2+} nanocrystals. J. Lumin. **111**, 105–111 (2005). https://doi.org/10.1016/j.jlumin.2004.06.008

87. X.J. Wang, D. Jia, W.M. Yen, Mn^{2+} activated green, yellow, and red long persistent phosphors. J. Lumin. **102**, 34–37 (2003). https://doi.org/10.1016/S0022-2313(02)00541-0
88. P. Feng, Y. Wei, Y. Wang, J. Zhang, H. Li, Z. Ci, Long persistent phosphor $CdSiO_3$:Gd^{3+}, B^{i3}+ and potential photocatalytic application of $CdSi_O3$:G^{d3}+, $^{Bi}3$+@T_iO2 in dark. J. Am. Ceram. Soc. **99**, 2368–2375 (2016). https://doi.org/10.1111/jace.14142
89. J. Kuang, Y. Liu, Luminescence properties of a Pb^{2+} activated long-afterglow phosphor. J. Electrochem. Soc. **153**, G245 (2006). https://doi.org/10.1149/1.2163780
90. J. Ueda, R. Maki, S. Tanabe, Vacuum referred binding energy (VRBE)-guided design of orange persistent $Ca_3Si_2O_7$:Eu^{2+} phosphors. Inorg. Chem. **56**, 10353–10360 (2017). https://doi.org/10.1021/acs.inorgchem.7b01214
91. T. Maldiney, A. Lecointre, B. Viana, A. Bessière, M. Bessodes, D. Gourier, C. Richard, D. Scherman, Controlling electron trap depth to enhance optical properties of persistent luminescence nanoparticles for *in vivo* imaging. J. Am. Chem. Soc. **133**, 11810–11815 (2011). https://doi.org/10.1021/ja204504w
92. K. Van den Eeckhout, P.F. Smet, D. Poelman, Luminescent afterglow behavior in the $M_2Si_5N_8$: Eu family (M = Ca, Sr, Ba). Materials **4**, 980–990 (2011). https://doi.org/10.3390/ma4060980
93. J. Li, B. Lei, J. Qin, Y. Liu, X. Liu, Temperature-dependent emission spectra of $Ca_2Si_5N_8$: Eu^{2+}, Tm^{3+} phosphor and its afterglow properties. J. Am. Ceram. Soc. **96**, 873–878 (2013). https://doi.org/10.1111/jace.12171
94. Y.Q. Li, A.C. Delsing, G. De With, H.T. Hintzen, Luminescence properties of Eu^{2+}-activated alkaline-earth silicon-oxynitride $MSi_2O_{2\text{-}\delta}N_{2+2/3\delta}$ (M = Ca, Sr, Ba): a promising class of novel LED conversion phosphors. Chem. Mater. **17**, 3242–3248 (2005). https://doi.org/10.1021/cm050175d
95. S.A. Fartode, V. Singh, S.J. Dhoble, Photoluminescence and thermoluminescence properties of Eu^{2+} and Ce^{3+} activated $BaAlSi_5O_2N_7$ phosphors. Defect. Diffus. Forum **361**, 177–193 (2015). https://doi.org/10.4028/www.scientific.net/DDF.361.177
96. H. Guo, Y. Wang, W. Chen, W. Zeng, S. Han, G. Li, Y. Li, Controlling and revealing the trap distributions of $Ca_6BaP_4O_{17}$:Eu^{2+}, ^{R3}+ (R = Dy, Tb, Ce, Gd, Nd) by codoping different trivalent lanthanides. J Mater Chem C **3**, 11212–11218 (2015). https://doi.org/10.1039/c5tc02283e
97. G. Ju, Y. Hu, L. Chen, X. Wang, Z. Mu, The influence of auxiliary codopants on persistent phosphor $Sr_2P_2O_7$:Eu^{2+}, ^{R3}+ (R = Y, La, Ce, Gd, Tb and Lu). Mater. Res. Bull. **48**, 4743–4748 (2013). https://doi.org/10.1016/j.materresbull.2013.08.011
98. M. Wu, W. Chen, Y. Zhang, J. Zhang, G. Chen, Z. Zheng, Defects enhanced photoluminescence of Mn^{2+}-doped ZrP_2O_7 blue LLP materials. J. Alloys Compd. **789**, 375–380 (2019). https://doi.org/10.1016/j.jallcom.2019.03.022
99. S.K. Sharma, M. Bettinelli, I. Carrasco, J. Beyer, R. Gloaguen, J. Heitmann, Dynamics of charges in superlong blacklight-emitting CaB_2O_4:Ce^{3+} persistent phosphor. J. Phys. Chem. C **123**, 14639–14646 (2019). https://doi.org/10.1021/acs.jpcc.9b02210
100. R. Stefani, A.D. Maia, C.A. Kodaira, E.E. Teotonio, M.C. Felinto, H.F. Brito, Highly enhanced luminescence of SrB_4O_7:Eu^{2+} phosphor prepared by the combustion method using glycine as fuel. Opt. Mater. **29**, 1852–1855 (2007). https://doi.org/10.1016/j.optmat.2006.10.012
101. Y. Zheng, D. Chen, Luminescence studies on $Al_4B_2O_9$:Eu^{2+} phosphor crystals. Luminescence **26**, 481–485 (2011). https://doi.org/10.1002/bio.1256
102. H. Yao, Z. Qiu, X. Xiong, Preparation and luminescence properties of $Ba_2Mg(BO_3)_2$:$Eu^{2+/3+}$phosphor. IOP Conf. Ser: Earth Environ. Sci. **714**, 022019 (2021). https://doi.org/10.1088/1755-1315/714/2/022019
103. T. Tsuboi, P. Silfsten, The lifetime of Eu^{2+} fluorescence in CaF_2: Eu^{2+} crystals. J. Phys. Condens. Matter **3**, 9163 (1991). https://doi.org/10.1088/0953-8984/3/46/017
104. S. Fujihara, Y. Kishiki, T. Kimura, Pyrolytic synthesis and $Eu^{3+}\rightarrow Eu^{2+}$ reduction process of blue-emitting perovskite-type $BaLiF_3$: Eu thin films. J. Solid State Chem. **177**, 1032–1036 (2004). https://doi.org/10.1016/j.jssc.2003.10.008
105. R.E. Rojas-Hernandez, F. Rubio-Marcos, M.Á. Rodriguez, J.F. Fernandez, Long lasting phosphors: $SrAl_2O_4$:Eu, Dy as the most studied material. Renew. Sustain. Energy Rev. **81**, 2759–2770 (2018). https://doi.org/10.1016/j.rser.2017.06.081

106. D. Singh, S. Sheoran, Synthesis and luminescent characteristics of M3Y2Si3O12:Eu3+ (M = Ca, Mg, Sr and Ba) nanomaterials for display applications. J. Mater. Sci. Mater. Electron. **27**, 12707–12718 (2016). https://doi.org/10.1007/s10854-016-5405-5
107. P. Kumar, D. Singh, I. Gupta, S. Singh, S. Nehra, R. Kumar, A study of phase evolution, crystallographic and down-conversion luminescent behaviour of monoclinic $Y_4Al_2O_9$:Dy^{3+} nanophosphors for white light applications. Opt. Mater. **138**, 113677 (2023). https://doi.org/10.1016/j.optmat.2023.113677
108. P. Kumar, D. Singh, I. Gupta, S. Singh, S. Nehra, R. Kumar, Realization of warm reddish-orange light emitter single phase $Y_4Al_2O_9$:Sm^{3+} nanophosphors for indoor lighting applications. J. Lumin. **257**, 119703 (2023). https://doi.org/10.1016/j.jlumin.2023.119703
109. D. Singh, V. Tanwar, A.P. Simantilleke, B. Mari, P.S. Kadyan, I. Singh, Rapid synthesis and enhancement in down conversion emission properties of $BaAl_2O_4$:Eu^{2+}, R^{E3}+ (R^{E3}+=Y, Pr) nanophosphors. J. Mater. Sci. Mater. Electron. **27**, 2260–2266 (2016). https://doi.org/10.1007/s10854-015-4020-1
110. D. Singh, V. Tanwar, A.P. Simantilleke, S. Bhagwan, B. Mari, P.S. Kadyan, K.C. Singh, I. Singh, Synthesis and enhanced luminescent characterization of $SrAl_4O_7$:Eu^{2+}, R^{E3}+ (RE = Nd, Dy) nanophosphors for light emitting applications. J. Mater. Sci. Mater. Electron. **27**, 5303–5308 (2016). https://doi.org/10.1007/s10854-016-4428-2
111. D. Zhou, Z. Wang, Z. Song, F. Wang, S. Zhang, Q. Liu, Enhanced persistence properties through modifying the trap depth and density in $Y_3Al_2Ga_3O_{12}$:$Ce3^{+,}$ $Yb3^{+}$phosphor by Co-doping $B3^{+}$. Inorg. Chem. **58**, 1684–1689 (2019). https://doi.org/10.1021/acs.inorgchem.8b03270
112. L. Li, V. Castaing, D. Rytz, A.D. Sontakke, Y. Katayama, S. Tanabe, M. Peng, B. Viana, Tunable trap depth for persistent luminescence by cationic substitution in Pr^{3+}:$K_{1-x}Na_xNbO_3$ perovskites. J. Am. Ceram. Soc. **102**, 2629–2639 (2019). https://doi.org/10.1111/jace.16116
113. B. Marí, K.C. Singh, P. Cembrero-Coca, I. Singh, D. Singh, S. Chand, Red emitting $MTiO_3$ (M = Ca or Sr) phosphors doped with Eu $^{3+}$ or Pr^{3+} with some cations as co-dopants. Displays **34**, 346–351 (2013). https://doi.org/10.1016/j.displa.2013.07.003
114. P. Kumar, D. Singh, I. Gupta, S. Singh, S. Nehra, R. Kumar, Combustion derived single phase $Y_4Al_2O_9$:Tb^{3+} nanophosphor: crystal chemistry and optical analysis for solid state lighting applications. RSC Adv. **13**, 7752–7765 (2023). https://doi.org/10.1039/d3ra00735a
115. P. Kumar, D. Singh, I. Gupta, S. Singh, S. Nehra, R. Kumar, Er^{3+}-doped $Y_4Al_2O_9$ nanophosphors for advance display applications: synthesis, crystal chemistry and down conversion photoluminescent investigation. Mater. Chem. Phys. **301**, 127610 (2023). https://doi.org/10.1016/j.matchemphys.2023.127610
116. H. Lin, J. Xu, Q. Huang, B. Wang, H. Chen, Z. Lin, Y. Wang, Bandgap tailoring *via* Si doping in inverse-garnet $Mg_3Y_2Ge_3O_{12}$:Ce^{3+} persistent phosphor potentially applicable in AC–LED. ACS Appl. Mater. Interf. **7**, 21835–21843 (2015). https://doi.org/10.1021/acsami.5b06071
117. T. Aitasalo, P. Dereń, J. Hölsä, H. Jungner, J.C. Krupa, M. Lastusaari, J. Legendziewicz, J. Niittykoski, W. Stręk, Persistent luminescence phenomena in materials doped with rare earth ions. J. Solid State Chem. **171**, 114–122 (2003). https://doi.org/10.1016/S0022-4596(02)00194-9
118. F. Clabau, X. Rocquefelte, S. Jobic, P. Deniard, M.H. Whangbo, A. Garcia, T. Le Mercier, On the phosphorescence mechanism in $SrAl_2O_4$:Eu^{2+} and its codoped derivatives. Solid State Sci. **9**, 608–612 (2007). https://doi.org/10.1016/j.solidstatesciences.2007.03.020
119. T. Aitasalo, J. Hassinen, J. Hölsä, T. Laamanen, M. Lastusaari, M. Malkamäki, J. Niittykoski, P. Novák, Synchrotron radiation investigations of the $Sr_2MgSi_2O_7$:Eu^{2+}, ^{R3}+ persistent luminescence materials. J. Rare Earths **27**, 529–538 (2009). https://doi.org/10.1016/S1002-0721(08)60283-5
120. L. Fernández-Rodríguez, A. Durán, M.J. Pascual, Silicate–based persistent phosphors. Open Ceram. **7**, 100150 (2021). https://doi.org/10.1016/j.oceram.2021.100150
121. S. Singh, A.P. Simantilleke, D. Singh, Synthesis, structural and photoluminescence behaviour of novel La_2SiO_5:Eu^{3+}/Tb^{3+} nanomaterials for UV-LEDs. Optik **221**, 165324 (2020). https://doi.org/10.1016/j.ijleo.2020.165324

122. S. Kadyan, S. Singh, A. Simantilleke, B. Mari, D. Singh, Rapid-gel combustion synthesis, structure and luminescence investigations of trivalent europium doped $MGdAlO_4$ (M = Mg^{2+}, Ca^{2+}, Sr^{2+} and Ba^{2+}) nanophosphors. Optik **200**, 163450 (2020). https://doi.org/10.1016/j.ijleo.2019.163450
123. S. Sheoran, V. Singh, S. Singh, S. Kadyan, J. Singh, D. Singh, Down-conversion characteristics of Eu^{3+} doped $M_2Y_2Si_2O_9$ (M = Ba, Ca, Mg and Sr) nanomaterials for innovative solar panels. Prog Nat Sci Mater Int **29**, 457–465 (2019). https://doi.org/10.1016/j.pnsc.2019.07.003
124. H.F. Brito, J. Hölsä, T. Laamanen, M. Lastusaari, M. Malkamäki, L.C. Rodrigues, Persistent luminescence mechanisms: human imagination at work. Opt. Mater. Express **2**, 371–381 (2012). https://doi.org/10.1364/ome.2.000371
125. K. Korthout, K. Van den Eeckhout, J. Botterman, S. Nikitenko, D. Poelman, P.F. Smet, Luminescence and x-ray absorption measurements of persistent $SrAl_2O_4$:Eu, Dy powders: evidence for valence state changes. Phys. Rev. B. Condens. Matter. Mater. Phys. **84**, 085140 (2011). https://doi.org/10.1103/PhysRevB.84.085140
126. Z. Qi, C. Shi, M. Liu, D. Zhou, X. Luo, J. Zhang, Y. Xie, The valence of rare earth ions in $R_2MgSi_2O_7$: Eu, Dy (R = Ca, Sr) long-afterglow phosphors. Phys. Status Solidi (a) **201**, 3109–3112 (2004). https://doi.org/10.1002/pssa.200406873
127. J. Ueda, P. Dorenbos, A.J. Bos, K. Kuroishi, S. Tanabe, Control of electron transfer between Ce^{3+} and Cr^{3+} in the $Y3Al_{5-x}Ga_xO_{12}$ host *via* conduction band engineering. J. Mater. Chem. C. **3**, 5642–5651 (2015). https://doi.org/10.1039/c5tc00546a
128. H. Luo, A.J. Bos, A. Dobrowolska, P. Dorenbos, Low-temperature VUV photoluminescence and thermoluminescence of UV excited afterglow phosphor $Sr_3Al_xSi_{1-x}O_5$:Ce^{3+}, $L^{n3}+$ (Ln = Er, Nd, Sm, Dy and Tm). Phys. Chem. Chem. Phys. **17**, 15419–15427 (2015). https://doi.org/10.1039/c5cp01710f
129. S. Singh, D. Singh, Down-conversion and structural characterizations of trivalent terbium-doped garnet nanocrystalline phosphors for lighting applications. J. Mater. Sci. Mater. Electron. **32**, 17674–17685 (2021). https://doi.org/10.1007/s10854-021-06303-9
130. V. Tanwar, S. Singh, I. Gupta, P. Kumar, H. Kumar, B. Mari, D. Singh, Preparation and luminescence characterization of Eu(III)-activated Forsterite for optoelectronic applications. J. Mol. Struct. **1250**, 131802 (2022). https://doi.org/10.1016/j.molstruc.2021.131802
131. T. Lyu, P. Dorenbos, Charge carrier trapping processes in lanthanide doped $LaPO_4$, $GdPO_4$, YPO_4, and $LuPO_4$. J. Mater Chem. C. **6**, 369–379 (2018). https://doi.org/ https://doi.org/10.1039/C7TC05221A
132. P. Dorenbos, The electronic level structure of lanthanide impurities in $REPO_4$, $REBO_3$, $REAlO_3$, and RE_2O_3 (RE = La, Gd, Y, Lu, Sc) compounds. J. Phys. Condens. Matter **25**, 225501 (2013). https://doi.org/10.1088/0953-8984/25/22/225501
133. H. Luo, A.J. Bos, P. Dorenbos, Controlled electron-hole trapping and detrapping process in $GdAlO_3$ by valence band engineering. J. Phys. Chem. C **120**, 5916–5925 (2016). https://doi.org/10.1021/acs.jpcc.6b00129
134. P. Dorenbos, Electronic structure engineering of lanthanide activated materials. J. Mater. Chem. **22**, 22344–22349 (2012). https://doi.org/10.1039/c2jm34252a
135. Y.F. Liu, P. Liu, L. Wang, C.E. Cui, H.C. Jiang, J. Jiang, A two-step solid-state reaction to synthesize the yellow persistent $Gd_3Al_2Ga_3O_{12}$:Ce^{3+} phosphor with an enhanced optical performance for AC-LEDs. Chem. Commun. **53**, 10636–10639 (2017). https://doi.org/10.1039/c7cc05041k
136. J. Xu, J. Ueda, Y. Zhuang, B. Viana, S. Tanabe, $Y_3Al_{5-x}Ga_xO_{12}$:Cr^{3+}: a novel red persistent phosphor with high brightness. Appl. Phys. Express **8**, 042602 (2015). https://doi.org/10.7567/APEX.8.042602
137. P. Dorenbos, Electronic structure and optical properties of the lanthanide activated $RE_3(Al_{1-x}Ga_x)_5O_{12}$ (RE=Gd, Y, Lu) garnet compounds. J. Lumin. **134**, 310–318 (2013). https://doi.org/10.1016/j.jlumin.2012.08.028
138. L. Wu, J. Hu, Q. Zou, Y. Lin, D. Huang, D. Chen, H. Lu, H. Zhu, Synthesis and optical properties of a $Y_3(Al/Ga)_5O_{12}$:Ce^{3+}, $C^{r3}+$, $^{Nd}3+$ persistent luminescence nanophosphor: a promising near–infrared–II nanoprobe for biological applications. Nanoscale **12**, 14180–14187 (2020). https://doi.org/10.1039/d0nr03269g

139. H. Lin, B. Wang, Q. Huang, F. Huang, J. Xu, H. Chen, Z. Lin, J. Wang, T. Hu, Y. Wang, $Lu_2CaMg_2(Si_{1-x}Ge_x)_3O_{12}:Ce^{3+}$ solid-solution phosphors: bandgap engineering for blue-light activated afterglow applicable to AC–LED. J. Mater. Chem. C. **4**, 10329–10338 (2016). https://doi.org/10.1039/c6tc03818b
140. H. Lin, B. Wang, J. Xu, R. Zhang, H. Chen, Y. Yu, Y. Wang, Phosphor-in-glass for high-powered remote-type white AC-LED. ACS Appl. Mater Interf. **6**, 21264–21269 (2014). https://doi.org/10.1021/am506251z
141. K.Y. Jung, H.W. Lee, H.K. Jung, Luminescent properties of (Sr, Zn)$A_l2_O4:E^{u2}+,^B3+$ particles as a potential green phosphor for UV LEDs. Chem. Mater. **18**, 2249–2255 (2006). https://doi.org/10.1021/cm060003w
142. J. Xu, J. Ueda, K. Kuroishi, S. Tanabe, Fabrication of Ce^{3+}-Cr^{3+} co-doped yttrium aluminium gallium garnet transparent ceramic phosphors with super long persistent luminescence. Scr. Mater. **102**, 47–50 (2015). https://doi.org/10.1016/j.scriptamat.2015.01.029
143. B.B. Srivastava, S.K. Gupta, S. Mohan, Y. Mao, Molten-salt-assisted annealing for making colloidal $ZnGa_2O_4$: Cr nanocrystals with high persistent luminescence. Chem. A. Eur. J. **27**, 11398–11405 (2021). https://doi.org/10.1002/chem.202101234
144. X. Wei, S.V. Kershaw, X. Huang, M. Jiao, C.C. Beh, C. Liu, M. Sarmadi, A.L. Rogach, L. Jing, Continuous flow synthesis of persistent luminescent chromium-doped zinc gallate nanoparticles. J. Phys. Chem. Lett. **12**, 7067–7075 (2021). https://doi.org/10.1021/acs.jpclett.1c01767
145. T. Maldiney, A. Bessière, J. Seguin, E. Teston, S.K. Sharma, B. Viana, A.J.J. Bos, P. Dorenbos, M. Bessodes, D. Gourier, D. Scherman, C. Richard, The *in vivo* activation of persistent nanophosphors for optical imaging of vascularization, tumours and grafted cells. Nat. Mater. **13**, 418–426 (2014). https://doi.org/10.1038/nmat3908
146. T. Itoh, Fluorescence and phosphorescence from higher excited states of organic molecules. Chem. Rev. **112**, 4541–4568 (2012). https://doi.org/10.1021/cr200166m
147. D.L. Fritzen, L. Giordano, L.C.V. Rodrigues, J.H.S.K. Monteiro, Opportunities for persistent luminescent nanoparticles in luminescence imaging of biological systems and photodynamic therapy. Nanomaterials **10**, 2015–2052 (2020). https://doi.org/10.3390/nano10102015
148. Y. Zhuang, Y. Katayama, J. Ueda, S. Tanabe, A brief review on red to near-infrared persistent luminescence in transition-metal-activated phosphors. Opt Mater (Amst) **36**, 1907–1912 (2014). https://doi.org/10.1016/j.optmat.2014.05.035
149. Y. Zheng, C. Shi, Y. Liang, Q. Lin, C. Guo, H. Zhang, Synthesis and electroluminescent properties of a novel terbium complex. Synth. Met. **114**, 321–323 (2000). https://doi.org/10.1016/s0379-6779(00)00265-4
150. C. Wang, Y. Jin, Y. Lv, G. Ju, D. Liu, L. Chen, Z. Li, Y. Hu, Trap distribution tailoring guided design of super-long-persistent phosphor $Ba_2SiO_4:Eu^{2+}$, $H^{o3}+$ and photostimulable luminescence for optical information storage. J Mater Chem C **6**, 6058–6067 (2018). https://doi.org/10.1039/c8tc01722k
151. A. Feng, J.J. Joos, J. Du, P.F. Smet, Revealing trap depth distributions in persistent phosphors with a thermal barrier for charging. Phys. Rev. B **105**, 205101 (2022). https://doi.org/10.1103/PhysRevB.105.205101
152. X. Zhang, L. Zhou, Q. Pang, J. Shi, M. Gong, Tunable luminescence and $Ce^{3+}\rightarrow Tb^{3+}\rightarrow Eu^{3+}$ energy transfer of broadband-excited and narrow line red emitting $Y_2SiO_5:Ce^{3+}$, Tb^{3+}, Eu^{3+} phosphor. J. Phys. Chem. C **118**, 7591–7598 (2014). https://doi.org/10.1021/jp412702g
153. Z. Pan, V. Castaing, L. Yan, L. Zhang, C. Zhang, K. Shao, Y. Zheng, C. Duan, J. Liu, C. Richard, B. Viana, Facilitating low-energy activation in the near-infrared persistent luminescent phosphor $Zn_{1+x}Ga_{2-2x}Sn_xO_4:Cr^{3+}$ *via* crystal field strength modulations. J. Phys. Chem. C **124**, 8347–8358 (2020). https://doi.org/10.1021/acs.jpcc.0c01951
154. R. Kabe, C. Adachi, Organic long persistent luminescence. Nature **550**, 384–387 (2017). https://doi.org/10.1038/nature24010
155. B. Valeur, M.N. Berberan-Santos, A brief history of fluorescence and phosphorescence before the emergence of quantum theory. J. Chem. Educ. **88**, 731–738 (2011). https://doi.org/10.1021/ed100182h

156. W. Zhao, Z. He, B.Z. Tang, Room-temperature phosphorescence from organic aggregates. Nat. Rev. Mater. **5**, 869–885 (2020). https://doi.org/10.1038/s41578-020-0223-z
157. W. Shao, J. Kim, Metal-free organic phosphors toward fast and efficient room-temperature phosphorescence. Acc. Chem. Res. **55**, 1573–1585 (2022). https://doi.org/10.1021/acs.accounts.2c00146
158. W. Nau, Modern molecular photochemistry of organic molecules. by N. J. Turro, V. Ramamurthy, J. C. Scaiano. ChemPhysChem **12**, 2496–2497 (2011). https://doi.org/10.1002/cphc.201000976
159. B. Xu, H. Wu, J. Chen, Z. Yang, Z. Yang, Y.C. Wu, Y. Zhang, C. Jin, P.Y. Lu, Z. Chi, S. Liu, J. Xu, M. Aldred, White-light emission from a single heavy atom-free molecule with room temperature phosphorescence, mechanochromism and thermochromism. Chem. Sci. **8**, 1909–1914 (2017). https://doi.org/10.1039/c6sc03038f
160. C. Zhou, S. Zhang, Y. Gao, H. Liu, T. Shan, X. Liang, B. Yang, Y. Ma, Ternary emission of fluorescence and dual phosphorescence at room temperature: a single-molecule white light emitter based on pure organic Aza-aromatic material. Adv. Funct. Mater. **28**, 1802407 (2018). https://doi.org/10.1002/adfm.201802407
161. S. Riebe, C. Vallet, F. van der Vight, D. Gonzalez-Abradelo, C. Wölper, C.A. Strassert, G. Jansen, S. Knauer, J. Voskuhl, Aromatic thioethers as novel luminophores with aggregation-induced fluorescence and phosphorescence. Chem. A. Eur. J. **23**, 13660–13668 (2017). https://doi.org/10.1002/chem.201701867
162. Y. Takeda, T. Kaihara, M. Okazaki, H. Higginbotham, P. Data, N. Tohnai, S. Minakata, Conformationally-flexible and moderately electron-donating units-installed D-A–D triad enabling multicolor-changing mechanochromic luminescence, TADF and room-temperature phosphorescence. Chem. Commun. **54**, 6847–6850 (2018). https://doi.org/10.1039/c8cc02365d
163. Y. Wen, H. Liu, S. Zhang, Y. Gao, Y. Yan, B. Yang, One-dimensional π-π Stacking induces highly efficient pure organic room-temperature phosphorescence and ternary-emission single-molecule white light. J. Mater. Chem. C. **7**, 12502–12508 (2019). https://doi.org/10.1039/c9tc04580e
164. J. Wang, X. Gu, H. Ma, Q. Peng, X. Huang, X. Zheng, S.H.P. Sung, G. Shan, J.W.Y. Lam, Z. Shuai, B.Z. Tang, A facile strategy for realizing room temperature phosphorescence and single molecule white light emission. Nat. Commun. **9**, 2963 (2018). https://doi.org/10.1038/s41467-018-05298-y
165. J. Yang, X. Zhen, B. Wang, X. Gao, Z. Ren, J. Wang, Y. Xie, J. Li, Q. Peng, K. Pu, Z. Li, The influence of the molecular packing on the room temperature phosphorescence of purely organic luminogens. Nat. Commun. **9**, 840 (2018). https://doi.org/10.1038/s41467-018-03236-6
166. Z. Yang, Z. Mao, X. Zhang, D. Ou, Y. Mu, Y. Zhang, C. Zhao, S. Liu, Z. Chi, J. Xu, Y.C. Wu, P.Y. Lu, A. Lien, M.R. Bryce, Intermolecular electronic coupling of organic units for efficient persistent room-temperature phosphorescence. Angew. Chemie. **128**, 2221–2225 (2016). https://doi.org/10.1002/ange.201509224
167. X. Chen, C. Xu, T. Wang, C. Zhou, J. Du, Z. Wang, H. Xu, T. Xie, G. Bi, J. Jiang, X. Zhang, J.N. Demas, C.O. Trindle, Y. Luo, G. Zhang, Versatile room-temperature-phosphorescent materials prepared from N-substituted naphthalimides: emission enhancement and chemical conjugation. Angew Chemie - Int Ed **55**, 9872–9876 (2016). https://doi.org/10.1002/anie.201601252
168. W.Z. Yuan, X.Y. Shen, H. Zhao, J.W.Y. Lam, L. Tang, P. Lu, C. Wang, Y. Liu, Z. Wang, Q. Zheng, J.Z. Sun, Y. Ma, B.Z. Tang, Crystallization-induced phosphorescence of pure organic luminogens at room temperature. J. Phys. Chem. C **114**, 6090–6099 (2010). https://doi.org/10.1021/jp909388y
169. O. Bolton, K. Lee, H.J. Kim, K.Y. Lin, J. Kim, Activating efficient phosphorescence from purely organic materials by crystal design. Nat. Chem. **3**, 205–210 (2011). https://doi.org/10.1038/nchem.984

170. G. Chen, S. Guo, H. Feng, Z. Qian, Anion-regulated transient and persistent phosphorescence and size-dependent ultralong afterglow of organic ionic crystals. J. Mater. Chem. C. **7**, 14535–14542 (2019). https://doi.org/10.1039/c9tc04951G
171. M.M. Fang, J. Yang, Z. Li, Recent advances in purely organic room temperature phosphorescence polymer. Chinese J. Polym. Sci. (English Ed) **37**, 383–393 (2019). https://doi.org/10.1007/s10118-019-2218-z
172. H. Ma, Q. Peng, Z. An, W. Huang, Z. Shuai, Efficient and long-lived room-temperature organic phosphorescence: theoretical descriptors for molecular designs. J. Am. Chem. Soc. **141**, 1010–1015 (2019). https://doi.org/10.1021/jacs.8b11224
173. A.D. Sontakke, A. Ferrier, P. Burner, V.F. Guimarães, M. Salaün, V. Maurel, I. Gautier-Luneau, A. Ibanez, B. Viana, Afterglow luminescence in wet-chemically synthesized inorganic materials: ultra-long room temperature phosphorescence instead of persistent luminescence. J. Phys. Chem. Lett. **8**, 4735–4739 (2017). https://doi.org/10.1021/acs.jpclett.7b01702
174. K. Jiang, Y. Wang, C. Cai, H. Lin, Activating room temperature long afterglow of carbon dots *via* covalent fixation. Chem. Mater. **29**, 4866–4873 (2017). https://doi.org/10.1021/acs.chemmater.7b00831
175. V.F. Guimarães, A.D. Sontakke, L.J.Q. Maia, M. Salaün, I. Gautier-Luneau, A. Ferrier, B. Viana, A. Ibanez, Photoluminescence properties of glassy yttrium aluminum borate powders: dopant-free phosphors for solid-state lighting. J. Lumin. **188**, 448–453 (2017). https://doi.org/10.1016/j.jlumin.2017.05.013
176. A. Dalal, K. Nehra, A. Hooda, D. Singh, S. Kumar, R.S. Malik, Synthesis, photophysical characteristics and geometry optimization of Tris(2-benzoylacetophenonate)europium complexes with 2, 2′-Bipyridine derivatives. J. Lumin. **247**, 118873 (2022). https://doi.org/10.1016/j.jlumin.2022.118873
177. K. Nehra, A. Dalal, A. Hooda, D. Singh, S. Kumar, Exploration of newly synthesized red luminescent material of samarium for display applications. Inorg. Chem. Commun. **139**, 109361 (2022). https://doi.org/10.1016/j.inoche.2022.109361
178. G. Baryshnikov, B. Minaev, H. Ågren, Theory and calculation of the phosphorescence phenomenon. Chem. Rev. **117**, 6500–6537 (2017). https://doi.org/10.1021/acs.chemrev.7b00060
179. A. Dalal, K. Nehra, A. Hooda, S. Singh, D. Singh, S. Kumar, Synthesis, Optoelectronic and photoluminescent characterizations of green luminous heteroleptic ternary terbium complexes. J. Fluoresc. **32**, 1019–1029 (2022). https://doi.org/10.1007/s10895-022-02920-7
180. Y. Liu, G. Zhan, Z.W. Liu, Z.Q. Bian, C.H. Huang, Room-temperature phosphorescence from purely organic materials. Chinese Chem. Lett. **27**, 1231–1240 (2016). https://doi.org/10.1016/j.cclet.2016.06.029
181. K. Nehra, A. Dalal, A. Hooda, K. Jakhar, D. Singh, S. Kumar, Preparation, optoelectronic and spectroscopic analysis of fluorinated heteroleptic samarium complexes for display applications. Inorganica Chim. Acta **537**, 120958 (2022). https://doi.org/10.1016/j.ica.2022.120958
182. C. Salas Redondo, P. Kleine, K. Roszeitis, T. Achenbach, M. Kroll, M. Thomschke, S. Reineke, Interplay of fluorescence and phosphorescence in organic biluminescent emitters. J. Phys. Chem. C **121**, 14946–14953 (2017). https://doi.org/10.1021/acs.jpcc.7b04529
183. Z. Wu, J. Nitsch, T.B. Marder, Persistent room-temperature phosphorescence from purely organic molecules and multi-component systems. Adv. Opt. Mater. **9**, 2100411 (2021). https://doi.org/10.1002/adom.202100411
184. A. Hooda, K. Nehra, A. Dalal, S. Singh, R.K. Saini, S. Kumar, D. Singh, Terbium complexes of an asymmetric β-diketone: preparation, photophysical and thermal investigation. Inorganica Chim. Acta. **536**, 120881 (2022). https://doi.org/10.1016/j.ica.2022.120881
185. H. Yuasa, S. Kuno, Intersystem crossing mechanisms in the room temperature phosphorescence of crystalline organic compounds. Bull. Chem. Soc. Jpn **91**, 223–229 (2018). https://doi.org/10.1246/bcsj.20170364
186. D. Singh, K. Nehra, R.K. Saini, A. Dalal, S. Bhagwan, K. Singh, A.P. Simantilleke, S. Kumar, Luminescence intensification of terbium(III) ion complexes with dipivaloylmethane (tmhd)

and monodentate auxiliary ligands. Optik **206**, 164338 (2020). https://doi.org/10.1016/j.ijleo.2020.164338
187. J. Yu, H. Ma, W. Huang, Z. Liang, K. Zhou, A. Lv, X.G. Li, Z. He, Purely organic room-temperature phosphorescence endowing fast intersystem crossing from through-space spin-orbit coupling. JACS Au **1**, 1694–1699 (2021). https://doi.org/10.1021/jacsau.1c00290
188. W. Zhu, H. Xing, E. Li, H. Zhu, F. Huang, Room-temperature phosphorescence in the amorphous state enhanced by copolymerization and host-guest complexation. Macromolecules **55**, 9802–9809 (2022). https://doi.org/10.1021/acs.macromol.2c00680
189. T. Theiss, S. Buss, I. Maisuls, R. López-Arteaga, D. Brünink, J. Kösters, A. Hepp, N.L. Doltsinis, E.A. Weiss, C.A. Strassert, Room-temperature phosphorescence from Pd(II) and Pt(II) complexes as supramolecular luminophores: the role of self-assembly, metal-metal interactions, spin-orbit coupling, and ligand-field splitting. J. Am. Chem. Soc. **145**, 3937–3951 (2023). https://doi.org/10.1021/jacs.2c09775
190. S. Kuila, S. Garain, G. Banappanavar, B.C. Garain, D. Kabra, S.K. Pati, S.J. George, Ambient room temperature phosphorescence and thermally activated delayed fluorescence from a core-substituted pyromellitic diimide derivative. J. Phys. Chem. B **125**, 4520–4526 (2021). https://doi.org/10.1021/acs.jpcb.1c02253
191. A. Hooda, A. Dalal, K. Nehra, D. Singh, S. Kumar, R.S. Malik, P. Kumar, Preparation and optical investigation of green luminescent ternary terbium complexes with aromatic β-diketone. Chem. Phys. Lett. **794**, 139495 (2022). https://doi.org/10.1016/j.cplett.2022.139495
192. A. Dalal, K. Nehra, A. Hooda, D. Singh, K. Jakhar, S. Kumar, Preparation and photoluminescent characteristics of green Tb(III) complexes with β-diketones and N donor auxiliary ligands. Inorg. Chem. Commun. **139**, 109349 (2022). https://doi.org/10.1016/j.inoche.2022.109349
193. T. Zhang, B. Chu, W. Li, Z. Su, Q.M. Peng, B. Zhao, Y. Luo, F. Jin, X. Yan, Y. Gao, H. Wu, F. Zhang, D. Fan, J. Wang, Efficient Triplet application in exciplex delayed-fluorescence oleds using a reverse intersystem crossing mechanism based on a Δ E S-T of around Zero. ACS Appl. Mater. Interf. **6**, 11907–11914 (2014). https://doi.org/10.1021/am501164s
194. D. Singh, S. Bhagwan, A. Dalal, K. Nehra, K. Singh, A. Simantilleke, S. Kumar, I. Singh, Intense red luminescent materials of ternary Eu^{3+} complexes of oxide ligands for electroluminescent display devices. Optik **208**, 164111 (2020). https://doi.org/10.1016/j.ijleo.2019.164111
195. E.L. Cates, M. Cho, J.H. Kim, Converting visible light into UVC: microbial inactivation by Pr^{3+}-activated upconversion materials. Environ. Sci. Technol. **45**, 3680–3686 (2011). https://doi.org/10.1021/es200196c
196. Y. Zhang, R. Huang, H. Li, D. Hou, Z. Lin, J. Song, Y. Guo, H. Lin, C. Song, Z. Lin, J. Robertson, Germanium substitution endowing Cr^{3+}-doped zinc aluminate phosphors with bright and super-long near-infrared persistent luminescence. Acta Mater. **155**, 214–221 (2018). https://doi.org/10.1016/j.actamat.2018.06.020
197. H. Tan, T. Wang, Y. Shao, C. Yu, L. Hu, Crucial breakthrough of functional persistent luminescence materials for biomedical and information technological applications. Front. Chem. **7**, 387 (2019). https://doi.org/10.3389/fchem.2019.00387
198. L.C. Mattheakis, J.M. Dias, Y.J. Choi, J. Gong, M.P. Bruchez, J. Liu, E. Wang, Optical coding of mammalian cells using semiconductor quantum dots. Anal. Biochem. **327**, 200–208 (2004). https://doi.org/10.1016/j.ab.2004.01.031
199. Y. Li, Y. Li, R. Chen, K. Sharafudeen, S. Zhou, M. Gecevicius, H. Wang, G. Dong, Y. Wu, X. Qin, J. Qiu, Tailoring of the trap distribution and crystal field in Cr^{3+}–doped non-gallate phosphors with near–infrared long-persistence phosphorescence. NPG Asia Mater **7**, e180 (2015). https://doi.org/10.1038/am.2015.38

Theoretical Aspects of Persistent Luminescence

M. Y. A. Yagoub, H. C. Swart, and E. Coetsee

Abstract This chapter presents the reader with a brief overview of the theoretical aspect of persistent luminescence, instead of presenting an extensive review of literature. The chapter focuses on the pathway of the theory of calculating the continuous trap distributions. The persistent luminescence of ultraviolet, red, white, and over 1000 nm were also discussed.

Keywords Persistent luminescence · Trap models · Trap distributions · Trapping kinetics · Thermoluminescence · Trap depth spectroscopy

1 Introduction

Persistent luminescence is prolonged luminescence in luminescent materials (phosphors) after the excitation source ceases. Due to this distinctive optical property, various suggestions to apply persistent luminescent materials to different applications, from luminous paint in the dark to phototherapy, were made [1]. $SrAl_2O_4:Eu^{2+}$, Dy^{3+} was the first commercial persistent phosphor developed in 1995 by Matsuzawa et al. [2]. Since then, many emerging persistent phosphors developed such as $Sr_2MgSi_2O_7$:Eu,Dy, Y_2O_2S:Eu,Ti,Mg, $Sr_4Al_{14}O_{25}$:Eu,Dy, and $CaAl_2O_4$:Eu,Nd that emitted in the blue, red, green, and violet regions (visible regions), respectively [3–5]. The lanthanide ions in these materials acted as excellent traps for carriers. In addition to the visible persistent phosphors, there were tremendous attempts to produce other persistent luminescence materials that would emit in the ultra-violet (UV)-A, B, and C and near-infrared (NIR)-I, II, and III regions [6–8].

M. Y. A. Yagoub (✉) · H. C. Swart · E. Coetsee
Department of Physics, University of the Free State, Free State, Bloemfontein 9300, South Africa
e-mail: yagoub.mya@ufs.ac.za

H. C. Swart
e-mail: swarthc@ufs.ac.za

E. Coetsee
e-mail: coetseee@ufs.ac.za

V. Kumar et al. (eds.), *Persistent Luminescence*, Progress in Optical Science and Photonics 30, https://doi.org/10.1007/978-981-97-4943-0_2

The energy levels inside the bandgap of the materials, introduced by defects, play a crucial role in persistent luminescence. These energy levels are so-called traps, that can capture the charge carriers. In most cases the charge carriers are electrons but some reports showed that holes can also be trapped by the traps [9]. The charge carriers will only be released if enough thermal energy is applied for emission to occur through recombination at the luminescent centers. The depth of the trap is the amount of energy needed to release the imprisoned charge carriers [10]. In the case of electron traps or hole traps, the energy between the trap and the material's valence band or conduction band is used to determine the depth of the trap [9, 11]. Thus, the shallow traps (close to the valance band or conduction band) will give short afterglows. No afterglow was observed at room temperature for traps being too deep, as there were no charge carriers to escape [9]. It was therefore stated that, for an afterglow to be observed at room temperature, a trap depth in the 0.6–0.7 eV range is ideal [2]. Although the general principle responsible for the afterglow mechanism was explained above, there are more complex details involved.

Many efforts have already been made by other researchers to explain persistent luminescence mechanisms [12–14]. It is clear that to understand the persistent luminescence mechanisms, the depth and the density of the energy levels (traps) must be known. The persistent luminescent materials often consist of traps in similar depths or in multiple levels or a continuous distribution [15]. This normally complicates the procedure to determine the traps' depths and will not give accurate values. Therefore, to investigate the persistent luminescence mechanisms, we need to understand the behavior of the persistent materials during the excitation phase and what is happening within the material during the afterglow phase. Thus, to understand the processes of the luminescent materials, we need to know the charge carrier's kinetics inside the persistent materials.

In this chapter, the detrapping and trapping kinetics of the charge carriers are reviewed. The procedure for estimating the trap distribution in persistent luminescence is given. An inside overview of some of the white, NIR, UV, and red persistent phosphors is presented as examples of persistent luminescence.

2 Kinetics of Trapping and Detrapping

There are two main types of phosphorescence. The first type relies on the time that the electron spends in the excited state of the luminescent center. This type of phosphorescence continues for only a few milliseconds and can be represented by an exponential decay curve, which is independent of temperature. The second type of phosphorescence depends on the time that the electron spends in the trap. This type may continue for hours or days. The decay curve of this type of phosphorescence is strongly dependent on temperature and is also an exponential decay curve by nature [13]. The phosphorescence emission after trapping is a result of the contribution of several trap depths. Therefore, the resulting decay curve is a contribution of each trap's depth. Thus, the form of the decay curve strongly depends on the distribution of the trap depths [15]. In this section, the trapping and detrapping kinetics as well as the distribution of traps are discussed and reviewed.

2.1 De-Trapping Kinetics

Persistent luminescence occurs when a detrapped electron or hole was transferred to a luminescent center. The detrapping of the carrier is determined by the carrier's trap depth. To understand the detrapping kinetics let us consider a one-center/trap model with only three energy levels, namely, the trap level, the excited state, which acts as an intermediate level for charge carriers, and the ground state of the luminescent center. For simplicity, the excited state is taken as the conduction band, that allows transportation of the charge carriers between the trap to the luminescence center. In a one-center model there are only three processes that can occur [16]:

1. Detrapping. In this step, the charge carriers are transferred from the trap level to the conduction band (excited state).
2. Recombination. The charge carriers recombine with holes in the luminescent center.
3. Retrapping. Some of the charge carriers could be transferred back to the trap level.

These three processes are schematically presented in Fig. 1. In this model, it is assumed that the electron acts as the charge carrier. Also in this model, the general formula of the emission intensity can be written within the following frame of aspects: charge neutrality, the charge carriers are independent of temperature, the time, and a quasi-equilibrium state [16].

$$I(t, T) = ns\exp\left(-\frac{\Delta E}{KT}\right)\left[1 - \frac{1-(N-n)\sigma_n}{(N-n)\sigma_n + m\sigma_{mn}}\right] \quad (1)$$

The cross-section for retrapping and that for recombination are represented by σ_n and σ_{mn}, respectively. The ratio between these two cross-sections strongly influences the afterglow's decay shape. N is the concentration of total traps (including filled and unfilled traps), n is the filled traps concentration and m is the concentration of available hole states for recombination. The constant s may slowly vary with temperature. The nature of the constant s can be understood if it is assumed that the trap resembles a potential box. In this case, the constant s can be expressed as how frequently the charge carrier would collide with the wall of the box/trap. In scientific language s therefore is known as the frequency factor for the detrapping electron, which is related to phonon-electron interactions [13, 17]. Therefore, the magnitude of s is in the order of the vibration of the studied crystal (~10^{20} s^{-1}). K is the Boltzmann constant and ΔE represents the trap's depth.

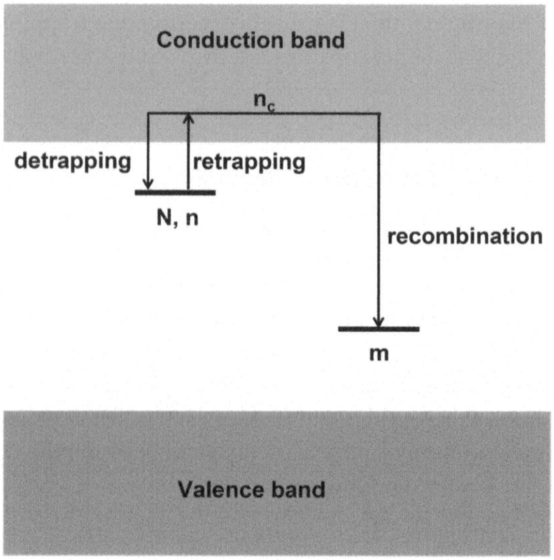

Fig. 1 A schematic diagram representing the one center model where only the transition of detrapping, recombination, and retrapping are allowed

The concentration of filled traps, *n*, strongly relies on the temperature and the time, which makes Eq. 1 a complex differential equation. This means that solving Eq. 1 requires some approximations.

2.1.1 First and Second-Order Kinetics

Randall and Wilkins have set the first theoretical model of a single afterglow peak [13, 17]. They have made several assumptions in order to interpret the afterglow curve of persistent luminescence. They assumed that the overlapping between the different trapping levels, that a direct transfer between the luminescent centers and the various traps and that the retrapping process was negligible. These assumptions implied that all the detrapping carriers will recombine within the luminescent center, which meant that the retrapping cross-section, σ_n, in Eq. 1 will be equal to zero. The Randall and Wilkins method; therefore, is known as the first-order kinetic and Eq. 1 can be rewritten as

$$I = ns\exp\left(-\frac{\Delta E}{KT}\right) \tag{2}$$

The detrapping of charge carriers occurs through several mechanisms, namely, photoionization, two photon ionization, and thermal ionization [10, 18]. In the case of photoionization, the trapped charge carrier received sufficient photon energy to allow detrapping and to reach the conduction band. Two photons are required during the two-photon ionization process to transfer the electron from the excited state of the luminescence center to the traps. This process could not trap enough electrons to produce persistent luminescence because of its weak nature [18]. The detrapping of charge carriers can also occur through a given temperature where the charge carriers can thermally be excited to the luminescent center or to the conduction band of the host. The gradual release of the trapped carriers would result in continuous luminescence for extremely long periods ranging from seconds to days. The trapped charge carriers can be distributed with a Maxwell distribution that possesses thermal energy; therefore, the probability for the charge carrier to escape from the traps or the detrapping can be expressed with a given temperature (T) [13].

$$p = s\exp(-\frac{\Delta E}{KT}) \tag{3}$$

The intensity of the afterglow curves depends on both time and temperature. In the case of constant temperature, the afterglow decay curve will be having an exponential appearance. Thus, the first-order kinetics in this case can be expressed as follows:

$$I(t) = I_0\exp\left(-\text{s}\exp\left(-\frac{\Delta E}{KT}\right)t\right) \tag{4}$$

Figure 2 shows the plot of the expected exponential decay for different trap depths, ΔE. The curves are characterized by a simple exponential decay, which is not a practical curve in actual persistent luminescent materials. This means that neglecting the retrapping process would not describe the afterglow decay curves accurately.

A search for a model that can better describe the observed afterglow decay was proposed by Garlick and Gibson [19]. They investigated the possibility of having equal retrapping and recombination probabilities ($\sigma_n = \sigma_{mn}$). Second-order kinetics refers to the Garlick and Gibson model, where the intensity of the afterglow decay

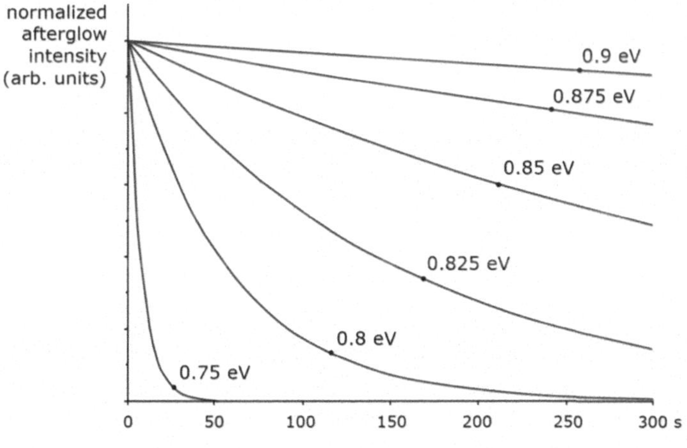

Fig. 2 A first-order kinetics afterglow decay curve for different trap depths. T = 293 K and s = 10^{12} s^{-1} [16]

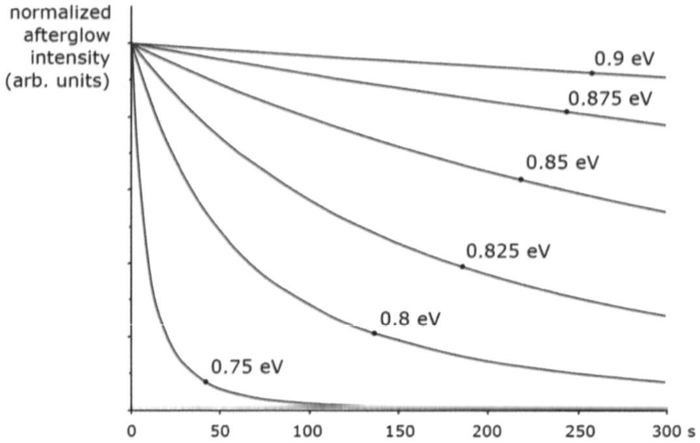

Fig. 3 The second-order kinetics afterglow curve for different trap depths. T = 293 K, s = 10^{12} s^{-1} and $n_0/N = ½$ [16]

is given as

$$I(t) = I_0\left[1 + s\,\frac{n_o}{N}\exp\left(-\frac{\Delta E}{KT}\right)t\right]^{-2} \tag{5}$$

The plot of the second-order kinetics of the afterglow decay, which exhibits a power law curve behavior, is shown in Fig. 3. The curve has a slope of -2 and approaches a straight line. In contrast to first-order kinetics, the decay curve for second-order kinetics takes longer to approach 0 intensity.

2.1.2 General Order Kinetics

The first- and second-order kinetics were built according to negligible retrapping and equal recombination and retrapping probability, which are very specific cases. An empirical expression for general order kinetics, therefore, was then developed based on the first- and second-order kinetics [20, 21].

$$I = \frac{n^b}{N^{1-b}}.s.\exp\left(-\frac{\Delta E}{KT}\right) \tag{6}$$

b represents the order of kinetics and can be determined from the curve fit. The general order kinetics is considered as a transition between the shapes of the afterglow decay of the first ($b = 1$) and second ($b = 2$) order kinetics and beyond. It should be noted that the order, b, has no direct physical meaning and that the general order kinetics expression is just a mathematical expression.

2.2 Trapping Kinetics

The trapping kinetics are associated with the capturing of an electron and a hole within the crystal defects after the excitation process. If the material is exposed to radiation with sufficient energy, the electrons will be promoted from the ground to the excited levels. Either the excited electron in the luminescent center can move to the trap level or recombine by going back to the luminescent center's ground level. Figure 4 shows the model of the trapping kinetics schematically. There are a few parameters to consider in this trapping kinetics model [15, 16]:

1. The trapping probability (p).
   Here, the assumption is that p depends on the fraction of the traps getting filled. This means that p is a maximum when all the traps are empty ($p = p_o$). Whereas $p = 0$ when all the traps are fully filled.
2. The detrapping probability (p′).
   p′ is assumed to be independent of the fraction of the traps getting filled. The trapping probability decreases with filling of the traps until it balances with the small value of the detrapping probability.
3. The excitation strength.
   This parameter strongly influences both the trapping and detrapping parameters.

The practical implementation of this model was done on $Sr_2MgSi_2O_7$:Eu,Dy [22]. A typical luminescence intensity behavior of the traps' kinetics of the $Sr_2MgSi_2O_7$:Eu,Dy persistent luminescent material during and after the excitation is shown in Fig. 5. Initially, the luminescence intensity started to increase gradually until a balance between the trapping and retrapping occurred. In this phase,

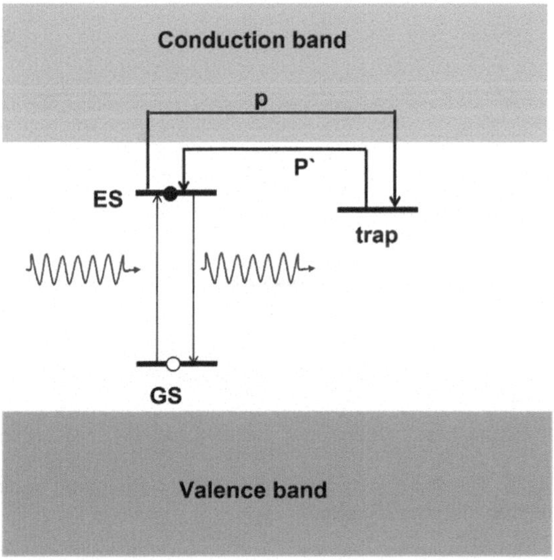

Fig. 4 A schematic diagram of a one-trap level explaining the trapping kinetics

the normal emission will dominate over the trap filling. The model fits well with the experimental data of $Sr_2MgSi_2O_7$:Eu,Dy's persistent luminescence. The model showed that p_0 is of around 80.3% and a negligible p'_o of 0.06% [22].

The initial photons reaching the material have a probability of 80% to induce trapping, which is the maximum trapping probability, p_o. Therefore, there is a recombination probability of $1 - p_o$, which is around 20%, where the electrons can return to the ground states before it can be trapped again. This observation can be seen in Fig. 5 where the luminescence intensity starts from 20%. The schematic diagram that explains the emission intensity in the beginning of Fig. 5, in the emission intensity of $Sr_2MgSi_2O_7$:Eu,Dy through the trapping and recombination probability, is shown in Fig. 6.

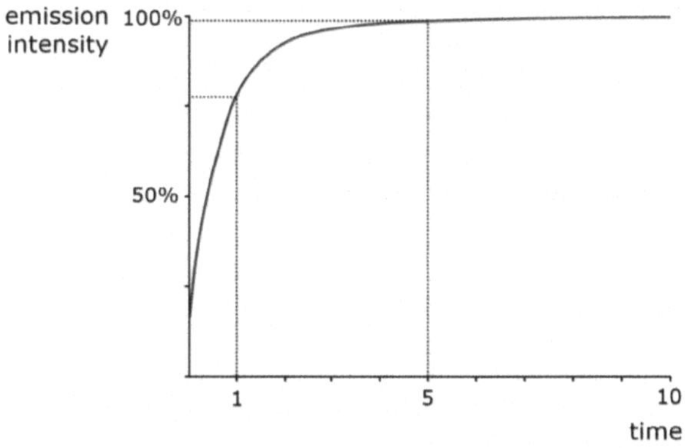

Fig. 5 $Sr_2MgSi_2O_7$:Eu,Dy emission intensity excited with a filtered Xe-lamp [16]

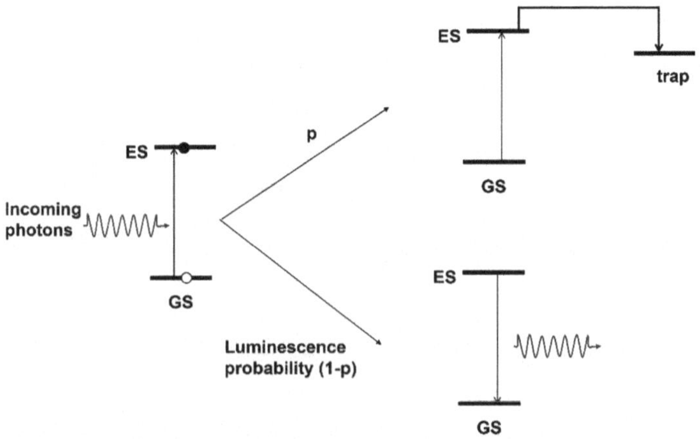

Fig. 6 Schematic diagram explaining the emission observed in $Sr_2MgSi_2O_7$:Eu,Dy emission intensity during excitation

2.2.1 Trapping Levels

The imperfection in the crystal lattice such as surfaces, impurity ions, or crystal defects can lead to additional levels inside the bandgap of the material. After the material receives sufficient excitation energy, the generated holes and electrons can be transferred to the hole traps and electron traps, respectively. The electrons and holes in both the excited states and ground states are localized around their parent atoms under normal conditions. The capturing duration of electrons and holes depends on the trap depths. The electron and hole depths are normally measured from the conduction and valence bands, respectively. The deep traps require high energy to release the carriers. Therefore, thermal energy with a relative temperature is not enough to release the carriers from deep traps. The carriers are then stored for a longer duration. The traps that usually result in persistent luminescence possess depths ranging from a few to several eV activation energy [2, 16].

2.2.2 Hole and Electron Traps

In general, for persistent luminescence to occur, the energy levels of an impurity ion must be located within the bandgap of the host. These ions can then act as holes or electron traps. For the impurity ion with M^{n+} to become an electron or a hole trap, its valance states must be altered to $M^{(n-1)+}$ or $M^{(n+1)+}$ for capturing an electron or a hole, respectively [9]. This implies that for the impurity ion to be an electron or a hole trap for persistent luminescence, it must possess the ability to have multi-valence states. Literature shows that all the luminescence centers (lanthanide ions) in persistent luminescence are stable in several valence states [23–26]. Table 1 lists the important impurity ions that are used as carrier traps in persistent luminescence. Although the changing valence states for electrons and holes are discussed thoroughly in literature to explain the hole and electron traps, it is controversial [9]. This is because it is difficult to obtain direct evidence of changing valence states from very low trap concentrations. However, recently the valence state change has been demonstrated using X-ray absorption fine structure (XAFS) in several persistent luminescent centers [22, 27–29].

Table 1 Hole and electron trapping centers

Trapping center	Ions
Electron ($M^{n+} + e^- \rightarrow M^{(n-1)+}$)	Dy^{3+}, Eu^{3+}, Sm^{3+}, Ho^{3+}, Nd^{3+}, Yb^{3+}, Er^{3+}, V^{3+}, Sc^{3+}, Fe^{3+}, Tb^{4+}, Bi^{3+}, Mn^{4+}, Cr^{3+}, V^{3+}, Tm^{3+}, Ti^{3+}
Hole ($M^{n+} + e^- \rightarrow M^{(n+1)+}$)	Tb^{3+}, Bi^{3+}, Pr^{3+}, Mn^{2+}, Ce^{3+}, Cr^{3+}, Yb^{2+}, Eu^{2+}

2.3 Number of Traps

Persistent luminescence occurs via a detrapping process of charge carriers from traps that were loaded by an excitation process. Thus, the number of traps (storage capacity of carriers) is a vital factor in determining the intensity of persistent luminescence. However, the light output of afterglow materials is proportional to the excitation intensity [22]. The light output will eventually reach a saturation point at a certain excitation energy. The saturation normally occurs at a certain excitation intensity. The saturation of light output is a result of all the available traps being completely filled. Therefore, the beginning of the light output saturation is a critical point to estimate the available traps in the material. Assuming 100% quantum efficiency of the light output of the afterglow, then the light output of the afterglow can also be used to estimate the total number of traps. Thus, the sample's quantum efficiency and known excitation source can be used to estimate:

1. The number of activators and then
2. The number of traps.

As mentioned previously, the lanthanide ions are normally utilized as activators (luminescent centers) for persistent luminescence. Estimating the number of activators; therefore, is determined from the doping ratio of the sample. Let's assume that the sample (ABC) is doped with 1% of sample X in the A site. This means that 1% of the A ions have been substituted by an X ion. If we know our sample contains 1.06×10^{19} of A ions per mm^3 with a layer thickness of 100 μm. The number of the dopant X is, therefore, 1.06×10^{17} ions per mm^3 within the layer. If the excitation source's spot size is 0.6 mm^2, the excitation beam is assumed to penetrate the sample deep enough to excite the entire thickness of the sample (100 μm). The schematic representation of the model is shown in Fig. 7. The material volume excited by the excitation beam is then about 0.06 mm^3. Thus, we can estimate that a total number of 6.4×10^{15} of X ions in the sample will be excited by the excitation beam.

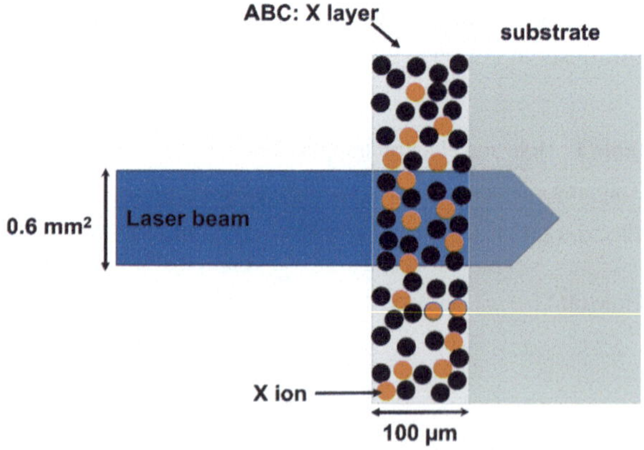

Fig. 7 Schematic geometry shows the method that can be used to estimate the number of available traps in any material

Consequently, estimating the number of traps requires concrete knowledge on the excitation beam used and the quantum efficiency of the sample. From the excitation beam we can calculate the number of photons that hit the sample. A laser with a known energy is usually used as an excitation beam. Whereas the number of photons that lead to a radiative transition can be estimated from the quantum efficiency of the sample.

On the other hand, there are only a fraction of the incoming photons that will assist in trapping. Thus, most of the incoming photons will result in recombination within the luminescent center without leading to charge carrier trapping. The fraction of photons that lead to trapping can be estimated from the emission intensity during the excitation. This fraction is very small compared to the photons leading to the emission. The practical estimation of the number of traps can be found in [16] which estimated the number of traps of the $SrAl_2O_4$:Eu,Dy persistent phosphor by using a 40 mW laser with 457 nm excitation wavelength.

As an illustration, using the same parameters as in [16], we can estimate the number of traps for our sample ABC: X (1%). A 40 mW laser with a blue wavelength can approximately carry 9.2×10^{16} incoming photons per second. The number of traps' saturation occurred approximately at 10% of the maximum laser intensity. Thus, the number of photons that caused the saturation was $9.2 \times \frac{10^{16}}{10} = 9.2 \times 10^{15}$ per second. The excitation duration was 30 s, which gave the total number of $9.2 \times 10^{15} \times 30 = 2.8 \times 10^{17}$ photons at 30 s.

The calculation of the number of photons that lead to radiative transition can be done from the quantum efficiency of the sample. If we assume the quantum efficiency is 24% of our ABC:X persistent phosphor sample, a total of 6.7×10^{16} photons lead to radiative transition.

The fraction of the number of photons that lead to the charge carrier trapping is shown in Fig. 8 [16]. The grey area in Fig. 8 illustrates this fraction of photons that lead to charge carrier trapping, that is approximately 0.97% of the total number of incoming photons. This gives $6.7 \times 10^{16} \times 0.97\%(\sim 1\%) = 6.7 \times 10^{14}$ photons that will create a trapped charge carrier. This implies the saturation of the number of traps that occurred at this excitation intensity. Thus, the maximum traps' number available in the material is 6.7×10^{14} traps. From this discussion we can conclude that.

- The available traps number is 6.7×10^{14} and
- The number of activators is 6.4×10^{15} ions.

This gives 16% of the number of activators that were converted to traps.

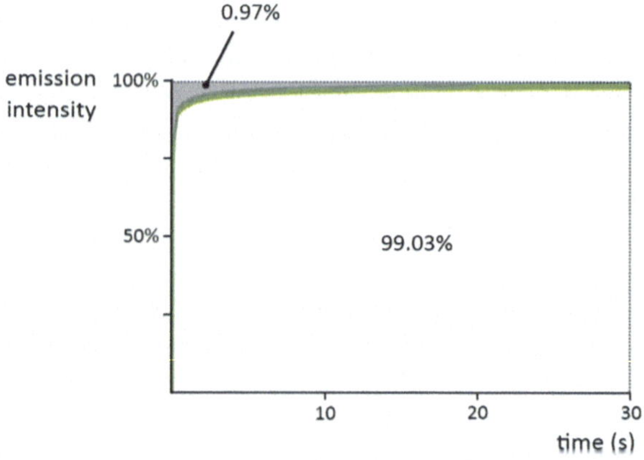

Fig. 8 The percentage of incoming photons that generated trapped electrons, which was around 1% from 10% of the maximum intensity [16]

3 Estimation of the Trap Depths

The trap depth (ΔE) is defined as the minimum energy that is needed to liberate a carrier from the trap. ΔE is the most important parameter in understanding the persistent luminescence mechanism. Unfortunately, an accurate and reliable evaluation of the trap depth in persistent materials is rather complicated [14]. Therefore, this section is dedicated to the estimation of the trap depths via the thermoluminescence (TL) route.

3.1 Determining Trap Depths

Different methods have been suggested for determining the ΔE value [30, 31]. The most convenient methods were based on the thermally stimulated luminescence experiment. Based on thermally stimulated luminescence, ΔE, can be determined from the luminescence intensity after a linear heating of the sample (known as a TL experiment). There have been many methods devoted for determining the trap depth, which rely on the TL experiment. The methods of determining ΔE can roughly be divided into these categories:

1. Peak position methods, which are based on the maximum temperature.
2. Shape methods, which are based on the curve's shape.
3. Area methods, which are based on the area of the glow curve shape.
4. Initial rise methods, which are based on the initial rise part of the afterglow curve.
5. Computerized methods, which are based on curve fitting using series expansion.

The Hoogenstraaten method that includes the shape method and the computerized fitting method are the most used method for determining the ΔE value [32–35]. The Hoogenstraaten method [32] is based on the peak position's shift as a function of heating rates. The shape method was first proposed by Chen [33], that was based

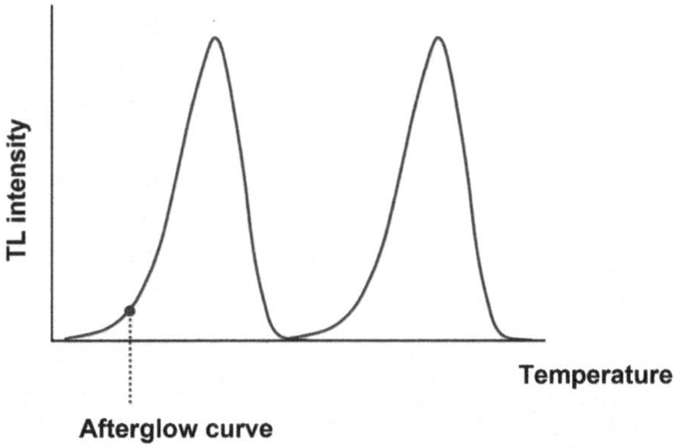

Fig. 9 A schematic representation of a TL curve

on the shape of the afterglow curve. In the computerized fitting method, specialized software was used to estimate the trap's depth [34, 35].

Figure 9 shows the schematic representation of a TL experiment. The maximum intensity of the glow curve corresponds to the maximum temperature (T_{max}). Different maximum temperature peaks result in different trap depths, which is the temperature of releasing the charge carrier from the traps and where they recombine in the ground state at the luminescent centers. ΔE can be determined from analyzing the individual peaks. A low temperature glow peak corresponds to a shallow trap, whereas higher temperature peaks correspond to deeper traps. The depth, number, and shape of the traps, as well as the detrapping and trapping dynamics, are all revealed by the height, position, shape, and location of the glow peak.

The intensity of the TL afterglow curve begins to drop at higher temperatures. This is because the non-radiative decay processes dominate over the radiative transition in this region. Therefore, both the radiative and non-radiative transitions should be included when analyzing the TL afterglow curves. To reflect the total number of liberated charge carriers, the TL experiment needs to be corrected for thermal quenching [30, 36, 37].

The different kinetics involved in the persistent luminescent dynamics affect the decay curve. In the case of the negligible retrapping, the first-order detrapping carrier transportation occurs where the charge carriers are released from the traps and do not get transferred back to the trap. In this case, the intensity of the persistent luminescence can be expressed from the detrapping rate using Randall and Wilkins model [17]. The shape of the glow curve is strongly affected with this kinetics. The assumption of negligible retrapping leads to an asymmetric TL glow peak as well as an exponential afterglow decay profile. The mathematical expression for the intensity and contour of the first-order glow peak is given below

$$I(T) = \frac{s}{\beta}.n_o.\exp\left(-\frac{\Delta E}{KT}\right).exp\left[-\frac{s}{\beta}\int_{T_o}^{T}\exp\left(-\frac{\Delta E}{K\theta}\right)d\theta\right] \qquad (7)$$

n_o and β represent the initial concentration of trapped electrons and the heating rate constant during the TL experiment, respectively. The detrapped electrons, however, can return to the trap defect. When taking the retrapping process into account, the detrapping process becomes complicated and implementation of a higher order model is required for understanding the detrapping kinetics.

The second-order kinetics was proposed by Garlick and Gibson [19] for the TL afterglow peak. The expression was derived for the case where the probability of retrapping and recombination are identical. The Garlick and Gibson expression of the TL intensity is given by

$$I(T) = \frac{s}{\beta}.\frac{{n_o}^2}{N}.\exp\left(-\frac{\Delta E}{KT}\right).\left[1+\frac{s}{\beta}.\frac{n_o}{N}\int_{T_o}^{T}\exp\left(-\frac{\Delta E}{K\theta}\right)d\theta\right]^{-2} \tag{8}$$

From the above discussion, the first- and second-order kinetics are valid for only special situations. Therefore, May and Partridge proposed an expression of an intermediate situation known as the general order kinetics, which is given by

$$I(T) = \frac{s}{\beta}.\frac{{n_o}^b}{N^{b-1}}.\exp\left(-\frac{\Delta E}{KT}\right).\left[1+\frac{s}{\beta}.\left(\frac{n_o}{N}\right)^{b-1}.(b-1)\int_{T_o}^{T}\exp\left(-\frac{\Delta E}{K\theta}\right)d\theta\right]^{\frac{-b}{b-1}} \tag{9}$$

These three expressions (7–9) are widely used nowadays for determining the trap depth, ΔE. The effect of the order of kinetics on the afterglow peak is summarized in Fig. 10. An asymmetric peak is obtained in the case of the first order (b = 1), whereas the second order (b = 2) gives nearly a symmetric TL peak. The general order kinetics showed an intermediate TL curve between the first and second-order kinetics.

3.2 Continuous Trap Depth Distributions

The first, second, or general order kinetics cannot explain most of the afterglow decay and glow curves. This is because the afterglow decay curves sometimes behave as a function of t^{-1}, which is complicated to be explained by these simple models [38]. On the other hand, the TL measurements often result in a broad afterglow peak in the high temperature region. Such peaks can consist of several discrete trap levels that can be described using the first-order kinetics. The same for the afterglow decay curve that can be regarded as discrete energy levels whereby each individual decay curve can be described by first-order kinetics [39–41]. The individual decay curves can then consist of three components that are known as slow, intermediate, and fast components, that can make the physical interpretation of the fitting of the decay curve procedures unreliable.

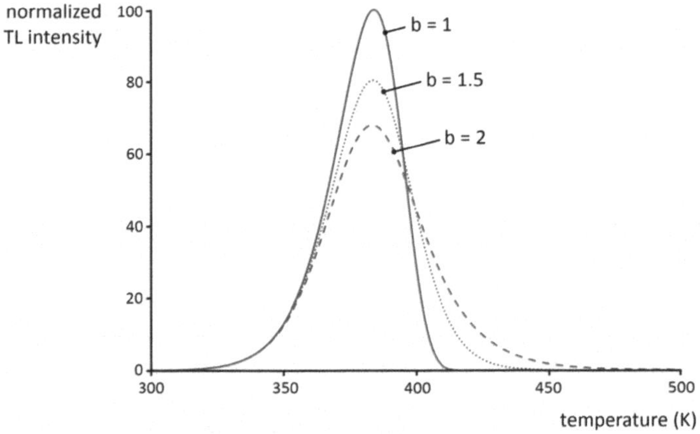

Fig. 10 The effect of the kinetic order on the afterglow curve. $E = 1$ eV, $s = 1 \times 1012\ \mathrm{s}^{-1}$, $n_0 = N = 1\ \mathrm{m}^{-3}$, and $\beta = 1$ K/s [15]. Copyright American Physical Society 2013

Similar to this, within the general order kinetics, the broad TL afterglow peaks are sometimes suited to numerous distinct glow peaks. This procedure of the TL afterglow peak requires around five independent parameters for each individual peak. Thus, the number of the fitting parameters is increased when fitting the broad TL afterglow peak. This makes the physical interpretation also impossible.

For these reasons, the continuous trap depth distribution was proposed in which the trap levels were distributed continuously rather than a discrete distribution in the bandgap of the host material [38, 42, 43]. There are three proposed types of continuous trap distributions: exponential, uniform, and Gaussian profiles, Fig. 11.

The traps in the uniform distribution have equal density. Whereas in the exponential distribution, the number of traps is distributed in shallow and deep traps where the shallow traps have the largest density. The Gaussian profile has a characteristic of a Gaussian peak centered around the value of the mean trap depth. The first-order kinetic of the continuous trap distribution can mathematically be expressed to integrate over the traps' depths [30].

Fig. 11 Schematic diagram showing the three continuous trap depth distributions: Gaussian, exponential, and uniform

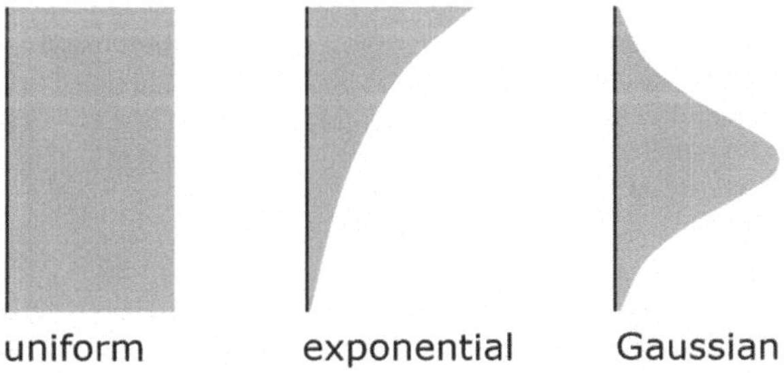

$$I(T) = \int_0^{\infty} \frac{s}{\beta}.N(\Delta E).\mathrm{f}_0(\Delta \mathrm{E}).\ \exp\left(-\frac{\Delta E}{KT}\right).exp\left[-\frac{s}{\beta}\int_{T_o}^{T}\exp\left(-\frac{\Delta E}{K\theta}\right)d\theta\right] \quad (10)$$

Here $N(\Delta E)$ represents the available trap depths' distribution that can be exponential, uniform, or Gaussian.

3.3 Initial Rise Analysis

One of the most helpful analysis methods is the first rise method analysis for figuring out the trap depth. This method makes the premise that the amount of trapped charge carriers in the low temperature area of the TL afterglow curve is essentially constant. This is so because only a small portion of the charge carriers can escape on the low temperature side due to the limited thermal energy present. The first, second, and general order kinetics equations can be roughly represented by the following equation under this supposition [15, 19].

$$I(T) \propto \mathrm{C.exp}\left(-\frac{\Delta E}{KT}\right) \quad (11)$$

The frequency factor (s), which is a component of the constant C, is unaffected by temperature. Figure 12 shows a plot between In(T) and 1/T that results in a straight line from which the trap depth can be calculated. The initial rise technique has two benefits:

1. The method is independent of the kinetic order within the detrapping and trapping processes.
2. There is a significant improvement in the issue of peak overlapping between the various TLs. This is because the approach only takes into account the TL glow curve's first phase (low temperature side).

As the initial rise employs only the low temperature part of the TL, this method is expected to reveal only the shallow traps' depths and the deeper traps remain filled

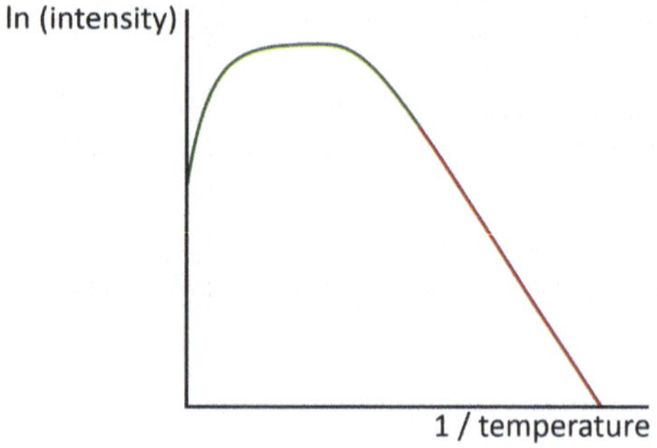

Fig. 12 A plot that demonstrates the In(intensity) as a function of 1/T as an illustration of the initial rise analysis

in the continuous trap distribution, which is considered as one of the disadvantages of using this method.

3.4 *Thermoluminescence and Initial Rise Analysis for Trap Depth Spectroscopy*

Thermoluminescence is the term used to describe the thermally stimulated emission of light emitted from materials. Electrons promoted to excited states during radiation with beta, gamma, or alpha particles and UV may lead to charge carriers getting trapped in the trap centers. The trapped charge carriers require sufficient energy to be released or to be stimulated from the traps. Upon thermal stimulation, the charge carriers recombine in the recombination centers from where luminescence can occur [30, 44]. The energy required to release the electrons from the traps measures the trap depths of the trap centers. The initial rising method has been successfully used to approximate the depth of the trapped electrons.

In the thermoluminescence experimental procedure, the heating rate of the sample is done at a linear heating rate ($\boldsymbol{\beta = dT/dt}$). According to the first-order kinetic, the thermoluminescence intensity is identical to the recombination rate. The TL intensity is then given by

$$I(T) = ns\exp\left(-\frac{\Delta E}{KT}\right) \tag{12}$$

From the assumption of the initial rise analysis, as discussed in Sect. 3.3, the trap concentrations at the low temperature side remain constant. Thus, the expression in Eq. 11 is only valid when n***(T)*** is independent of temperature (T) in the low temperature region. This assumption is correct to a temperature in which the thermoluminescence intensity is about 15% of the maximum intensity (I_M). From Fig. 13, this temperature is marked as temperature (***Tc***), which corresponds to the thermoluminescence intensity (I_C) [24]. From Eq. 11, a linear curve of ln(I(T)) vs 1/KT can be plotted where the trap depth (ΔE) is represented by the slope of this plotted curve.

4 Trap Depth Distribution in $CaAl_2O_4$:Eu,Nd

As discussed in the previous sections, the trap distribution in materials can be in different types including discrete or continuous distributions. The physical interpretation of the discrete trap depth distribution is rather impossible as many parameters are needed to fit only one broad TL peak. The continuous trap depth distribution assumes a continuous distribution of the traps in the bandgap of the material.

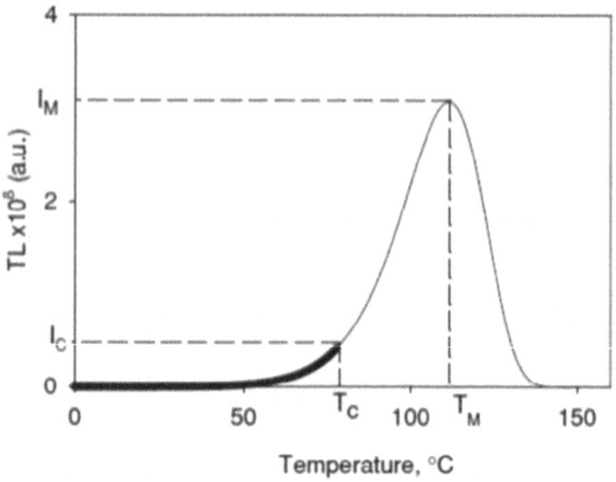

Fig. 13 An illustration of the initial rise method

$CaAl_2O_4$ material crystallizes into a monoclinic structure [45]. The crystal structure of the $CaAl_2O_4$ material is also dependent on the preparation method. When $CaAl_2O_4$ is prepared by the sol–gel or combustion methods, orthorhombic or hexagonal phases are obtained [46, 47]. Eu^{2+} ions act as trap centers when co-doped in this material. There are three possible cation sites in the monoclinic structure of the $CaAl_2O_4$ material. The Eu^{2+} ions sometimes possess larger ionic radii than Ca^{2+} ions, and it is only incorporated into the antiprism trigonal site of the Ca^{2+} ion [48]. This makes the Eu^{2+} ions occupy only one site in the monoclinic structure of $CaAl_2O_4$. The presence of only one site of Eu^{2+} in the monoclinic structure makes it a suitable material that simplify the analysis of the persistent luminescence data [48].

$CaAl_2O_4$:Eu,Nd material can display various numbers of TL afterglow peaks, which depend on the excitation intensity [49, 50]. Aitasalo et al. [49] reported two overlapping afterglow peaks from the TL measurement of $CaAl_2O_4$:Eu,Nd using relatively low excitation intensity. The main afterglow peak was assigned to the $CaAl_2O_4$:Eu-related defect traps and the second peak to the Nd-related trap defect. Whereas Eeckhout et al. [15] found only a single afterglow peak from this material, which was attributed to the Nd-related defect traps. The two tests were conducted at various levels of excitation, which suggests that the excitation level may have an impact on the relative filling of both types of traps.

There are different methods for probing the trap depth distribution of the persistent materials. Eeckhout et al. [15] provided a systematic procedure of probing the continuous trap depth distribution that can be applied to the persistent materials. This procedure involves two principles. (1) A series of TL measurements with varying the excitation temperature and duration and (b) using the initial rise method for analyzing the TL experiment.

Upon exciting the $CaAl_2O_4$:Eu,Nd persistent material with different excitation intensities at various temperatures, a gradual deepening of the trap depths was observed, which proved that the $CaAl_2O_4$:Eu,Nd material consisted of a continuous trap distribution [15, 47]. Figure 14 shows the TL glow curve of the $CaAl_2O_4$:Eu,Nd material for various excitation temperatures [15]. By employing the initial rise method discussed above, the minimum and maximum estimated trap depths were found to be 0.69 and 1.2 eV at room temperature and at 483 K, respectively. It was found that the trap depths deeper than 1.2 eV in this $CaAl_2O_4$:Eu,Nd persistent material were difficult to observe as the TL intensity of the afterglow peaks was weak for these high excitation temperatures. Figure 15 represents the trap depth distributions with the traps' shape in the $CaAl_2O_4$:Eu,Nd persistent material. The trap depths' energies consisted of a continuous distribution with the highest trap's intensity available at 0.9 eV with more and less a Gaussian shape.

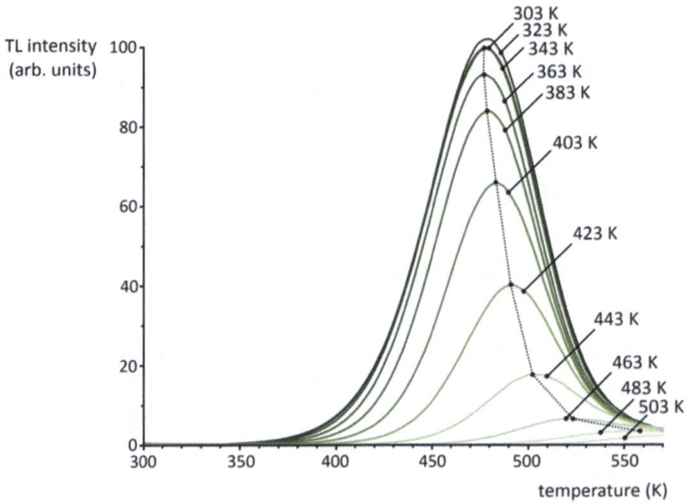

Fig. 14 The TL afterglow curve of $CaAl_2O_4$:Eu,Nd. The samples were excited by 365 nm light for 60 s at various excitation temperatures [15]. Copyright American Physical Society 2013

Fig. 15 $CaAl_2O_4$:Eu,Nd trap distribution [15]. Copyright American Physical Society 2013

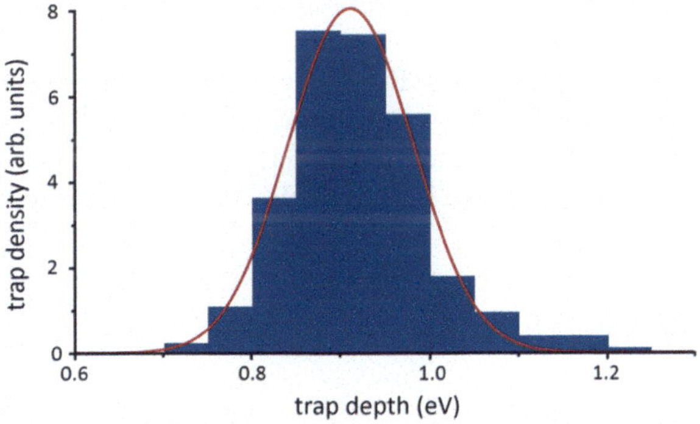

Table 2 Some of the NIR and over 1000 nm emission persistent phosphors

Host	Activator	Emission peak	Duration	Refs.
$La_3Ga_5GeO_{14}$	Cr^{3+}	~ 1000	> 8 h	Jia et al. [72]
$SrAl_2O_4$	$Eu^{2+} \rightarrow Er^{3+}$	1530	> 10 min	Yu et al. [73]
	$Eu^{2+} \rightarrow Nd^{3+}$	1064	> 15 min	Teng et al. [74]
$Sr_3Sn_2O_7$	Nd^{3+}	1079	> 1 h	Kamimura et al. [75]
$GdY_2Al_3Ga_2O_{12}$	$Cr^{3+} \rightarrow Nd^{3+}$	1060	> 15 min	Qin et al. [76]
(Y/Lu/La/Gd)PO_4	$Bi^{3+} \rightarrow Nd^{3+}$	1062	> 1 h	Lyu and Dorenbos [77]
$Zn_3Ga_2Ge_2O_{10}$	Ni^{2+}	1290	> 12 h	Liu et al. [78]
$ZnGa_2O_4$	Ni^{2+}	1275	> 5 min	Pellerin et al. [79]
	Co^{2+}	~ 1000	> 40 min	Pellerin et al. [79]
$ZnGa_2O_4$	$Cr^{3+} \rightarrow Yb^{3+}$	~ 1000	> 10 h	Castaing et al. [80]
$MgGeO_3$	Pr^{3+}	1100	> 120 h	Liang et al. [81]

5 Ultraviolet and Red Persistent Phosphors

5.1 *Ultraviolet Persistent Phosphors*

Ultraviolet (UV) persistent phosphors with an emission wavelength >400 nm bought much attention within the scientific research community due to their different applications in microbial inactivation water sterilization, drug delivery, photocatalysis, and phototherapy [7]. Many UV persistent luminescence materials, mainly rare-earth-based materials, have already been investigated [51–53]. Table 2 shows some of the published research on UV persistent phosphors in the three different UV regions, UV-A, UV-B, and UV-C. The inorganic materials based Ce^{3+}, Gd^{3+}, and Pr^{3+} were the most widely utilized for UV persistent luminescence [7]. Ce^{3+} doped materials are known as UV-A persistent luminescence. Gd^{3+} ion emission convers the UV-B region, while Pr^{3+} ion exhibit persistent luminescence in different regions including UV-A, B, and C.

Besides these ions, Bi^{3+} in the $LiScGeO_4$ persistent phosphor was reported with a long duration persistent luminescence of over 120 h that peaked at 365 nm using a 254 nm excitation wavelength [54]. The proposed mechanism of the UV-A persistent luminescence of the $LiScGeO_4$:Bi^{3+} phosphor is shown in Fig. 16.

5.2 *Red Persistent Phosphors*

The most commercially used persistent phosphors are blue, green, and yellow emitting phosphors. In contrast, red persistent phosphors are still developing as there are a few red persistent phosphors that can be acceptable for practical applications.

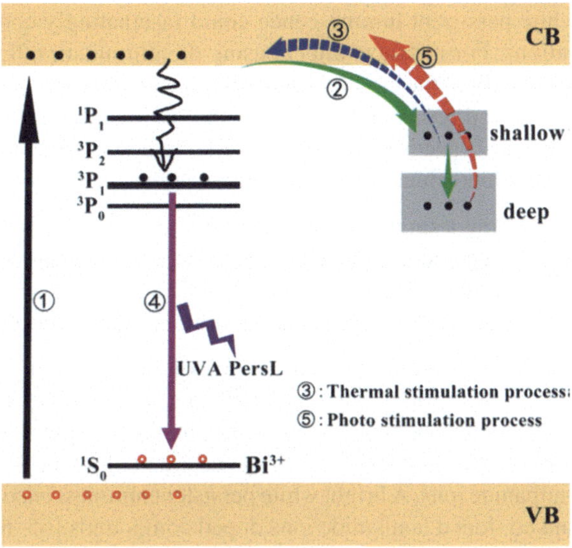

Fig. 16 Proposed persistent luminescence mechanism in the Bi^{3+} doped $LiScGeO_4$ phosphor [54]. Copyright American Chemical Society 2020

In this subsection, the best-known red persistent phosphors are introduced. The red persistent phosphors are considered to emit in the range of 590–680 nm. Because the red color is unpleasant to the sensitivity of the human eye, most research on this topic involves improving the red phosphor emission through choosing the best luminescent ions [55]. Eu^{3+} doped compounds are the most extensively studied red persistent materials due to the best red color band situated between 593 and 650 nm [56, 57]. Eu^{2+} doped nitrides are also red persistent phosphors that emit bright red luminescence with good chemical and thermal stability [58]. The drawback of the Eu^{2+} doped nitrides is their complex synthesis procedures. Sm^{3+} doped compounds are considered as orange-red persistent luminescence ions [59]. The sulfides-based lanthanide ions showed bright red afterglow but due to their moisture sensitivity and thermal instability, their applications are restricted [60]. Pr^{3+} doped compounds also showed red afterglow but the afterglow's duration is still limited [61, 62]. The Mn^{2+} and Mn^{4+} ions doped compounds were also proposed as red-emitting persistent phosphors. For instance, fluoride-based Mn^{2+} persistent phosphors exhibit red luminescence in the range of 600–630 nm [63].

6 White Persistent Phosphors

White light is a combination of three primary colors of red, green, and blue (RGB). Despite the advantages of these RGB colors, the research on white persistent luminescence is limited [64]. One of the advantages of the white persistent luminescence is that it can protect the cone and rod retina cells' sensitivity. The development of the white persistent luminescence could fascinatingly contribute to illumination applications. For instance, after ceasing the illumination light in the dark, the maximum sensitivity of the eye changes from green to blue due to the different adaptation times for the rod retina and cone cells [64, 65]. Under a bright white persistent luminescence an object's color can correctly be observed. However, there are various routes to generate white persistent luminescence:

(1) Using a single luminescent ion.
(2) Using energy transfer mechanisms in composite materials with RGB persistent luminescence.
(3) Using the blue, green and red luminescent ions doped compounds.

The most prominent route is utilizing the lanthanide ions doped compounds such as Dy^{3+}, Eu^{2+}, Pr^{3+}, Tb^{3+}, and Bi^{3+} to generate strong white persistent luminescence. Lanthanide ions work as good luminescent centres and excellent traps. Excellent persistent luminescence with the desired color can be achieved by choosing proper lanthanide ions. A bright white persistent luminescence has been reported from singly and co-doped lanthanide ions doped compounds [65–67]. Huang et al. [66] reported white persistent luminescence from a Dy^{3+} doped Y_2O_2S phosphor under 357 nm UV excitation. Huimin et al. [65] generated bright white persistent luminescence from a Pr^{3+} doped $Ba_2Ga_2GeO_7$ phosphor excited with 255 nm excitation wavelength. The observed white persistent luminescence was a result of an appropriate red and blue-greenish color combination with a duration of three hours. The schematic diagram responsible for the white persistent luminescence of Pr^{3+} doped $Ba_2Ga_2GeO_7$ is presented in Fig. 17.

The white persistent luminescence has also been produced using co-doped or tri-doped compounds [64]. The idea is to co-dope or tri-dope two or three different lanthanide ions with the three primary RGB colors. With an appropriate excitation

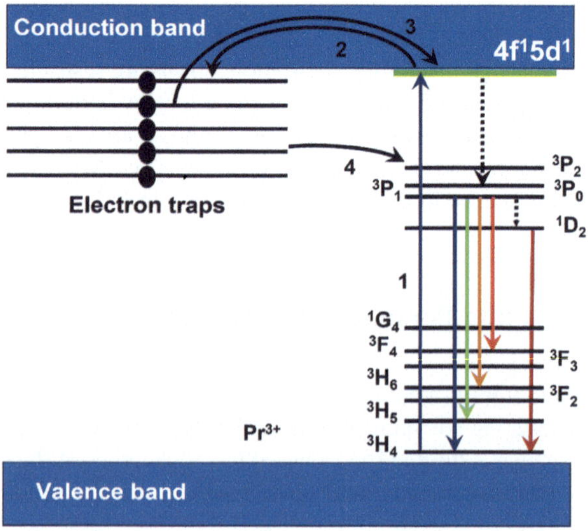

Fig. 17 Schematic diagram showing the persistent luminescence mechanism of Pr^{3+} doped $Ba_2Ga_2GeO_7$ phosphors [65]

energy, white persistent luminescence can be obtained. By this way, Ueda et al. [64] managed to produce bright white light from an Eu^{3+} co-doped or tri-doped $Gd_3Ga_5O_{12}$: Tb^{3+}, Pr^{3+} phosphor where Eu^{3+} acted as an electron trap.

A bright white persistent luminescence can also be obtained from composite phosphors using the energy transfer process [68]. At least two types of the composite phosphor's particles are required to produce white persistent luminescence for example a yellow component and a blue persistent component. The blue component can transfer some of the energy to the yellow component within the composite accompanied by the process of blue and yellow color mixing to produce the white persistent luminescence. The chosen composite should be designed in a way that the afterglow emission of the blue persistent component overlaps with the excitation of the yellow-fluorescent component.

7 Persistent Luminescence Over 1000 nm

Masne de Chermont et al. were the first to show that an NIR persistent phosphor is a promising optical tool for in vivo bioimaging [8, 69]. It was shown that using the NIR emission in the NIR-II (1000–1400 nm) and NIR-III (1500–1800 nm) wavelength ranges led to higher tissue penetration and increased contrast quality [6, 8, 69]. Despite this promising application of the persistent phosphor over 1000 nm in the visualization of deep anatomical features, only a few studies on persistent luminescence with long duration and intense emission have been reported [6, 8]. From the table, the choice of the persistent luminescence phosphors may be divided into two types based on the emitting/trapping centers: (1) Materials activated by transition metal ions, and (2) materials activated by rare-earth ions. For the transition metal ion-activated materials, Cr^{3+} is the most popular ion that can emit in the NIR-I window but when doped into $La_3Ga_5GeO_{14}$ it can emit a broad peak that can extend to the NIR-II optical window [70, 71]. It is possible to successfully shift the primary Cr^{3+} persistent emission peak in $La_3Ga_5GeO_{14}$ to a wavelength of more than 1000 nm. The spin-allowed $^4T_2(^4F) \rightarrow {}^4A_2(^4F)$ transition was blamed for the widespread emission of Cr^{3+} ions over 1000 nm. However, there are also other transition metal ions that can become better choices for persistent luminescence over 1000 nm such as the Ni^{2+} ion [6]. The main peak of Ni^{2+} persistent luminescence is located at around 1290 nm. For the group (II), the lanthanide ions with their sharp 4f-4f transition bands in the NIR-II optical window such as Er^{3+}, Pr^{3+}, Nd^{3+}, Tm^{3+}, and Ho^{3+} are potential candidates for this purpose.

7.1 Materials-Based Lanthanide Ions

The emission of the 4f-4f transition of the lanthanide ions is almost independent of the hosts' crystal fields [82–84]. The singly doped rare-earth ions such as Yb^{3+}, Er^{3+},

Pr^{3+}, Nd^{3+}, Tm^{3+}, and Ho^{3+} doped materials were reported to have good persistent luminescence over 1000 nm. The Yb^{3+} ion has only one excited state that produces an emission around 1000 nm. Although the partial overlap of Yb^{3+} emission with the vibrational overtone of water, some reports still considered the emission in the region of 900–1000 nm to be considered for in vivo optical imaging [85].

Liang et al. [86] have reported persistent luminescence from a $MgGeO_3$:Yb^{3+} phosphor that exhibited long persistent luminescence for more than 100 h. The material consisted of two types of traps: a continuous trap distribution as shallow traps and a photochromic center related to the deep traps. Figure 18 schematically shows the proposed model to explain the persistent luminescence approach in $MgGeO_3$:Yb^{3+}. Moreover, persistent luminescence from a $CaTiO_3$:Yb^{3+} phosphor with sunlight excitation was reported by Y. Wang et al. [87]. The efficiency of the $CaTiO_3$:Yb^{3+} persistent luminescence and duration of more than 80 h increased through co-doping with Bi^{3+} ions.

Although tremendous attention has been paid to the dominant red persistent luminescence of the Pr^{3+} ion, it could be a feasible persistent luminescence (over 1000 nm) candidate owing to its $^1D_2 \rightarrow {}^3H_6$ (900 nm), $^3F_{3,4}$ (1080 nm), or $^1G4 \rightarrow {}^3H_5$ (1310 nm) transitions that falls in the NIR-II and III spectral regions [6, 81]. Liang et al. [81] reported persistent luminescence over 1000 nm for more than 120 h from Pr^{3+} doped $MgGeO_3$ and $CdSiO_3$ multiband persistent phosphors that peaked at 900 and 1080 nm in addition to the well-pronounce red peak at 612 nm. The practical demonstration of this persistent phosphor in an ex vivo experiment in chicken breast showed the penetration power of these three-emission wavelengths in the order of 1080 > 900 > 612 nm, which was in line with the knowledge of interaction of light-tissue.

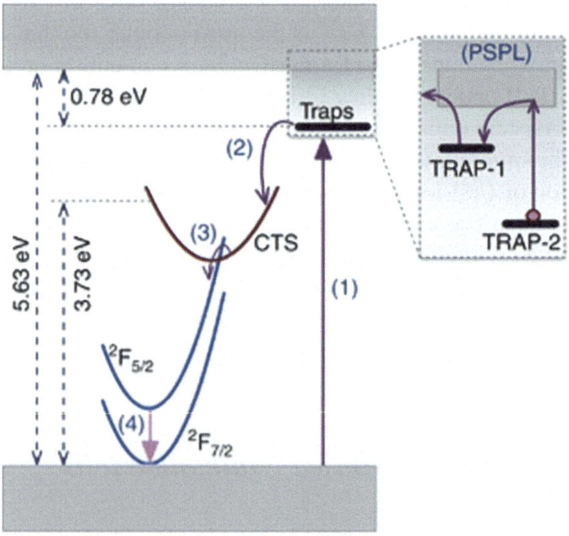

Fig. 18 The schematic diagram of the persistent luminescence mechanism of a $MgGeO_3$:Yb^{3+} phosphor [86]. Copyright Springer Nature 2016

Nd^{3+} has been considered as the best candidate for persistent luminescence over 1000 nm for the NIR-II optical imaging due to the transitions of the $^4F_{3/2}$ excited state to the ^4I_J (J = 9/2, 11/2, 13/2) ground states that exhibit emission bands that peaked at 900, 1060, and 1340 nm, respectively [88, 89]. It is well-known that Nd^{3+} ions possess a relatively low absorption coefficient. Therefore, it is often utilized as electron trapping centers (co-dopant) for improving the persistent luminescence for other rare-earth ions activated persistent phosphors such as in the case of $CaAl_2O_4$:Eu^{2+},Nd^{3+} [15]. However, C. Xu et al. [75] explored the possibility of using the Nd^{3+} ion doped $Sr_3Sn_2O_7$ with a double layer perovskite structure as a potential new class of NIR-II persistent luminescence. They reported that Nd^{3+} intensively emitted between 850 and 1450 nm. As a practical demonstration, the NIR-II persistent luminescence of the $Sr_3Sn_2O_7$:Nd^{3+} phosphor was clearly detectable when embedded under the skin of a chicken and recorded by using an InGaAs camera. This indicated its usefulness for high image contrast detection. The persistent luminescence duration of Nd^{3+} doped NIR-II phosphor was found to be around 1 h, which is considered as a problem for biomedical imaging application.

7.2 Co-Doping Energy Transfer Persistent Luminescence

It was shown that the NIR-II/III persistent luminescence can be obtained by using single rare-earth activated materials. There are difficulties for some lanthanide ions to be used as single ions for persistent luminescence over 1000 nm. Although lanthanide ions such as Ho^{3+}, Er^{3+}, and Tm^{3+} possess transition levels that can give rise to the line emission in the NIR-II/III regions, the trivalent states of these ions are either very close or embedded inside the valance band [6]. Furthermore, these ions are difficult to form tetravalent (+4) states upon photoexcitation, which cannot be utilized as effective recombination centers for inducing NIR persistent luminescence. Therefore, the energy transfer between two or more different lanthanide ions was suggested to overcome these restrictions. In this case, the energy from other activator centers can be absorbed in the UV/visible regions and then directly transferred to the NIR emitted centers that generated the persistent luminescence over 1000 nm.

The persistent luminescence over 1000 nm via energy transfer was effectively demonstrated by Xu et al. in the $LaAlO_3$ and $LaGaO_3$, Ho^{3+}–Cr^{3+}–Sm^{3+} tri-doped materials in which the Cr^{3+} ion acted as the donor for Ho^{3+} ion with emission around 1200 nm [90]. Besides Ho^{3+} and Nd^{3+}, the Er^{3+} ion was considered as a potential candidate for application in brain imaging owing to its Er^{3+}:$^4I_{13/2} \rightarrow {}^4I_{15/2}$ transition that is located around 1500 nm [73, 91]. Z. Yu et al. [73] firstly reported the persistent luminescence from Eu^{2+} and Er^{3+} co-doped $SrAl_2O_4$ via the persistent energy transfer from Eu^{2+} to Er^{3+} ions. This compound exhibited weak persistent luminescence intensity from Er^{3+} ions that peaked at 1530 nm with a short duration of less than 10 min after ceasing the UV excitation energy compared to the super-long persistent luminescence of the green emission of the Eu^{2+} ion. Xu et al. also reported long persistent luminescence over 1000 nm with a duration greater than 10 h

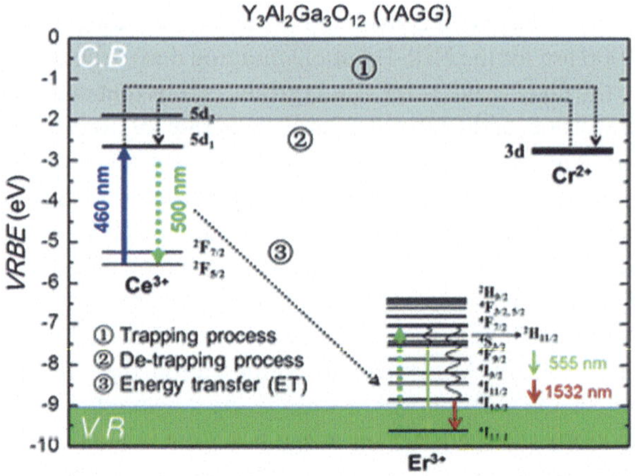

Fig. 19 Schematic diagram showing the persistent luminescence mechanism of the tri-doped $Y_3Al_2Ga_3O_{12}$:Er^{3+}, Ce^{3+}, Cr^{3+} phosphor [93]. Copyright Royal Society of Chemistry 2016

from Er^{3+}, Ce^{3+} and Cr^{3+} in YAGG garnets and $LaAlO_3$ perovskites after ceasing the excitation [92]. The $Y_3Al_2Ga_3O_{12}$:Er, Ce, Cr exhibited NIR photon energy within the NIR-II window was twice the height than that of the widely used $ZnGa_2O_4$:Cr^{3+} deep-red persistent phosphor for the NIR-I optical window [93]. This implied that the $Y_3Al_2Ga_3O_{12}$:Er^{3+}, Ce^{3+}, Cr^{3+} persistent phosphor showed excellent persistent luminescence in the NIR-II optical window by the Er^{3+} ion. Figure 19 shows the trapping-detrapping mechanism of $Y_3Al_2Ga_3O_{12}$:Er^{3+}, Ce^{3+}, Cr^{3+} persistent phosphor along with the obtained persistent luminescence using the vacuum referred binding energy (VRBE) diagram. Various studies have been reported on persistent luminescence via energy transfer using the above-mentioned lanthanide ions, which can be found elsewhere [6].

8 Summary and Conclusion

Considering the fast development of persistent luminescence in various applications from phototherapy to optical imaging, in this chapter, the fundamental mechanisms of persistent luminescence were discussed. The one center model was built to describe the detrapping and trapping kinetics. The behavior of the luminescence intensity during and after the excitation intensity and how the shape of the afterglow decay affected detrapping kinetics was described. The influence of the excitation intensity on the detrapping kinetics and then on the persistent luminescence output was investigated.

The trap depth is the most important parameter in understanding the persistent luminescence mechanism. Two approaches can be considered in trap depth distribution: discrete distribution or continuous trap depth distribution. The continuous trap depth distribution was discussed in which the trap levels were distributed continuously rather than a discrete distribution in the bandgap of the host material. To study the material through continuous trap depth distribution, thermoluminescence analysis with a constant heating rate is very important. The initial rise method can be used to gauge the trap depth. This strategy is predicated on the notion that the concentration of trapped charge carriers on the low-temperature section of the TL afterglow curve is essentially constant. One can determine the trap depth and the form of the continuous trap depth distribution by taking a series of TL measurements while adjusting the excitation temperature and time and using the initial rise method. The $CaAl_2O_4$:Eu,Nd persistent material's continuous trap depth distribution was evaluated. It was possible to generate a continuous Gaussian-shaped trap depth distribution between 0.69 eV and 1.2 eV.

Persistent luminescence can be obtained in all regions of the electromagnetic spectrum depending on the luminescent center. UV persistent luminescence was reviewed. The lanthanide ions were extensively used as luminescent centers for producing the UV persistent luminescence. For visible persistent luminescence, the Cr^{3+} ion is however, extensively utilized for generating the red persistent luminescence besides the Pr^{3+} ion. The development of white persistent luminescence could fascinatingly contribute to illumination applications. This is because after ceasing the illumination light in the dark, the maximum sensitivity of the eye changes from green to blue due to the different adaptation times for the rod retina and cone cells. Therefore, under a bright white persistent luminescence, an object's color can correctly be observed.

The development of persistent luminescence over 1000 nm for optical imaging operation was summarized. Substantial progress on the persistent luminescence over 1000 nm has been achieved but there is a further demand on the defect identification optimization and evaluation of NIR persistent luminescence in the nanoscale.

Acknowledgements This work is based on the research supported by the South African Research Chairs Initiative of the Department of Science and Technology and the National Research Foundation of South Africa (grant 84415). The financial assistance of the National Research Foundation (NRF) and the Cluster program of the University of the Free State towards this research is hereby acknowledged.

References

1. Y. Li, M. Gecevicius, J. Qiu, Long persistent phosphors - From fundamentals to applications. Chem. Soc. Rev. **45**, 2090–2136 (2016). https://doi.org/10.1039/c5cs00582e
2. T. Matsuzawa, Y. Aoki, N. Takeuchi, Y. Murayama, A new long phosphorescent phosphor with high brightness, $SrAl_2O_4$: Eu^{2+}, Dy^{3+}. J. Electrochem. Soc. **143**, 2670–2673 (1996). https://doi.org/10.1149/1.1837067
3. Y. Lin, Z. Tang, Z. Zhang, Preparation of long-afterglow $Sr_4Al_{14}O_{25}$-based luminescent material and its optical properties. Mater. Lett. **51**, 14–18 (2001). https://doi.org/10.1016/S0167-577X(01)00257-9
4. X. Wang, Z. Zhang, Z. Tang, Y. Lin, Characterization and properties of a red and orange Y_2O_2S-based long afterglow phosphor. Mater. Chem. Phys. **80**, 1–5 (2003). https://doi.org/10.1016/S0254-0584(02)00097-4

5. J. Hölsä, H. Jungner, M. Lastusaari, J. Niittykoski, Persistent luminescence of Eu^{2+} doped alkaline earth aluminates, MAl_2O_4:Eu^{2+}. J. Alloys Compd. **323–324**, 326–330 (2001). https://doi.org/10.1016/S0925-8388(01)01084-2
6. J. Xu, M. Back, T. Setsuhisa, Near-infrared phosphor with persistent luminescence over 1000 nm for optical images, in: Phosphor Handb. 2nd Addit. pp. 363–418 (2021)
7. S.K. Sharma, J. James, S.K. Gupta, S. Hussain, UV-A, B, C Emitting persistent luminescent materials. Materials (Basel). **16**, 1–16 (2023). https://doi.org/10.3390/ma16010236
8. Y. Zhuang, Y. Katayama, J. Ueda, S. Tanabe, A brief review on red to near-infrared persistent luminescence in transition-metal-activated phosphors. Opt. Mater. (Amst) **36**, 1907–1912 (2014). https://doi.org/10.1016/j.optmat.2014.05.035
9. J. Ueda, How to design and analyze persistent phosphors? Bull. Chem. Soc. Jpn **94**, 2807–2821 (2021). https://doi.org/10.1246/bcsj.20210255
10. V. Shanker, D. Haranath, G. Swati, Persistence mechanisms and applications of long afterglow phosphors, defect diffus. Forum **361**, 69–94 (2015). https://doi.org/10.4028/www.scientific.net/DDF.361.69
11. Y. Lin, Z. Tang, Z. Zhang, X. Wang, J. Zhang, Preparation of a new long afterglow blue-emitting $Sr_2MgSi_2O_7$-based photoluminescent phosphor. J. Mater. Sci. Lett. **20**, 1505–1506 (2001). https://doi.org/10.1023/A:1017930630889/METRICS
12. K. Van den Eeckhout, P.F. Smet, D. Poelman, Persistent luminescence in Eu^{2+}-doped compounds: a review. Mater. **3**, 2536–2566 (2010). https://doi.org/10.3390/MA3042536
13. J.T. Randall, M.H.F. Wilkins, Phosphorescence and electron traps - I. The study of trap distributions. Proc. R. Soc. London. Ser. A. Math. Phys. Sci. **184**, 365–389 (1945). https://doi.org/10.1098/rspa.1945.0024
14. C.S. Shalgaonkar, A.V. Narlikar, Review: a review of the recent methods for determining trap depth from glow curves. J. Mater. Sci. **7**, 1465–1471 (1972). https://doi.org/10.1007/BF00574938
15. K. Van Den Eeckhout, A.J.J. Bos, D. Poelman, P.F. Smet, Revealing trap depth distributions in persistent phosphors. Phys. Rev. B - Condens. Matter Mater. Phys. **87**, 1–11 (2013). https://doi.org/10.1103/PhysRevB.87.045126
16. K. van den Eeckhout, Persistent luminescence: kinetics and compounds, Gent University (2013)
17. B.J.T. Randall, M.H.F. Wilkins, Phosphorescence and electron traps II. The interpretation of long-period phosphorescence. Proc. R. Soc. London. Ser. A. Math. Phys. Sci. **184**, 390–407 (1945). https://doi.org/10.1098/rspa.1945.0025
18. S.A. Basun, T. Danger, A.A. Kaplyanskii, D.S. Mcclure, K. Petermann, W.C. Wong, Optical and photoelectrical studies of charge-transfer processes in $YAlO_3$: Ti crystals, **54** , 6141–6149 (1996)
19. G.F.J. Garlick, A.F. Gibson, The electron trap mechanism of luminescence in sulphide and silicate phosphors. Proc. Phys. Soc. **60**, 574 (1948). https://doi.org/10.1088/0959-5309/60/6/308
20. C.E. May, J.A. Partridge, Thermoluminescent kinetics of alpha-irradiated Alkali Halides. J. Chem. Phys. **40**, 1401–1409 (1964). https://doi.org/10.1063/1.1725324
21. M.S. Rasheedy, On the general-order kinetics of the thermoluminescence glow peak. J. Phys. Condens. Matter **5**, 633 (1993). https://doi.org/10.1088/0953-8984/5/5/014
22. K. Korthout, K. Van Den Eeckhout, J. Botterman, S. Nikitenko, D. Poelman, P.F. Smet, Luminescence and x-ray absorption measurements of persistent $SrAl_2O_4$: Eu , Dy powders : Evidence for valence state changes, **085140** , 1–7 (2011). https://doi.org/10.1103/PhysRevB.84.085140
23. A.J.J. Bos, P. Dorenbos, A. Bessière, B. Viana, Lanthanide energy levels in YPO_4. Radiat. Meas. **43**, 222–226 (2008). https://doi.org/10.1016/J.RADMEAS.2007.10.042
24. P. Dorenbos, Modeling the chemical shift of lanthanide 4f electron binding energies, Phys. Rev. B - Condens. Matt. Mater. Phys. **85** , 165107 (2012). https://doi.org/10.1103/PHYSREVB.85.165107/FIGURES/10/MEDIUM
25. Y. Zhuang, Y. Lv, L. Wang, W. Chen, T.L. Zhou, T. Takeda, N. Hirosaki, R.J. Xie, Trap depth engineering of $SrSi_2O_2N_2$:Ln^{2+}, L^{n3}+ (L^{n2}+ = Yb, Eu; L^{n3}+ = Dy, Ho, Er)

persistent luminescence materials for information storage applications. ACS Appl. Mater. Interf. **10**, 1854–1864 (2018). https://doi.org/10.1021/ACSAMI.7B17271/ASSET/IMAGES/LARGE/AM-2017-17271Z_0006.JPEG
26. J. Ueda, S. Miyano, S. Tanabe, Formation of deep electron traps by Yb^{3+} Codoping leads to super-long persistent luminescence in Ce^{3+}-doped yttrium aluminum gallium garnet phosphors. ACS Appl. Mater. Interf. **10**, 20652–20660 (2018). https://doi.org/10.1021/ACSAMI.8B02758/ASSET/IMAGES/LARGE/AM-2018-02758J_0005.JPEG
27. J.J. Joos, K. Korthout, L. Amidani, P. Glatzel, D. Poelman, P.F. Smet, Identification of Dy^{3+}/Dy^{2+} as electron trap in persistent phosphors. Phys. Rev. Lett. **125**, 033001 (2020). https://doi.org/10.1103/PHYSREVLETT.125.033001/FIGURES/4/MEDIUM
28. S. Carlson, J. Hölsä, T. Laamanen, M. Lastusaari, M. Malkamäki, J. Niittykoski, R. Valtonen, X-ray absorption study of rare earth ions in $Sr_2MgSi_2O_7$:Eu^{2+}, ^{R3}+ persistent luminescence materials. Opt. Mater. (Amst) **31**, 1877–1879 (2009). https://doi.org/10.1016/J.OPTMAT.2008.12.020
29. J. Ueda, M. Katayama, J. Xu, Y. Inada, K. Asami, S. Tanabe, Evidence of valence state change of Ce^{3+} and Cr^{3+} during UV charging process in $Y_3Al_2Ga_3O_{12}$ persistent phosphors. Opt. Mater. Express **7**, 2471–2476 (2017). https://doi.org/10.1364/OME.7.002471
30. R. Chen, S.W.S. McKeever, Theory of thermoluminescence and related phenomena. Theory Thermolumin. Relat. Phenom. (1997). https://doi.org/10.1142/2781
31. V. Pagonis, G. Kitis, C. Furetta, Numerical and practical exercises in thermoluminescence. Numer. Pract. Exerc. Thermolumin. 1–208 (2006). https://doi.org/10.1007/0-387-30090-2/COVER.
32. W. Hoogenstraaten, Philips research reports (1958)
33. R. Chen, Glow curves with general order kinetics. J. Electrochem. Soc. **116**, 1254 (1969). https://doi.org/10.1149/1.2412291
34. A.J.J. Bos, T.M. Piters, J.M. Gomez Ros, A. Delgado, An intercomparison of glow curve analysis computer programs: I; Synthetic glow curves. Radiat. Prot. Dosimetry. **47** , 473–477 (1993). https://doi.org/10.1093/oxfordjournals.rpd.a081789
35. A.J.J. Bos, T.M. Piters, J.M. Gomez Ros, A. Delgado, An intercomparison of glow curve analysis computer programs: II. Measured glow curves, Radiat. Prot. Dosimetry. **51**, 257–264 (1994). https://doi.org/10.1093/oxfordjournals.rpd.a082143
36. A.J.J. Bos, Theory of thermoluminescence. Radiat. Meas. **41**, 45–56 (2006). https://doi.org/10.1016/j.radmeas.2007.01.003
37. A. Vedda, M. Nikl, M. Fasoli, E. Mihokova, J. Pejchal, M. Dusek, G. Ren, C.R. Stanek, K.J. McClellan, D.D. Byler, Thermally stimulated tunneling in rare-earth-doped oxyorthosilicates, Phys. Rev. B - Condens. Matter Mater. Phys. **78**, 195123 (2008). https://doi.org/10.1103/PHYSREVB.78.195123/FIGURES/9/MEDIUM
38. W.F. Hornyak, R. Chen, Thermoluminescence and phosphorescence with a continuous distribution of activation energies. J. Lumin. **44**, 73–81 (1989). https://doi.org/10.1016/0022-2313(89)90023-9
39. F. Gutiérrez-Martín, F. Fernández-Martinez, P. Díaz, C. Colón, A. Alonso-Medina, Persistent UV phosphors for application in photo catalysis. J. Alloys Compd. **501**, 193–197 (2010). https://doi.org/10.1016/J.JALLCOM.2010.04.178
40. F.B. Dejene, D.B. Bem, H.C. Swart, Synthesis and characterization of $CaAl_xO_y$:Eu^{2+} phosphors prepared using solution-combustion method. J. Rare Earths **28**, 272–276 (2010). https://doi.org/10.1016/S1002-0721(10)60380-8
41. W. Pan, G. Ning, X. Zhang, J. Wang, Y. Lin, J. Ye, Enhanced luminescent properties of long-persistent $Sr_2MgSi_2O_7$:Eu^{2+}, Dy^{3+} phosphor prepared by the co-precipitation method. J. Lumin. **128**, 1975–1979 (2008). https://doi.org/10.1016/J.JLUMIN.2008.06.009
42. A. Bosacchi, B. Bosacchi, S. Franchi, L. Hernandez, Optical properties of a 'quasi-disordered' semiconductor: $ZnIn_2S_4$. Solid State Commun. **13**, 1805–1809 (1973). https://doi.org/10.1016/0038-1098(73)90734-5
43. H. Gobrecht, D. Hofmann, Spectroscopy of traps by fractional glow technique. J. Phys. Chem. Solids **27**, 509–522 (1966). https://doi.org/10.1016/0022-3697(66)90194-6

44. Y. Kirsh, Kinetic analysis of thermoluminescence. Phys. Status Solidi **129**, 15–48 (1992). https://doi.org/10.1002/PSSA.2211290102
45. W. Hörkner, H. Müller-Buschbaum, Zur kristallstruktur von $CaAl_2O_4$. J. Inorg. Nucl. Chem. **38**, 983–984 (1976). https://doi.org/10.1016/0022-1902(76)80011-5
46. T. Aitasalo, J. Hölsä, H. ögne Jungner, M. Lastusaari, J. Niittykoski, Sol–gel processed Eu^{2+}-doped alkaline earth aluminates. J. Alloys Compd. 341, 76–78 (2002). https://doi.org/10.1016/S0925-8388(02)00068-3
47. T. Aitasalo, J. Hölsä, H. Jungner, M. Lastusaari, J. Niittykoski, M. Parkkinen, R. Valtonen, Eu^{2+} doped calcium aluminates prepared by alternative low temperature routes. Opt. Mater. (Amst) **26**, 113–116 (2004). https://doi.org/10.1016/J.OPTMAT.2003.11.007
48. T. Nakamura, T. Matsuzawa, C.C. Rowlands, V. Beltrán-López, G.M. Smith, P.C. Riedi, EPR investigations on europium(II)-doped aluminates. J. Chem. Soc. Faraday Trans. **94**, 3009–3012 (1998). https://doi.org/10.1039/A804494E
49. T. Aitasalo, J. Hölsä, H. Jungner, M. Lastusaari, J. Niittykoski, Thermoluminescence study of persistent luminescence materials: Eu^{2+}- and R^{3+}-doped calcium aluminates, $CaAl_2O_4$:Eu^{2+}, ^{R3}+. J. Phys. Chem. B **110**, 4589–4598 (2006). https://doi.org/10.1021/JP057185M/ASSET/IMAGES/LARGE/JP057185MF00016.JPEG
50. T. Aitasalo, J. Hölsä, H. Jungner, M. Lastusaari, J. Niittykoski, Mechanisms of persistent luminescence in Eu^{2+}, RE^{3+} doped alkaline earth aluminates. J. Lumin. **94–95**, 59–63 (2001). https://doi.org/10.1016/S0022-2313(01)00279-4
51. X. Qu, L. Cao, W. Liu, G. Su, P. Wang, I. Schultz, Sol–gel synthesis of long-lasting phosphors $CdSiO_3$: Mn^{2+}, RE^{3+} (RE = Tb, Eu, Nd) and luminescence mechanism research. Mater. Res. Bull. **47**, 1598–1603 (2012). https://doi.org/10.1016/J.MATERRESBULL.2011.08.012
52. J. Liu, Y. Liang, S. Yan, D. Chen, S. Miao, W. Wang, Narrowband ultraviolet-B persistent luminescence from (Y, Gd$_)$3G$_a$5$_O$$_1$2:B^{i3}+ phosphors for optical tagging application. Dalt. Trans. **50**, 15413–15421 (2021). https://doi.org/10.1039/D1DT02568F
53. X. Wang, Y. Chen, P.A. Kner, Z. Pan, Gd^{3+}-activated narrowband ultraviolet-B persistent luminescence through persistent energy transfer. Dalton Trans. **50**, 3499–3505 (2021). https://doi.org/10.1039/D1DT00120E
54. Z. Zhou, P. Xiong, H. Liu, M. Peng, Ultraviolet-A persistent luminescence of a Bi^{3+}-activated $LiScGeO_4$ material. Inorg. Chem. **59**, 12920–12927 (2020). https://doi.org/10.1021/ACS.INORGCHEM.0C02007/ASSET/IMAGES/LARGE/IC0C02007_0007.JPEG
55. C. Chiatti, C. Fabiani, A.L. Pisello, Long persistent luminescence: a road map toward promising future developments in energy and environmental science. Annu. Rev. Mater. Res. **51**, 409–433 (2021). https://doi.org/10.1146/annurev-matsci-091520-011838
56. N.C. Strandwitz, G.D. Stucky, Hollow microporous cerium oxide spheres templated by colloidal silica. Chem. Mater. **21**, 4577–4582 (2009). https://doi.org/10.1021/CM901516B/SUPPL_FILE/CM901516B_SI_001.PDF
57. S. Som, S. Dutta, V. Kumar, A. Pandey, V. Kumar, A.K. Kunti, J. Priya, S.K. Sharma, J.J. Terblans, H.C. Swart, $CaTiO_3$:Eu^{3+}, a potential red long lasting phosphor: energy migration and characterization of trap level distribution. J. Alloys Compd. **622**, 1068–1073 (2015). https://doi.org/10.1016/J.JALLCOM.2014.11.001
58. X. Teng, Y. Liu, Y. Liu, Y. Hu, H. He, W. Zhuang, Luminescence properties of Tm^{3+} co-doped $Sr_2Si_5N_8$:Eu^{2+} red phosphor. J. Lumin. **130**, 851–854 (2010). https://doi.org/10.1016/J.JLUMIN.2009.12.012
59. V.R. Prasad, S. Damodaraiah, S. Babu, Y.C. Ratnakaram, Structural, optical and luminescence properties of Sm^{3+} and Eu^{3+} doped calcium borophosphate phosphors for reddish-orange and red emitting light applications. J. Lumin. **187**, 360–367 (2017). https://doi.org/10.1016/J.JLUMIN.2017.03.050
60. D. Jia, X. jun Wang, Alkali earth sulfide phosphors doped with Eu^{2+} and Ce^{3+} for LEDs, Opt. Mater. (Amst). **30** , 375–379 (2007). https://doi.org/10.1016/j.optmat.2006.11.061
61. L.L. Noto, O.M. Ntwaeaborwa, M.Y.A. Yagoub, H.C. Swart, Enhancement of persistent luminescence of $ZnTa_2O_6$:Pr^{3+} by addition Li^+, Na^+, K^+ and Cs^+ ions. Mater. Res. Bull. **70**, 545–552 (2015). https://doi.org/10.1016/J.MATERRESBULL.2015.05.023

62. X. Wang, Y. Mao, Recent advances in Pr^{3+} -activated persistent phosphors. J. Mater. Chem. C. **10**, 3626–3646 (2022). https://doi.org/10.1039/D2TC00208F
63. D. Sekiguchi, S. Adachi, Synthesis and photoluminescence spectroscopy of $BaGeF_6$:Mn^{4+} red phosphor. Opt. Mater. (Amst) **42**, 417–422 (2015). https://doi.org/10.1016/j.optmat.2015.01.039
64. J. Ueda, S. Miyano, J. Xu, P. Dorenbos, S. Tanabe, Development of white persistent phosphors by manipulating lanthanide ions in gadolinium gallium garnets. Adv. Photonics Res. **2**, 2000102 (2021). https://doi.org/10.1002/adpr.202000102
65. H. Li, R. Pang, Y. Luo, H. Wu, S. Zhang, L. Jiang, D. Li, C. Li, J. Feng, H. Zhang, Commendable Pr^{3+}-activated $Ba_2Ga_2GeO_7$ phosphor with high-brightness white long-persistent luminescence. J. Mater. Chem. C. **7**, 6698–6705 (2019). https://doi.org/10.1039/c9tc01735f
66. P. Huang, F. Yang, C. Cui, L. Wang, X. Lei, Luminescence properties of white-light long-lasting phosphor Y_2O_2S:Dy^{3+}, Mg^{2+}, Si^{4+}. Ceram. Int. **39**, 7193–7197 (2013). https://doi.org/10.1016/j.ceramint.2013.02.064
67. Z. Wang, J. Xu, Z. Ju, S. Zhao, P. Pei, X. Ma, W. Liu, Novel white-emitting afterglow phosphor $Na_2CaSn_2Ge_3O_{12}$:Dy^{3+}: preparation, photoluminescence, and phosphorescence properties. J. Alloys Compd. **856**, 157230 (2021). https://doi.org/10.1016/J.JALLCOM.2020.157230
68. H. Nath Luitel, Design of long persistent white phosphorescence in a composite phosphor via energy transfer mechanism. Int. J. Mater. Sci. Appl. **4**, 107 (2015). https://doi.org/10.11648/j.ijmsa.20150402.16
69. Q.L.M. De Chermont, C. Chanéac, J. Seguin, F. Pellé, S. Maîtrejean, J.P. Jolivet, D. Gourier, M. Bessodes, D. Scherman, Nanoprobes with near-infrared persistent luminescence for in vivo imaging. Proc. Natl. Acad. Sci. U. S. A. **104**, 9266–9271 (2007). https://doi.org/10.1073/pnas.0702427104
70. W. Yan, Y.-Y. Lu, X.-J. Wang, M. Yin, Z. Pan, F. Liu, Near infrared long-persistent phosphorescence in $La_3Ga_5GeO_{14}$:Cr^{3+} phosphor. Opt. Express **18**(19), 20215–20221 (2010). https://doi.org/10.1364/OE.18.020215
71. Z. Zhou, Y. Li, M. Peng, Near-infrared persistent phosphors: Synthesis, design, and applications. Chem. Eng. J. **399**, 125688 (2020). https://doi.org/10.1016/j.cej.2020.125688
72. D. Jia, L.A. Lewis, X.J. Wang, Cr^{3+} -doped lanthanum gallogermanate phosphors with long persistent IR emission. Electrochem. Solid-State Lett. **13**, J32 (2010). https://doi.org/10.1149/1.3294520/XML
73. N. Yu, F. Liu, X. Li, Z. Pan, Near infrared long-persistent phosphorescence in $SrAl_2O_4$: Eu^{2+}, Dy^{3+}, Er^{3+} phosphors based on persistent energy transfer. Appl. Phys. Lett. **95**, 231110 (2009). https://doi.org/10.1063/1.3272672/120942
74. Y. Teng, J. Zhou, Z. Ma, M.M. Smedskjaer, J. Qiu, Persistent near infrared phosphorescence from rare earth ions co-doped strontium aluminate phosphors. J. Electrochem. Soc. **158**, K17 (2011). https://doi.org/10.1149/1.3518767/XML
75. S. Kamimura, C.N. Xu, H. Yamada, G. Marriott, K. Hyodo, T. Ohno, Near-infrared luminescence from double-perovskite $Sr_3Sn_2O_7$:Nd^{3+}: a new class of probe for in vivo imaging in the second optical window of biological tissue. J. Ceram. Soc. Japan. **125**, 591–595 (2017). https://doi.org/10.2109/JCERSJ2.17051
76. X. Qin, Y. Li, R. Zhang, J. Ren, M. Gecevicius, Y. Wu, K. Sharafudeen, G. Dong, S. Zhou, Z. Ma, J. Qiu, Hybrid coordination-network-engineering for bridging cascaded channels to activate long persistent phosphorescence in the second biological window, Sci. Reports **61. 6** , 1–9 (2016). https://doi.org/10.1038/srep20275
77. T. Lyu, P. Dorenbos, Designing thermally stimulated 1.06 μm Nd^{3+} emission for the second bio-imaging window demonstrated by energy transfer from Bi^{3+} in La-, Gd-, Y-, and $LuPO_4$. Chem. Eng. J. **372** , 978–991 (2019). https://doi.org/10.1016/J.CEJ.2019.04.125
78. F. Liu, Y. Liang, Y. Chen, Z. Pan, Divalent nickel-activated Gallate-based persistent phosphors in the short-wave infrared. Adv. Opt. Mater. **4**, 562–566 (2016). https://doi.org/10.1002/ADOM.201500656
79. M. Pellerin, V. Castaing, D. Gourier, C. Chanéac, B. Viana, Persistent luminescence of transition metal (Co, Ni...)-doped ZnGa2O4 phosphors for applications in the near-infrared range **10533**, 196–203 (2018). https://doi.org/10.1117/12.2294986

80. V. Castaing, A.D. Sontakke, J. Xu, A.J. Fernández-Carrión, C. Genevois, S. Tanabe, M. Allix, B. Viana, Persistent energy transfer in ZGO:Cr^{3+}, Y^{b3}+: a new strategy to design nano glass-ceramics featuring deep red and near infrared persistent luminescence. Phys. Chem. Chem. Phys. **21**, 19458–19468 (2019). https://doi.org/10.1039/C9CP02927C
81. Y. Liang, F. Liu, Y. Chen, X. Wang, K. Sun, Z. Pan, Red/near-infrared/short-wave infrared multi-band persistent luminescence in Pr^{3+}-doped persistent phosphors. Dalt. Trans. **46**, 11149–11153 (2017). https://doi.org/10.1039/C7DT02271A
82. M.Y.A. Yagoub, H.C. Swart, E. Coetsee, The role of Li^{+} interstitial ions in up-conversion intensity of CaF2:Yb^{3+}, Tb^{3+} phosphors. Mater. Res. Bull. **156**, 111986 (2022). https://doi.org/10.1016/J.MATERRESBULL.2022.111986
83. J.C.G. Bünzli, C. Piguet, Taking advantage of luminescent lanthanide ions. Chem. Soc. Rev. **34**, 1048–1077 (2005). https://doi.org/10.1039/B406082M
84. J.G. Bünzli, S.V. Eliseeva, Lanthanide NIR luminescence for telecommunications, bioanalyses and solar energy conversion. J. Rare Earths **28**, 824–842 (2010). https://doi.org/10.1016/S1002-0721(09)60208-8
85. G. Deng, S. Li, Z. Sun, W. Li, L. Zhou, J. Zhang, P. Gong, L. Cai, Near-infrared fluorescence imaging in the largely unexplored window of 900–1,000 nm. Theranostics. **8**, 4116–4128 (2018). https://doi.org/10.7150/THNO.26539
86. Y.J. Liang, F. Liu, Y.F. Chen, X.J. Wang, K.N. Sun, Z. Pan, New function of the Yb^{3+} ion as an efficient emitter of persistent luminescence in the short-wave infrared, Light Sci. Appl. **57. 5**, e16124–e16124 (2016). https://doi.org/10.1038/lsa.2016.124
87. D. Wang, H. Li, J. Zhang, Z. Zou, X. Li, C. Wu, Y. Wang, Near-infrared persistent luminescence of Yb^{3+} in perovskite phosphor. Opt. Lett. **42** , 4510–4512 (2017). https://doi.org/10.1364/OL.42.004510
88. M. Back, E. Casagrande, E. Trave, D. Cristofori, E. Ambrosi, F. Dallo, M. Roman, J. Ueda, J. Xu, S. Tanabe, A. Benedetti, P. Riello, Confined-melting-assisted synthesis of bismuth silicate glass-ceramic nanoparticles: formation and optical thermometry investigation. ACS Appl. Mater. Interf. **12**, 55195–55204 (2020). https://doi.org/10.1021/acsami.0c17897
89. M. Back, J. Ueda, J. Xu, D. Murata, M.G. Brik, S. Tanabe, Ratiometric luminescent thermometers with a customized phase-transition-driven fingerprint in perovskite oxides. ACS Appl. Mater. Interf. **11**, 38937–38945 (2019). https://doi.org/10.1021/acsami.9b13010
90. J. Xu, D. Murata, B. So, K. Asami, J. Ueda, J. Heo, S. Tanabe, 1.2 μm persistent luminescence of Ho^{3+} in $LaAlO_3$ and $LaGaO_3$ perovskites. J. Mater. Chem. C. **6**, 11374–11383 (2018). https://doi.org/10.1039/C8TC04393K
91. J. Xu, S. Tanabe, A.D. Sontakke, J. Ueda, Near-infrared multi-wavelengths long persistent luminescence of Nd^{3+} ion through persistent energy transfer in Ce^{3+}, Cr^{3+} co-doped Y3Al2Ga3O12 for the first and second bio-imaging windows. Appl. Phys. Lett. **107**, 81903 (2015). https://doi.org/10.1063/1.4929495/30916
92. J. Xu, D. Murata, Y. Katayama, J. Ueda, S. Tanabe, Cr^{3+}/Er^{3+} co-doped $LaAlO_3$ perovskite phosphor: a near-infrared persistent luminescence probe covering the first and third biological windows. J. Mater. Chem. B. **5**, 6385–6393 (2017). https://doi.org/10.1039/C7TB01332A
93. J. Xu, D. Murata, J. Ueda, S. Tanabe, Near-infrared long persistent luminescence of Er^{3+} in garnet for the third bio-imaging window. J. Mater. Chem. C. **4**, 11096–11103 (2016). https://doi.org/10.1039/C6TC04027F

Different Persistent Luminescent Materials and Suitable Adoptions for Improving the Efficiency

Yuhua Wang, Xilin Ma, Songsong Ding, Hongzhe Zhang, and Runlin Liu

Abstract In this chapter, we focus on the research results of persistent luminescent materials (PLMs) based on our laboratory research and other related reports, systematically elaborate, and classify the types of PLMs from different host materials, and emission wavelengths, and summarize the main issues of PLMs. Then, the suitable means to improve the efficiency (such as afterglow duration and brightness) of PLMs are discussed and summarized, such as doping effect, solid solution, energy transfer, surface modification, and so on. Finally, we draw a conclusion and a perspective of the PLMs, to extend and provide a reference for the researchers in this field.

Keywords Multicolor persistent luminescent materials · Defect · Trap · Afterglow mechanism

Y. Wang (✉) · X. Ma · S. Ding · H. Zhang · R. Liu
National and Local Joint Engineering Laboratory for Optical Conversion Materials and Technology of National Development and Reform Commission, Department of Materials Science, School of Materials and Energy, Cross-Strait Research and Technology Transfer Center for Optoelectronic Materials and Devices of Lanzhou University and Taiwan Kun Shan University, Lanzhou University, No. 222, South Tianshui Road, Lanzhou, Gansu 730000, P. R. China
e-mail: wyh@lzu.edu.cn

X. Ma
e-mail: maxilin20@lzu.edu.cn

S. Ding
e-mail: dingss20@lzu.edu.cn

H. Zhang
e-mail: 220220900431@lzu.edu.cn

R. Liu
e-mail: lrl2023@lzu.edu.cn

V. Kumar et al. (eds.), *Persistent Luminescence*, Progress in Optical Science and Photonics 30, https://doi.org/10.1007/978-981-97-4943-0_3

1 Introduction

Persistent luminescent materials (PLMs) are characterized as green, clean, and pollution-free new energy materials, which can absorb and store the energy of an external excitation source (ultraviolet light, electron beam, X-ray, et al.), and a long afterglow is produced after removing the irradiation source for several minutes or hours. Hence, PLMs are extensively employed in daily production and life due to their "light storage" characteristics, such as night lighting, surface coating, interior decoration, information storage, bioimaging, and many other fields. Up to now, a variety of novel PLMs with different colors and properties have developed in the laboratory, however, only a yellow-green persistence phosphor $SrAl_2O_4$: Eu^{2+}, Dy^{3+} (520 nm, >30 h) has been commercially spread and widely used [1].

The process of persistent luminescence can be explained by a simplified diagram in Fig. 1, including the ground state and the excited state of the emission center, and the metastable trap level of the carriers. The fluorescence process is mainly originated from the electron transition between the ground state and the excited state (Step I and II) and determines the emission peak shape, location, and decay lifetime. For persistent luminescence, the long afterglow emission mainly depends on the process of trapping and de-trapping of the carriers (Step III), and the parameters of interest can be classified into the following categories:

(1) Afterglow emission color, which is generally related to the doping centers, which act not only as the fluorescence emission center but also as the afterglow emission center, where the transition electrons come from the trap level.
(2) Afterglow duration, for human visible PLMs, is defined as the duration of the afterglow emission intensity attenuation to 0.032 mcd/m^2, which is about 100 times of the dark perception limit of the naked eye.
(3) Afterglow brightness, that is, the brightness of the PLMs during the decay process when the pre-irradiation is stopped. As the long afterglow properties are mainly dependent on the trap properties caused by co-doping ions and other factors that are related to the captured carriers of the trap level, some viable adoptions have emerged for improving the PLMs efficiency.

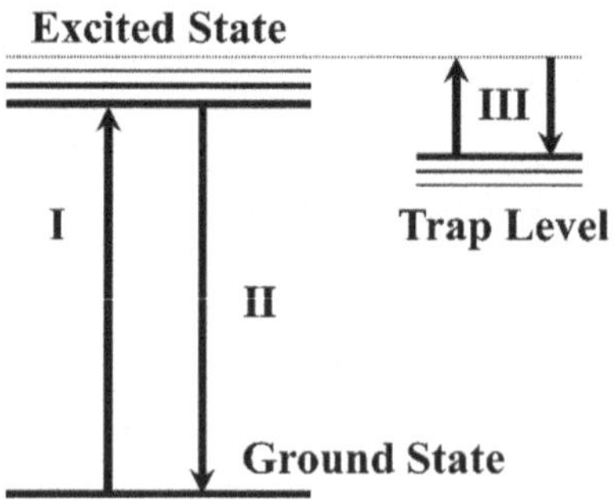

Fig. 1 The simplified diagram of fluorescence and persistent luminescence process

This review is primarily based on our works and other related research. Firstly, classify some of the PLMs, according to the classification principle of the type of host materials and emission bands. On this basis, the relevant strategies to improve the properties of long afterglow materials are concerned, including host composition regulation, doping effect, synthesis condition improvement, surface modification, and other strategies.

2 Persistent Luminescent Materials (PLMs) and Their Categories

Since the appearance of PLMs, it has been paid attention to be studied for more than 100 years [2]. Up to now, a wide variety of PLMs have been developed, and their luminescent properties have been continuously improved [3]. In the following sections, PLMs are classified, and their characteristics are summarized from two perspectives: the host materials and the emission wavelengths.

2.1 Host Materials

PLMs are generally composed of a host material and doping centers. The former is the main component of PLMs and can provide suitable lattice sites for dopant ions. The latter form the emission centers and trap centers, where the emission center determines the emission wavelength of the materials in photoluminescence (PL) and persistence luminescence, while the trap centers tend to form the metastable energy traps inside the host materials for trapping and de-trapping carriers to produce long afterglow. According to the different types of hosts, PLMs have gradually formed several systems from the first discovered sulfides which will be described below.

2.1.1 Sulfide, Sulfur-Oxide, and Binary Oxide-Based

The earliest sulfide-based PLMs with practical application value is ZnS: Cu^{+} (with a green emission peak at 510 nm), which has been used in watches, instrument panels, and other military departments [4]. Since then, many PLMs containing sulfur compounds have been discovered and can be divided into two categories, the transition metal and alkaline earth metal sulfide PLMs, and sulfur-oxide PLMs, as shown in Table 1. The typical type is Y_2O_2S: Eu^{3+}, Mg^{2+}, Ti^{4+}, which can achieve a yellow-to-red afterglow by adjusting the doping ratio of Eu^{3+}, Mg^{2+}, and Ti^{4+} [5]. In addition, persistent luminescence was also found in some doped simple binary oxide compounds. However, the best initial afterglow brightness of the above materials is only about 40 mcd/m^2 with a rapid decrease, and the effective afterglow duration is

Table 1 Several typical sulfide, sulfur-oxide PLMs

Materials	Afterglow color	Emission peak (nm)	Long afterglow duration (h)	Refs.
CaS: Ce^{3+}	Cyan	507	0.2	[6]
CaS: Eu^{3+}, Tm^{3+}	Red	650	0.75	[7]
CaS: Eu^{3+}, Ce^{3+}	Red	650	1	[8]
SrS: Eu^{3+}, Sm^{3+}	Red	610	0.5	[9]
Y_2O_2S: Eu^{3+}, Mg^{2+}, Ti^{4+}	Red	625	>2	[5]
Y_2O_2S: Dy^{3+}, Mg^{2+}, Si^{4+}	White	488, 579	>1	[10]

Table 2 Several typical binary oxide PLMs

Host material	Activator	Co-doping ions	Emission peak (nm)	Long afterglow duration	Refs.
SnO_2	Sm^{3+}	–	567,607,625	40 min	Zhang et al. [11]
CaO	Eu^{3+}	–	594,616	>2 h	Fu [12]
Ga_2O_3	Cr^{3+}	–	720	>4 h	Lu et al. [13]
Lu_2O_3	Eu^{3+}	–	611	>3 min	Zych and Trojan-Piegza [14]
SrO	Eu^{3+}	–	594,616	>1 h	Fu [15]
	Pb^{2+}	–	390	>1 h	
Y_2O_3	Eu^{3+}	–	612	90 min	Lin et al. [16]
ZrO_2	Sm^{3+}	–	570,614	15 min	Zhao et al. [17], Carvalho et al. [18]
	Ti^{4+}	–	470–500	1 h	

generally only 0.5 h. Meanwhile, the poor chemical stability in a humid and acidic environment also restricts the extensive applications and is gradually replaced by other more efficient PLMs (as well as the binary oxide compounds in Table 2).

2.1.2 Aluminate

In 1996, yellow-green long afterglow material $SrAl_2O_4$: Eu^{2+}, Dy^{3+} were identified, and the decay time and afterglow brightness are 5 to 10 times higher than that of traditional sulfide materials [19]. Subsequently, a series of lanthanide elements doped aluminate-based PLMs were developed, as shown in Table 3. We prepared the $CaAl_2O_4$: Eu^{2+}, Nd^{3+}, a blue emitting persistent phosphors by a high-temperature state reaction with the higher afterglow brightness (>1238 mcd/m^2) and longer duration (>20 h) [20, 21].

Table 3 Several typical aluminate PLMs

Host material	Activator	Co-doping ions	Emission peak (nm)	Long afterglow duration	Refs.
$SrAl_2O_4$	Eu^{2+}	Dy^{3+}	520	>60 h	Matsuzawa et al. [19], Yamamoto and Matsuzawa [22]
$CaAl_2O_4$	Eu^{2+}	Nd^{3+}	450	>20 h	Sakai et al. [23]
	Eu^{2+}	Dy^{3+}	500	>10 h	
$BaAl_2O_4$	Ce^{3+}	–	450,402	10 h	Jia et al. [24]
$MgAl_2O_4$	Ce^{3+}	–	520	10 h	Jia et al. [25]
$ZnAl_2O_4$	Mn^{2+}	–	512	2 h	Matsui et al. [26]
$SrAl_4O_7$	Eu^{2+}	Dy^{3+}	475	>10 h	Katsumata et al. [27]
$CaAl_4O_7$	Tb^{3+}	Ce^{3+}	543	10 h	Jia et al. [28]
$Sr_3Al_2O_6$	Eu^{2+}	Dy^{3+}	612	>10 min	Zhang et al. [29]
$Sr_4Al_{14}O_{25}$	Eu^{2+}	Dy^{3+}	486	>30 h	Lin et al. [30]
$Ca_{12}Al_{14}O_{33}$	Eu^{2+}	Nd^{3+}	440	1 h	Zhang et al. [31]
$CaYAl_2O_7$	Ce^{3+}	–	425	>1 h	Kodama et al. [32]

As can be seen, the aluminate PLMs have a longer duration and are relatively stable than sulfide-base materials. However, the afterglow color is limited to the short-wavelength region such as blue and green, and due to the poor water resistance, hydrolysis effect will greatly reduce the afterglow performance (Taken $SrAl_2O_4$ as an example, $4SrAl_2O_4 + 16H_2O \rightarrow Sr_3Al_2(OH)_{12} + Sr^{2+} + 2OH^- + 6Al(OH)_3$ will happen), which limits its application and storage.

2.1.3 Silicate

Silicate PLMs have good chemical stability and thermal stability, and the main raw material is silicate sand as a host, which is cheap and easily available. Eu^{2+} has been selected as the main activator in silicate PLMs, whose long afterglow color is in the blue and green region, among which the best one is $Sr_2MgSi_2O_7$: Eu^{2+}, Dy^{3+} (cyan, > 20 h). Moreover, some achieve long wavelength emission, such as Sr_3SiO_5: Eu^{2+}, Dy^{3+} (yellow) [33], Li_2SrSiO_4: Eu^{2+}, Dy^{3+} (orange yellow) [34], $MgSiO_3$: Eu^{2+},Dy^{3+} (red) [35] and so on (in Table 4). The afterglow color in the silicate system can form a good complement with aluminate PLMs. In the meantime, we have also developed a variety of novel silicate PLMs, such as $Ba_2Gd_2Si_4O_{13}$: Eu^{2+}, RE^{3+} ($RE^{3+} = Dy^{3+}$, Ho^{3+}, Tm^{3+}, Nd^{3+} and Tb^{3+}) [36], $Ca_2ZnSi_2O_7$: Eu^{2+} [37], $Ca_3SnSi_2O_9$: Re^{3+} (Re = Pr, Tb, Sm) [38], and $BaLu_2Si_3O_{10}$: Eu^{2+}, Nd^{3+} [39]. In addition, $BaZrSi_3O_9$:

Eu^{2+}, Pr^{3+} (cyan) [40], $Ba_4(Si_3O_8)_2$: Eu^{2+}, Dy^{3+} (bluish green) [41] have ultra-long afterglow durations, which can reach to 15 h and 24 h, respectively. However, due to the absorption only in the near-ultraviolet region of silicate, it is often difficult to achieve daylight charging for silicate PLMs, and the long afterglow brightness is relatively low compared with the aluminate.

In addition, aluminosilicate PLMs have also been reported, as listed in Table 5. The afterglow emission color is mainly blue and green, and its afterglow performance is not as good as aluminate or silicate.

Table 4 Several typical silicate PLMs

Host material	Activator	Co-doping ions	Emission peak (nm)	Long afterglow duration	Refs.
$Sr_3MgSi_2O_8$	Eu^{2+}	Dy^{3+}	475	5 h	Alvani et al. [42]
$Sr_2MgSi_2O_7$	Eu^{2+}	Dy^{3+}	466	>20 h	Lin et al. [43]
$Sr_2ZnSi_2O_7$	Eu^{2+}	Dy^{3+}	457	1 h	Jiang et al. [44]
$Ca_3MgSi_2O_8$	Eu^{2+}	Dy^{3+}	471	5 h	Lin et al. [45]
$Ca_2MgSi_2O_7$	Eu^{2+}	Dy^{3+}	545	>20 h	Fei et al. [46]
$CaMgSi_2O_6$	Eu^{2+}	Dy^{3+}	447	>20 h	Jiang et al. [47]
$Ba_3MgSi_2O_8$	Eu^{2+}	Dy^{3+}	439	1 h	Lin et al. [48]
$BaMg_2Si_2O_7$	Mn^{2+}	Eu^{2+}, Dy^{3+}	660	2 h	Ye et al. [49]
Sr_3SiO_5	Eu^{2+}	Dy^{3+}	570	6 h	Sun et al. [33]
Sr_2SiO_4	Dy^{3+}	–	479	5 min	Lakshminarasimhan and Varadaraju [50]
$MgSiO_3$	Mn^{2+}	Dy^{3+}	661	10 min	Lin et al. [35]
Mg_2SiO_4	Sm^{3+}	–	603	5 h	Lei et al. [51]
$MgSiO_3$	Pr^{3+}	–	reddish orange	5 h	Kuang and Liu [52]

Table 5 Several typical aluminosilicate PLMs

Host material	Activator	Co-doping ions	Emission peak (nm)	Long afterglow duration	Refs.
$Ca_2Al_2SiO_7$	Ce^{3+}	–	417	>1 h	Kodama et al. [32]
$Ca_2Al_2Si_2O_8$	Eu^{2+}	Dy^{3+}	430	1 h	Wang et al. [53]
$Sr_2Al_2SiO_7$	Eu^{2+}	Dy^{3+}	484	>7 min	Ding et al. [54]

Table 6 Several typical Cr^{3+} activated gallate and germanate PLMs

Host material	Activator	Emission range (nm)	Emission peak (nm)	Transition	Refs.
$La_3Ga_5GeO_{14}$	Cr^{3+}	700–1300	785	$^4T_2 \rightarrow {^4A_2}$	Wu et al. [59]
$ZnGa_2O_4$	Cr^{3+}	600–800	696	$^2E \rightarrow {^4A_2}$	Zhou et al. [60]
$MgGa_2O_4$	Cr^{3+}	650–770	695	$^2E \rightarrow {^4A_2}$	Mondal et al. [61]
$LiGa_5O_8$	Cr^{3+}	650–800	712	$^2E \rightarrow {^4A_2}$	Liu et al. [62]
$Zn_3Ga_2Ge_2O_{10}$	Cr^{3+}	650–950	696	$^2E \rightarrow {^4A_2}, {^4T_2} \rightarrow {^4A_2}$	Wu et al. [63]
La_2MgGeO_6	Cr^{3+}	640–750	710	$^2E \rightarrow {^4A_2}$	Wang et al. [64]
$SrGa_{12}O_{19}$	Cr^{3+}	600–950	750	$^2E \rightarrow {^4A_2}$	Xu et al. [65]

2.1.4 Gallate and Germanate

The valence shell electron configurations of gallium and germanium are $4s^2\,4p^1$ and $4s^2\,4p^2$, respectively, which feature similar chemical properties. The basic composed polyhedron of gallates, germanates, or gallium-germanates are [Ga/GeO_4] or [Ga/GeO_6], and this is mainly suitable for the transition metal ions to occupy, due to the similar ionic radius Ga^{3+}(CN = 4, 0.47 Å, CN = 6, 0.62 Å), Ge^{4+}(CN = 4, 0.39 Å, CN = 6, 0.53 Å), Mn^{2+}(CN = 4, 0.57 Å, CN = 6, 0.72 Å), and Cr^{3+}(CN = 6, 0.615 Å). Table 6 exhibits some reported near-infrared PLMs. Among them, $Zn_3Ga_2Ge_2O_{10}$: Cr^{3+} has the better performance, for its long afterglow duration can reach 400 h [55]. We also devoted some new germanate-based PLMs with outstanding performance, like $CaGa_2O_4$: Pr^{3+} (yellow, 6 h) [56], $Na_2CaGe_5SiO_{14}$: Cr^{3+} (600–900 nm, 10 h), $Na_2CaGe_6O_{14}$: Cr^{3+} (600–1200 nm, 10 h) [57, 58]. Nevertheless, the required raw materials (Ga_2O_3, GeO_2) for the preparation of gallium, and germanate are often expensive, not suitable for low-cost application, and these types of materials cannot be synthesized in a reduction atmosphere, so they can only for dopant ions that do not require reduction (Cr^{3+}, Mn^{4+}, Eu^{3+}, etc.).

2.1.5 Nitride and Phosphate

Nitride-base materials are usually characterized by a closely packed three-dimensional network structure composed of nitride polyhedrons, such as SiN_4, AlN_4, [Si, Al]N_4, et al. The afterglow in $Ba_2Si_5N_8$: Eu^{2+} phosphor was first found in 2000, with a red emission at 600 nm [66]. After that, solid solution $M_2Si_5N_8$: Eu^{2+}(M = Ca, Sr, Ba) phosphors were studied, and co-doped Tm^{3+} could effectively improve the long afterglow duration to 2650 s under 420 nm pre-irradiation [67]. In 2015, the afterglow performance of $CaAlSiN_3$: Eu^{2+} was discovered, and the afterglow time can reach to 2700 s, and the duration of solid solution $Sr_{0.8}Ca_{0.2}AlSiN_3$: Eu^{2+} is enhanced to 9600 s [68, 69]. However, nitride PLMs need ultra-high temperatures and pressure to be synthesized, and the price of raw materials is expensive, these factors restrict the development and application, in addition, the afterglow performance is not usable compared with aluminate and silicate PLMs (in Table 7).

Phosphate has been widely studied as a host material for vacuum ultraviolet and phosphor-converted light-emitting diode (pc-LED) [75]. While, there are relatively few works about phosphate PLMs, except for our discovered $Ca_6BaP_4O_{17}$ (yellow, >20 h) [76], the afterglow properties of other phosphate PLMs are not ideal. Based on the related studies published so far (in Table 8), their afterglow performance is not outstanding.

Table 7 Several typical nitride PLMs

Host material	Activator	Co-doping ions	Emission peak (nm)	Long afterglow duration (s)	Refs.
$CaAlSiN_3$	Eu^{2+}	–	635	2700	Wang et al. [68]
$Sr_{0.8}Ca_{0.2}AlSiN_3$	Eu^{2+}	–	628	9600	Liu et al. [69]
$Ca_2Si_5N_8$	Eu^{2+}	–	610	150	Miyamoto et al. [67]
	Eu^{2+}	Tm^{3+}	610	2650	
$Sr_2Si_5N_8$	Eu^{2+}	–	620	80	Teng et al. [70]
	Eu^{2+}	Tm^{3+}	620	270	
$Ba_2Si_5N_8$	Eu^{2+}	–	580	400	Hoppe et al. [66]
	Eu^{2+}	Tm^{3+}	580	495	
$SrCaSi_5N_8$	Eu^{2+}	Tm^{3+}	647	4800	Wang et al. [71]
$CaSi_2O_2N_2$	Eu^{2+}	–	556	400	Botterman et al. [72]
$SrSi_2O_2N_2$	Eu^{2+}	–	539	6000	Zhuang et al. [73]
	Eu^{2+}	Ho^{3+}	539	21,600	
$BaSi_2O_2N_2$	Eu^{2+}	–	493	2800	Botterman et al. [72]
$SrSi_2AlO_2N_3$	Eu^{2+}	–	500	6000	Zhuang et al. [74]
	Eu^{2+}	Dy^{3+}	500	7800	

Table 8 Several typical phosphate PLMs

Host material	Activator	Co-doping center	Emission peak (nm)	Long afterglow duration	Refs.
$Ca_2P_2O_7$	Eu^{2+}	Y^{3+}	416	10 min	Pang et al. [77]
$SrMg_2(PO_4)_2$	Eu^{3+}	–	588	1.5 h	Liu et al. [78]
$Zn_3(PO_4)_2$	Mn^{2+}	Al^{3+} or Ga^{3+}	616	>2 h	Wang et al. [79]
	Hf^{4+}	–	470	40 min	
$Ca_9Gd(PO_4)_7$	Mn^{2+}	–	602,628,660	>20 min	[80]
$Ca_9Lu(PO_4)_7$	Mn^{2+}	–	660	>20 min	Bessiere et al. [80]
$Ca_3(PO_4)_2$	Mn^{2+}	–	645	1 h	Dikhtyar et al. [81]
$SrZn_2(PO_4)_2$	Eu^{2+}	Mn^{2+}	421,547	1 min	Jeong et al. [82]
	Mn^{2+}	–	547	1 min	
YPO_4	Pr^{3+}	–	600,620	>30 min	Lecointre et al. [83]

2.1.6 Other Options

In recent years, some novel PLMs have been reported, such as stannate, which can only be synthesized in air atmosphere. In addition, $YTaO_4$: Tb^{3+} (green, 2 h) [84], $CaTiO_3$: Pr^{3+} (red, 1 h) [85], CaWO4: Sm^{3+} (red, 0.5 h) [86], etc. are also reported. However, deficiencies still exist among them, such as lower charging efficiency under visible light, shorter afterglow duration, or weaker afterglow brightness. As a supplementary, we obtained a novel borate persistent phosphor, Ca_2BO_3Cl: Eu^{2+}, Dy^{3+}, which peaked at 580 nm with more than 48 h long afterglow [87].

In addition, some host materials have self-activation accompanied by the afterglow emission, this phenomenon relies on the intrinsic defects of the materials as the emission centers rather than the doping centers. Such as in Mg_2SnO_4, which can be excited by 291 nm ultraviolet light, and emits a 500 nm blue-green afterglow, similarly, our research of $SrZrSi_2O_7$ also presents a blue afterglow emission of 497 nm after excitation, and the sample prepared under the reduction atmosphere has a higher afterglow intensity, indicating that the increase of intrinsic defects (such as oxygen vacancy) in the material, thus improving the afterglow performance [88, 89].

To sum up, distinct host materials will have different effects on the luminescence behaviors of PLMs. This gives us the following implications when designing and synthesizing novel PLMs.

(1) Suitable band gap. It determines the optical absorption and efficient charging wavelengths. Different bandgap structures can affect the absolute position of the energy level of doping centers; hence the trapping and de-trapping of trap centers are also affected.

(2) Suitable coordination environment. Ce^{3+}, Eu^{2+} (4f-5d transition), and Mn^{2+}, Cr^{3+} (3d-3d transition) need to occupy suitable lattice sites with appropriate crystal fields for demanded emission wavelengths. For Tb^{3+}, Yb^{3+}, Pr^{3+}, etc. (4f-4f transition), their emission is not affected by crystal fields, however, the ion radius and the position relationship between rare earth energy level and band structure should also be considered when doping.
(3) Intrinsic or extrinsic defects. defects can lead to certain trap properties of the PLMs, which is useful for the generation of afterglow, and the PLMs with outstanding properties are mainly lanthanide ion co-doped. The doping effect in PLMs will be described in detail later.

2.2 *Emission Wavelengths*

2.2.1 Visible Wavelengths

Through the above summary, the earlier discovered blue PLMs are mainly sulfide-based, due to their weak afterglow performance and poor chemical properties, which cannot be widely applied. Since then, aluminate and silicate PLMs have been developed successively, such as high-brightness aluminate $CaAl_2O_4$: Eu^{2+}, Nd^{3+} (430 nm, >6 h), $Sr_4Al_{14}O_{25}$: Eu^{2+}, Dy^{3+} (490 nm, >20 h) and silicate $Sr_2MgSi_2O_7$: Eu^{2+}, Dy^{3+} (470 nm, >10 h), and other PLMs systems. Eu^{2+} and Ce^{3+} are selected as the main emission centers for blue-emitting PLMs, and Ti^{4+}, Bi^{3+}, Hf^{4+}, and Pb^{2+} doped materials can also emit long afterglow in the blue region, but their emission intensity and quantum efficiency are not as good as the former. In addition, some undoped materials, such as $BaZrSi_3O_9$ [90] and $BaZrO_3$ [91], due to their intrinsic defects, can also generate blue afterglow.

Yellow-green emitting long afterglow material $SrAl_2O_4$: Eu^{2+}, Dy^{3+} has been successfully commercialized because of its high daylight charging efficiency and ultra-long afterglow decay time (more than 30 h). Other samples doped with Tb^{3+} and Mn^{2+} ions can also emit green afterglow; however, the performance is still not comparable to $SrAl_2O_4$: Eu^{2+}, Dy^{3+}.

The red PLMs are mainly concentrated in the simple compounds at first, such as CaS: Eu^{2+}, Tm^{3+} (650 nm, >1 h), Y_2O_2S: Eu^{3+}, Mg^{2+}, Ti^{4+} (625 nm, >2 h) and CaO: Eu^{3+} (616 nm, >2 h), then comes to $CaZnGe_2O_6$: Mn^{2+} (648 nm, >3 h) [92], $Zn_3(PO_4)_2$: Mn^{2+} (616 nm, >2 h), Ca_2SnO_4: Sm^{3+} (653 nm, >1 h) [93], and $CaMgGe_2O_6$: Mn^{2+}, Sm^{3+} (our work, 675 nm, >1 h) [94]. As the red afterglow can not only need to overcome the larger Stokes shift of down-conversion but also needs a suitable energy trap, it is challenging to develop new materials, whose performances are still far inferior to blue or green PLMs.

In addition, white PLMs (full spectrum PLMs) have also been reported, mainly in some Dy^{3+} activated materials, like Mg_2SiO_4 (>5 h), as well as in some co-doping materials, such as $CaAl_2Si_2O_8$: Eu^{2+}, Mn^{2+}, Dy^{3+}, where Eu^{2+} and Mn^{2+} produce cyan and orange afterglow, respectively [95]. As the durations of different emission centers are different, the color of the afterglow can be changed, which is detrimental to practical applications.

2.2.2 Near-Infrared Wavelengths

Near-infrared light can be divided into the first near-infrared window (NIR-I, 650–1000 nm) and the second near-infrared window (NIR-II, 1000–1700 nm) by wavelengths. Due to higher spatiotemporal resolution and lower signal-to-noise (SNR) to biological tissues, the optical imaging technology based on near-infrared light can realize the in situ detection of cells, tissues, and organisms, which is of great significance for the diagnosis and treatment of human diseases. The near-infrared PLMs can slowly emit light after pre-irradiation, which can significantly improve the SNR ratio due to the avoidance of tissue autofluorescence. Therefore, the study of near-infrared PLMs has become a hot spot in the field of bioimaging.

The emission centers of near-infrared PLMs can be classified into rare earth ions (Eu^{2+}, Eu^{3+}, Sm^{3+}, and Pr^{3+}) and transition metal ions (mainly Mn^{2+} and Cr^{3+}). Combined with the previous tables, it can be found that the emission peaks of Eu^{2+}, Eu^{3+}, Sm^{3+}, Pr^{3+}, and Mn^{2+} are mainly located in the red or deep red regions, which is not conducive for application in biology. Therefore, Cr^{3+} is the main emission center of near-infrared PLMs, however, the emission peak still cannot exceed 1000 nm and with the risk of oxidation into toxic Cr^{4+} and Cr^{6+}. Therefore, it is still necessary to develop novel non-toxic PLMs with larger Stokes shifts located in the NIR II region.

2.2.3 Ultra-Violet Wavelengths

The conventional long afterglow emission spectrum overlaps with the natural light spectrum, and the emission intensity can be overwhelmed. In recent years, some ultraviolet (UV) PLMs have been studied, because of higher energy than visible light, which can be utilized for sterilization, solar blind marking, and so on [96]. According to the different emission bands, it can be divided into: UVA (320–380 nm), UVB (280–320 nm), and UVC (200–280 nm) regions, which are described below.

(1) UVA PLMs are well examined, and the emission centers that can produce UVA long afterglow emission can be selected as Pb^{2+}, Bi^{3+}, Ce^{3+}, and Tb^{3+}. Pb^{2+} activated UVA PLMs such as SrO: Pb^{2+} (254 nm excited, 390 nm, >1 h) [15], $Sr_2MgGe_2O_7$: Pb^{2+} (370 nm, >12 h) [97]. The optically stimulated luminescence is also observed in $Sr_2MgGe_2O_7$: Pb^{2+}, whose afterglow can be significantly enhanced as a white light irradiated for a short time. However, the toxicity of Pb^{2+} is not conducive to humans and the environment, so it has been gradually abandoned. Bi^{3+} activated UVA PLMs can be list as follows: $CdSiO_3$: Bi^{3+} (221 nm activated, 360 nm, > a few minutes) [98], $NaLuGeO_4$: Bi^{3+}, Eu^{3+} (400 nm, >63 h) [99], $CaZnGe_2O_6$: Bi^{3+} (300–700 nm, >12 h) [100], $LiScGeO_4$: Bi^{3+} (365 nm,

Table 9 Suitable doping ions for demanded PLMs emission colors

Emission wavelength		Doping ions
Visible regions	Blue afterglow	Eu^{2+}, Ce^{3+}, Tm^{3+}, defects
	Green afterglow	Eu^{2+}, Tb^{3+}, Ce^{3+}, Mn^{2+}
	Red afterglow	Eu^{3+}, Eu^{2+}, Yb^{2+}, Sm^{3+}, Pr^{3+}, Mn^{2+}
	White afterglow	Dy^{3+}
Near-infrared region		Nd^{3+}, Bi^{2+}, Yb^{3+}, Mn^{2+}, Mn^{4+}, Cr^{3+}, Ni^{2+}, Er^{3+}, Fe^{3+}
Ultra-violet region		Ce^{3+}, Pr^{3+}, Bi^{3+}, Pb^{2+}, Tb^{3+}

>120 h) [101], $Sr_3Y_2Ge_3O_{12}$: Bi^{3+} (354 nm, >60 h) [102], and so on. Moreover, some Ce^{3+} or Tb^{3+} doped UVA PLMs such as $SrAl_2O_4$: Ce^{3+} (385 nm, >1000 min) [103], CaB_2O_4: Ce^{3+} (340–400 nm, >15 h) [104], $Sr_4Al_{14}O_{25}$: Tb^{3+} (380 nm) [105] are also reported.

(2) There are relatively few studies and reports on UVB PLMs. Currently, UVB luminescent materials can be used in the medical field to treat skin diseases [106, 107]. Sun et al. [108], reported that $Y_3Ga_5O_{12}$: Bi^{3+} and $Y_3Al_5O_{12}$: Bi^{3+} emit UVB long afterglow, whose emission is located at 316 nm and 303 nm, respectively, and the afterglow duration exceeds 60 min.

(3) Because of the shorter wavelength and higher energy, UVC light can destroy DNA structure and effectively kill viruses and bacteria, so it is widely applied in sterilization and disinfection [109]. Due to the atmospheric absorption, UVC light is absent in the natural environment, so it also has an application potential for outdoor sun-blind imaging. Some have been reported are Cs_2NaYF_6: Pr^{3+} (X-ray activated, 250 nm, >2 h) [110], Li_2CaGeO_4: Pr^{3+} (X-ray activated, 240–330 nm, >2 h) [111], etc.

However, considering the above, when obtaining UV afterglow emission, higher energy irradiation (such as UVC or X-ray) is often needed for charging, which is not conducive to the widespread use of the materials.

As a conclusion, during the design and synthesis of the PLMs, the desired wavelengths can be realized by purposefully doping the appropriate activator ions. The relevant emission colors and corresponding doping ions are summarized in Table 9.

3 Effect of Doping and Co-doping on Persistent Luminescence

Through the above summary, it can be recognized that the long afterglow is closely related to the different doping ions. Doping and co-doping refer to introducing a very small amount (generally no more than 5%) of other element ions in the host material in addition to generating appropriate emission centers and trap centers. It generally contains rare earth ions, transition metal ions, and some main group ions, their specific role will be discussed below.

3.1 Doping as Emission Centers

The electron configuration of lanthanide ions is [Xe] $4f^i$ $5d^j$ (i = 1–14, j = 0–1), and the electron configuration of transition metal ions is [Ar] $3d^i$ (i = 1–10). Ions of this electron configuration will produce electron transition under irradiation, thus forming a colorful afterglow emission.

3.1.1 Rare Earth Elements

The rare earth element is the general name of 17 elements of lanthanide (Ln) elements, scandium, and yttrium. Their similar outer electron structure shows unique luminescence characteristics in different doped materials. So far, Eu^{2+} is the most reported doping ion of PLMs, Tb^{3+}, Pr^{3+} followed by Ce^{3+}, Sm^{3+}, and Eu^{3+} less.

Among the Ln^{2+} ions, Eu^{2+} has a high absorption cross-section and effective excitation intensity due to parity allowed $4f^65d^1$ to $4f^7$ transition. The spectrum character is broadband with high intensity, and the peak position can be modulated by controlling the Nephelauxetic effect and the crystal field splitting effect, which emits different colors of the afterglow in different hosts. Among them, longer wavelength emission of yellow and red (or near-infrared) light requires lowering the lowest excitation level of the $4f^65d^1$ configuration of Eu^{2+}. Therefore, this tends to occur in some materials with a stronger crystal field, such as Ca_2BO_3Cl: Eu^{2+}, Dy^{3+} [87], Sr_3SiO_5: Eu^{2+}, Lu^{3+} [112], $Ca_2ZnSi_2O_7$: Eu^{2+}, Dy^{3+} [113], CaS: Eu^{2+} [8], where Eu^{2+} is occupied in low coordination number site. In addition, in some nitride compounds with strong bond covalence, longer wavelength afterglow emission can also be generated due to the larger Nephelauxetic effect, such as $Ca_2Si_5N_8$: Eu^{2+}, Tm^{3+} [114]. Ce^{3+} has a $4f^1$ outer electron configuration (Ce is $4f^15d^16s^2$) and can also realize a wide emission band due to a 5d-4f transition, which is similar to Eu^{2+}. The typical persistent emission band of Ce^{3+} is generally located in the blue to green light region, and only in $Gd_3Al_2Ga_3O_{12}$ has been found a yellow afterglow emission [115].

Among the trivalent Ln^{3+} ions, the Eu^{3+} ion does not have such a variable afterglow emission as Eu^{2+}, as the Eu^{3+} emission mainly comes from the 4f-4f transition of the $4f^6$ inner shell, whose spectrum character features a sharp line from the transition of $^5D_0$-^7F_J (J = 0, 1, 2, 3, 4) level. Eu^{3+} is more commonly doped in sulfide PLMs and the long afterglow is concentrated in the red region such as Gd_2O_2S: Eu^{3+} [116], Y_2O_2S: Eu^{3+} [117]. Sm^{3+} is known for realizing a reddish-orange afterglow. In fact, as an important red emission center, Sm^{3+} is found in a variety of materials from oxide compounds (CaO: Sm^{3+} [118], $La_2Zr_2O_7$: Sm^{3+}, Ti^{4+} [119], $Sr_2ZnSi_2O_7$: Sm^{3+} [120]) to sulfides (La_2O_2S: Sm^{3+} [121]) and fluoride (KY_3F_{10}: Sm^{3+} [122]). Similarly, the Pr^{3+} ion is another ion that produces red afterglow resulting from transitions of $^1D_2$-$^3H_4$, $^3P_0$-$^3H_6$ in our study of $LiTaO_3$: Pr^{3+} and $BaGa_2O_4$: Pr^{3+} [123, 124]. Tb^{3+} doped PLMs always present an effective green afterglow due to the transition from $^5D_4$-^7F_J ($J = 3, 4, 5, 6$) and has been reported in Lu_2O_3: Tb^{3+} [125], $CdSiO_3$: Tb^{3+} [126], $CaZnGe_2O_6$: Tb^{3+} [127], $Sr_3Al_2O_5Cl_2$: Tb^{3+} (our results) [128], and so on. Dy^{3+}, due to its blue emission (470–500 nm) of $^4F_{9/2}$-$^6H_{15/2}$ and yellow emission (570–600 nm) of $^4F_{9/2}$-$^6H_{13/2}$ transition, can emit white afterglow in a single host, such as $Sr_2Al_2SiO_7$: Dy^{3+} [129], $CaAl_2O_4$: Dy^{3+} [130], Y_2O_2S: Dy^{3+}, Mg^{2+}, Si^{4+} [10]. Few other Ln ions have been reported as persistent luminescence emission centers, which may be due to the absence of suitable hosts for generating afterglow.

3.1.2 Transition Metal Elements

Transitional metals are a series of elements in the d region of the periodic table. With the variation of valence states, the 3d electrons also show different configurations. Unlike rare earth element ions, the 3d orbital is exposed, and their outer 4 s electrons have no shielding effect on them, so they are easily affected by crystal fields. The transition behavior can be regulated through a crystal field. The transition of 3d electrons is parity forbidden, however, this can be relaxed by the change of crystal structure, hence the 3d transition metal ions have rich spectral properties in different types of hosts. The following is a detailed introduction of Cr^{3+}, Mn^{2+}, Fe^{3+}, and Ni^{2+} doping ions for PLMs.

Cr^{3+} is one of the most selected ions of the PLMs to produce near-infrared, whose 3d electron configuration is $3d^3$, which is sensitive to the changes in the crystal field. Cr^{3+} can occupy the tetrahedral or octahedral site of the crystal lattice, and the Dq/B value can be distinct as can be seen from the energy level diagram in Fig. 2a. Therefore, Cr^{3+} has a narrow band emission (peaked at ~700 nm) caused by the ^2E-$^4A_2$ transition. And broaden band emission (peaked at 650–1000 nm) due to $^4T_2$-$^4A_2$ transitions (in Fig. 3). As Table 6 shows, Cr^{3+}-activated PLMs with good performances are mainly gallate or germanate compounds [55]. This is because Cr^{3+} ions tend to replace Ga^{3+} or Ge^{4+} ions in the octahedron or tetrahedron as a suitable luminescence site, and the energy traps can largely generate to contribute to the persistent luminescence [131]. Moreover, due to the transition of $^4A_2$-$^4T_1$ (^4P), $^4A_2$-$^4T_1$ (^4F), and $^4A_2$-$^4T_2$ (^4F), Cr^{3+} has a strong and wide absorption band in the visible region, which can sensitize Nd^{3+}, Ho^{3+}, Er^{3+}, Tb^{3+}, Tm^{3+}, and so on [132].

Mn^{2+} has a $3d^5$ electron configuration, whose transition can generate a wide emission band ranging from blueish green (490 nm) to near-infrared (750 nm). In general, most substitutions of Mn^{2+} ions occur in octahedral and tetrahedral crystal structures. The emission of the Mn^{2+} transition is attributed to the spin-forbidden 3d-3d transition from the lowest excited energy level $^4T_1$ (^4G) to the ground state $^6A_1$ (^6S) (correspond to Fig. 2b). Tetrahedron-coordinated Mn^{2+} ions emit green, while octahedron-coordinated Mn^{2+} ions emit orange red. Thus, the emission color is strongly dependent on the coordination number (CN) of Mn^{2+} in the host. A distinct

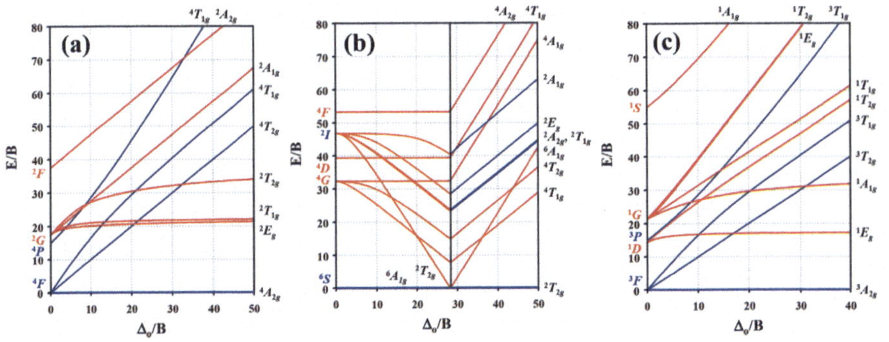

Fig. 2 The energy level diagram of some transition metals: **a** Cr^{3+} ($3d^3$), **b** Mn^{2+} and Fe^{3+} ($3d^5$ configuration), **c** Ni^{2+} ($3d^8$ configuration)

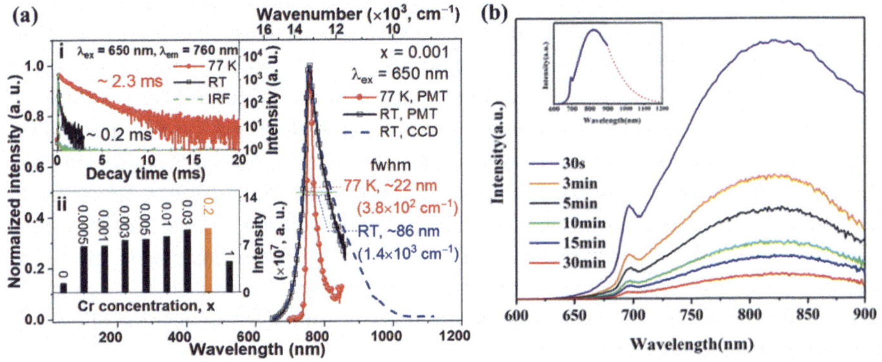

Fig. 3 The emission of some Cr^{3+} activated PLMs: **a** the emission ($\lambda_{ex} = 650$ nm) spectra of $Na_{0.5}Gd_{0.5}TiO_3$: x%Cr (x = 0, 0.001, 1), the inset of (**a**) present (i) the luminescence decay curves of the x = 0.001 sample and (ii) the Cr^{3+} concentration-dependent integrated emission intensity. (Reprinted with permission from Chem. Mater. 2020, 32, 13, 5579–5588. Copyright 2020 American Chemical Society.) **b** The afterglow spectra of $Na_2CaGe_6O_{14}$: Cr^{3+} from 30 s to 10 h after 10 min of irradiation with a 254 nm UV lamp, the inset is the simulated completion spectrum of the afterglow spectrum at 30 s. (Reprinted with permission from ACS Appl. Mater. Interfaces 2022, 14, 39, 44,622–44,631. Copyright 2022 American Chemical Society.)

sample is shown in the afterglow spectra of β-$Zn_3(PO_4)_2$: Mn^{2+}, Ga^{3+} [133] and γ-$Zn_3(PO_4)_2$: Mn^{2+}, Ga^{3+} [134]. The former is a red-emitting PLM, while the latter shows a green and red afterglow. That is because there are two coordination sited (CN = 4 and 6) in γ-$Zn_3(PO_4)_2$, while only one site (CN = 6) in β-$Zn_3(PO_4)_2$.

Fe^{3+} has the same electron configuration as Mn^{2+}, and their luminescence behavior is also caused by the $^4T_1$ (^4G)-$^6A_1$ (^6S) transition, but Fe^{3+} ions are often treated as a killer to emission centers, hence avoid mixing in the preparation of luminescent materials. However, some Fe^{3+} activated PLMs have been examined, such as Mg_2SnO_4 [135], $LiGaO_2$ [136], and $SrAl_{12}O_{19}$ [137], which are beneficial to the human body for the application of near-infrared biological probes.

Ni^{2+} ions with $3d^8$ electron configuration, when used as activators of luminescent materials, usually replace cations in matrix compounds and occupy octahedral sites. Their luminescence properties are more complex than those of manganese and iron ions. The absorption of Ni^{2+} ions is formed by the transition from the ground state $^3A_2$ to the excited state $^3T_1$, followed by the absorption of the transition $^3T_1$-$^3T_2$ (as well as other transitions to ^1E, $^1A_1$ cause weaker absorption peaks). Meanwhile, the emission spectrum of Ni^{2+} consists of multiple peaks, which are $^1T_2(^1D) \rightarrow {}^3A_2(^3F)$ transitions located at the visible wavelength of about 500-600 nm. The $^1T_2$ $(^1D) \rightarrow {}^3T_2(^3F)$ transition in the red or near-infrared band of 700–800 nm and the $^3T_2(^3F) \rightarrow {}^3A_2(^3F)$ transition in the infrared band greater than 1500 nm (in Fig. 2c).

3.1.3 Main Group Elements

As for luminescent materials, excitation and emission spectra of Sn^{2+} [138], Sb^{3+} [139], Tl^+ [140], Pb^{2+} [141], and Bi^{3+} ions [142] doped oxides have been studied. Among them, bismuth (Bi) is one of the most widely studied main group doped ions in PLMs. The outer electron configuration of Bi is $4f^{14}5d^{10}6s^26p^3$, and the 6p orbital is half-full, thus forming a variety of valence states, and +3 valence state is stable and common, which can produce luminescence in deep ultraviolet, visible, and even near-infrared regions. For example, in Fig. 4, the afterglow spectrum of $Sr_3Y_2Ge_3O_{12}$: Bi^{3+} almost completely covers the ultraviolet to near-infrared band (320–830 nm) [143]. Under the white LED charging, it still has strong afterglow emission up to 15 h, and the spectral signal can be detected after 60 h. In addition, Bi^{3+} can also be utilized as a sensitizer to improve the blue light absorption efficiency to promote afterglow intensity [144].

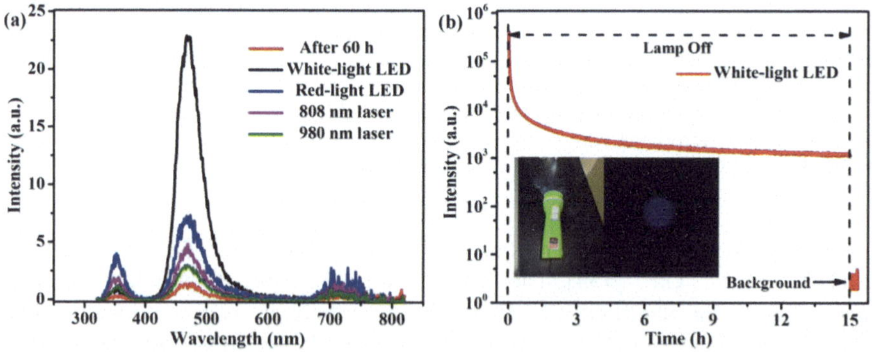

Fig. 4 **a** Long afterglow emission spectra of $Sr_3Y_2Ge_3O_{12}$:Bi^{3+} after irradiation by the LED (white and red light) or the NIR laser (808 and 980 nm; 500 mW) for 30 s. **b** The decay curve of $Sr_3Y_2Ge_3O_{12}$:Bi^{3+} after removing the white-light LED (~9.5 mW/cm^2; the excitation time was set as 15 min). The inset shows the photograph of the white-light LED for the experiment. (Reprinted with permission from Chemical Engineering Journal 421 (2), 127,820. Copyright 2021 ELSEVIER.)

3.2 *Co-doping Ions for Traps*

In many works, it can be found that the host material is doped with a single ion (called single doping), and although the afterglow can be emitted, the afterglow intensity and afterglow duration are poor. On this basis, as a second ion doped (called co-doped), the afterglow performance of some materials can be greatly improved. This is attributed to the introduction of co-doped ions leading to an increase in traps involved in the afterglow process. In this case, co-doping ions is considered to be the simplest and most effective adoption, which has become the mainstream method to modify the PLMs, and Ln ions co-doping realizes several excellent PLMs, and it is generally attributed to the valence change of Ln ions [145]. This can be attributed to the electron configurations of lanthanide ions, for instance, the La^{3+} [empty, $(4f)^0$], Gd^{3+} [half full, $(4f)^7$], and Lu^{3+} [full, $(4f)^{14}$] are stable configurations, and the rest are either prone to losing electrons (Ce^{3+}, Pr^{3+}, and Tb^{3+} can be oxidized to +4 state) or obtaining electrons (Sm^{3+}, Eu^{3+}, Dy^{3+}, and Yb^{3+} can be altered to +2-valence state). This has been detected and confirmed in PLMs activated by Eu^{2+} with Ln^{3+} co-doping, the photo-stimulated compensation for the excitation in $Sr_4Al_{14}O_{25}$: Eu^{2+}, Dy^{3+} during an X-ray scan, and monitored the existence of Dy^{2+} (in Fig. 5b and c) [146]. Similarly, in our results of $BaZrSi_3O_9$: Eu^{2+}, Sm^{3+} material of Fig. 5a, the Sm^{2+} is detected during the excitation by X-ray absorption near-edge structure (XANES), and there is a decrease of Sm^{2+} coincides with the increase in Sm^{3+}. The decay time τ_1 and τ_2 of PL ($^5D_0 \rightarrow {}^7F_0$) of Sm^{2+} coincides with the two-evolution time of PL ($5d \rightarrow 4f$) of Eu^{2+}. Sm^{2+} can return to Sm^{3+} by transferring its electron to the 5d band of Eu^{2+} [147].

Figure 5d reveals this trend, the short vertical lines represent the trend of valence state changing, which also reflects the intrinsic ability of lanthanide 4f level to bind carriers (as an available trap level in long afterglow). Therefore, the selection of co-doping ions from other rare earth ions with variable valence as the active ions has a certain guiding role in enhancing present PLMs.

4 Impact of Design on the Efficiency of Persistent Luminescence

From the above analysis, doping and co-doping have become simple and effective ways to improve the properties of PLMs. In the process of exploring PLMs, there are other useful and meaningful designs to optimize afterglow performance, which are described below.

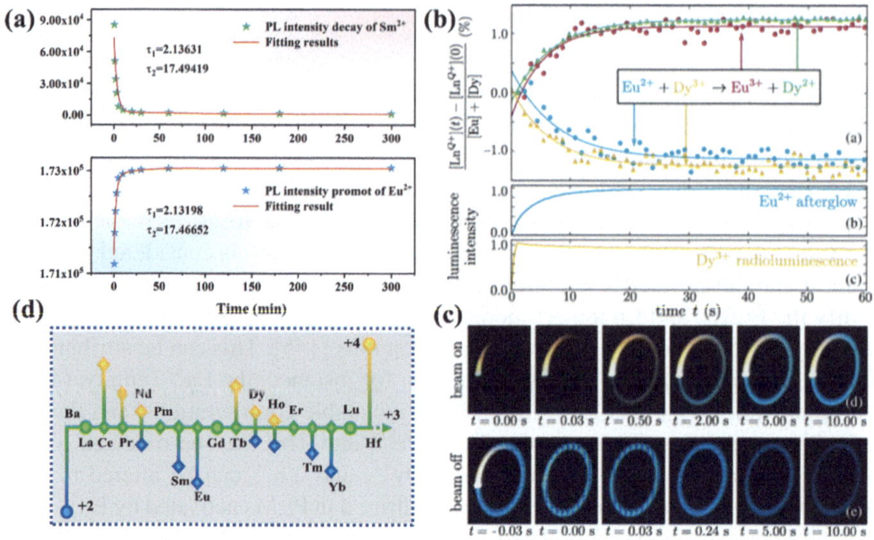

Fig. 5 **a** Decay lifetime fitting result of Eu^{2+} and Sm^{2+} in $BaZrSi_3O_9$: Eu^{2+}, Sm^{3+} material. (Reprinted with permission from Adv. Funct. Mater. 2023, 33, 2,300,721. Copyright 2023 John Wiley & Sons.) **b** and **c** are the time evolution of the lanthanide oxidation states in $Sr_4Al_{14}O_{25}$: Eu^{2+}, Dy^{3+}, relative to the total [Eu] + [Dy] content from HERFD XANES: (**a**) The X-ray beam is switched on at time t = 0, (**b**) and Dy^{3+} (line emission in 550–600 nm range), (**c**) camera images of a pristine rotating sample illustrate the charging and persistent luminescence of the blue Eu^{2+} emission when the beam is switched on (**d**) and off (**e**), respectively. (Reprinted with permission from Phys. Rev. Lett. 125, 033,001. Copyright 2020 American Physical Society.) (**d**) The tendency of the valence state change of rare earth elements

4.1 Solid Solution

Except for ion doping or co-doping, another effective method to improve the afterglow efficiency is solid solution, which refers to gradually dissolving a second phase into the host material and keeping the main phase unchanged. The specific practice is to replace the corresponding cations or anions in the crystal lattice with other different elements, compared to doping, the proportion of solid solution is higher (generally $\geq$10%). Through this method, the crystal structure of the hosts can be modulated, as well as the location of the conduction bands and valence bands of the hosts, and then affect the afterglow performance of the PLMs. In $Ca_2Ga_2(Si, Ge)O_7$: Pr^{3+}, Yb^{3+}, the solid solution between Si and Ge is used to control the afterglow properties. On the one hand, with the increase of Ge^{4+} solid solubility (from 0 to 100%), the band gap of the materials gradually decreases from the initial 5.047–4.809 eV. On the other hand, the decay rate of Pr^{3+} is inconsistent, resulting in the afterglow color change from orange to bluish-white, as shown in Fig. 6. And the afterglow duration can be strongly enhanced by adjusting the contents of Si^{4+}/Ge^{4+} [148]. Figure 7 exhibits the change of the TL spectra of $BaSi_{2-x}Ge_xO_5$: Eu^{2+}, Pr^{3+} under the substitution of the

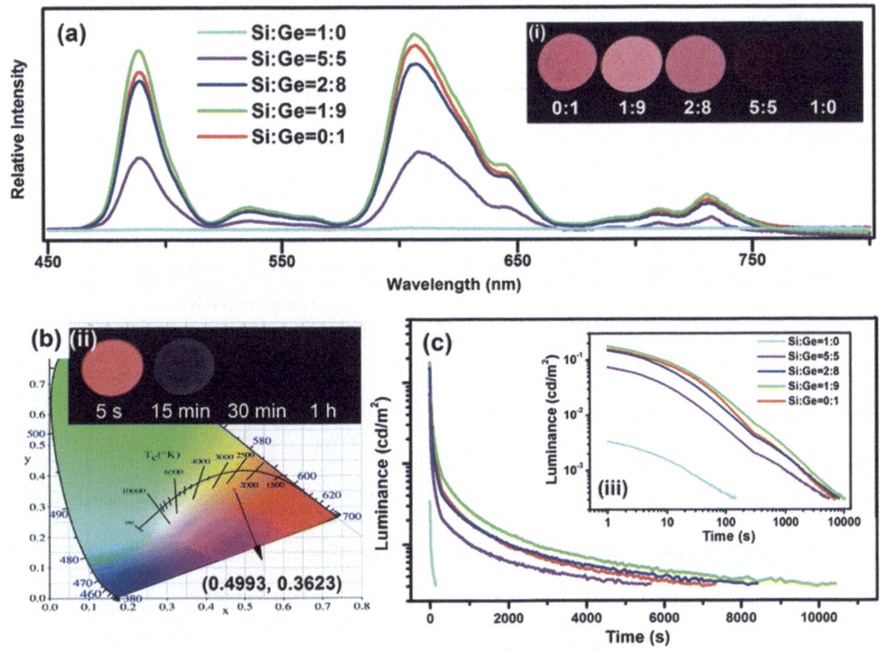

Fig. 6 Changes of afterglow properties with solid solubility: **a** afterglow emission spectra of $Ca_2Ga_2(Si, Ge)O_7$: Pr^{3+}, Yb^{3+} under different Ge^{4+} solubility, inset is the photographs of the corresponding samples, **b** CIE coordinates of $Ca_2Ga_2(Si, Ge)O_7$: Pr^{3+}, Yb^{3+}, and inset is the afterglow photos, **c** the afterglow duration of $Ca_2Ga_2(Si, Ge)O_7$: Pr^{3+}, Yb^{3+}, and inset is the relation under dual-log treatment. (Reprinted with permission from J. Mater. Chem. C, 2016,4, 10,026–10,031. Copyright 2016 ROYAL SOC CHEMISTRY.)

same valence ions, and the intensity of the TL spectra is significantly increased, indicating that the concentration of trapped carriers is greatly increased, which ensures the substantial improvement of the afterglow performance [149].

In addition, if the different valence state ions replace the host in a solid solution, it can form different types of non-charge balance defects in the lattice, such as cationic and anion vacancies. These defects can form trap levels, which are conducive to the generation of afterglow. For example, our research of (Ba, Li) $(Si, Ge, P)_2O_5$: Eu^{2+}, Pr^{3+} blue PLMs is conducted via the solid solution of Ba^{2+}-Li^+ and Si^{4+}-P^{5+}, and oxygen vacancies and barium vacancies can be newly introduced. The trap levels caused by these defects greatly increase the trapped carrier's concentration for generating afterglow, so that the afterglow duration and brightness are greatly improved (in Fig. 8).

In short, the effect of the solid solution method of the PLMs can be divided as follows.

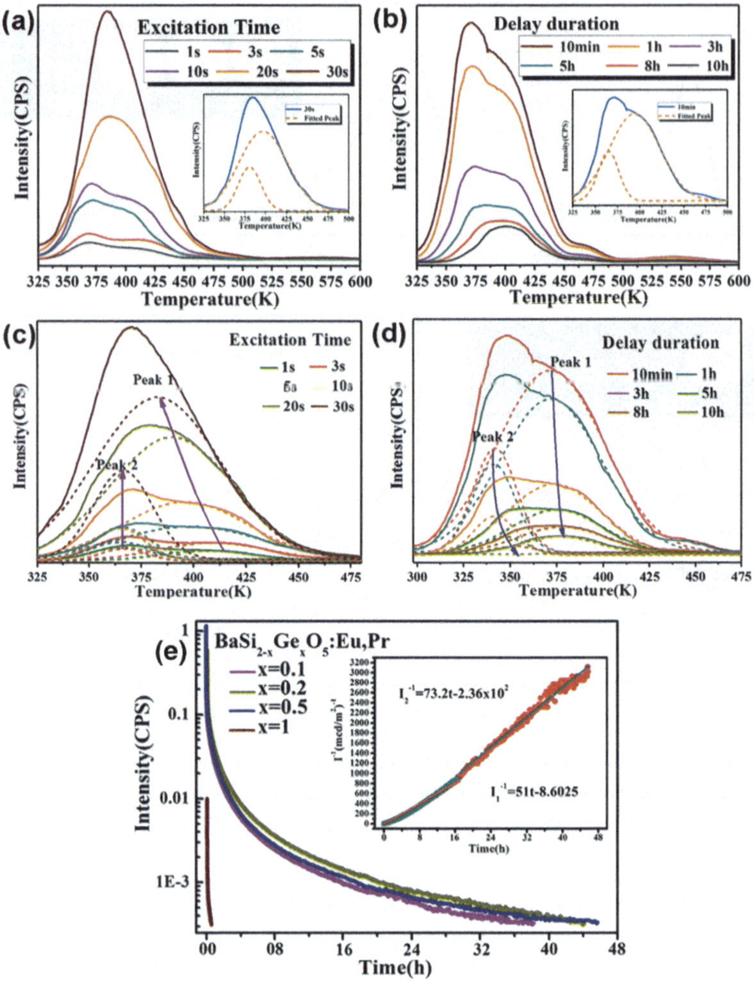

Fig. 7 Changes of TL intensity under solid solution: **a** TL curves of the sample $BaSi_{1.5}Ge_{0.5}O_5$: Eu^{2+}, Pr^{3+} recorded after UV light (254 nm) excitation for various durations (the inset shows the fitting results after 30 s excitation), **b** decay duration dependent TL curves after UV light (254 nm) excitation for 10 s (the inset shows the fitting results of delay duration in 10 min), **c** fitting results of excitation time-dependent TL glow curves, **d** fitting results of decay duration-dependent TL curves, **e** persistent decay curves of $BaSi_{2-x}Ge_xO_5$: Eu^{2+}, Pr^{3+}, and inset is reciprocal afterglow intensity (I^{-1}) versus time (t) of $BaSi_{1.5}Ge_{0.5}O_5$: Eu^{2+}, Pr^{3+} sample excited by UV light (254 nm) for 10 min. (Reprinted with permission from J. Phys. Chem. C 2019, 123, 5, 3102–3109. Copyright 2019 American Chemical Society.)

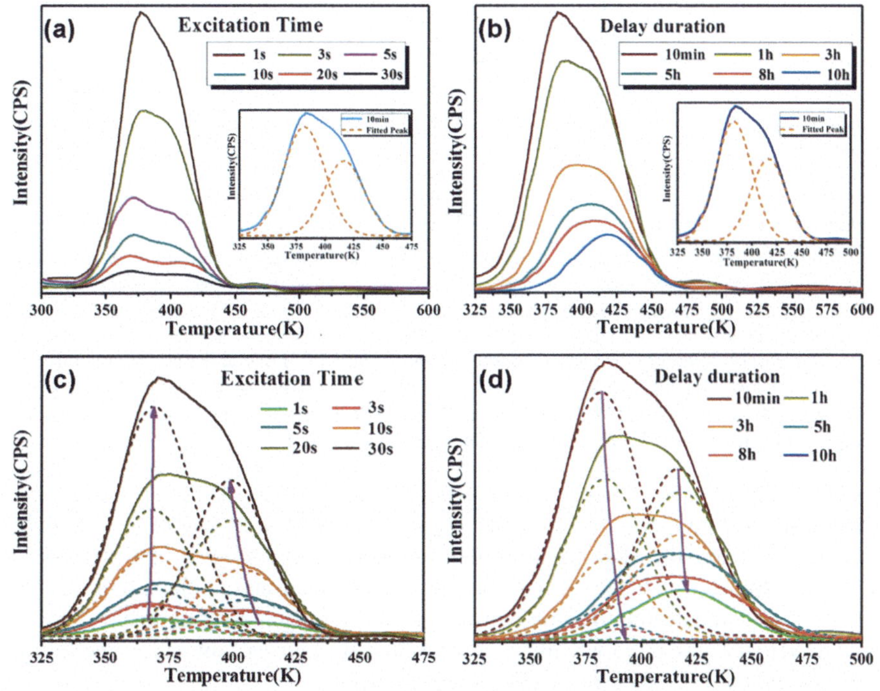

Fig. 8 Changes of afterglow properties under the solid solution of doping ions with various valences: **a** TL curves of the sample $Ba_{0.92}Li_{0.08}Si_{1.92}P_{0.08}O_5$: Eu^{2+}, Pr^{3+} recorded after UV light (254 nm) excitation with various durations (the inset shows the fitting results after 30 s excitation duration), **b** decay duration dependent TL curves after UV light (254 nm) excitation for 10 s (the inset shows the fitting results of delay duration in 10 min), **c** fitting results of excitation time-dependent TL glow curves, **d** fitting results of decay duration-dependent TL curves. (Reprinted with permission from J. Phys. Chem. C 2019, 123, 5, 3102–3109. Copyright 2019 American Chemical Society.)

(1) The introduction of new ions can affect the energy level of the material, thus changing the band gap structure and controlling the color of the afterglow emission.
(2) The defects caused by the solid solution can produce a large number of traps in the material, thereby increasing the number of captured electrons and extending the afterglow properties.

4.2 *Energy Transfer*

In some cases, the desired afterglow color cannot be obtained by doping and solid solution alone, and the wavelength of PLMs is still located outside the demanded colors, or the afterglow intensity is too weak to be effectively applied. Therefore, it

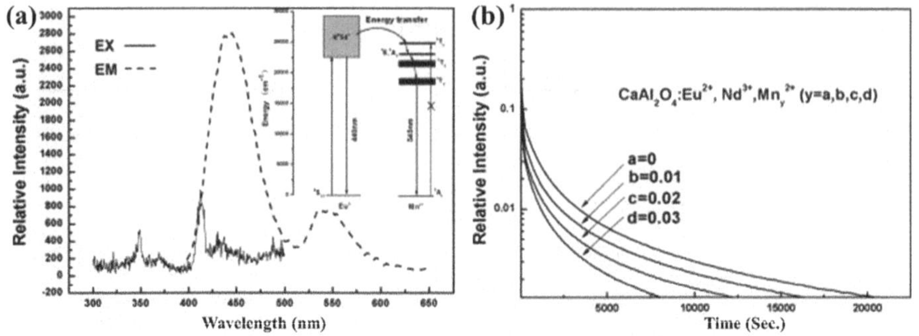

Fig. 9 The persistent energy transfer from the doping centers in PLMs: **a** the afterglow emission spectrum of $CaAl_2O_4$: Eu^{2+}, Nd^{3+}, and the PL excitation spectrum of $CaAl_2O_4$: Mn^{2+}, **b** the afterglow duration curves of $CaAl_2O_4$: Eu^{2+}, Nd^{3+}, Mn^{2+}. (Reprinted with permission from Journal of Applied Physics 105, 083,502. Copyright 2009 AIP Publishing.)

is necessary to find a way to move the original emission wavelength to the goals, or greatly increase the existing afterglow intensity. The energy in the process of afterglow is transferred in the form of charge carriers (electrons or holes). If the charge carriers can be transferred to some emission center ions that meet the conditions, or transferred to the activator ion with weak afterglow, the above problems can be solved well. This method is called persistent energy transfer.

In general, there are two feasible approaches to achieve the persistent energy transfer, one is co-doping in the same host, through the energy transfer between two (or more) doped ions to regulate the afterglow performance. As shown in Fig. 9, $CaAl_2O_4$: Eu^{2+}, Nd^{3+} with good properties is chosen, which have strong afterglow emission in blue regions. Meanwhile, the emission band covers most of the excitation bands of Mn^{2+}. Therefore, the co-doping of Mn^{2+} in the $CaAl_2O_4$: Eu^{2+}, Nd^{3+} can absorb the emitted light of Eu^{2+}, resulting in the yellow emission of Mn^{2+} [150]. Similarly, this also happens in $SrAl_2O_4$: Ce^{3+}, Mn^{2+} [151], $CaAl_2O_4$: Ce^{3+}, Tb^{3+} [152].

Another way is to alter the energy transfer mode to "PLMs + phosphors" to get the afterglow emission of the required wavelength, or even get the full wavelength afterglow emission. As shown in Fig. 10a, the afterglow emission spectrum of CaAl2O4: Eu^{2+}, Nd^{3+} is perfectly covered by the excitation spectrum of $Y_3Al_5O_{12}$: Ce^{3+}, so the PL of $Y_3Al_5O_{12}$: Ce^{3+} can be produced by the absorption of $CaAl_2O_4$: Eu^{2+}, Nd^{3+}. Our works exhibit that there are two spectra components respectively originating from a blue emission of $CaAl_2O_4$: Eu^{2+}, Nd^{3+} and a broadband yellow emission of $Y_3Al_5O_{12}$: Ce^{3+}. This indicates that the energy transfer process from $CaAl_2O_4$: Eu^{2+}, Nd^{3+} to $Y_3Al_5O_{12}$: Ce^{3+} strongly occurs. With the different mix contents, the color of the afterglow can also be changed from blue to white, as shown in Fig. 10b–d [153].

In summary, the emission bands of PLMs can be designed by various means. Therefore, the concept of "multicolor long persistent phosphors" and corresponding

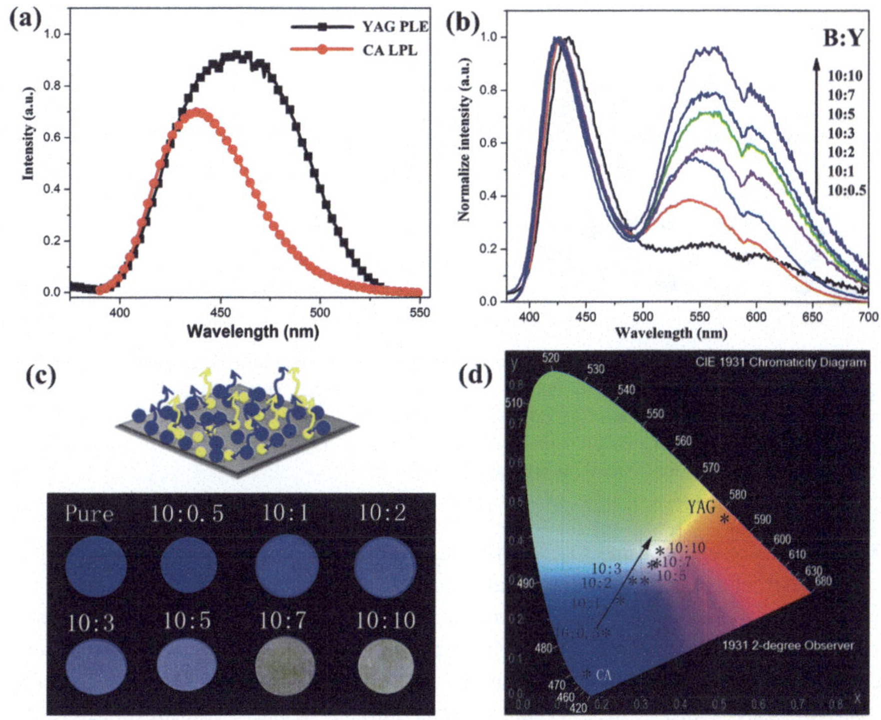

Fig. 10 The persistent energy transfer between PLMs particles: **a** the afterglow emission spectrum of $CaAl_2O_4$: Eu^{2+}, Nd^{3+} and the PL excitation spectrum of $Y_3Al_5O_{12}$: Ce^{3+}, **b** the afterglow emission spectra of the mixed $CaAl_2O_4$: Eu^{2+}, Nd^{3+}&$Y_3Al_5O_{12}$: Ce^{3+} sample, **c** the afterglow emission color in different proportions of $CaAl_2O_4$: Eu^{2+}, Nd^{3+} and $Y_3Al_5O_{12}$: Ce^{3+}, **d** CIE coordinates of $CaAl_2O_4$: Eu^{2+}, Nd^{3+}&$Y_3Al_5O_{12}$: Ce^{3+} at different proportions. (Reprinted with permission from New J. Chem., 2016,40, 485–491. Copyright 2015 ROYAL SOC CHEMISTRY.)

systematic research methods are summarized and proposed [154]. Studying efficient PLMs with different emissions will remain a challenging task and of great significance.

4.3 Synthesis Atmosphere Improvement

Persistent luminescence is closely related to the concentration and distribution of the trap center. Whether solid-state synthesis or liquid-phase synthesis, high-temperature treatment is required, which is to promote the production of defects in the materials. The most common is the oxygen vacancy defect, which can temporarily trap and store electrons for afterglow. In Fig. 11a, $Gd_{9.33}(SiO_4)_6O_2$: Sm^{3+}, the silicon-oxygen tetrahedrons in the crystal structure are separated, and the oxygen atoms contained in $Gd_{9.33}(SiO_4)_6O_2$ are divided into bond oxygen atoms (O_1, O_2, and

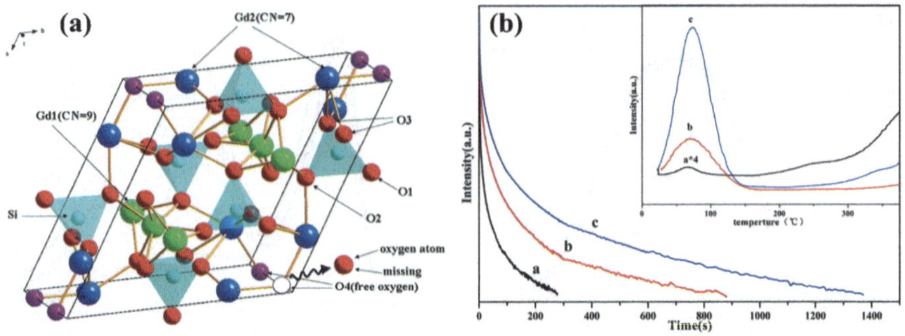

Fig. 11 The persistent energy transfer from the PLMs to the phosphors: **a** schematic drawing of oxygen vacancy, **b** afterglow decay curves of (**a**): the air-sintered, **b**: the vacuum-sintered and **c**: reduction-sintered $Gd_{9.33}(SiO_4)_6O_2$: 1.5%Sm^{3+} samples (the inset shows their TL spectra, respectively). (Reprinted with permission from Gen Li et al. 2013 ECS J. Solid State Sci. Technol. 2 R161. Copyright 2013 The Electrochemical Society.)

O_3) and free oxygen atoms O_4. Therefore, it is relatively easy to produce oxygen vacancy defects under certain synthetic conditions. Among synthesized under the air, vacuum, and reduction atmosphere, the afterglow duration and TL peak intensity are gradually longer and stronger with the air atmosphere, vacuum atmosphere, and reduction atmosphere, indicating that for the long-afterglow luminescent material $Gd_{9.33}(SiO_4)_6O_2$: Sm^{3+}, the oxygen vacancy plays a decisive role in its afterglow performance in Fig. 11b [155]. We point out that oxygen vacancy is most easily formed in the reducing atmosphere ($O_O \rightarrow V_O^{\cdot\cdot} + O_i''$, $O_i'' + 2h^{\cdot} \rightarrow \frac{1}{2}O_2$), while it is difficult to form oxygen vacancy in the air atmosphere due to enough partial pressure of oxygen [124].

In short, atmosphere improvement has been applied in some PLMs with low melting points and boiling points constituents to improve afterglow performance. This is closely related to the crystal structure itself, and it is difficult to achieve this method when the coordination polyhedral of the component material is closely connected (such as edge-sharing or coplanar connected) [156].

4.4 Persistent Luminescence Under Extreme Conditions

To search for new techniques to improve the efficiency of PLMs in extreme conditions, the afterglow behaviors, and mechanisms under extreme conditions (ultra-high pressure) are deeply studied. Such research is the first exploration in the field of PLMs. Figure 12a shows the afterglow emission and mechanism model of $Ca_6BaP_4O_{17}$: Eu^{2+}, Tb^{3+} under ultra-high pressure. It can be seen that during the process of pressure increasing from 0 to 11 GPa, the emission intensity of the material first increases and then decreases, and the emission color changes from green to orange and then to red (as shown in Fig. 12b). The spectrum consists of two parts: broadband emission and peak emission, which are attributed to the characteristic emission of Eu^{2+} and Eu^{3+} (not completely reduced), respectively. In the process of pressure change, the emission of Eu^{2+} weakens, while the emission of Eu^{3+} does not significantly change, so the above appears in the different emission colors. In addition, the afterglow duration time of $Ca_6BaP_4O_{17}$: Eu^{2+}, Tb^{3+} is also first extended and then shortened under pressure (see Fig. 12c). The reason is as below: under normal pressure, the transition pathways of electrons are (1) being excited to the excited state energy level, (2) entering into the conduction band, (3) being trapped by the trap levels, (4) return to emission center to produce afterglow. However, under high pressure, the band gap of the material increases, and the process (2) is limited during excitation, resulting in more electrons participating in the luminescence process instead of entering the conduction band, thus reducing the electrons captured by the traps. At the same time, the depth of the original trap becomes deeper, which is no longer suitable for participating in the afterglow process, and after the excitation stops, the difficulty of electrons escaping from the traps to return to the energy level of the emission center, as shown in Fig. 12d [157].

5 Surface Modification of PLMs

During the specific use, the performance of PLMs is closely connected with the application environments. Even if applied in plastics, inks, coatings, fibers, and other aspects (including nano PLMs in biological applications), the particles cannot be separated from the environment and exist alone in the process of use. They will be affected by temperature field, moisture, acid, alkali and salt corrosion, radiation, and other external factors to varying degrees. It requires PLMs must have a certain stability (mainly including water resistance, thermal stability, and chemical stability). Therefore, to ensure the light storage characteristics of PLMs, the simplest way is to start from the surface of the PLMs powders, that is, to reduce or isolate their contact with the environment or chemical substances. The surface modification technique refers to uniformly wrapping a dense film on the surface of the PLMs particles by physical or chemical means, to effectively improve the stability and the functionality of the PLMs, which will be discussed separately below.

5.1 Surface Modification for Improving Stability

For the surface modification to improve the stability of PLMs, it is usually wrapped with inorganic or organic film on the surface, some mainly selected options for inorganic coating and organic coating are listed in Table 10, both types can improve the weather resistance of PLMs to a certain extent, such as water resistance and acid resistance. However, the melting point of the organic film is lower than that of the inorganic type in improving the thermal stability of PLMs.

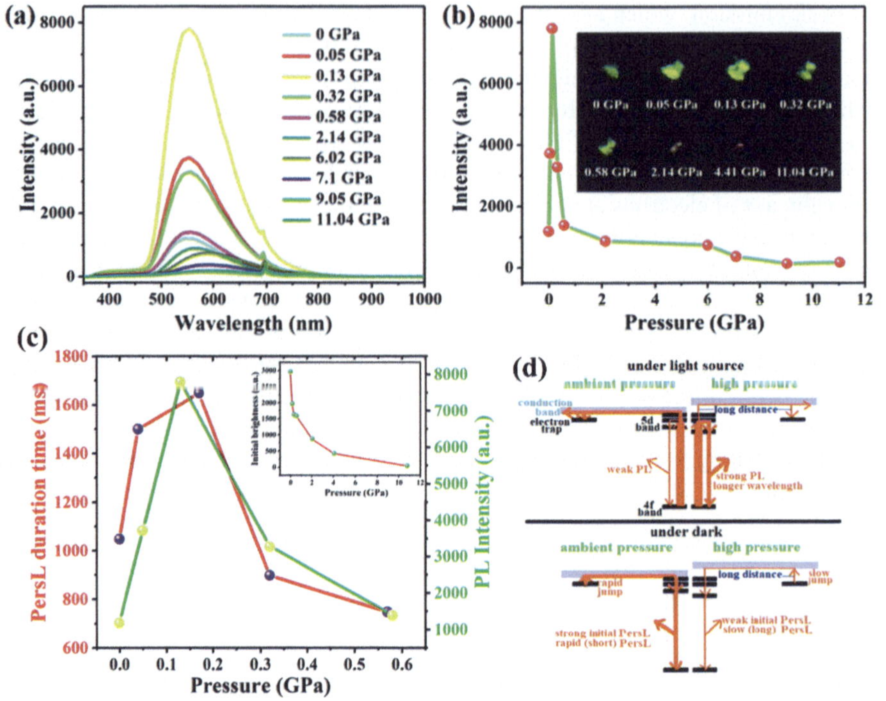

Fig. 12 Properties and mechanism of the afterglow material under ultra-high pressure: **a** emission spectra of $Ca_6BaP_4O_{17}$: Eu^{2+}, Tb^{3+} during pressure increase (0 GPa to 11 GPa), **b** emission intensity changes of $Ca_6BaP_4O_{17}$: Eu^{2+}, Tb^{3+} during pressure increase (0 GPa to 11 GPa), **c** curves of emission intensity and afterglow decay time with pressure, **d** possible mechanism model of high-pressure afterglow. (Reprinted with permission from Dalton Trans., 2020,49, 8056–8059. Copyright 2020 ROYAL SOC CHEMISTRY.)

Table 10 Several typical options for the surface modification of PLMs

Type	Materials	Refs.
Inorganic films	SiO_2	Jung et al. [158]
	Al_2O_3	Seo et al. [159]
	MgF_2	Guo et al. [160]
	TiO_2	Lee and Yoo [161]
	In_2O_3	Chang et al. [162]
	Y_2O_3	Kim et al. [163]
Organic films	PVP, SA	Li et al. [164]

In implementation, the surface modification is mainly carried out by high-temperature solid-state reaction or liquid reactions. The former is to mix the phosphors with the substance that needs to be coated in a certain proportion and calcined at low temperatures. For example, SrF_2 coated $SrA1_2O_4$: Eu^{2+}, Dy^{3+} material is prepared by calcined at 700 °C for 30 min in a reducing atmosphere after mixing NH_4HF_2 and $SrA1_2O_4$: Eu^{2+}, Dy^{3+} evenly according to the mass ratio of 1:4 [160]. This method can improve the water-resistance of $SrA1_2O_4$: Eu^{2+}, Dy^{3+}, however, the solid-state treatment needs a higher temperature, and the layer is not evenly distributed among the particles. The latter can be achieved by the homogeneous precipitation method, sol–gel method, heterogeneous nucleation method, etc. The basic principle is that the liquid precursor is deposited on the phosphor particles to achieve coating. For example, SiO_2 coated $SrA1_2O_4$: Eu^{2+}, Dy^{3+} is prepared by heterogeneous nucleation method [165]. The $Si(OH)_4$ is first generated by hydrolysis in solution and reacts with the hydroxyl group on the surface of the $SrA1_2O_4$: Eu^{2+}, Dy^{3+} to form M–O-Si (M is Al and Sr) bond. With the further progress of the reaction, the silica gel is formed by the polycondensation of $Si(OH)_4$ (In Fig. 13). The precursor of the coating takes the surface of the PLMs as the heterogeneous nucleation matrix. With the gradual change of the pH value of the solution, the coating layer can be deposited on the surface of the PLMs better, finally forming a continuous dense coating.

In summary, for the surface coating, the appropriate choice should be made according to the characteristics of the long-lingering phosphor and the performance that needs to be promoted after coating. The principles can be summarized as follows:

(1) The coating materials should have good stability under the action of high temperature and ultraviolet radiation.
(2) The coating material is easy to form a firm bond with the surface of the PLMs.
(3) Good transmittance to the excitation light, to reduce the absorption effect.
(4) Relatively simple preparation process, and easy to achieve large-scale industrialization with the low cost of raw materials.

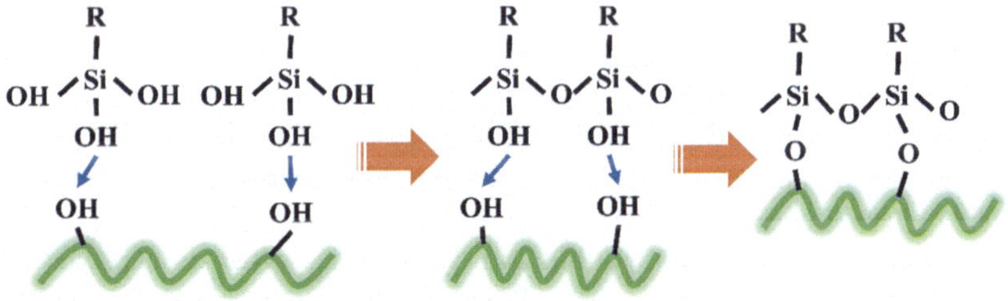

Fig. 13 The diagram of the preparation of SiO_2 coated $SrA1_2O_4$: Eu^{2+}, Dy^{3+} by heterogeneous nucleation method (R refers to $CH_2 = C\ (CH_3)COO(CH_2)_3$)

5.2 *Surface Modification for Better Biocompatibility*

Surface modification for biocompatibility refers to the modulation of the particle surface to obtain a high degree of specificity and thus improve its functionality applied in biology. PLMs do not have a surface functional ligand, therefore, in biological applications, they need to have a functional ligand attached to their surface, which is optimized to have a specific biological function to make PLMs better for use in living tissues. It is well known that the stability, suspension, and biocompatibility of PLMs become the most important factors to making LMs used in the biological field. Currently, there are three main methods for surface modification of PLMs to enhance biocompatibility.

5.2.1 Assembling and Ordering Surfaces

To give different functions to the surface of the material, modeled on the construction of cell membranes, the immobilization of polylysine, polyornithine, etc. on the surface of the material can enhance the attachment properties of the cells. Specifically, hemi-myosin, glutathione, etc., are assembled into monolayers on the surface of the material as a way to be used to improve the feasibility of binding the material to biomolecules. Aranit et al. [166] functionalized $LiGa_5O_8$: Cr^{3+} nanoparticles with excellent biocompatibility and stability in biological media by biocompatible lipid coatings. In vitro experiments showed no cytotoxicity and no adverse effect on the maturation potential of dendritic cells even at very high particle concentrations. Dendritic cells labeled with lipid-encapsulated $LiGa_5O_8$: Cr^{3+} nanoparticles could be imaged within a few days after injection into mice, confirming the effectiveness of dendritic cell migration to popliteal lymph nodes in vivo. The results suggest that lipid-encapsulated $LiGa_5O_8$: Cr^{3+} nanoparticles have good prospects for application in the research and development of dendritic cell vaccines.

5.2.2 SiO_2 Surface Modification

SiO_2 surface modification was used to form silica shells on the surface of PLMs. SiO_2 surface modification has excellent anti-photo-bleaching properties and also reduces the effect of heavy metal ions released from PLMs, reduces the biotoxicity of PLMs, and makes the material biocompatible and hydrophilic, which is expected to be applied in the field of biological imaging. Wu et al. [167] firstly coated hydroxy PLMs with a mesoporous silica shell layer to obtain PLM@SiO_2.Then, polypyrrole (PPy) was introduced onto PLM@SiO_2 to obtain PLM@SP.PLM@SP with a core–shell structure was prepared by introducing dopamine polyethylene glycol (DA-PEG) on the surface of PLM@SP, improved biocompatibility. Wang et al. [168] encapsulated zinc gallogermanate (ZGGO) materials with SiO_2 and then further modified them with ethyl orthosilicate to obtain silica-encapsulated ZGGO materials. At this time,

ZGGO nanomaterials have the advantages of ultra-long persistent luminescence, red light renewability, monodisperse nano-size, and good biocompatibility, which are suitable for in vivo long-cycle bioimaging and concomitant drug delivery. In addition, L. reuteri lactobacillus biofilm was coated on PLMs nanocomposites, making them useful for oral PL bioimaging and delivery of colorectal cancer chemotherapeutic drugs [169]. This novel mode of drug delivery avoids digestion of the drug by gastric acid and localizes the colon, which may provide new avenues for oral drug delivery.

5.2.3 Polymer Surface Modification Method

Polyethylene glycol (PEG) and its derivatives were coupled to the surface of PLMs by in situ polymerization to obtain a polymer shell layer, which makes the biocompatibility of PLMs better and at the same time provides a conduit for the binding of PLMs to biomolecules. Recently, He et al. [170] developed an organic afterglow diagnostic nano-assembly using PEG derivatives to surface-modify organic afterglow nanoparticles, giving the material properties such as afterglow imaging and drug release in response to the tumor microenvironment (excess H_2O_2). This molecular structure combines the properties of passive tumor targeting, specific drug release, and spontaneous afterglow generation, providing a design guide for activatable cancer therapy.

5.3 Surface Modification for Multifunctional Platform

As diagnosis and treatment are two separate parts of conventional medicine, it can lead to difficulties in achieving simultaneous and co-targeted diagnosis and treatment in vivo with imaging probes and therapeutic reagents. Therefore, the development of multifunctional nanoplatforms for diagnostic and therapeutic integration is very promising [171]. The ability of PLMs to avoid tissue autofluorescence and excellent signal-to-noise ratio in optical detection make these optical probes well suited for biosensing, cell tracking, and in vivo deep tissue imaging [172, 173]. Combining PLMs with various imaging techniques, such as computed tomography (CT) [174], magnetic resonance (MR) imaging [175], and single-photon emission computed tomography (SPECT) [176], construct multifunctional nanoplatforms with multiple therapeutic modalities provides a new approach to the treatment of inflammation or tumors. Zou et al. [177] designed and prepared multifunctional hybrid mesoporous nanoparticles (MHMN) with a mixture of PLMs, 68 Ga, and magnetic nanoparticles by using the mesoporous silica nanoparticle template method (macroporous MSN template method), in which PLNP, 68 Ga, and magnetic nanoparticles were used as the contrast agents for NIR-PL imaging, magnetic resonance imaging (MRI), and positron emission tomography (PET), respectively. These multifunctional nanoparticles can be used in vivo for long-term NIR-PL imaging, MRI, and

PET. Of interest, the application of macroporous MSN templates maintains the mesoporous structure of the MHMNs, which gives them excellent drug-carrying capacity. In addition, they rationally designed MHMN with high doses of DOX and Si-Pc for chemotherapy and NIR-PL-sensitized photodynamic therapy, respectively. Studies on mouse tumor models demonstrated that the DOX/Si-Pc-loaded HMNP had good tumor cell-killing ability and significant tumor inhibition, demonstrating the great potential of MHMN as an all-in-one nano-diagnostic platform for multi-modal NIR-PL/MR/PET imaging-guided chemotherapy and NIR-PL-sensitized photodynamic cancer therapy, and providing an opportunity for the development of a NIR-PLMs based nanoplatform in cancer therapy.

6 Conclusion and Future Perspective

Due to the unique light storage ability, PLMs are widely used in human daily life, production, and some special night occasions. On the one hand, this chapter reviews some typical reported afterglow materials. From the sulfide system to the present aluminate system, silicate system, and other novel systems like borate, stannate, and so on, the host materials of PLMs are diversified, and the doping elements are constantly enriched. On the other hand, how to improve the afterglow performance with suitable adoptions is summarized, when selecting the host materials, the cation radius, coordination environment and the fit of combination of activators should be considered, as well as the band gap of host materials. The activators are preferentially Eu^{2+} and Mn^{2+} with broadband emission, while the other trivalent rare earth ions are more suitable as co-doped ions to enhance the afterglow properties of PLMs. Through some simple and effective schemes, such as doping/co-doping, solid solution, energy transfer, and other means, the performance of long afterglow materials has also been improved.

Although PLMs have been studied for more than 100 years, the obstacles in afterglow emission and afterglow mechanisms have not been completely solved. In the future, the research of PLMs will continue to be explored, including but not inadequate to

(1) Discover new PLMs. For the visible regions, there is still a lack of high-brightness yellow and red PLMs, particularly, higher brightness red PLMs are demanded due to their lower insensitivity to the human eyes. For the invisible regions, the development of the UV band and the near-infrared over 1000 nm for biological imaging are also needed.
(2) Perfect the existing mechanism to guide performance improvement. For lanthanide co-doped PLMs, Ln^{2+}/Ln^{3+} valence changing behavior reveals the important role of electron traps in the performance improvement of co-doped PLMs. It can be seen that it is very necessary to study the mechanism of such a complex persistent luminescence process, such as how to achieve a synchronous enhancement of afterglow brightness and afterglow duration, and

how to improve the efficiency of visible light excitation and charging, how to adjust the ratio of deep and shallow traps, and so on.

(3) As well as new applications. At present, the practical applications of PLMs are only in the field of weak-light lighting. The combination with interdisciplinary should be actively promoted, seeking more important application fields based on PLMs, such as information storage, biological imaging, and all-weather stress monitoring, to improve the practical application value of the charming PLMs.

Acknowledgements The authors gratefully acknowledge the financial support from the Gansu Provincial Science and Technology Department Key R&D Projects (No. 22YF7FA017).

References

1. D. Van der Heggen, et al., Adv. Funct. Mater. **32** (2022)
2. J. Xu, S. Tanabe, J. Lumin. **205**, 581 (2019)
3. Y. Li, M. Gecevicius, J. Qiu, Chem. Soc. Rev. **45**, 2090 (2016)
4. L. Ma, X.J. Zou, B. Bui, W. Chen, K.H. Song, T. Solberg, Appl. Phys. Lett. **105**(5), 013702 (2014)
5. Z.P. Yang, Z. Guo, S.C. Zhu, W.J. Wang, Spectrosc. Spectral Anal. **24**, 1506 (2004)
6. F. Okamoto, K. Kato, J. Electrochem. Soc. **130**, 432 (1983)
7. D.D. Jia, W.Y. Jia, D.R. Evans, W.M. Dennis, H.M. Liu, J. Zhu, W.M. Yen, J. Appl. Phys. **88**, 3402, Pii [s0021–8979(00)05115-x] (2000)
8. D.D. Jia, J. Electrochem. Soc. **153**, H198 (2006)
9. L.H. Robins, J.A. Tuchman, Phys. Rev. B **57**, 12094 (1998)
10. P. Huang, F. Yang, C.E. Cui, L. Wang, X. Lei, Ceram. Int. **39**, 7193 (2013)
11. J.C. Zhang, X.L. Ma, Q.S. Qin, L.R. Shi, J.Y. Sun, M.J. Zhou, B.T. Liu, Y.H. Wang, Mater. Chem. Phys. **136**, 320 (2012)
12. J. Fu, Electrochem. Solid State Lett. **3**, 350 (2000)
13. Y.Y. Lu, F. Liu, Z.J. Gu, Z.W. Pan, J. Lumin. **131**, 2784 (2011)
14. E. Zych, J. Trojan-Piegza, J. Lumin. **122**, 335 (2007)
15. J. Fu, J. Am. Ceram. Soc. **85**, 255 (2002)
16. Y.H. Lin, C.W. Nan, N. Cai, X.S. Zhou, H.F. Wang, D.P. Chen, J. Alloy. Compd. **361**, 92 (2003)
17. Z.Y. Zhao, Y.H. Wang, J. Lumin. **132**, 2842 (2012)
18. J.M. Carvalho, L.C.V. Rodrigues, J. Holsa, M. Lastusaari, L.A.O. Nunes, M. Felinto, O.L. Malta, H.F. Brito, Opt. Mater. Express **2**, 331 (2012)
19. T. Matsuzawa, Y. Aoki, N. Takeuchi, Y. Murayama, J. Electrochem. Soc. **143**, 2670 (1996)
20. L. Wang, Y. Wang, Physica B-Condensed Matter **393**, 147 (2007)
21. Y. Wang, L. Wang, J. Appl. Phys. **101**, 053108 (2007)
22. H. Yamamoto, T. Matsuzawa, J. Lumin. **72–4**, 287 (1997)
23. R. Sakai, T. Katsumata, S. Komuro, T. Morikawa, J. Lumin. **85**, 149 (1999)
24. D. Jia, X.J. Wang, E. van der Kolk, W.M. Yen, Opt. Commun. **204**, 247, Pii s0030–4018(02)01300–7 (2002)
25. D. Jia, W.M. Yen, J. Lumin. **101**, 115, Pii s0022–2313(02)00394–0 (2003)
26. H. Matsui, C.N. Xu, H. Tateyama, Appl. Phys. Lett. **78**, 1068 (2001)
27. T. Katsumata, K. Sasajima, T. Nabae, S. Komuro, T. Morikawa, J. Am. Ceram. Soc. **81**, 413 (1998)
28. D. Jia, W. Jia, X.J. Wang, W.M. Yen, Solid State Commun. **129**, 1 (2004)

29. P. Zhang, M.X. Xu, Z.T. Zheng, B. Sun, Y.H. Zhang, Mater. Sci. Eng. B-Solid State Mater. Adv. Technol. **136**, 159 (2007)
30. Y.H. Lin, Z.L. Tang, Z.T. Zhang, C.W. Nan, Appl. Phys. Lett. **81**, 996 (2002)
31. J.Y. Zhang, Z.T. Zhang, T.M. Wang, W.C. Hao, Mater. Lett. **57**, 4315 (2003)
32. N. Kodama, Y. Tanii, M. Yamaga, J. Lumin. **87–9**, 1076 (2000)
33. X.Y. Sun, J.H. Zhang, X. Zhang, Y.S. Luo, X.J. Wang, J. Phys. D-Appl. Phys. **41**(4), 195414 (2008)
34. S.A. Cheng, X.H. Xu, J. Han, J.B. Qiu, B.H. Zhang, Powder Technol. **276**, 129 (2015)
35. L. Lin, C.S. Shi, Z.F. Wang, W.P. Zhang, M. Yin, J. Alloy. Compd. **466**, 546 (2008)
36. G. Li, Y. Wang, D. Liu, H. Guo, J. Liu, RSC Adv. **6**, S7024 (2016)
37. W. Zeng, Y. Wang, X. Xu, Y. Gong, Funct. Mater. Lett. **4**, 289 (2011)
38. X. Xu, Y.H. Wang, W. Zeng, Y. Gong, B.T. Liu, J. Am. Ceram. Soc. **94**, 3632 (2011)
39. J. Liu, G. Li, H. Guo, D. Liu, P. Feng, Y. Wang, RSC Adv. **8**, 10246 (2018)
40. H. Guo, Y. Wang, G. Li, J. Liu, P. Feng, D. Liu, J. Mater. Chem. C **5**, 2844 (2017)
41. Y. Gong, Y.H. Wang, Y.Q. Li, X.H. Xu, W. Zeng, Opt. Express **19**, 4310 (2011)
42. A.A.S. Alvani, F. Moztarzadeh, A.A. Sarabi, J. Lumin. **114**, 131 (2005)
43. Y.H. Lin, Z.L. Tang, Z.T. Zhang, X.X. Wang, J.Y. Zhang, J. Mater. Sci. Lett. **20**, 1505 (2001)
44. L. Jiang, C.K. Chang, D.L. Mao, B. Zhang, Mater. Lett. **58**, 1825 (2004)
45. Y.H. Lin, Z.T. Zhang, Z.L. Tang, X.X. Wang, J.Y. Zhang, Z.S. Zheng, J. Eur. Ceram. Soc. **21**, 683 (2001)
46. Q. Fei, C.K. Chang, D.L. Mao, J. Alloy. Compd. **390**, 133 (2005)
47. L. Jiang, C.K. Chang, D.L. Mao, J. Alloy. Compd. **360**, 193 (2003)
48. Y.H. Lin, Z.L. Tang, Z.T. Zhang, C.W. Nan, J. Alloys Compd. **348**, 76, Pii s0925–8388(02)00796-x (2003)
49. S. Ye, J.H. Zhang, X. Zhang, S.Z. Lu, X.G. Ren, X.J. Wang, J. Appl. Phys. **101**(6), 063545 (2007)
50. N. Lakshminarasimhan, U.V. Varadaraju, Mater. Res. Bull. **43**, 2946 (2008)
51. B.F. Lei, Y.L. Liu, H. Liu, Z.R. Ye, C.S. Shi, J. Solid State Chem. **177**, 1333 (2004)
52. J.Y. Kuang, Y.L. Liu, Chem. Phys. Lett. **424**, 58 (2006)
53. Z. Wang, Y.H. Wang, P.Y. Zhang, X.P. Fan, G.D. Qian, J. Lumin. **124**, 140 (2007)
54. Y.L. Ding, Y.X. Zhang, Z.Y. Wang, W. Li, D.L. Mao, H.B. Han, C.K. Chang, J. Lumin. **129**, 294 (2009)
55. Z.W. Pan, Y.Y. Lu, F. Liu, Nat. Mater. **11**, 58 (2012)
56. Y. Wang, P. Feng, S. Ding, S. Tian, Y. Wang, Inorg. Chem. Front. **8**, 3748 (2021)
57. S. Ding, H. Guo, P. Feng, Q. Ye, Y. Wang, Adv. Opt. Mater. **8**, 2000097 (2020)
58. S. Ding, P. Feng, J. Cao, X. Ma, Y. Wang, ACS Appl. Mater. Interfaces (2022)
59. Y.L. Wu, Y. Li, X.X. Qin, R.C. Chen, D.K. Wu, S.J. Liu, J.R. Qiu, J. Alloy. Compd. **649**, 62 (2015)
60. Z.H. Zhou, W. Zheng, J.T. Kong, Y. Liu, P. Huang, S.Y. Zhou, Z. Chen, J.L. Shi, X.Y. Chen, Nanoscale **9**, 6846 (2017)
61. A. Mondal, S. Das, J. Manam, RSC Adv. **6**, 82484 (2016)
62. F. Liu, W.Z. Yan, Y.J. Chuang, Z.P. Zhen, J. Xie, Z.W. Pan, Sci. Rep. **3**(9), 1554 (2013)
63. Y.L. Wu, Y. Li, X.X. Qin, R.C. Chen, D.K. Wu, S.J. Liu, J.R. Qiu, Spectrochimica Acta Part A-Mol. Biomol. Spectrosc. **151**, 385 (2015)
64. Q. Wang, Z.F. Mu, S.A. Zhang, Q.T. Zhang, D.Y. Zhu, J.Q. Feng, Q.P. Du, F.G. Wu, J. Lumin. **206**, 618 (2019)
65. J. Xu, D.Q. Chen, Y.L. Yu, W.J. Zhu, J.C. Zhou, Y.S. Wang, Chem.-An Asian J. **9**, 1020 (2014)
66. H.A. Hoppe, H. Lutz, P. Morys, W. Schnick, A. Seilmeier, J. Phys. Chem. Solids **61**, 2001 (2000)
67. Y. Miyamoto, H. Kato, Y. Honna, H. Yamamoto, K. Ohmi, J. Electrochem. Soc. **156**, J235 (2009)
68. J. Wang, H.R. Zhang, B.F. Lei, Z.G. Xia, H.W. Dong, Y.L. Liu, M.T. Zheng, Y. Xiao, J. Mater. Chem. C **3**, 4445 (2015)
69. X.L. Liu, Z. Song, S.X. Wang, Q.L. Liu, J. Lumin. **208**, 313 (2019)

70. X.M. Teng, Y.H. Liu, Y.Z. Liu, Y.S. Hu, H.Q. He, W.D. Zhuang, J. Lumin. **130**, 851 (2010)
71. J. Wang, H.R. Zhang, B.F. Lei, H.W. Dong, H.M. Zhang, Y.L. Liu, N.L. Lai, Y. Fang, Z.J. Chen, Opt. Mater. **36**, 1855 (2014)
72. J. Botterman, K. Van den Eeckhout, A.J.J. Bos, P. Dorenbos, P.F. Smet, Opt. Mater. Express **2**, 341 (2012)
73. Y.X. Zhuang, Y. Lv, L. Wang, W.W. Chen, T.L. Zhou, T. Takeda, N. Hirosaki, R.J. Xie, A.C.S. Appl, Mater. Interfaces **10**, 1854 (2018)
74. Y.X. Zhuang, Y. Lv, Y. Li, T.L. Zhou, J. Xu, J. Ueda, S. Tanabe, R.J. Xie, Inorg. Chem. **55**, 11890 (2016)
75. W. Sun et al., ACS Appl. Mater. Interfaces **7**, 25219 (2015)
76. H. Guo, W. Chen, W. Zeng, G. Li, Y. Wang, Y. Li, Y. Li, X. Ding, J. Mater. Chem. C **3**, 5844 (2015)
77. R. Pang, C.Y. Li, S. Zhang, Q. Su, Mater. Chem. Phys. **113**, 215 (2009)
78. L.Y. Liu, C.Y. Li, S.B. Wang, Q.A. Su, Appl. Phys. Lett. **88**(3), 241107 (2006)
79. J. Wang, S.B. Wang, Q. Su, J. Mater. Chem. **14**, 2569 (2004)
80. A. Bessiere, R.A. Benhamou, G. Wallez, A. Lecointre, B. Viana, Acta Mater. **60**, 6641 (2012)
81. Y.Y. Dikhtyar, D.V. Deyneko, K.N. Boldyrev, O.V. Baryshnikova, A.A. Belik, V.A. Morozov, B.I. Lazoryak, J. Lumin. **236**(10), 118083 (2021)
82. J. Jeong, M. Jayasimhadri, H.S. Lee, K.W. Jang, S.S. Yi, J.H. Jeong, C. Kim, Physica B-Condensed Matter **404**, 2016 (2009)
83. A. Lecointre, A. Bessiere, A.J.J. Bos, P. Dorenbos, B. Viana, S. Jacquart, J. Phys. Chem. C **115**, 4217 (2011)
84. T. Takayama, T. Katsumata, S. Komuro, T. Morikawa, J. Cryst. Growth **275**, E2013 (2005)
85. X.M. Zhang, J.H. Zhang, X. Zhang, L. Chen, S.Z. Lu, X.J. Wang, J. Lumin. **122**, 958 (2007)
86. F.W. Kang, Y.H. Hu, H.Y. Wu, Z.F. Mu, G.F. Ju, C.J. Fu, N.N. Li, J. Lumin. **132**, 887 (2012)
87. W. Zeng, Y. Wang, S. Han, W. Chen, G. Li, Y. Wang, Y. Wen, J. Mater. Chem. C **1**, 3004 (2013)
88. J.C. Zhang, M.H. Yu, Q.S. Qin, H.L. Zhou, M.J. Zhou, X.H. Xu, Y.H. Wang, J. Appl. Phys. **108**(7), 123518 (2010)
89. X.C. Li, Z.H. Zou, Z.L. Wang, C.Q. Wu, J.C. Zhang, Y.H. Wang, J. Rare Earths **33**, 37 (2015)
90. Y. Takahashi, H. Masai, T. Fujiwara, K. Kitamura, S. Inoue, J. Ceram. Soc. Jpn. **116**, 357 (2008)
91. D.L. Sun, D.C. Li, Z.W. Zhu, J. Xiao, Z.T. Tao, W. Liu, Opt. Mater. **34**, 1890 (2012)
92. G.B. Che, C.B. Liu, X.Y. Li, Z.L. Xu, Y. Liu, H. Wang, J. Phys. Chem. Solids **69**, 2091 (2008)
93. X.H. Xu, Y. Gong, W. Zeng, Y.H. Wang, in 16th International Conference on Luminescence and Optical Spectroscopy of Condensed Matter (ICL) (Elsevier Science Bv, Univ Michigan, Cent Campus, Ann Arbor, MI, 2011), p. 62
94. Q. Ye, Y. Wang, H. Guo, X. Zhou, P. Feng, S. Ding, Dalton Trans. **48**, 11052 (2019)
95. J.S. Zhang, B.J. Chen, J.S. Sun, X.P. Li, L.H. Cheng, H.Y. Zhong, J. Phys. D-Appl. Phys. **45**(6), 325105 (2012)
96. S.K. Sharma, J. James, S.K. Gupta, S. Hussain, Materials **16**, 236 (2023)
97. Y.J. Liang, F. Liu, Y.F. Chen, K.N. Sun, Z.W. Pan, Dalton Trans. **45**, 1322 (2016)
98. Z.W. Yang, J.Y. Liao, T.Y. Wang, H.J. Wu, J.B. Qiu, Z.G. Song, D.C. Zhou, Mater. Express **4**, 172 (2014)
99. W.X. Wang, Z.Y. Sun, X.Y. He, Y.D. Wei, Z.H. Zou, J.C. Zhang, Z.F. Wang, Z.Y. Zhang, Y.H. Wang, J. Mater. Chem. C **5**, 4310 (2017)
100. X.J. Dou, H.W. Xiang, P.L. Wei, S.A. Zhang, G.F. Ju, Z.M. Meng, L. Chen, Y.H. Hu, Y. Li, Mater. Res. Bull. **105**, 226 (2018)
101. Z.H. Zhou, P.X. Xiong, H.L. Liu, M.Y. Peng, Inorg. Chem. **59**, 12920 (2020)
102. Z.H. Zhou, X. Wang, X.D. Yi, H. Ming, Z.J. Ma, M.Y. Peng, Chem. Eng. J. **421**(11), 127820 (2021)
103. D.D. Jia, X.J. Wang, W. Jia, W.M. Yen, J. Lumin. **122**, 311 (2007)
104. S.K. Sharma, M. Bettinelli, I. Carrasco, J. Beyer, R. Gloaguen, J. Heitmann, J. Phys. Chem. C **123**, 14639 (2019)

105. S. Zhang, R. Pang, C.Y. Li, Q.A. Su, J. Lumin. **130**, 2223 (2010)
106. A.J. Kanwar, S. Dogra, Clin. Exp. Dermatol. **30**, 332 (2005)
107. M.D. Njoo, J.D. Bos, W. Westerhof, J. Am. Acad. Dermatol. **42**, 245 (2000)
108. H.X. Sun, Q.Q. Gao, A.Y. Wang, Y.C. Liu, X.J. Wang, F. Liu, Opt. Mater. Express **10**, 1296 (2020)
109. P. Caillet-Fauquet, M. Di Giambattista, M.L. Draps, F. Sandras, T. Branckaert, Y. de Launoit, R. Laub, J. Virol. Methods **118**, 131 (2004)
110. Y.M. Yang et al., Light-Sci. Appl. **7**(11), 88 (2018)
111. X.Q. Zhou, J.W. Qiao, Y.F. Zhao, K. Han, Z.G. Xia, Sci. China-Mater. **65**, 1103 (2022)
112. X. Yu, T. Wang, X.H. Xu, D.C. Zhou, J.B. Qiu, ECS Solid State Lett. **3**, R4 (2014)
113. L.L. Jiang, S.G. Xiao, X.L. Yang, X.A. Zhang, X.H. Liu, B.Y. Zhou, X.L. Jin, Mater. Sci. Eng. B-Adv. Funct. Solid-State Mater. **178**, 123 (2013)
114. T. Maldiney et al., Opt. Mater. Express **2**, 261 (2012)
115. Y.F. Liu, P. Liu, L. Wang, C.E. Cui, H.C. Jiang, J. Jiang, Chem. Commun. **53**, 10636 (2017)
116. J. Huang, Y.H. Song, Y. Sheng, K.Y. Zheng, H.B. Li, H.G. Zhang, Q.H. Huo, X.C. Xu, H.F. Zou, J. Alloy. Compd. **532**, 34 (2012)
117. Z.L. Wang, Y.H. Wang, J.C. Zhang, Y.H. Lu, Y.H. Lu, Mater. Res. Bull. **44**, 1183 (2009)
118. Y.H. Jin, Y.H. Hu, L. Chen, X.J. Wang, G.F. Ju, Z.F. Mou, F. Liang, Opt. Commun. **311**, 266 (2014)
119. B.C. Duan, J.C. Zhang, X. Liu, Q. Yuan, Y.H. Wang, J. Alloy. Compd. **587**, 318 (2014)
120. Y.Y. Zhang, R. Pang, C.Y. Li, C.Y. Zang, Q.A. Su, J. Rare Earths **28**, 705 (2010)
121. G.D. Liu, Q.H. Zhang, H.Z. Wang, Y.G. Li, Mater. Sci. Eng. B-Adv. Funct. Solid-State Mater. **177**, 316 (2012)
122. J.S. Zhang, H.Y. Zhong, J.S. Sun, L.H. Cheng, X.P. Li, B.J. Chen, Chin. Phys. Lett. **29**, 017101 (2012)
123. S. Tian, P. Feng, S. Ding, Y. Wang, Y. Wang, J. Alloy. Compd. **899**, 163325 (2022)
124. X. Ma, P. Feng, Y. Wang, S. Ding, S. Tian, Y. Wang, J. Mater. Chem. C **10**, 1105 (2022)
125. V.N. Makhov, C. Lushchik, A. Lushchik, M. Kirm, Z.F. Wang, W.P. Zhang, M. Yin, J.T. Zhao, J. Lumin. **129**, 1711 (2009)
126. L.C.V. Rodrigues, H.F. Brito, J. Holsa, R. Stefani, M. Felinto, M. Lastusaari, T. Laamanen, L.A.O. Nunes, J. Phys. Chem. C **116**, 11232 (2012)
127. C.B. Liu, G. Che, Z.L. Xu, Q.W. Wang, J. Alloy. Compd. **474**, 250 (2009)
128. W.B. Chen, Y.H. Wang, W. Zeng, S.C. Han, G. Li, Opt. Mater. **36**, 1850 (2014)
129. A.V. Saraswathi, K. Naregundi, M.S. Murari, M.I. Sayyed, N. Almousa, S.D. Kamath, Ceram. Int. **48**, 36110 (2022)
130. B.G. Zhai, M.M. Chen, Y.M. Huang, RSC Adv. **12**, 31757 (2022)
131. S.K. Singh, RSC Adv. **4**, 58674 (2014)
132. Y. Liu et al., Nanoscale Adv. **4**, 4972 (2022)
133. Y.H. Song, H.F. Zou, S.C. Gan, Y.F. Deng, G.Y. Hong, J. Meng, J. Mater. Sci. **42**, 4899 (2007)
134. R.X. Zhong, Z.R. Liu, X.P. Meng, M.Y. Li, X.Y. Zhang, in 2nd International Conference on Advanced Engineering Materials and Technology (AEMT) Zhuhai, Peoples R China, 2012), p. 2396
135. M. Li, Y. Jin, L. Yuan, B. Wang, H. Wu, Y. Hu, F. Wang, A.C.S. Appl, Mater. Interfaces **15**, 13186 (2023)
136. Z. Zhou, X. Yi, P. Xiong, X. Xu, Z. Ma, M. Peng, J. Mater. Chem. C **8**, 14100 (2020)
137. J. Zeng, G. Peng, S. Zhang, G.R. Krishnan, G. Ju, Y. Li, Y. Hu, J. Lumin. **232**, 117810 (2021)
138. A. Antuzevics, G. Krieke, G. Doke, B. Berzina, Materialia **21**, 101374 (2022)
139. E. Oomen, L.C.G. Vangorkom, W.M.A. Smit, G. Blasse, J. Solid State Chem. **65**, 156 (1986)
140. T. Ohata, T. Hayashi, S. Koshino, J. Phys. Soc. Jpn. **48**, 1193 (1980)
141. H.F. Folkerts, M.A. Hamstra, G. Blasse, Chem. Phys. Lett. **246**, 135 (1995)
142. Z.W. Long et al., Adv. Opt. Mater. **7**, 1900006 (2019)
143. C.Y. Liu, Z.G. Xia, M.S. Molokeev, Q.L. Liu, J. Am. Ceram. Soc. **98**, 1870 (2015)
144. S.X. Yan, J.H. Zhang, X. Zhang, S.Z. Lu, X.G. Ren, Z.G. Nie, X.J. Wang, J. Phys. Chem. C **111**, 13256 (2007)

145. P. Feng, Y. Wang, S. Ding, Chem. Eng. J. **450**, 137820 (2022)
146. J.J. Joos, K. Korthout, L. Amidani, P. Glatzel, D. Poelman, P.F. Smet, Phys. Rev. Lett. **125**, 033001 (2020)
147. F. Peng, T. Seto, Y.H. Wang, Adv. Funct. Mater. **13** (2023)
148. Z.H. Zou, H. Zhou, W.X. Wang, J.C. Zhang, C. Cao, T. Zhang, Z.P. Ci, Z.Y. Zhao, Y.H. Wang, J. Mater. Chem. C **4**, 10026 (2016)
149. P. Feng, G. Li, H.J. Guo, D.W. Liu, Q.F. Ye, Y.H. Wang, J. Phys. Chem. C **123**, 3102 (2019)
150. X. Xu, Y. Wang, Y. Li, Y. Gong, J. Appl. Phys. **105**, 083502 (2009)
151. X. Xu, Y. Wang, X. Yu, Y. Li, Y. Gong, J. Am. Ceram. Soc. **94**, 24 (2011)
152. D. Jia, R.S. Meltzer, W.M. Yen, W. Jia, X. Wang, Appl. Phys. Lett. **80**, 1535 (2002)
153. W. Chen, Y. Wang, W. Zeng, S. Han, G. Li, H. Guo, Y. Li, Q. Qiang, New J. Chem. **40**, 485 (2016)
154. Y.H. Wang, Y. Gong, X.H. Xu, Y.Q. Li, J. Lumin. **133**, 25 (2013)
155. G. Li, Y.H. Wang, S.C. Han, W. Zeng, W.B. Chen, Y. Gong, ECS J. Solid State Sci. Technol. **2**, R161 (2013)
156. H. Li, Y. Wang, Inorg. Chem. **56**, 10396 (2017)
157. H.J. Guo, T. Seto, T. Geng, B. Zou, G.S. Li, Y. Uwatoko, Z.B. Tang, Z.B. Li, Y.H. Wang, Dalton Trans. **49**, 8056 (2020)
158. I.Y. Jung, Y. Cho, S.G. Lee, S.H. Sohn, D.K. Kim, D.K. Lee, Y.M. Kweon, Appl. Phys. Lett. **87**(3), 191908 (2005)
159. J.H. Seo, S.M. Lee, H.S. Moon, S.J. Han, S.H. Sohn, Mol. Cryst. Liquid Cryst. **531**, 82 (2010)
160. C.F. Guo, L. Luan, D.X. Huang, Q. Su, Y.H. Lv, Mater. Chem. Phys. **106**, 268 (2007)
161. H.S. Lee, J.W. Yoo, Appl. Surf. Sci. **257**, 8355 (2011)
162. C.H. Chang, B.S. Chiou, K.S. Chen, C.C. Ho, J.C. Ho, Ceram. Int. **31**, 635 (2005)
163. D. Kim, S. Jeong, J. Moon, S.H. Cho, J. Colloid Interface Sci. **297**, 589 (2006)
164. J. Li, J.F. Zhao, H.F. Zhou, J.A. Liang, X.G. Liu, B.S. Xu, Spectrochimica Acta Part A-Mol. Biomol. Spectrosc. **78**, 1310 (2011)
165. X.D. Lu, X.H. Liu, G.S. Gai, in *International Forum on Powder Technology and Application* (Trans Tech Publications Ltd, Tsinghua Univ, Beijing, PEOPLES R CHINA, 2008), p. 169
166. A. Harizaj, O.Q. De Clercq, B. Descamps, C. Vanhove, S.C. De Smedt, D. Poelman, I. Lentacker, K. Braeckmans, Part. Part. Syst. Charact. **36**(9), 1900371 (2019)
167. S.Q. Wu, Y. Li, R.F. Zhang, K.L. Fan, W.H. Ding, L.T. Xu, L.B. Zhang, Talanta **221**(10), 121435 (2021)
168. Z.H. Wang, J.M. Liu, C.Y. Li, D. Wang, H. Lv, S.W. Lv, N. Zhao, H. Ma, S. Wang, A.C.S. Appl, Mater. Interfaces **11**, 36409 (2019)
169. J.M. Liu, D.D. Zhang, G.Z. Fang, S. Wang, Biomaterials **165**, 39 (2018)
170. S.S. He, C. Xie, Y.Y. Jiang, K.Y. Pu, Adv. Mater. **31**(7), 1902672 (2019)
171. L.J. Chen, S.K. Sun, Y. Wang, C.X. Yang, S.Q. Wu, X.P. Yan, A.C.S. Appl, Mater. Interfaces **8**, 32667 (2016)
172. Y. Lv, D.D. Ding, Y.X. Zhuang, Y.S. Feng, J.P. Shi, H.W. Zhang, T.L. Zhou, H.M. Chen, R.J. Xie, A.C.S. Appl, Mater. Interfaces **11**, 1907 (2019)
173. S.Q. Wu, Z.H. Qian, Y. Li, S.P. Hu, Y. Ma, S.Y. Wei, L.B. Zhang, A.C.S. Appl, Mater. Interfaces **12**, 25572 (2020)
174. J.M. Liu, Y.Y. Liu, D.D. Zhang, G.Z. Fang, S. Wang, A.C.S. Appl, Mater. Interfaces **8**, 29939 (2016)
175. E. Teston et al., Small **14**(11), 1800020 (2018)
176. Q. Wang, N. Liu, Z.Y. Hou, J.P. Shi, X.H. Su, X.L. Sun, Adv. Healthc. Mater. **10**(6), 2000802 (2021)
177. R. Zou, Y.F. Gao, Y. Zhang, J. Jiao, K.L. Wong, J. Wang, A.C.S. Appl, Mater. Interfaces **13**, 9667 (2021)

NIR Persistent Luminescence

Yang Li, Shaoan Zhang, and S. Saravanakumar

Abstract The allure of long-afterglow materials lies in their ability to emit light for extended durations, ranging from minutes to even hours, after being exposed to external energy sources such as UV–VIS radiation, high-energy rays, or electron beams. In comparison to other fluorescent labels, near-infrared fluorescent nanomaterials possess a unique delayed detection mode that eliminates the requirement for additional light sources during image acquisition. This exceptional property not only safeguards biological tissue but also reduces background signals while enhancing the signal-to-noise ratio—a valuable asset in the field of biomedical imaging technology. This paper extensively discusses the development process of red and near-infrared afterglow nanomaterials, elucidates the mechanism evolution of afterglow nanomaterials, presents the synthesis scheme of nanomaterials along with key achievements, highlights major challenges encountered during the development process and provides corresponding solutions. We firmly believe that these distinctive long-afterglow materials will continue to radiate infinite vitality through their captivating luminescent phenomenon.

1 Introduction

An unusual optical phenomenon known as persistent luminescence occurs when a material is energized by high-energy radiation such as ultraviolet rays, X-rays, and other external sources [1–5]. After stopping the excitation source, the afterglow

Y. Li (✉) · S. Zhang · S. Saravanakumar
Institute of Light+X Science and Technology, Faculty of Electrical Engineering and Computer Science, Ningo University, Ningbo 315211, China
e-mail: lychris@sina.com

Y. Li
School of Biomedical Engineering, Guangzhou Medical University, Guangzhou 510006, China

S. Saravanakumar
Department of Physics, Kalasalingam University, Krishnan Koil, Viridhunagar, Tamil Nadu 626126, India

V. Kumar et al. (eds.), *Persistent Luminescence*, Progress in Optical Science and Photonics 30, https://doi.org/10.1007/978-981-97-4943-0_4

decays slowly over a significant duration of up to several hours. It is called long-persistent luminescence or long-lasting phosphorescence [1–4]. Persistent luminescence is characterized by the delayed release of trapped charge carriers due to thermal excitation [4, 5]. The persistent luminescence is realized and evaluated by afterglow intensity and lifetime parameters [6, 7]. In persistent luminescence, the afterglow lifetime refers to the duration starting from the end of excitation until the luminescence intensity decreases to 0.32 mcd/m^2. This intensity is 100 times higher than the dark-adapted sensitivity threshold of the human eye [6]. Commercial persistent phosphors are known for their ability to produce strong and long-lasting luminescence. Numerous long-persistent materials have been developed, with new materials increasing annually [8–11]. For instance, in the past twenty years, the persistent phosphors $CaAl_2O_4:Eu^{2+}$, Nd^{3+} [8], $SrAl_2O_4:Eu^{2+}$, Dy^{3+} [9], Y_2O_2S: Eu^{3+}, Mg^{2+}, Ti^{4+} [10], and $ZnGa_2O_4:Cr^{3+}$ [11], have been created in the colors blue, green, red, and NIR.

Since 2007, bioluminescent nanoparticles have been utilized to demonstrate the various bio-applications [12–15]. These nanoparticles are used in biological systems to avoid autofluorescence and can offer autofluorescence-free signals for ultrasensitive and bio-imaging applications. Hence, the optical research society has to focus on the persistent luminescent nanoparticles (PLNPs) by uniform size & shape, excitation and emission wavelength to meet diverse application requirements [12–15]. The uniform size and shape of the nanoparticles are advantages for the surface functionalization of PLNPs, whereas irregular size and morphology produce less performance [14, 15]. The uniform size and reproducible functionalization of PLNPs are crucial for ensuring robustness and reproducibility in complex biological systems [16–20]. Numerous studies have indicated that the biodistribution and cellular uptake of PLNPs can be significantly influenced by their size, shape, and surface properties in vivo. [16–18]. So, uniform-sized PLNPs are vital for the next generation of clinical applications [19, 20]. In addition to the uniform-sized nanoparticles, the enhancement of afterglow intensity is essential for improving image quality in biomedical applications [21–23]. Furthermore, developing PLNPs by controlling excitation and emission wavelength enables their utilization in applications that demand multiple signal readouts and excitation sources [21–23]. Hence, tuning the excitations and emissions to help efficiently apply PLNPs in animal or human bodies to accomplish deep-tissue penetration is essential.

2 Persistent Mechanism of NIR PLNPs

Trapping and de-trapping processes involve several steps, including excitation, carriers trapping, carriers storing, carriers releasing, and afterglow emission. Initially, charge carriers undergo excitation to higher energy levels of activators. Following this, these carriers are ensnared by traps consisting of isolated or continuously distributed energy levels within the forbidden bandgap. The captured carriers remain stored in these traps until they are released due to sufficient thermal disturbance. This release of trapped charge carriers leads to the emission of photons, resulting in an afterglow [1, 2]. The optical energy storage by trapping electrons is very short under UV, X-rays, etc. excitation. However, the emission of electrons caused by thermal fluctuations generates a longer-lasting afterglow that can persist for several hours or even longer.

The persistent phosphorescence is due to a trap-controlled process in inorganic phosphors [3, 4]. The defect existence in the materials is strongly responsible for the afterglow mechanism [5]. Contrarily, the lengthy luminescence of organic phosphors is due to the excitation of triplet energy level. The organic afterglow mechanisms are entirely different and can be regulated by adjusting the surface structure, molecular chain and hydrogen bond [5]. In this chapter, the afterglow mechanisms are discussed for inorganic phosphors. Many different models have been proposed to explain the mechanisms involved, but we still need a complete understanding of the trapping and de-trapping process. Hence, this chapter contributes more to understanding the proposed models, which are discussed as follows.

2.1 Conduction Band-Valence Band Model

In the mid-1990s, there was a detailed formulation of the afterglow mechanism. The fundamental mechanism for energy storage is the trapping process. The energy storage is attained due to the spatial localization of the excited carriers. The trapped electron/hole near CB/VB can freely move until captured by another trap or recombination trap. It is recommended based on the participation of CB/VB in the energy release process, migrations, excitation, etc. This model includes the electron, hole, recombination, and two-photon absorption models [25–30].

2.1.1 Recombination Model

The recombination model entails the simultaneous involvement of electron traps in proximity to the CB and hole traps near the VB. The hole is created in VB by stimulating UV light and pumps an electron into the CB. This excited electron could be trapped near CB while holes could be trapped, which are considered electron traps (E-trap) and hole traps (H-trap), respectively. Once the irradiation ceases, the trapped electrons and holes have the ability to reenter the CB and VB, resulting in a process called recombination. Also, there is a possibility to re-trap the released electrons/holes by E-trap and H-trap, producing the re-trapping effect [24, 25].

The recombination model is proposed to understand the process of trap and de-trap. Later, this model is modified and presented as an energy-transfer recombination model for further understanding. This proposed recombination model accounts for the energy transfer between the active centers. The carriers could be captured by E-trap and H-trap, which may combine to produce emissions afterwards. It is observed in $CaSnSiO_5$ [26], $Ca_3SnSi_2O_9$ [27] and $CdSiO_3$ [28] phosphors. The activator-free $CaSnSiO_5$, $Ca_3SnSi_2O_9$ and $CdSiO_3$ phosphors exhibit the afterglow band at 430 nm [26–28]. The presence of a blue afterglow in this case is considered unusual and is attributed to trap-mediated recombination. To study and understand this phenomenon, a range of persistent phosphors doped with Er^{3+}, Pr^{3+}, Dy^{3+}, Tb^{3+}, Tm^{3+}, Sm^{3+}, and Nd^{3+} have been synthesized. These phosphors were specifically chosen to demonstrate a typical afterglow performance and provide insights that align with the trap-mediated recombination model.

2.1.2 Hole Model

As shown in Fig. 1a, Matsuzawa and his team reported that $SrAl_2O_4$:Eu^{2+}, Dy^{3+} material exhibits the persistence of length afterglow around 10 h [9]. Also, the report extends to elucidate the cause of the origination of this extraordinary persistent luminescence. As shown in Fig. 1a, this model considers that holes are the majority charge carriers due to photoconductive behaviours of $SrAl_2O_4$: Eu^{2+}, Dy^{3+}. It is also highlighted that the hole could escape to the VB when Eu^{2+} ions are excited by a photon, producing Eu^{+} ions. The released hole could then be trapped by Dy^{3+}, forming a Dy^{4+} ion. After a while, thermal energy facilitates the release of the trapped hole back into the valence band (VB), leading to the regeneration of Eu^{2+} ions and the subsequent emission of afterglow. These findings indicate that Dy^{3+} ions are essential for generating photoconductivity as they serve as hole-trapping centers. The introduction of Dy^{3+} ions also create a new trap level [29].

2.1.3 Electron Model

In 2003, Aitasalo et al. documented the observation of a direct electron excitation from the valence band (VB) to trap levels. However, the specific origin of these trap levels was unspecified [30]. As depicted in Fig. 1b, the CB position is significantly higher than the energy level of the oxygen vacancy trap. Furthermore, the transition to the conduction band induced by thermal effects poses a significant challenge. Hence, they assumed that the recombination energy was directly transferred to the Eu ions. This electron model has also elucidated the doping effect on trivalent rare earth co-dopants.

Holsa et al. reported the instability of Eu^{2+} ions in aluminates [9, 29]. Furthermore, they suggested that Matsuzawa's model virtually ignored the appearance of persistent phosphorescence in $SrAl_2O_4$: Eu^{2+} phosphors and R^{3+} could play a role in improving more trap or trap properties [31]. In addition, they synthesized $SrAl_2O_4$: Eu^{2+}, Sm^{3+} persistent phosphors and elucidated the inclusion of Sm^{3+} in $SrAl_2O_4$: Eu^{2+} and its afterglow emission. In a reducing atmosphere consisting of a mixture of N_2-H_2, Sm^{3+} ions can be effectively reduced to Sm^{2+} ions, leading to a decrease in trap concentrations. Clabau and his team did not encourage to accept the aforementioned Matsuzawa model [32]. The results obtained from Electron Paramagnetic Resonance (EPR) measurements indicate a decrease in the concentration of Eu^{2+}

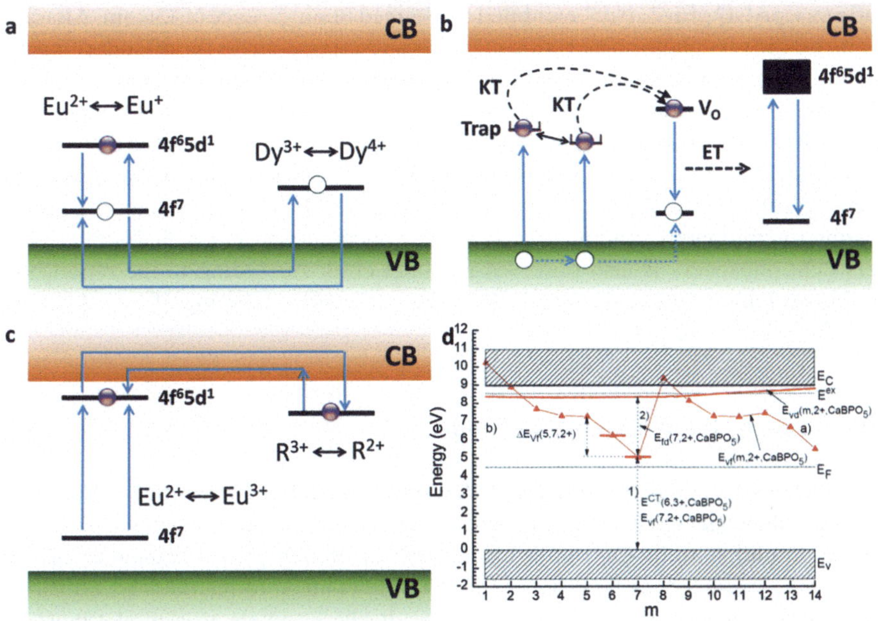

Fig. 1 **a** Hole model: Matsuzawa et al. proposed a persistent phosphorescence mechanism for $SrAl_2O_4:Eu^{2+}$, Dy^{3+}. Reproduced with permission from Ref. [5], © The Royal Society of Chemistry 2016. **b** Electron model: Aitasalo et al. proposed persistent phosphorescence mechanism for $CaAl_2O_4:Eu^{2+}$, Dy^{3+}. Reproduced with permission from Ref. [5], © The Royal Society of Chemistry 2016. **c** Dorenbos et al. proposed afterglow mechanism for aluminate and silicate compounds. Reproduced with permission from Ref. [5], © The Royal Society of Chemistry 2016. **d** Energy level scheme of the divalent lanthanides in $CaBPO_5$. Reproduced with permission from Ref.[22], © Electrochemical Society 2005

during the photoexcitation process. This decrease is followed by an increase in Eu^{2+} concentration after the excitation ceases and continues until the end of persistent luminescence. Based on these findings, it has been proposed that Eu^{2+} is involved in the trapping and de-trapping processes, which contradicts the previous proposal suggesting energy transfer to Eu^{2+} after trapping.

2.1.4 Two-Photon Absorption Model

In the persistent phosphors $CaAl_2O_4:Eu^{2+}$, Dy^{3+}, an unanticipated afterglow excitation band, approximately at 530 nm, has been observed. This indicates the existence of a novel excitation mechanism for producing afterglow [9]. Aitasalo et al. presented a novel energy-transfer model involving the absorption of two photons at a wavelength of 530 nm. This mechanism initiates deep energy level excitation, followed by the subsequent capture of two electrons from the shallow energy level [33]. They then proved that when stimulated with light of wavelengths exceeding

530 nm, $CaAl_2O_4:Eu^{2+}$, Nd^{3+} exhibited a sustained luminescence at 430 nm. Aitasalo proposed the two-photon absorption model as an explanation for the afterglow mechanism. According to their research, this mechanism involves the absorption of two photons, which directly excite an e^- from the VB into a long-lived intermediate trap level. After undergoing two steps of optical stimulation, Aitasalo believed that defects in the material acted as electron–hole recombination centers within the bandgap. The energy resulting from this recombination process was then transferred to Eu2+ ions, leading to continuous illumination. A similar phenomenon in $Ca_2Si_5N_8:Eu^{2+}$, Tm persistent phosphors is also observed by Smet et al. According to the report, the double excitation process is explained by the consecutive trapping of two electrons by a Tm^{3+}-Eu^{2+} pair [34, 35].

2.1.5 Energy Band Engineering Model

Dorenbos and colleagues supported Aitasalo's model in their study (Fig. 1c). However, they later showed that the reasoning behind the assumption that the carriers on the ground state of Eu^{2+} after excitation were faulty [36]. Lanthanide energy levels were localized, as he proposed in 2005, in contrast to the delocalized Bloch states of the valence and CBs [37]. As a result, the basic electron model must be revised. The significant change in his new model involved the energization and assignment of electrons to divalent europium ions instead of the VB. The updated electron model, however, did not account for the occurrence of PP in $SrAl_2O_4$: Eu^{2+} phosphors. Numerous researchers have dedicated substantial efforts to calculate the energy levels of lanthanides in persistent phosphors [38, 39]. Holsa et al. published numerous papers on this topic. They proposed a great method for understanding this afterglow mechanism based on the experimental data, including the (i) bandgap energy of hosts, (ii) energy level positions of R^{2+} and R^{3+} ions, and (iii) trap level locations about the bottom of the CB of hosts. In their study, they also provided a demonstration of how to calculate the ground and excitation level energies for R^{2+} and R^{3+} ions within the host band structure of $CdSiO_3:Tb^{3+}$ [40–42]. The previous empirical model was justified by the observation that the ground-state energy levels of both R^{2+} and R^{3+} ions were independent of the host. For instance, while the energy level of Eu^{3+} ($^7F_0$) was only experimentally determined, we could infer the ground level location of Tb^{3+} ($^7F_6$) in $CdSiO_3$ by referencing the ground level of Eu^{3+} ($^7F_0$).

Persistent luminescence applications typically prioritize excitation in the visible spectral region, either by using daylight or artificial lighting sources. If visible excitation is not feasible, the next preference is near-UV radiation. In the rare earth series, Ce^{3+}, Eu^{2+}, Pr^{3+}, and Tb^{3+} ions exhibit the lowest 5d energies, enabling them to be excited by longer wavelengths. This characteristic is advantageous as it allows for efficient electron transfer from the CB of the host material to trapping sites. The significant overlap between the excited 5d levels and the CB enhances the potential for high-efficiency electron transfer to occur. Though UV radiation usually is required to excite trivalent ions (Tb^{3+}, Ce^{3+}, and Pr^{3+}), blue light can excite Eu^{2+} in many hosts [42–47]. Bos et al. and Zych et al. conducted Thermoluminescence (TL) measurement, a powerful technique widely recognized for identifying the trap level locations [47, 48].

2.2 *Quantum Tunneling Model*

Quantum tunneling is a fascinating phenomenon in quantum mechanics where a particle has the ability to pass through a barrier that it would not be able to overcome using classical mechanics. Electrons have the ability to pass through thin potential barriers demonstrated by Oppenheimer in 1928 and other researchers. This phenomenon can be explained by perpendicular carrier transport in two-dimensional systems. It is simple to understand by recognizing an electron's wave function. Instead of abruptly stopping at a barrier, electrons gradually lose their energy into it in an exponential manner, with the rate of energy loss. When the barrier is thin, there is a non-zero amplitude of the electron's wave function that remains at the end of the barrier. This implies that the electron has a non-zero probability of penetrating through the barrier.

The barrier's shape governs the probability of penetration. In semiconductor devices, a triangular form, like the simplified metal–semiconductor heterojunction interface and the band-to-band transition example, represents the potential barrier. Tunnelling through such a barrier is crucial when the band edges are tilted at high field strength sufficiently. Furthermore, the tunnelling current's dependence on the level-density distribution allows for this current to gather information regarding carrier distribution. This technique employs a similar mechanism to manipulate the Fermi or quasi-Fermi level on one side of a barrier compared to the Fermi level on the other side. This enables carriers to tunnel through the barrier at specific energies.

According to the findings presented in Fig. 2a, Hoogeustraten et al. conducted pioneering research on the presence of quantum tunneling in compounds (sulfide) [49]. Tanabe and his team later utilized a similar approach to investigate $SrAl_2O_4:Eu^{2+}$, Dy^{3+} PPs [50]. Photocurrent excitation spectra (Fig. 2b) observed that the majority of the 5d levels of Eu^{2+} at the Sr site, which experienced a weak crystal field, localized within the CB. These findings led the researchers to propose a hypothesis regarding the perfect route for trapping and de-trapping processes. Instead of relying on the CB or valence band (VB), they suggested that quantum tunneling could play a significant role in these processes [10].

Pan et al. conducted a study in Cr^{3+}: zinc gallogermanate exhibited NIR emissions. They stated that these phosphors demonstrate the emissions within the 650–1000 nm range and feature an afterglow persisting for over 360 h [10]. The ionization threshold energy in the $Zn_3Ga_2Ge_2O_{10}$ host closely aligned with the CB because of the effective activation of afterglow by UV light. This observation was made using the CB/VB model. However, it is worth noting that visible light in the range of 400–630 nm, which has photon energy below the ionization threshold, can still generate a faint yet persistent luminescence. Consequently, an alternative model known as the quantum tunneling model was put forward to account for the processes of electron trapping

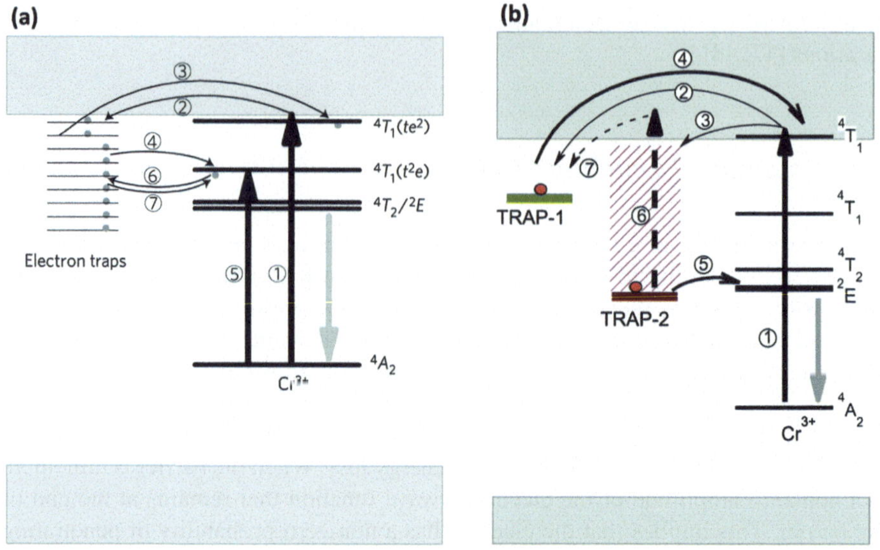

Fig. 2 **a** Representation of persistent NIR luminescence mechanism. Reproduced with permission from Ref. [10], © Nature Publishing Group 2012. **b** Schematic representation of the NIR persistent phosphorescence and photo stimulated NIR persistent phosphorescence mechanisms Reproduced with permission from Ref. [4], © Nature Publishing Group 2013

and de-trapping. The electrons in the ground state of Cr^{3+} emitters were stimulated to deep trap energy levels below the ionization threshold by short-wavelength visible light. Subsequently, the traps were filled through a tunneling mechanism from the accessible Cr^{3+} energy levels. The reverse tunneling recombination produced a faint yet long-lasting light. To further validate this concept, they also conducted experiments to measure the degradation of long-lasting phosphors at a temperature of 77 K. These experiments involved using different excitation wavelengths, which ranged from 320 to 600 nm. They then observed a similar event in a $LiGa_5O_8$ phosphor doped with Cr^{3+} ions, which exhibited a persistent duration exceeding 1000 h [51].

2.3 *Oxygen Vacancy Model and Coordination Model*

Qiu et al. studied the photo-stimulated luminescence properties of fluoroaluminate glasses doped with Eu^{2+} [52], strontium aluminosilicate glasses doped with Eu^{2+} [53], alkali borate and silicate glasses doped with Ce^{2+} and gold nanoparticle-doped glasses [54, 55]. A novel afterglow model known as the oxygen vacancy model (Introduced in the 1990s. See Dorenbos' study on LSO: Ce) was offered to describe the trapping and detrapping mechanism using EPR spectra analysis. This model was constructed to identify and confirm the trap types to understand the afterglow mechanism.

In their previous study, Aitasalo et al. introduced an electron model where they proposed an oxygen vacancy as the main trap for capturing electrons. Clabau et al. also proposed that oxygen vacancies in $SrAl_2O_4:Eu^{2+}$, Dy^{3+} served as an excellent electron trap. The afterglow curves of $SrAl_2O_4$: Eu^{2+} and Eu^{2+}/Dy^{3+}-codoped $SrAl_2O_4$ were compared, revealing the presence of two TL peaks. These peaks exhibited different intensities and positions, but shared a similar spectral shape. In addition, it was discovered that the process of co-doping had no impact on the chemical composition of the traps. Interestingly, it was observed that only oxygen vacancies retained their ability to store and release electrons. Lim and his colleagues conducted a study on the afterglow properties of spinel powders doped with titanium. Pure spinel materials have been found to exhibit numerous intrinsic defects, including vacancies of Mg^{2+} and O^{2+}, which tend to occur in pairs. These paired defects are commonly referred to as Schottky defects. It has been observed that these defects are closely linked to the emission of red light at a wavelength of 620 nm [56].

3 Red Emitting Persistent Materials

As known, the persistent luminescence of red-emitting materials mainly originates from emitters, such as Eu^{2+}, Eu^{3+}, Sm^{3+}, Yb^{2+} ions, etc. [57–61]. The defect traps determine the persistent intensity and duration, which are closely related to co-dopants, synthetic methods, and inherent properties of the host materials. Therefore, in designing red-emitting persistent luminescence materials, it is essential to consider both a well-suited emitter center and an appropriate host matrix capable of generating appropriate traps for the persistent luminescence (Fig. 3).

In Eu^{2+}-based phosphors, the broad emission bands are commonly observed. These bands are caused by transitions from the $4f^65d^1$ excited states to the $4f^7$ ground states. (parity allowed electron transition). Emission wavelengths of Eu^{2+} strongly depend on their local structure, such as the anionic-Eu bond lengths, the surrounding coordination number of Eu^{2+}, the symmetry, and the nephelauxetic effect. Thus, Eu^{2+} ions can emit broad luminescence from blue to red emission, even to NIR light. As is known to us, red persistent sulfide phosphors activated by Eu^{2+} ions, such as $CaS:Eu^{2+}$, Dy^{3+}/Tm^{3+} [63] or $SrS:Eu^{2+}$, Dy^{3+} [64], exhibit irreparable shortcomings, such as chemical instability, i.e., the sulfide material is easily degraded in the presence of water. Until now, only yellow-orange persistent phosphors in the oxide system, such as $Sr_3SiO_5:Eu^{2+}$ (yellow) [64, 65], $Ca_3Si_2O_7:Eu^{2+}$ (orange) [66], and $LiSr_4(BO_3)_3:Eu^{2+}$ (orange) [67], were found. However, red-emitting persistent materials of Eu^{2+} doped oxides were barely reported. In the nitride system, several high-performance red-emitting persistent materials, such as $Ca_2Si_5N_8:Eu^{2+}$, Tm^{3+} [68], $SrCaSi_5N_8:Eu^{2+}$, Tm^{3+} [69], had been developed. Typically, red-emitting persistent luminescence of $Ca_2Si_5N_8:Eu^{2+}$, Tm^{3+} was independently founded by Miyamoto et al. and Van den Eeckhout et al. [68]. It was discovered that $Ca_2Si_5N_8:Eu^{2+}$,

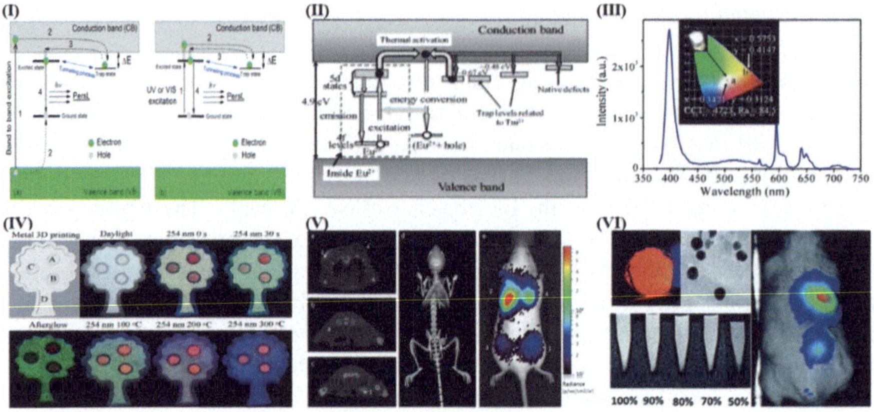

Fig. 3 Red emitting persistent materials and its applications. (I) The proposed scheme of the electron trapping-detrapping model for **a** high energy band-to-band excitation **b** low energy UV or visible light excitation only pumping the emission ions [57]. (II) The afterglow decay curves of different amount of Eu^{2+}, Tm^{3+}doped $Ca_2Si_5N_8$ sample [58]. (III) Electroluminescence spectrum of the W-LEDs using a phosphors blend of $Sr_2P_2O_7$:$x$$Sm^{3+}$ ($x = 6$.mol%) and (Sr, Ba)$_2$$SiO_4$:$Eu^{2+}$ phosphors. The inset: picture of W-LEDs and CIE chromaticity coordinates of (**a**) W-LEDs and(b) $Sr_2P_2O_7$:Sm^{3+} phosphor. CIE, color-rendering index [59]. (IV). Appearances of the signals for the"apple tree" pattern with four typical samples placed in the designed container fabricated by metal 3D printing. Reproduced with permission from Ref. [60] © Elsevier Ltd and Techna Group S.r.l 2022. **(V)** Bioimaging of CaWO4: 1% Eu3 + , 0.5% Sm^{3+} nanoparticles. The nanoparticles were embedded into a sacrificed mouse by different ways and the positions of the droplets were visualized with μCT (sagittal sections (**a–c**), 3D projection (**d**)). Bioluminescence of the nanoparticles was measured 25 min after UV excitation (**e**) [61]. **(VI)** Internalization of Gd_2O_2S:Eu^{3+} NPs in MDA-MB231 cancer cells after exposure to Gd_2O_2S:Eu^{3+} nanoparticles (overnight, 0.1 mg mL^{-1}). **a** The bright field images with UV excitation and the time gated detection of Gd_2O_2S:Eu^{3+} nanoparticles under UV excitation [62]

Tm^{3+} demonstrate a potent, persistent reddish-orange luminescence between 500 and 750 nm [58]. What's noteworthy is that the afterglow duration of $Ca_2Si_5N_8$:Eu^{2+} without co-dopants lasts for 150 s with a threshold of 0.32 mcd/m^2; however, $Ca_2Si_5N_8$:Eu^{2+}, Tm^{3+} emitted strong red persistent luminescence for more than 1 h afterglow time [68]. At the proper co-doping concentration of Tm^{3+}, the afterglow length increases to 2500 s. Besides, additional traps are created when Dy^{3+} or Nd^{3+} are added to $Ca_2Si_5N_8$:Eu^{2+}, but they are too shallow to be useful at room temperature.

Divalent Yb^{2+} ions have an electronic structure similar to that of Eu^{2+}, [Xe] $4f^{13}$ $6s^2$, producing a broad emission band because of the significant crystal field dependence of its 5d-4f transitions. The red emitting persistent phosphorescence observed in $Na_2CaGe_2O_6$:Yb^{2+} [70], $Ba_{0.8}Sr_{0.2}AlSi_5O_2N_7$:$Yb^{2+}$ [71], $Ba_{1-x}Ca_xS$:Yb^{2+} [72], and $SrSi_2O_2N_2$: Yb^{2+}, Ho^{3+} [73]. The developed $SrSi_2O_2N_2$:Yb^{2+}, Ho^{3+} persistent phosphors showed excellent potential as an optical information storage medium by exhibiting 2.5 h of effective red sustained phosphorescence. Others Eu^{2+} and Yb^{2+} doped red persistent phosphors are listed in Table 1.

Table 1 Detail of host, emitter (Eu^{2+}), co-dopants, persistent region, and afterglow time of red-emitting persistent materials

Compounds	Emitter	Co-dopants	Persistent region (nm)	Afterglow time	Year	Refs.
Ca_2SiS_4	Eu^{2+}	Nd^{3+}	660	30 min	2010	Van den Eeckhout et al. [25]
$Sr_3Al_2O_5Cl_2$	Eu^{2+}	Tm^{3+}	620	220 min	2010	Li et al. [74]
$LiSr_4(BO_3)_3$	Eu^{2+}		630		2013	Li et al. [67]
$Sr_5(BO_3)_3Cl$	Eu^{2+}		610	15 min	2014	Jin et al. [75]
$Ca_2Si_5N_8$	Eu^{2+}	Tm^{3+}	615	2500 s	2014	Miyamoto et al. [68]
$SrCaSi_5N_8$	Eu^{2+}	Tm^{3+}	647	30 min	2014	Wang et al. [69]
CaS	Eu^{2+}	Dy^{3+}	650	18 min	2015	Rodriguez Burbano et al. [63]
$Sr_{0.8}Ca_{0.2}AlSiN_3$	Eu^{2+}		628	9600 s	2019	Liu et al. [76]
CaS	Eu^{2+}	Tm^{3+}	650	700 s	2021	Suda et al. [77]
$Sr_3Al_{2-x}B_xO_5Cl_2$	Eu^{2+}	Dy^{3+}	625		2021	Li et al. [78]
$LiSr_{4-x}Ba_x(BO_3)_3$	Eu^{2+}	Dy^{3+}	611	8 min	2022	Li et al. [79]

Among these above red persistent emitters, the strong red-emitting activator ion is the Eu^{3+} ion due to the inner-shell 4f-4f electron transitions. The activators such as Dy^{3+}, Mg^{2+}, Zr^{4+}, and Ti^{4+} are trapping centers in the oxides and oxysulfide-based persistent phosphors. Besides, Eu^{3+} replaces Y^{3+}/Gd^{3+} in doped oxy-sulfides, such as $Gd_2O_2S:Eu^{3+}$, Ti^{4+}, Mg^{2+}, [80] and $Y_2O_2S:Eu^{3+}$, Ti^{4+}, Mg^{2+} [81] phosphors. The nonequivalent substitution, Ti^{4+} and Mg^{2+}, occupy the sites of Y^{3+}/Gd^{3+} to create traps to enhance the afterglow with high efficiency in PPs. Unfortunately, the afterglow duration of Eu^{3+} activated oxysulfide compounds remains relatively short (<5 h), failing to meet the demands of optical applications. Additionally, oxysulfides have a shorter afterglow time than persistent phosphors with Eu^{3+} doping in oxides, such as $Na_2CaSn_2Ge_3O_{12}:Eu^{3+}$ and Dy^{3+}, which has a 5-h afterglow period [82]. There have also been reports of other Eu^{3+} doped with oxides, including CaO: Eu^{3+} [83], $Lu_2O_3:Eu^{3+}$ [84], $SrMg_2(PO_4)_2:Eu^{3+}$, Zr^{4+} [85], $Ca_2SnO_4:Eu^{3+}$ [86], and $SnO_2:Eu^{3+}$ [87].

Like Eu^{3+} ions, Sm^{3+} ions were widely used as an activator of red-emitting persistent materials due to its $^4G_{5/2} \rightarrow {}^6H_J$ (J = 5/2, 7/2, 9/2 and 11/2) transitions ranging from orange to red emission. Until now, red-emitting persistent materials doped with Sm^{3+} have been mainly in oxides. Sm^{3+}-doped phosphors such as $BaGa_2O_4:Sm^{3+}$, Sr^{2+} [88], $Na_2CaSn_2Ge_3O_{12}:Sm^{3+}$ [89], $Ca_2SnO_4:Sm^{3+}$ [90], $Sr_2SnO_4:Sm^{3+}$, Dy^{3+} [91], $Zn_2P_2O_7:Sm^{3+}$, Li^+ [92], Y_2O_2S: Sm^{3+}, Mg^{2+}, Ti^{4+} [93], and $Sr_3Sn_2O_7:Sm^{3+}$ [94] phosphors have been thoroughly investigated for their afterglow properties, which gives intense reddish-orange afterglow up to 7 h due to $^4G_{5/2}$ to ^6H_J (J = 5/2, 7/2, 9/2) transitions [95–99]. Among the persistent phosphors, the afterglow of $Ca_2SnO_4:Sm^{3+}$ for more than 7 h is the most extended red or reddish-orange emission [90]. The experiments of thermoluminescence and positron annihilation lifetime were conducted to provide evidence for the persistent mechanism in the materials. The findings indicated that the defects and structural defects like Sn^{3+} and Sn^{2+} play a role in trapping holes, whereas the defects Sm_{Ca} serve as centers for trapping electrons. In $Ca_2SnO_4:Sm^{3+}$ red persistent phosphors, these trapping sites are crucial for storing photo energy. Others Eu^{3+} and Sm^{2+} doped red persistent phosphors are listed in Table 2.

4 Transition Metal Ions Based NIR PLNPs

4.1 Cr^{3+}-doped NIR PLNPs

The Cr^{3+} ion has been extensively studied for persistent phosphors. This is primarily due to its role as a typical NIR activator [109–111]. The valence electron configuration of Cr^{3+} is $[Ar]3d^54s^1$. In the context of ligand field theory, d^3 spectral components will split into $^4A_{2g}$, ^2E_g, $^2T_{1g}$, and $^2T_{2g}$ in an octahedral site with a high crystal field. Consequently, the ^2E and $^4T_2$ energy levels emerge as the lowest excited states, influenced by the intensity of the crystal field. Cr^{3+} emissions usually display a distinctive sharp R-line because of the $^2E \rightarrow {}^4A_2$ transition in a strong crystal field. Meanwhile, the transition $^4T_2 \rightarrow {}^4A_2$ will result in broad-band emission in a weak crystal field. Both $^2E \rightarrow {}^4A_2$ (R-line) and $^4T_2 \rightarrow {}^4A_2$ bands are inhomogeneously broadened due to site-to-site disordering. The development of persistent phosphors with visible light emission that has high brightness and a lengthy afterglow lifespan has been quite successful during the past two decades [12, 110–112]. The exploration of NIR emitting persistent phosphors has steadily increased recently, mainly because of their prospective uses in industries including labelling, security, and defense technology [83, 95–100]. A recent demonstration of in vivo imaging utilizing persistent phosphors that generate NIR light sparked a huge interest in persistent material research and imaging technique improvement [12, 113–117]. Optical imaging is a promising tool for observing biological tissues using the radioactive-free technique.

NIR persistent phosphors find utility due to their capacity for effective signal transmission through biological oxide hosts, characterized by their high transmittance. Figure 4 shows the timeline of the evolution of uniform persistent luminescence of Cr^{3+} doped nanoparticles. The sustained phosphorescence with a peak at 695 nm

Table 2 Detail of host, emitter (Eu^{3+} and Sm^{3+}), co-dopants, persistent region, and afterglow time of red-emitting persistent materials

Compounds	Emitter	Co-dopants	Persistent region (nm)	Afterglow time	Year	Refs.
CaO	Eu^{3+}		616	2 h	2000	Fu [83]
Y_2O_3	Eu^{3+}		612.3	1.5 h	2003	Lin et al. [100]
Lu_2O_3	Eu^{3+}		611	1 h	2010	Trojan-Piegza and Zych [84]
$CaMoO_4$	Eu^{3+}		616	5 min	2011	Kang et al. [101]
$CaWO_4$	Eu^{3+}		616	35 min	2012	Kang et al. [102]
$BaZrO_3$	Eu^{3+}	Ti^{4+}	594	10 min	2012	Sun et al. [103]
$GdSr_2AlO_5$	Eu^{3+}		703	375 s	2015	Li et al. [104]
$Na_2CaSn_2Ge_3O_{12}$	Eu^{3+}	Dy^{3+}	588	5 h	2018	Xu et al. [82]
$Bi_4Ge_3O_{12}$	Eu^{3+}		612	5400 s	2023	Liu et al. [105]
$Sr_2ZnSi_2O_7$	Sm^{3+}		598	30 min	2010	Zhang et al. [99]
$Ca_3SnSi_2O_9$	Sm^{3+}		600	1200 s	2011	Xu et al. [106]
SnO_2	Sm^{3+}	Zr^{4+}	607	940 s	2012	Zhang et al. [107]
ZrO_2	Sm^{3+}	Sn^{4+}	618	1000 s	2012	Zhao and Wang [96]
$Gd_{9.33}(SiO_4)_6O_2$	Sm^{3+}		615	20 min	2013	Li et al. [98]
$SrSnO_3$	Sm^{3+}		596	150 s	2015	Wang et al. [108]
Sr_2ScGaO_5	Sm^{3+}		619	300 s	2018	Li et al. [97]
$BaGa_2O_4$	Sm^{3+}		602	300 s	2020	Zhou et al. [88]
$BaGa_2O_4$	Sm^{3+}	Sr2 +	650	2 h	2020	Zhou et al. [88]
$SrZrO_3$	Sm^{3+}	Mg^{2+}	644	10 min	2023	Yen et al. [110]

in $ZnGa_2O_4$:Cr^{3+} crystals was initially described by Bessiere et al. [118–120]. A super-long NIR afterglow emission period of 360 h was then accomplished by using $Zn_3Ga_2Ge_2O_{10}$ as the host by Pan et al., which can be excited by solar spectrum in 2012 [120–122]. Later, Li and Qiu et al. enhanced the afterglow qualities by partially replacing Ga with Zn and Sn to create a solid solution of Zn-Ga-Sn-O. The researchers confirmed that $Zn_3Ga_2SnO_8$:0.5Cr^{3+} displays intense and long-lasting phosphorescence for up to 300 h in the NIR region. This characteristic could potentially facilitate real-time, prolonged, repeatable, and reliable structural imaging of deep tissues [123].

Significant efforts have been undertaken in recent years to create a controllable "bottom-up" synthesis of Cr^{3+} doped PLNPs to overcome the uniformity of the

Table 3 Detail of host, afterglow time and synthetic method of Cr^{3+} doped NIR persistent materials

Compounds	Size	Afterglow time	Synthetic method	Year	Refs.
$La_3Ga_5GeO_{14}$	/	>1 h	Solid-state reaction method	2010	Yan et al. [137]
ß-Ga_2O_3	360 ± 87 nm	72 h	The hydrothermal process followed by calcination	2012	Wang et al. [138]
$LiGa_5O_8$	2–10 μm	>1000 h	Solid-state reaction method	2013	Liu et al. [139]
$SrGa_{12}O_{19}$	/	>2 h	A high-temperature solid-state reaction method	2014	Allix et al. [140]
$LiGa_5O_8$	50–100 nm	~120 h	Sol–gel method	2014	Xu et al. [141]
$ZnAl_2O_4$	/	30 min	Solid state method	2014	Chuang et al. [142]
$Ca_3Ga_2Ge_3O_{12}$	/	>7000 s	High-temperature solid-state reaction method	2014	Basavaraju et al. [143]
$Ca_{14}Zn_6Al_{10}O_{35}$	5 μm	1 h	The Pechini method	2015	Chen et al. [144]
$La_3Ga_5GeO_{14}$	/	40 min	Solid-state reaction method	2015	Zhuang et al. [119]
Zn_2SnO_4	/	>35 h	Solid-state reaction method	2015	Li et al. [145]
$Zn_3Ga_2Ge_2O_{10}$	/	1200 s	Solid-state reaction method	2015	Xue et al. [146]
$Zn_{1.2}Al_{1.6}Ge_{0.2}O_4$	/	>120 h	High-temperature solid-state sintering	2018	Liu et al. [147]
$Mg_4Ga_4Ge_3O_{16}$	/	>10 h	High-temperature solid-state reaction method	2018	Du and Poelman [148]
$Mg_{1.4}Zn_{0.6}SnO_4$	1–5 μm	180 min	The hydrothermal process followed by calcination	2021	Du et al. [149]
Zn_2SnO_4	/	440 h	The solid state reaction method	2021	Liang et al. [150]
$Cs_2AgNaInCl_6$	10–18 nm	168 h	Hot-injection method	2022	Yu et al. [151]

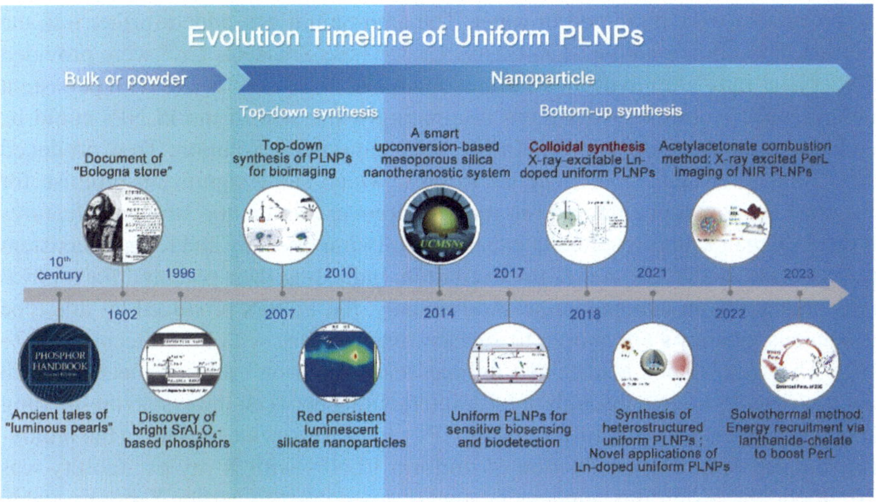

Fig. 4 The timeline of the evolution of uniform persistent luminescence nanoparticles (Ln = lanthanide elements). Insertion panels (from left to right) were reproduced with permission from Ref. [110] © Taylor &Francis Group, LLC; Ref. [29, 111] © The Electrochemical Society 2009; Ref. [112] © The National Academy of Sciences of the USA 2007; Ref. [12] © Macmillan Publishers Limited 2014; Ref. [113] © Elsevier B.V 2010; Ref. [114] © Elsevier B.V 2014; Ref. [115] © The Royal Society of Chemistry 2018; Ref. [116] © Nature Nanotechnology 2021; Ref. [117] ©The Royal Society of Chemistry 2022; Ref. [118] © Elsevier B.V 2023, respectively

size and shape in "top-down" synthesis. In 2015, mesoporous $SiO_2/ZnGa_2O_4:Cr^{3+}$ (mZGC) NIR PLNPs with in vivo rechargeable NIR emission was demonstrated by Han et al. [124]. The temperature required for synthesising $ZnGa_2O_4:Cr^{3+}$ (ZGC) NIR PLNPs was only 600 °C when the mesopores of mesoporous silica nanoparticles (MSNs) were used as a reaction template. The distinctive mesoporous characteristics and the uniform size with uniform morphology of the MSNs were observed in PLNPs at such low temperatures. The experimental results prove that MSNs are nanoreactors to synthesize ZGC nanoparticles with controlled morphology, and such porous nanostructure opens a new path to developing Cr^{3+}-doped $ZnGa_2O_4$ NIR PLNPs. In 2018, Zhang and his team proposed the uniform carbon nanosphere as a space-limited template which can absorb the target ions such as Zn^{2+}, Ga^{3+}, and Cr^{3+} in a precursor solution [125]. Such a space-limited template strategy keeps the target material $ZnGa_2O_4:Cr^{3+}$ PLNPs with a uniform spherical shape. Unlike the MSNs template, the carbon nano-sphere template was incinerated during the low-temperature annealing phase, fabricating a hollow structure of $ZnGa_2O_4:Cr^{3+}$ PLNPs. In mice, this will be useful to serve for theranostic purposes by filling the hollow cavity by photosensitisers and chemo drugs.

Recently, there has been significant attention directed towards both templated synthesis and direct synthesis approaches for PLNPs in aqueous solutions. $ZnGa_2O_4:Cr^{3+}$ PLNPs were synthesized hydrothermally in 2015 by Han et al. [126]. For the purpose of achieving uniform distribution in the reaction solution, the precursors were dissolved in water. The ammonium was added further into the solution to get precipitation of $ZnGa_2O_4:Cr^{3+}$. These $ZnGa_2O_4:Cr^{3+}$ were provided high crystallinity after hydrothermal treatment and became effective with persistent luminescence emission. Even more fascinatingly, the size of the PLNPs could be readily altered by modifying the Zn/Ga molar ratio in the solution. They produced sub-10 nm ultrasmall, homogeneous PLNPs, which show significant promise for sustained luminescence imaging in mice. Moreover, the unique structural features of the ZGC nanocubes resulted in enhanced emission intensity and longer persistent luminescence under UV and X-ray excitation, indicating their potential for innovative applications in nanomedicine. Additionally, the PLNPs produced by this type of aqueous synthesis are simple to surface-functionalize with biomolecules [127–130]. They established a sustained luminescence "turn-on" sensing mechanism in raw biological samples through precise interactions between the lysozyme-binding aptamer conjugated on the surface of PLNPs and the lysozyme present in the biological samples [84]. The production of uniform PLNPs with X-ray excitability was also applicable using this aqueous synthesis, in contrast to earlier X-ray excitable PLNPs, which required a high-temperature annealing process [132, 133].

The study conducted by Hao et al. in 2017 demonstrated the potential application of a novel class of X-ray-activated $ZnGa_2O_4:Cr^{3+}$ PLNPs for in vivo bioimaging through NIR emission [133]. The PEG-modified X-PLNPs exhibited 700 nm NIR emission under X-ray, originating from the Cr^{3+} transitions of $^2E \rightarrow {}^4A_2$. This enabled deep-tissue bioimaging in vivo, even at a depth of 20 mm, following intravenous injection and oral administration. Furthermore, the bioimaging process was repeatable through recharging with X-ray light. These findings provide the development of novel NIR-emitting PLNPs for long-term tracking and very sensitive in vivo bioimaging. In 2019, Yeh et al. synthesized $ZnGa_2O_4:Cr^{3+}$ (ZGC) with a concave morphology via a prolonged 72-h solvothermal process to achieve complete crystallization [134]. The resulting nanoparticles exhibited excellent dispersibility in aqueous solution and biocompatibility after modification with APTES and PEG, as well as highly passive targeting of orthotopic hepatic tumors. In 2020, Liu et al. developed a top-down composition strategy that utilizes size-mismatched Sb^{3+} co-doping to reasonably optimize the oxygen vacancy concentration of $LaGaO_3$, accordingly enhancing the NIR persistent luminescence (PerL) performance of Cr^{3+} [135]. Based on this theory, they designed the PLNPs ($LaGaO_3:Sb^{3+}$, Cr^{3+}) with ultra-long NIR-PerL emission (>500 h) at ~750 nm during low-dose X-ray excitation. This nanoparticle demonstrates the benefits of PLNPs in deep imaging and therapy while meeting the biosafety of X-rays and understanding the necessity for a high signal-to-noise ratio (SNR) of biological imaging. However, the designed nanoparticles have problems such as different particle sizes, uneven distribution, and limited stability, which limit their practical biological applications. In the year 2022, a photosensitiser rose bengal (RB) modified PLNPs ($Zn_3Ga_2Ge_2O_{10}:Cr^{3+}$, Mn^{2+}), which exhibited the peaks at 698 nm and 532 nm under X-ray source, was developed by Jiang et al. [117]. Among them, the Cr^{3+} emission peak (698 nm) has excellent performance for NIR PerL imaging, and the green emitting Mn^{2+} peak (532 nm) can effectively activate RB and produce reactive oxygen species (ROS), thus achieving anti-tumour function. This dual-functional nano platform, which enables both PerL imaging and continuous photodynamic therapy, has significant advantages in tumour diagnosis and treatment. In 2023, Serge Correales et al. optimized the design of monodisperse size-controlled (7–12 nm) ZGO PLNP in oleic acid (OA) [136]. On the one hand, they have increased their size to enhance their ability in persistent luminescence. On the other hand, surface modification is made to make it hydrophilic. The surface modification involving BF4- gives the nanoparticles ultra-long colloidal stability in water (>60 days). This design strategy has the advantages of controllable nanoparticle size, increased hydrophilicity and colloidal stability, and prolonged PerL performance, which makes it have broad biological application prospects. Meanwhile, Yan et al. successfully developed nanoscale ZGC ($ZnGa_2O_4$:Cr) PLNPs by employing MOF as a template [137]. These PLNPs demonstrated superior luminescence intensity and an exceptionally long persistent luminescence (PersL) life, allowing for accurate delineation of tumours for over 3 h. With their outstanding luminescent properties and long-lasting afterglow, these PLNPs hold tremendous promise for many applications. These remarkable advancements in PLNPs synthesis and their unique properties have opened new opportunities in biophotonics, cancer theranostics, and other biomedical applications [117, 118, 129–137]. The continuous efforts of researchers in this area promise even further breakthroughs in the future.

4.2 *Mn^{2+}-doped NIR PLNPs*

The divalent Mn^{2+} ([Ar] $3d^5$ $4s^2$) emission transition attributed to the parity-forbidden intra-configuration transition. The divalent manganese ion produces a broad emission due the $^4T_1$-$^6A_1$ transition which is strongly dependent on the crystal field strength. By adjusting the crystal field strength, it is possible to tune the wavelength of Mn^{2+}, spanning from green (in a strong crystal field) to orange/red (in a weak crystal field). Persistent phosphors activated by Mn^{2+} are widely recognized for their enduring and sustained long-lasting light emission. As a result, the Mn^{2+} ion was also regarded as a good emitter of persistent phosphors that generate red-infrared light. As typical examples, $CaMgSi_2O_6$:Mn^{2+} [152], $Ca_3(PO_4)_2$:Mn^{2+}/Dy^{3+} [153], $Ca_{9.5}Mn(PO_4)_7$:Mn^{2+} [115], $MgGeO_3$:Mn^{2+} [154] and $CaZnGe_2O_6$:Mn^{2+} [155], phosphors show good red-infrared broad persistent phosphorescence, while $Zn_3(PO_4)_2$:Mn^{2+}/Ga^{3+} [156] phosphor displays green and red persistent phosphorescence.

The first work using Mn^{2+} doped NIR PLNPs was reported by Gourier et al. in 2009 [158]. Then, his team reported that the red-infrared persistent luminescence exhibited at 680 nm in $Ca_{0.2}Zn_{0.9}Mg_{0.9}Si_2O_6$: Dy^{3+}, Eu^{2+}, Mn^{2+} PLNPs for one hour [159]. In this persistent luminescence process, Eu^{2+} acted as a sensitizer. Dy^{3+} creates a trap center where Mn^{2+} serves as the luminescent center. Viana et al. concentrated on the effect of trivalent lanthanide Ln^{3+} (Ln = Pr, Dy, Nd, Ce) electron traps on the optical characteristics of Mn^{2+} doped NIR PLNPs to enhance the persistent intensity and duration for *in-vivo* long-term monitoring and diagnosis applications [160]. This study demonstrates that Pr^{3+} serves as the most effective Ln^{3+} electron trap within the $CaMgSi_2O_6$ lattice. It discusses about an optimal trap depth. $CaMgSi_2O_6$:Eu^{2+}, Mn^{2+}, and Pr^{3+} PLNPs display the strongest NIR persistent luminescence at 690 nm. $CaMgSi_2O_6$:Eu^{2+}, Mn^{2+}, Pr^{3+} NIR PLNPs were effectively used in very sensitive real-time in vivo bioimaging in mice. In conclusion, the innovative work of Daniel Scherman and colleagues has substantially aided the development of Mn^{2+}-doped NIR PLNPs and their use in the biomedical field [161, 162].

4.3 *Mn^{4+}-doped NIR PLNPs*

Mn^{4+} is another notable activator that emits NIR radiation. Its electronic configu ration ([Ar] $3d^7$ $4s^2$) enables strong NIR emission spanning from 600 to 800 nm across various host materials. The effectiveness of this NIR emission is closely associated with the specific crystal field environment of the host lattices. In host systems featuring octahedral sites like GaO_6 or AlO_6, Cr^{3+} ions incorporate Ga^{3+} or Al^{3+} ions due to their closely matched valence states and similar ionic radii. Conversely, Mn^{4+} exhibits the versatility to substitute for ions such as Ga^{3+}, Ge^{4+}, Al^{3+}, Ti^{4+}, Si^{4+}, and Zr^{4+} ions [145–149]. This inherent capability of Mn^{4+} produces a more promising choice due to the variety and the feasibility of its substitution (Fig. 5).

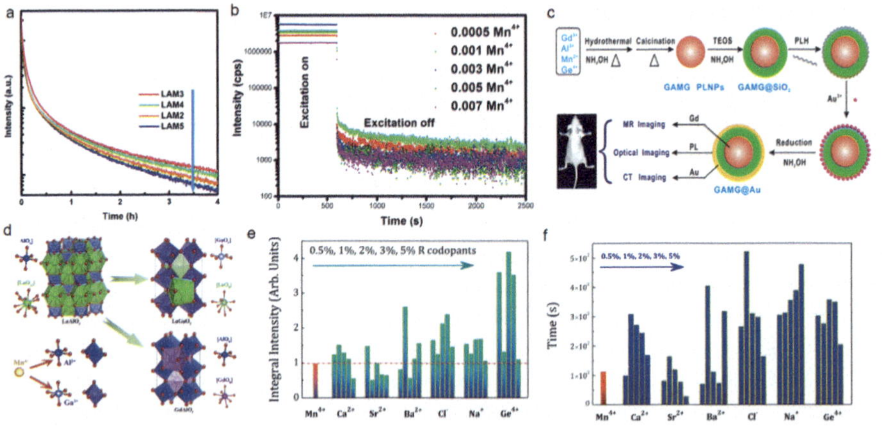

Fig. 5 **a** Afterglow curve of $La_1Al_{0.999}O_{2.9985}$: xMn^{4+} ($x = 0.001, 0.003, 0.005, 0.01$). Reproduced with permission from Ref. [145], © The Royal Society of Chemistry 2014. **b** Afterglow curve of $Mg_2Ge_{1\text{-}x}O_4$: xMn^{4+} ($x = 0.0005, 0.001, 0.003, 0.005, 0.007$). Reproduced with permission from Ref. [146], © Elsevier Ltd 2017. **c** Schematic illustration of the preparation of the $GdAlO_3$:Mn^{4+}, Ge^{4+}@Au core–shell multifunctional nanoprobe for in vivo trimodality bioimaging. Reproduced with permission from Ref. [147], © American Chemical Society 2016. **d** Schematic illustration of the doping positions and crystal structure transformation among the $LaAlO_3$, $LaGaO_3$ and $GdAlO_3$ hosts. Reproduced with permission from Ref. [148], © Elsevier Ltd 2019. The comparisons of **e** the afterglow duration time and **f** the integral intensity of $LaAlO_3$:Mn^{4+}, R (R = Ba^{2+}, Ca^{2+}, Sr^{2+}, Cl^-, Na^+, Ge^{4+}) phosphors with different co-dopant concentrations [149]

The small ionic radius of Mn^{4+} makes it generally a challenge to identify an appropriate host that induces lattice defects, consequently posing a challenge in achieving persistent phosphorescence. There are thus a limited number of persistent phosphors doped with Mn^{4+}. A Mn^{4+}-doped $MAlO_3$ (M = Gd, La) persistent phosphors were synthesized by Qiu et al. with an emission wavelength of 730 nm and 20-h persistent emissions [145]. Mn^{4+} emits NIR light for less than 3 h in Mg_2GeO_4 and La_2MgGeO_6 phosphors [145, 146]. Persistent phosphors made of $MAlO_3$ (M = La, Gd) with Ge^{4+} and Mn^{4+} codoped in place of Al^{3+}. As luminous and trapping centers, respectively, Ge^{4+} and Mn^{4+} are used. Consequently, effective generation of lattice defects occurred, leading to the observation of NIR persistent phosphorescence lasting for 20 h. The replacement of Mg^{2+} by Yb^{3+} in Mn^{4+}-doped Mg_2GeO_4 does not provide a considerable improvement in the NIR persistent luminescence, and the afterglow duration is also <3 h.

Therefore, a more effective strategy involves substituting the central ions within tetrahedral sites in the host lattice with trap center ions to enhance the depth of the trap.

Motivated by the recently developed NIR persistent luminescent phosphors (LPPs) doped with Mn^{4+}, Wang et al. indigenously designed a multifunctional nanoprobe with a $GdAlO_3$:Mn^{4+}, Ge^{4+}@Au core–shell structure. This design incorporates plasmon-enhanced NIR persistent luminescence for in vivo trimodality bioimaging [147]. In this nanoplatform, NIR persistent luminescence (PersL) was harnessed for optical imaging, while the incorporation of gadolinium (Gd) facilitated magnetic resonance imaging (MRI). Additionally, the integration of gold (Au) served the purpose of computed tomography imaging (CT). The application of a gold nano-coating on the NIR PLNPs notably enhanced PersL efficiency through plasmon resonance. Poelman and their team have introduced strategies to control the PL and afterglow performance to overcome the insufficiency of PersL in $LaAlO_3$:Mn^{4+} NIR LPPs [148, 149]. For example, the co-dopants, Cl^-/Na^+/Sr^{2+}/Ca^{2+}/Ba^{2+}/Ge^{4+}, are used in charge compensation strategy; the cation substitution approach is used to synthesize $LaAl_{1\text{-}x}Ga_xO_3$ and $La_{1\text{-}x}Gd_xAlO_3$ solid solutions. Manganese ion (Mn^{4+}) and germanium ion (Ge^{4+}) co-doped gadolinium aluminate ($GdAlO_3$:Mn^{4+}, Ge^{4+}) PLNPs served as the emission center for NIR persistent luminescence.

5 Lanthanide Ions-Doped NIR PLNPs

The valence electron shell structure of rare earth atoms is complicated. The valence electron layer structure of rare earth atoms has the following characteristics: because the system is stable when the 4f orbit is close to and in the whole, half full or empty state, the lanthanides have more stable valence states of +4 and +2 in addition to +3 valence [115, 150–154]. For example, Ce often behaves in the +4-valence state because Ce^{4+} has a $4f^0$ electron configuration [115]; Pr, Tb, and Dy can also be represented as +4 valence states, Pr^{4+} as $4f^1$, Tb^{4+} as $4f^7$, and Dy^{4+} as $4f^8$ [154]. Sm, Eu, Tm, and Yb often appear as +2 valence states because Sm^{2+} is $4f^6$, Eu^{2+}

is $4f^7$, Tm^{2+} is $4f^{13}$, and Yb^{2+} is $4f^{14}$. Thus, Lanthanides can emit light in a range of wavelengths, from the UV to the NIR [115, 154]. The lanthanide luminescence is predominantly driven by electron redistribution in the 4f orbitals, as the effective shielding of the 6 s and 5p orbitals diminishes the ligand field effects. Therefore, the manipulation of both the lanthanide emitters and the host materials enables the regulation of the emission in nanocrystals doped with lanthanide ions. Lanthanides demonstrate significantly prolonged luminescent behaviors, owing to the intersystem crossing and energy transfer mechanisms between the host and lanthanide ions. Nevertheless, the persistent luminescence (PersL) emitted by most lanthanide ions is observed in the visible spectrum rather than the NIR region. It is mainly because the lanthanide's 5d state is too high to be charged at room temperature, and it lacks carrier traps with adequate defect kinds, depths, distributions, etc. In an earlier work, Bunzli et al. detailed various NIR PLNPs incorporating lanthanide ions and elucidated their distinctive transitions within the NIR spectrum. Key lanthanides such as Eu, Pr, Nd, Yb, and Er were identified for their ability to generate robust NIR persistent luminescence (PersL) (Fig. 6) [155].

On the one hand, an idea of energy transfer between two various emission centers has been put out. If two emitting centers within the same matrix exhibit overlapping emission spectra during the energy transfer process, the subsequent emitting center can be triggered through persistent energy transfer to emit a persistence emission at a longer wavelength. On the other hand, another well-developed method uses adequate auxiliary dopants to produce highly dense trapping levels. For instance, the PersL of Eu^{2+}-doped Ca_2SiS_4 co-doped with the rare-earth elements (Dy, Nd, Sm, or Tm) was examined by Smet et al. [152]. Due to the co-doping of Nd, which resulted in the insertion of infrared emission lines arising from 4f-4f transitions within Nd^{3+} into the luminescence spectrum, only Ca_2SiS_4:Eu, Nd phosphor exhibits the greatest PersL at 660 nm.

Similar to Eu^{2+}, divalent Yb^{2+} ions can generate a broad emission band, attributed to the notable crystal field dependence of their 5d-4f transitions. Yb^{2+} ion was also chosen as the NIR emitter due to the parity-allowed 5d-4f transition and tunable emissions. $Gd_2O_2CO_3$: Yb^{3+} has demonstrated remarkable NIR persistent luminescence (NIR PersL) at 970 nm, producing a luminescence decay time of over 6 days [156]. It is attributed to the $^2F_{7/2}$–$^2F_{5/2}$ transition of Yb^{3+}. Ions. Furthermore, Xie and colleagues have reported that Yb^{2+} and Yb^{3+} ions are considered NIR PersL emitters, simultaneously showcasing robust PersL in the visible spectral range centred at 664 nm and in the NIR region at 980 nm [153]. Trap centers in the $Ba_{0.8}Sr_{0.2}AlSi_5O_2N_7$ phosphor can be traced to intrinsic flaws in the host material caused by slight fluctuations in the N/O ratio during manufacture. Thermal assistance mechanisms release captured electrons to the lowest $4f^{13}5d^1$ excited state of Yb^{2+} ions. Because the energy gap between the Yb^{2+} excited state ($4f^{13}5d^1$) and the energy level of the host CB is quite close, the ensuing $4f^{13}5d$-$4f^{14}$ transition of Yb^{2+} yields a broad emission band centred around 664 nm. Simultaneously, a resonance energy transfer from Yb^{2+} to Yb^{3+} causes Yb^{3+} to undergo the $^2F_{5/2}$-$^2F_{7/2}$ transition, resulting in a characteristic strong emission peak at 980 nm. As a result, Yb-doped

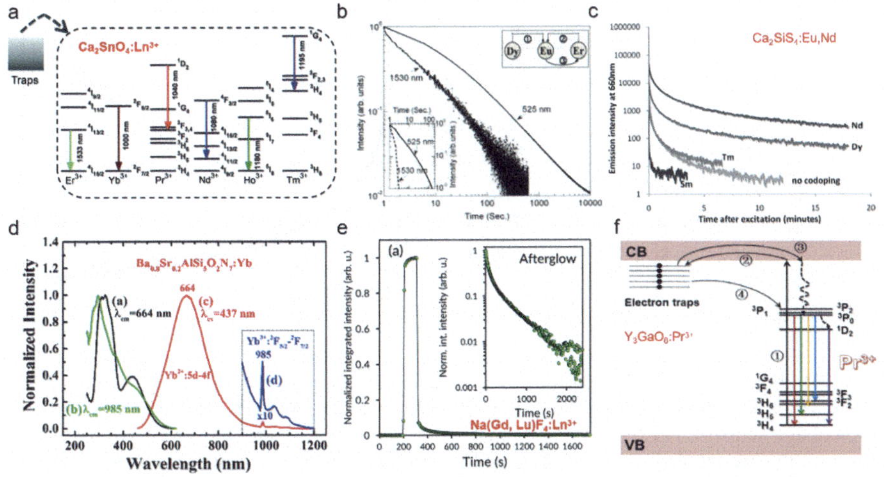

Fig. 6 **a** The possible electron transfer processes of Ca_2SnO_4:Ln^{3+} persistent phosphors in short-wave infrared persistent emissions. Reproduced with permission from Ref. [150], © The Royal Society of Chemistry 2017. **b** PersL decay curves of $SrAl_2O_4$: 1.0% Eu^{2+}, 1.5% Dy^{3+}, and 2% Er^{3+} phosphor at 525 and 1530 nm emissions. The upper right inset shows the charge trapping and energy transfer processes among Eu, Dy, and Er ions. The lower left inset shows PersL decay curves of $SrAl_2O_4$: 1.0% Eu^{2+}, 2% Er^{3+} persistent phosphors for 525 and 1530 nm emissions. Reproduced with permission from Ref. [151], © American Institute of Physics 2009. **c** PersL decay curves measured at 660 nm for Ca_2SiS_4: Eu with Dy, Nd, Sm, or Tm cooping and no codoping. Reproduced with permission from Ref. [152], © The Electrochemical Society 2009. **(d)** Normalized photoluminescence excitation (a, $\lambda_{em} = 664$ nm; b, $\lambda_{em} = 985$ nm) and photoluminescence (c, $\lambda ex = 437$ nm) spectra of $Ba_{0.8}Sr_{0.2}AlSi_5O_2N_7$: Yb. The blue solid line **d** shows the amplified tenfold photoluminescence spectrum in the wavelength of 900–1200 nm. Reproduced with permission from Ref. [153], © The Royal Society of Chemistry 2017 **e** Radioluminescence of colloidal oleate-capped β-Na (Gd, Lu) F_4: Ln^{3+} nanoparticles (Ln = Ce, Eu, Tb). Reproduced with permission from Ref. [115], © The Royal Society of Chemistry 2018. **(f)** The schematic mechanism in the afterglow process of YGO: Pr^{3+}. Reproduced with permission from Ref. [154], © Elsevier Inc 2023

NIR PLNPs can be chosen as suitable optical probes for deep-tissue bio-applications, successfully eliminating autofluorescence interference.

6 Organic Molecule-Based NIR PLNPs

Until now, most of the PLNPs were made of metal-containing inorganic compounds. However, certain limitations are frequently connected with these inorganic PLNPs [1–5]. As an example, their manufacture involves high temperatures, and the light scattering of inorganic hosts prevents the transparency of materials. Under thermal stimulation, the emission is regulated by the gradual liberation of trapped charge carriers from isolated traps, such as, defects, impurities. This process leads to low luminescence efficiency. The organic long persistent luminous (OLPL) systems have

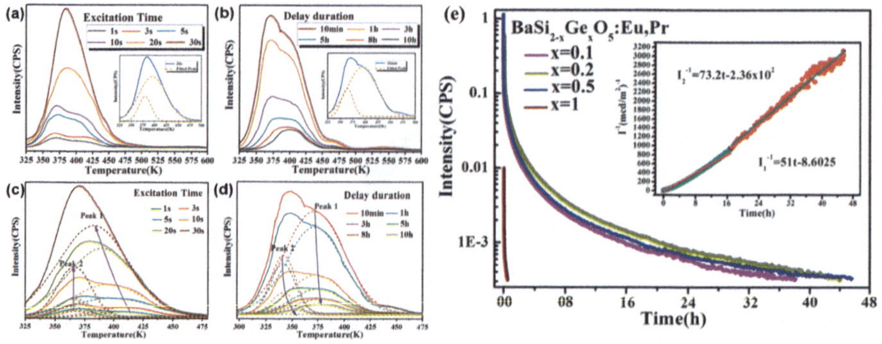

Fig. 7 (I) Photograph of OLPL (**a**) and the optical characterization of TMB and PPT films (**b–d**). Reproduced with permission from Ref. [157], ©Advanced Materials 2018. (II) The schematic illustration of dye-sensitized PLNPs for PersL imaging guided tumorectomy (**a**), the following illustrations show the overlapping spectra of ZGC and TAMRA and the mechanism of PersL enhancement (**b** and **c**). Reproduced with permission from Ref. [158], ©Angewandte Chemie International Edition 2021. (III) Long-lasting NIR sustained luminescence MOF (PLNPs@ZIF-8) activates tumours in an acidic microenvironment for imaging purposes. Reproduced with permission from Ref. [159], © Biomaterials 2019. (IV) After imaging in mice, the drug delivery strategy of PLNPs@MIL-100 (Fe) and the organ processing diagram. Reproduced with permission from Ref. [160], © Biomaterials Science 2023

advantages over inorganic PLNPs regarding solubility, transparency, flexibility, and colour tunability (Fig. 7) [157–160]. The recombination of long-lasting charge-separated states causes OLPL emission within an exciplex, produced by the interaction of acceptor radical anions and donor radical cations. Upon light excitation, acceptor and donor molecules generate charge-transfer states. The radical ions generated through these states disperse through charge hopping, giving rise to the formation of stable charge-separated states. When the excitation is turned off, the dispersed charge carriers within the exciplex gradually reunite, resulting in the emission of long-lasting light.

In 2018, Jinnai K et al. made charge carriers slowly recombine by doping emitter molecules in an organic long-persistent luminescent (OLPL) matrix to act as electron trapping sites [157]. This strategy extends the launch time and preserves the integrity of the charge separation step characteristic of OLPL. The researchers used this method to achieve greenish-blue to red emission but also improved colour purity, emission efficiency and LPL duration. This technology has led to the development of various organic glow-in-the-dark coatings. The long afterglow excitation spectra of $ZnGa_2O_4{:}Cr^{3+}$ (ZGC) at 702 nm overlapped with the fluorescence emission spectra of the organic Rhodamine dye (TAMAR). In 2021, Li and coworkers used the above findings to introduce a dye sensitization strategy [158]. They combined TAMRA with ZGC functionalized by bovine serum albumin (BSA) to get dye-sensitized PLNPs. The results show that this strategy enhanced the renewable sustained emission signal of NIR light under long-wave LED light, and was successfully used for microscopic imaging detection of tumor cells. This opens up new opportunities for a variety of bio-photonic and photonic applications. In addition, continuous luminescent metal–organic framework nanocomposites also show great potential applications in drug delivery. Zhao et al. prepared PLNPs by a one-pot method and then porous PLMOFPLNPs@ZIF-8 by surface adsorption-induced self-assembly [159]. This method could emit NIR light continuously for several days without external excitation. The results of biological experiments show that the activated sites can continue to glow in the acidic tumor microenvironment, promoting the material's application in deep tissue and long-term imaging. In 2023, Shu and his team members synthesized the core–shell structure PLNPs@MIL-100 (Fe) using a typical layer-by-layer strategy to reduce autofluorescence interference [160]. The CCD camera detected that the PLNPs solution still had continuous NIR luminescence after 168 h and then added DOX in various concentrations for drug loading. The treatment of the major organs of mice showed that the drug-carrying material had no obvious toxicity and could be used for imaging and guiding drug delivery. This paves the way for molecular imaging to guide the development of cancer interventions.

7 Semiconducting Polymer Self-assembled NIR PLNPs

A brand-new class of photonic nanomaterials called semiconductor polymer nanoparticles (SPNs) is made mainly of semiconducting polymers with π-electron delocalized back bones (Fig. 8). SPNs have the potential to generate bright fluorescence, with their optical properties primarily used by the polymer nanoparticles' molecular structure. Additionally, SPNs are broadly adopted as a very sensitive bioluminescent and chemical probe to satisfy imaging requirements since monomers used for polymerisation may alter the band gap.

In 2015, Rao et al. made the initial report on semiconducting conjugated polymer nanoparticles designed for in vivo optical imaging. These nanoparticles exhibited persistent luminescence upto one hour [161]. After this, other research group aims at highly sensitive and biodegradable semiconducting polymer nanoparticles SPNs. In 2017, Pu et al. described biodegradable SPNs as molecular afterglow probes in living mice [162]. The signal can be effectively detected in a live mouse body by amplifying and red-shifting the afterglow of SPNs. The afterglow intensity is brighter 100-fold than inorganic afterglow agents. In live mice, high-contrast imaging of lymph nodes and tumors using this approach demonstrated a 127-fold higher signal-to-noise ratio compared to near-infrared fluorescence imaging. Furthermore, Pu et al. designed self-assembled amphiphilic poly(p-phenylenevinylene) (PPV) derivatives nano reagents (SPPVN) with NIR afterglow luminescence for in vivo imaging of metastatic tumours in living mice [163]. SPPVN accumulates well in tumours and can detect xenograft tumours as small as 1 mm^3 and microscopic peritoneal metastatic tumours nearly invisible to the naked eye. Such organic compounds enable ultrasensitive in vivo imaging and serve as a platform for developing afterglow imaging techniques.

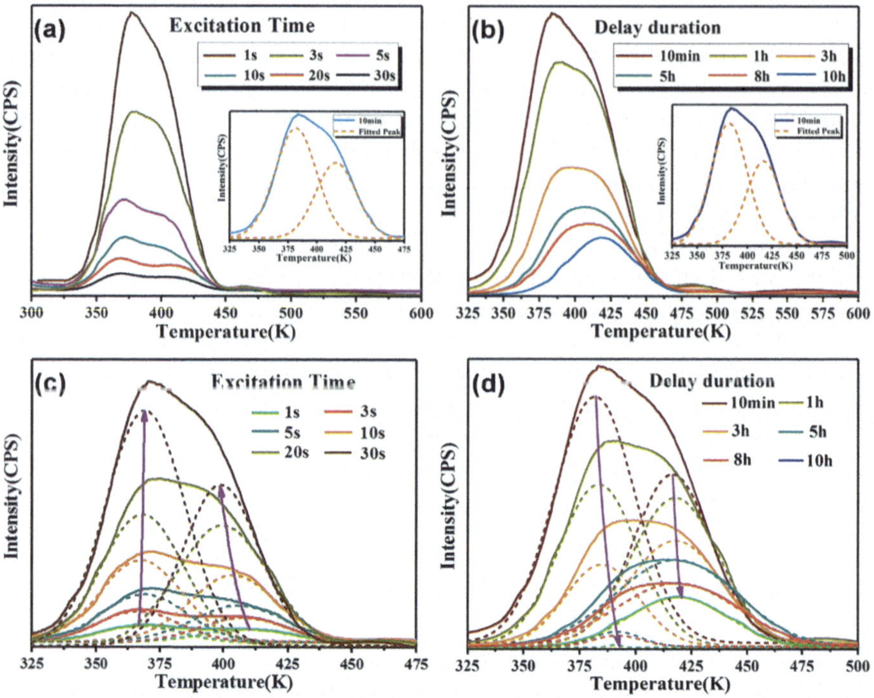

Fig. 8 **a** Synthesis of the polymer nanoparticles doped with NIR775 persistent luminescent. **b** Persistent luminescence spectra and time-dependent persistent luminescence intensity at 4, 28 and 35 °C of polymer nanoparticles doped with NIR775. **c** In vivo optical persistent luminescence imaging of a mouse before (Bkg), after 0 h and 24 h subcutaneous injection of nanoparticles that had been excited before injection. Reproduced with permission from Ref. [161], Angewandte Chemie International Edition 2015. **d** Schematic illustration of the preparation of SPNs through nanoprecipitation. **e** Average hydrodynamic diameters of SPNs. **f** Quantifying the absolute fluorescence and afterglow luminescence intensities of SPNs at different doping. **g** Fluorescence (left) and afterglow luminescence (right) imaging of a lymph node in a living mouse. Reproduced with permission from Ref. [162], Nature Biotechnology 2017. **h** Schematic illustration of the preparation of SPPVN. **i** Fluorescence and afterglow luminescence images of SPPVN solution through a living mouse. Fluorescence **j** and afterglow **k** intensities of the liver for 5 and 1 mm^3 tumour-bearing mice treated with SPPVN as a function of post-injection time. Reproduced with permission from Ref. [163], Advanced Materials 2018

8 Phosphors with Luminescence Centers of Complex Ions

The persistent phosphorescence in the NIR region can be induced due to transition of energy from one activator to another (e.g., Eu^{2+} and Ce^{3+}). After the UV light irradiation at 365 nm, a strong afterglow at 882 was observed in $SrAl_2O_4$:Eu^{2+}, Dy^{3+}, and Nd^{3+}. This phenomenon was attributed to the persistent energy transfer from excited Eu^{2+} ions to Nd^{3+} ions [164]. In $Y_3Al_2Ga_3O_{12}$:Ce^{3+}, Nd^{3+}, Nd^{3+} provides NIR persistent phosphorescence (808 nm $^4F_{5/2}$-$^4I_{13/2}$, 1064 nm $^4F_{3/2}$-$^4I_{11/2}$, 1335 nm $^4F_{3/2}$-$^4I_{13/2}$) with a lifetime at around 10 h [165]. Based on these findings, Nd^{3+} and its hosts are unsuitable for producing ultralong visible or NIR persistent phosphorescence for practical applications in *vivo* bioimaging. Upon activation with 455 nm light, the energy transfer between Cr^{3+} and Er^{3+}, produces NIR emission at 1533 nm which extended for 10 h in the $Y_3Al_2Ga_3O_{12}$ phosphor [166]. Finally, rare earth ion-doped persistent phosphors have a high potential for visible persistent luminescence with long afterglow periods. This is because of the prospects from matching trap depth and the luminous excited state of rare earth ions, as well as the possibility of creating lattice defects with a high trap density.

Nonetheless, rare earth ion-activated persistent phosphors encounter constraints in generating NIR persistent phosphorescence with longer afterglow periods due to the issue of extending and deepening trap depth. Significant efforts have been made toward creating transition metal ion-doped materials to obtain significantly prolonged NIR afterglow emission.

9 Summary and Outlook

There has been a flourishing moment in uniform PLNPs over the last five years, marked by advancements in controlled synthesis techniques, more systematic nanoparticle designs, and fascinating applications. This achievement, however, is accompanied by several unresolved obstacles and promising potential that remain the subject of significant research efforts.

One fundamental problem is enhancing the brightness of PLNPs. Despite several attempts, viable alternatives in this area have remained restricted. As a result, additional research efforts are required to realise these techniques' potential fully. Exploring beyond the approach of nonequivalent ion doping for energy trap engineering, a reassessment of these strategies in the context of persistent phosphors could open up new avenues for enhancing the optical properties of PLNPs. Furthermore, in the case of dye-sensitized PLNPs, optimizing dye selection could increase their brightness even further. Another possibility is to shift the excitation wavelength for charging PLNPs towards the deep tissue-penetrating NIR region. This might be accomplished using NIR excitable dye molecules to charge PLNPs within the NIR spectrum. These efforts will likely contribute to the continued improvement of uniform PLNPs as the field evolves.

References

1. S. Shionoya, W.M. Yen, H. Yamamoto, *Phosphor Handbook* (CRC Press, 2006)
2. G. Blasse, B.C. Grabmaier, *Luminescent Materials* (Springer, Berlin, Heidelberg, 1994)

3. R.S. Becker, R.S. Becker, *Theory and Interpretation of Fluorescence and Phosphorescence* (Wiley Interscience, New York, 1969)
4. J.R. Qiu, Y. Li, Y.C. Jia, *Persistent Phosphors: From Fundamentals to Applications* (Woodhead Publishing, 2020)
5. Y. Li, M. Gecevicius, J.R. Qiu, Long persistent phosphors–from fundamentals to applications. Chem. Soc. Rev. **45**, 2090–2136 (2016)
6. P.F. Smet, J. Botterman, K. Van den Eeckhout, K. Korthout, D. Poelman, Persistent luminescence in nitride and oxynitride phosphors: a review. Opt. Mater. **36**, 1913–1919 (2014)
7. Y. Zhuang, Y. Katayama, J. Ueda, S. Tanabe, A brief review on red to near–infrared persistent luminescence in transition–metal–activated phosphors. Opt. Mater. **36**, 1907–1912 (2014)
8. A.H. Wako, B.F. Dejene, H.C. Swart, Roles of doping ions in afterglow properties of blue $CaAl_2O_4$: Eu^{2+}, Nd^{3+} phosphors. Physica B **439**, 153–159 (2014)
9. H. Yamamoto, T. Matsuzawa, Mechanism of long phosphorescence of $SrAl_2O_4$: Eu^{2+}, Dy^{3+} and $CaAl_2O_4$: Eu^{2+}, Nd^{3+}. J. Lumin. **72**, 287–289 (1997)
10. Z. Pan, Y.Y. Lu, F. Liu, Sunlight–activated long–persistent luminescence in the near–infrared from Cr^{3+}–doped zinc gallogermanates. Nat. Mater. **11**, 58–63 (2012)
11. Y. Wang, Z. Wang, Characterization of Y_2O_2S:Eu^{3+}, Mg^{2+}, Ti^{4+} long–lasting phosphor synthesized by flux method. J. Rare Earths **24**, 25–28 (2006)
12. T. Maldiney, A. Bessière, J. Seguin, E. Teston, S.K. Sharma, B. Viana, A.J.J. Bos, P. Dorenbos, M. Bessodes, D. Gourier, D. Scherman, C. Richard, The in vivo activation of persistent nanophosphors for optical imaging of vascularization, tumours and grafted cells. Nat. Mater. **13**, 418 (2014)
13. P. Pei, Y. Chen, C. Sun, Y. Fan, Y. Yang, X. Liu, L. Lu, M. Zhao, H. Zhang, D. Zhao, X. Liu, F. Zhang, X–ray–activated persistent luminescence nanomaterials for NIR–II imaging. Nat. Nanotechnol. **16**, 1011 (2021)
14. X. Ou, X. Qin, B. Huang, J. Zan, Q. Wu, Z. Hong, L. Xie, H. Bian, Z. Yi, X. Chen, Y. Wu, X. Song, J. Li, Q. Chen, H. Yang, X. Liu, High–resolution X–ray luminescence extension imaging. Nature **590**, 410 (2021)
15. B.Y. Wu, H.F. Wang, J.T. Chen, X.P. Yan, Fluorescence resonance energy transfer inhibition assay for α–fetoprotein excreted during cancer cell growth using functionalized persistent luminescence nanoparticles. J. Am. Chem. Soc. **133**, 686 (2011)
16. A. Abdukayum, J.T. Chen, Q. Zhao, X.P. Yan, Functional near infrared-emitting Cr^{3+}/Pr^{3+} co–doped zinc gallogermanate persistent luminescent nanoparticles with superlong afterglow for in vivo targeted bioimaging. J. Am. Chem. Soc. **135**, 14125 (2013)
17. J. Shi, X. Sun, J. Li, H. Man, J. Shen, Y. Yu, H. Zhang, Multifunctional near infrared–emitting long–persistence luminescent nanoprobes for drug delivery and targeted tumor imaging. Biomaterials **37**, 260 (2015)
18. Z.Y. Zhou, N. Tian, J.T. Li, I. Broadwell, S.G. Sun, Nanomaterials of high surface energy with exceptional properties in catalysis and energy storage. Chem. Soc. Rev. **40**, 4167 (2011)
19. Q. Kuang, X. Wang, Z. Jiang, Z. Xie, L. Zheng, High–energy–surface engineered metal oxide micro–and nanocrystallites and their applications. Acc. Chem. Res. **47**, 308 (2014)
20. F. Benfenati, G. Lanzani, Clinical translation of nanoparticles for neural stimulation. Nat. Rev. Mater. **6**, 1 (2021)
21. L. Liang, N. Chen, Y. Jia, Q. Ma, J. Wang, Q. Yuan, W. Tan, Recent progress in engineering near–infrared persistent luminescence nanoprobes for time–resolved biosensing/bioimaging. Nano Res. **12**, 1279 (2019)
22. Z. Zhou, Y. Li, M. Peng, Near–infrared persistent phosphors: Synthesis, design, and applications. Chem. Eng. J. **399**, 125688 (2020)
23. K. Huang, N. Le, J.S. Wang, L. Huang, L. Zeng, W.C. Xu, Z. Li, Y. Li, G. Han, designing next generation of persistent luminescence: recent advances in uniform persistent luminescence nanoparticles. Adv. Mater. **34** (2022)
24. Z. Li, Y. Zhang, X. Wu, X. Wu, R. Maudgal, H. Zhang, G. Han, Adv. Sci. **2**, 1500001 (2015)
25. K. Van den Eeckhout, P.F. Smet, D. Poelman, Persistent luminescence in Eu^{2+}–doped compounds: a review. Materials **3**, 2536–2566 (2010)

26. X. Xu, J. Qiu, X. Yu, Multicolor long persistent luminescence phosphors $CaSnSiO_5:R^{3+}$ (R= Pr, Tb, Sm, Dy). ECS Solid State Lett. **2**, R9 (2012)
27. X. Xu, Y. Wang, W. Zeng, Y. Gong, B. Liu, Luminescent properties of the multicolor afterglow phosphors $Ca_3SnSi_2O_9:Re^{3+}$ (Re = Pr, Tb, Sm). J. Am. Ceram. Soc. **94**, 3632 (2011)
28. Y. Liu, B. Lei, C. Shi, Luminescent properties of a white afterglow phosphor $CdSiO_3:Dy^{3+}$. Chem. Mater. **17**, 2108 (2005)
29. T. Matsuzawa, Y. Aoki, N. Takeuchi, Y. Murayama, A new long phosphorescent phosphor with high brightness, $SrAl_2O_4$: Eu^{2+}, Dy^{3+}. J. Electrochem. Soc. **143**, 2670 (1996)
30. J. Holsa, T. Aitasalo, H. Jungner, M. Lastusaari, J. Niittykoski, G. Spano, Role of defect states in persistent luminescence materials. J. Alloys Compd. **374**, 56–59 (2004)
31. P. Gluchowski, W. Strek, M. Lastusaari, J. Holsa, Optically stimulated persistent luminescence of europium–doped $LaAlO_3$ nanocrystals. Phys. Chem. Chem. Phys. **17**, 17246–17252 (2015)
32. F. Clabau, X. Rocquefelte, S. Jobic, P. Deniard, M.H. Whangbo, A. Garcia, T. Le Mercier, Mechanism of phosphorescence appropriate for the long–lasting phosphors Eu^{2+}–doped $SrAl_2O_4$ with codopants Dy^{3+} and B^{3+}. Chem. Mater. **17**, 3904–3912 (2005)
33. T. Aitasalo, J. Hölsä, H. Jungner, M. Lastusaari, J. Niittykoski, Mechanisms of persistent luminescence in Eu^{2+}, RE^{3+} doped alkaline earth aluminates. J. Lumin. **94**, 59–63 (2001)
34. W. Chen, S. Zhang, M. Zhou, T. Zhao, X. Qin, X. Liu, M. Liu, P. Duan, Two–photon absorption–based upconverted circularly polarized luminescence generated in chiral perovskite nanocrystals. J. Phys. Chem. Lett. **10**, 3290–3295 (2019)
35. K. Eeckhout, P.F. Smet, D. Poelman, Persistent luminescence in rare–earth codoped $Ca_2Si_5N_8$: Eu^{2+}. J. Lumin. **129**, 1140–1143 (2009)
36. P. Dorenbos, Energy of the first 4f7→4 f65d transition of Eu+ in inorganic compounds. J. Lumin. **104**, 239–260 (2003)
37. P. Dorenbos, The Eu^{3+} charge transfer energy and the relation with the band gap of compounds. J. Lumin. **111**, 89–104 (2005)
38. D.C. Rodriguez Burbano, S.K. Sharma, P. Dorenbos, B. Viana, J.A. Capobianco, Persistent and photostimulated red emission in CaS: Eu^{2+}, Dy^{3+} nanophosphors. Adv. Opt. Mater. **3**, 551–557 (2015)
39. F.T. You, A.J.J. Bos, Q.F. Shi, S.H. Huang, P. Dorenbos, Thermoluminescence investigation of donor (Ce^{3+}, Pr^{3+}, Tb^{3+}) acceptor (Eu^{3+}, Yb^{3+}) pairs in $Y_3Al_5O_{12}$. Phys. Rev. B: Condens. Matter Mater. Phys. **85**, 115101 (2012)
40. L.C.V. Rodrigues, J. Holsa, M. Lastusaari, M.C.F.C. Felinto, H.F. Brito, Defect to R^{3+} energy transfer: colour tuning of persistent luminescence in $CdSiO_3$. J. Mater. Chem. C **2**, 1612–1618 (2014)
41. L.C.V. Rodrigues, H.F. Brito, J. Holsa, R. Stefani, M.C.F.C. Felinto, M. Lastusaari, T. Laamanen, L.A.O. Nunes, Discovery of the persistent luminescence mechanism of $CdSiO_3:Tb^{3+}$. J. Phys. Chem. C **116**, 11232–11240 (2012)
42. L.C.V. Rodrigues, H.F. Brito, J. Holsa, M. Lastusaari, Persistent luminescence behavior of materials doped with Eu^{2+} and Tb^{3+}. Opt. Mater. Express **2**, 382–390 (2012)
43. M. Lastusaari, A.J.J. Bos, P. Dorenbos, T. Laamanen, M. Malkamaki, L.C.V. Rodrigues, J. Holsa, Wavelength–sensitive energy storage in $Sr_3MgSi_2O_8$: Eu^{2+}, Dy^{3+}. J. Therm. Anal. Calorim. **121**, 29–35 (2015)
44. V. Kale, M. Lastusaari, J. Holsa, T. Soukka, Intense UV upconversion through highly sensitized $NaRF_4$: tm (R: Y, Yb) crystals. RSC Adv. **5**, 35858–35865 (2015)
45. L.U. Khan, H.F. Brito, J. Holsa, K.R. Pirota, D. Muraca, M.C.F.C. Felinto, E.E.S. Teotonio, O.L. Malta, Red–green emitting and superparamagnetic nanomarkers containing Fe_3O_4 functionalized with calixarene and rare earth complexes. Inorg. Chem. **53**, 12902–12910 (2014)
46. A. Lecointre, A. Bessiere, A.J.J. Bos, P. Dorenbos, B. Viana, S. Jacquart, Designing a red persistent luminescence phosphor: the example of YPO_4: Pr^{3+}, Ln^{3+} (Ln = Nd, Er, Ho, Dy). J. Phys. Chem. C **115**, 4217–4227 (2011)
47. A.J.J. Bos, N.R.J. Poolton, J. Wallinga, A. Bessiere, P. Dorenbos, Energy levels in YPO_4: Ce^{3+}, Sm^{3+} studied by thermally and optically stimulated luminescence. Radiat. Meas. **45**, 343–346 (2010)

48. A. Lecointre, A. Bessiere, A.J.J. Bos, P. Dorenbos, B. Viana, S. Jacquart, Designing a red persistent luminescence phosphor: the example of $YPO_4:Pr^{3+}$, Ln^{3+} (Ln = Nd, Er, Ho, Dy). J. Phys. Chem. C **115**, 4217–4227 (2011)
49. W. Hoogeustraten, Electron traps in zinc sulphide phosphors. Philips Res. Rep. **13**, 515–693 (1958)
50. S. Tanabe, S. Watanabe, F. Saif, M. Matsuzawa, Survival probability of a truncated radial oscillator subject to periodic kicks. Phys. Rev. A: At. Mol. Opt. Phys. **65**, 033420 (2002)
51. F. Liu, W.Z. Yan, Y.J. Chuang, Z.P. Zhen, J. Xie, Z.W. Pan, Photostimulated NIR persistent luminescence as a new optical read-out from Cr^{3+}-doped $LiGa_5O_8$. Sci. Rep. **3**, 01554 (2013)
52. J.H. Si, K. Kitaoka, J.R. Qiu, T. Mitsuyu, K. Hirao, Optically encoded second-harmonic generation in germanosilicate glass by a femtosecond laser. Opt. Lett. **24**, 911–913 (1999)
53. J.R. Qiu, K. Kojima, K. Miura, T. Mitsuyu, K. Hirao, Infrared femtosecond laser pulseinduced permanent reduction of Eu^{3+} to Eu^{2+} in a fluorozirconate glass. Opt. Lett. **24**, 786–788 (1999)
54. Y. Shimotsuma, P.G. Kazansky, J.R. Qiu, K. Hirao, Self-organized nanogratings in glass irradiated by ultrashort light pulses. Phys. Rev. Lett. **91**, 247405 (2003)
55. J.R. Qiu, X.W. Jiang, C.S. Zhu, M. Shirai, J. Si, N. Jiang, K. Hirao, Manipulation of gold nanoparticles inside transparent materials. Angew. Chem. Int. Ed. **43**, 2230–2234 (2004)
56. J.H. Lim, B.N. Kim, Y. Kim, S. Kang, R.J. Xie, I.S. Chong, K. Morita, H. Yoshida, K. Hiraga, Non-rare earth white emission phosphor: Ti-doped $MgAl_2O_4$. Appl. Phys. Lett. **102**, 031104 (2013)
57. J. Xu, S. Tanabe, Persistent luminescence instead of phosphorescence: history, mechanism, and perspective. J. Lumin. **205**, 581–620 (2019)
58. B. Lei, K. Machida, T. Horikawa, H. Hanzawa, N. Kijima, Y. Shimomura, H. Yamamoto, Reddish-orange long-lasting phosphorescence of $Ca_2Si_5N_8:Eu^{2+}$, Tm^{3+} Phosphor. J. Electrochem. Soc. **157**, J196–J201 (2010)
59. R. Cao, C. Cao, X. Yu, W. Li, J. Qiu, Photoluminescence of $Sr_2P_2O_7$:Sm3+phosphor as reddish orange emission for white light-emitting diodes. Int. J. Appl. Ceram. Tec. **12**, 755–759 (2015)
60. A. Guo, Q. Zhu, S.M. Zhang, X.D. Sun, J.G. Li, A building-block strategy for dynamic anti-counterfeiting by using (Ba, Sr)G_a2_O4: Sm new red persistent luminescent phosphor as an important component. Ceram. Int. **49**, 4622–4630 (2023)
61. A. Mattner, L. Zimmermann, S. Lienenklaus, S. Weiss, C. Wickleder. Tailoring long lasting luminescence of red-emitting $CaWO_4:Eu^{3+}$, Sm^{3+} nanoparticles with enhanced crystallinity for improved bio-imaging. Ceram. Int. **46**, 26295–26298 (2020)
62. S.A. Osseni, S. Lechevallier, M. Verelst, P. Perriat, J. Dexpert-Ghys, D. Neumeyer, R. Garcia, F. Mayer, K. Djanashvili, J.A. Peters, E. Magdeleine, H. Gros-Dagnac, P. Celsis, R. Mauricot, Gadolinium oxysulfide nanoparticles as multimodal imaging agents for T2-weighted MR, X-ray tomography and photoluminescence. Nanoscale **6**, 555–564 (2014)
63. C. Rodriguez Burbano, S.K. Sharma, P. Dorenbos, B. Viana, J.A.J. Capobianco, Persistent and photostimulated red emission in CaS: Eu^{2+}, Dy^{3+} nanophosphors. Adv. Opt. Mater. **3**, 551 (2015)
64. K. Szczodrowski, A. Chruscinska, J. Barzowska, K. Przegietka, K. Anders, R. Piramidowicz, M. Grinberg, Influence of Ti^{4+} on the long lasting luminescence of $Sr_2SiO_4:Eu^{2+}$. RSC Adv. **5**, 65236–65244 (2015)
65. Y. Li, B.H. Li, C.C. Ni, S.X. Yuan, J. Wang, Q. Tang, Q. Su, Synthesis, persistent luminescence, and thermoluminescence properties of yellow $Sr_3SiO_5:Eu^{2+}$, RE^{3+} (RE=Ce, Nd, Dy, Ho, Er, Tm, Yb) and orange–Red $S_{r3}xB_axSi_O5:E^{u2}$+, D^{y3}+ phosphor. Chem. Asian J. **9**, 494–499 (2014)
66. J. Ueda, R. Maki, S.J.I.C. Tanabe, Vacuum referred binding energy (VRBE)-guided design of orange persistent $Ca_3Si_2O_7:Eu^{2+}$ phosphors. Inorg. Chem. **56**, 10353 (2017)
67. G. Li, Y. Wang, W. Zeng, S. Han, W. Chen, Y. Li, H.J.O.M. Li, Enhanced long persistence of $LiSr_4(BO_3)_3:Eu^{2+}$ orange phosphors by co-doping with Dy^{3+}. Opt. Mater. **36**, 1808–1813 (2014)
68. Y. Miyamoto, H. Kato, Y. Honna, H. Yamamoto, K. Ohmi, An orange-emitting, long-persistent phosphor, $Ca_2Si_5N_8$: Eu^{2+}, Tm^{3+}. J. Electrochem. Soc. **156**, J235 (2009)

69. J. Wang, H. Zhang, B. Lei, H. Dong, H. Zhang, Y. Liu, N. Lai, Y. Fang, Z. Chen, Red persistent and photo-stimulated luminescence properties of SrCaSi5N8: Eu2+, Tm3+ solid solution. Opt. Mater. **36**, 1855–1858 (2014)
70. Z. Ma, S. Tian, H. Wu, J. Zhang, H. Li, Self-reduction of Yb^{3+} to Yb^{2+} and red persistent luminescence of $Na_2CaGe_2O_6$:Yb^{2+} phosphor. Ceram. Int. **44**, 14582–14586 (2018)
71. Y. Lv, L. Wang, Y. Zhuang, T.L. Zhou, R.J. Xie, Discovery of the Yb^{2+}–Yb^{3+} couple as red-to-NIR persistent luminescence emitters in Yb-activated ($Ba_{1\text{-}x}Sr_x$) $AlSi_5O_2N_7$ phosphors. J. Mater. Chem. C **5**, 7095–7101 (2017)
72. J. Wei, L. Chen, W. Zhang, Y. Yang, Z. Li, The orange–red persistent luminescence of $Ba_{1\text{-}x}Ca_xS$: Yb^{2+}. J. Lumin. **181**, 427–432 (2017)
73. Y. Zhuang, Y. Lv, L. Wang, W. Chen, T.L. Zhou, T. Takeda, N. Hirosaki, R.J. Xie, Trap depth engineering of $SrSi_2O_2N_2$:Ln^{2+}, Ln^{3+} (Ln^{2+}=Yb, Eu; Ln^{3+}=Dy, Ho, Er) persistent luminescence materials for information storage applications. ACS Appl. Mater. Interfaces **10**, 1854–1864 (2018)
74. Y.Q. Li, Y.H. Wang, Y. Gong, X.H. Xu, M.J. Zhou, Design, synthesis and characterization of an orange-yellow long persistent phosphor: $Sr_3Al_2O_5Cl_2$:Eu^{2+}, Tm^{3+}. Opt. Express **18**, 24853–24858 (2010)
75. Y.H. Jin, Y.H. Hu, L. Chen, X.J. Wang, Synthesis and persistent luminescence mechanism of a novel orange emitting persistent phosphor $Sr_5(BO_3)_3Cl$:Eu^{2+}. J. Am. Ceram. Soc. **97**, 2573–2579 (2014)
76. X. Liu, Z. Song, S. Wang, Q. Liu, The red persistent luminescence of (Sr, Ca)$AlSi_N3$:E^{u2}+ and mechanism different to SrA_l2_O4:E^{u2}+, $^{Dy}3$+. J. Lumin. **218**, 313–321 (2019)
77. Y. Suda, Y. Tamura S. Yamaguchi, Y. Nanai, T. Okuno, Red afterglow and luminescence arising from defects in CaS:Eu^{2+}, Tm^{3+}. J. Phys. D Appl. Phys. **54**, 415103 (2021)
78. H.H. Li, R.Z. Li, C.K. Chang, Red-shift and improved afterglow of $Sr_3Al_{2\text{-}x}B_xO_5Cl_2$:$Eu^{2+}$, Dy^{3+} phosphors via a cation substitution strategy. Ceram. Int. **48**, 1814–1819 (2022)
79. R.Z. Li, H. Cao, H.H. Li, D.Y. Zhang, C.K. Chang, Polychromatic luminescence of $LiSr_4(BO_3)_3$:Eu^{2+}, Dy^{3+} persistent phosphors. Ceram Int. **48**, 20546–20554 (2022)
80. T. Hang, Q. Liu, D.L. Mao, C.K. Chang, Long lasting behavior of Gd_2O_2S:Eu^{3+} phosphor synthesized by hydrothermal routine. Mater. Chem. Phys. **107**, 142–147 (2008)
81. J. Hölsä, T. Laamanen, M. Lastusaari, M. Malkamäki, J. Niittykoski, E.J.O.M. Zych, Effect of Mg^{2+} and TiIV doping on the luminescence of Y_2O_2S: Eu^{3+}. Opt. Mate. **31**, 1791–1793 (2009)
82. J. Xu, Z. Ju, X. Yang, X. Tang, W. Liu, Considerable improvement of $Na_2CaSn_2Ge_3O_{12}$:Eu^{3+} reddish orange long-persistent phosphorescence with co-doping of Dy^{3+}. Dyes Pigments **162**, 951–958 (2019)
83. J. Fu, Orange and red emitting long-lasting phosphors MO:Eu^{3+} (M = Ca, Sr, Ba). Electrochem. Solid-State Lett. **3**, 350–351 (2000)
84. J. Trojan-Piegza, E. Zych, Afterglow luminescence of Lu_2O_3: Eu ceramics synthesized at different atmospheres. J. Phys. Chem. C **114**, 4215–4220 (2010)
85. X.G. Wang, F.P. Du, D.L. Wei, Y.L. Huang, H.J. Seo, A new long-lasting phosphor Zr^{4+} and Eu^{3+} co-doped $SrMg_2(PO_4)_2$. Sens. Actuators B Chem. **158**, 171–175 (2011)
86. X.P. Gao, Z.Y. Zhang, C. Wang, J. Xu, Z.H. Ju, Y.Q. An, W.S. Liu, The persistent energy transfer and effect of oxygen vacancies on red long-persistent phosphorescence phosphors Ca_2SnO_4: Gd^{3+}, Eu^{3+}. J. Electrochem. Soc. **158**, J405–J408 (2011)
87. J. Kong, W. Zheng, Y. Liu, R. Li, E. Ma, H. Zhu, X. Chen, Persistent luminescence from Eu^{3+} in SnO_2 nanoparticles. Nanoscale **7**, 11048–11054 (2015)
88. X. Zhou, G. Ju, T. Dai, Y. Li, H. Wu, Y. Jin, Y. Hu, Strontium substitution enhancing a novel Sm^{3+}-doped barium gallate phosphor with bright and red long persistent luminescence. J. Lumin. **218**, 116820 (2020)
89. J. Xu, Z.H. Ju, X.P. Gao, Y.Q. An, X.L. Tang, W.S. Liu, $Na_2CaSn_2Ge_3O_{12}$: a novel host lattice for Sm^{3+}-doped long-persistent phosphorescence materials emitting reddish orange light. Inorg. Chem. **52**, 13875–13881 (2013)

90. Z.H. Ju, R.P. Wei, J.R. Zheng, X.P. Gao, S.H. Zhang, W.S. Liu, Synthesis and phosphorescence mechanism of a reddish orange emissive long afterglow phosphor Sm^{3+}-doped Ca_2SnO_4. Appl. Phys. Lett. **98**, 1906 (2011)
91. X.H. Xu, Y.H. Wang, Y. Gong, W. Zeng, Y.Q. Li, Effect of oxygen vacancies on the red phosphorescence of Sr_2SnO_4:Sm^{3+} phosphor. Opt. Express **18**, 16989–16994 (2010)
92. R. Pang, W.Z. Sun, J.P. Fu, H.F. Li, Y.L. Jia, D. Li, L.H. Jiang, S. Zhang, C.Y. Li, Luminescence properties of a novel reddish orange long-lasting phosphorescence phosphor $Zn_2P_2O_7$:Sm^{3+}, Li^+. RSC Adv. **5**, 82704–82710 (2015)
93. Y.J. Li, M.W. Wang, L.D. Zhang, D. Gao, S.X. Liu, Soft chemical synthesis and luminescence properties of red long-lasting phosphors Y_2O_2S:Sm^{3+}. Int. J. Miner. Metall. Mater. **20**, 972–977 (2013)
94. B.F. Lei, S.Q. Man, Y.L. Du, S. Yue, Luminescence properties of Sm^{3+}-doped $Sr_3Sn_2O_7$ phosphor. Mater. Chem. Phys. **124**, 912–915 (2010)
95. J.C. Zhang, X.L. Ma, Q.S. Qin, L.R. Shi, J.Y. Sun, M.J. Zhou, B.T. Liu, Y.H. Wang, The synthesis and afterglow luminescence properties of a novel red afterglow phosphor: SnO_2:Sm^{3+}, Zr^{4+}. Mater. Chem. Phys. **136**, 320–324 (2012)
96. Z. Zhao, Y. Wang, The synthesis and afterglow luminescence properties of a novel red afterglow phosphor: ZrO_2:Sm^{3+}, Sn^{4+}. J. Lumin. **132**, 2842–2846 (2012)
97. G. Li, W. Chen, Y. Wang, B. Duhan, Electronic structure, photoluminescence and phosphorescence properties in Sr_2ScGaO_5:Sm^{3+}. Dyes Pigments **157**, 259–266 (2018)
98. G. Li, Y.H. Wang, S.C. Han, W. Zeng, W.B. Chen, Y. Gong, A red long-lasting phosphorescence material $Gd_{9.33}(SiO_4)_6O_2$:Sm^{3+} and effect of oxygen vacancies on its performance. ECS J. Solid State Sci. Technol. **2**, R161–R164 (2013)
99. Y.Y. Zhang, R. Pang, C.Y. Li, C.Y. Zang, Q.A. Su, Reddish orange long lasting phosphorescence of Sm^{3+} in $Sr_2ZnSi_2O_7$:Sm^{3+} phosphors. J. Rare Earths **28**, 705–708 (2010)
100. Y. Lin, C.W. Nan, N. Cai, X. Zhou, H. Wang, D. Chen, Anomalous afterglow from Y_2O_3-based phosphor. J. Alloys Compd. **361**, 92–95 (2003)
101. F.W. Kang, Y.H. Hu, H.Y. Wu, G.F. Ju, Red afterglow properties of Eu^{3+} in CaMoO4 phosphor. Chin. Phys. Lett. **28**, 107201–107204 (2011)
102. F.W. Kang, Y.H. Hu, H.Y. Wu, Z.F. Mu, G.F. Ju, C. Fu, N. Li, Luminescence and red long afterglow investigation of Eu^{3+}–Sm^{3+} co-doped $CaWO_4$ phosphor. J. Lumin. **132**, 887–894 (2012)
103. D. Sun, Z. Li, J. Zhu, Z. Xiao, W. Tao, Liu, Photoluminescence properties of europium and titanium Co-doped $BaZrO_3$ phosphors powders synthesized by the solid-state reaction method. Opt. Mater. **34**, 1890–1896 (2012)
104. G. Li, Y. Wang, W. Zeng, W. Chen, S. Han, H. Guo, Luminescent and magnetic properties of the afterglow phosphors $GdSr_2AlO_5$:RE^{3+} ($RE^{3+} = Eu^{3+}$, Sm^{3+}, P^{r3}+ and D^{y3}+). RSC Adv. **5**, 20884–20889 (2015)
105. Z. Liu, T. Lin, N. Chen et al., From scintillator to phosphor: X-ray induced long afterglow red luminescence from $Bi_4Ge_3O_{12}$:Eu^{3+}. J. Lumin. **263**, 120002 (2023)
106. X. Xu, Y. Wang, W. Zeng, Y. Gong, B. Liu, Luminescent properties of the multicolor afterglow phosphors $Ca_3SnSi_2O_9$:Re^{3+} (Re ¼ Pr, Tb, Sm). J. Am. Ceram. Soc. **94**, 3632–3635 (2011)
107. J.C. Zhang, X.L. Ma, Q.S. Qin, L.R. Shi, J.Y. Sun, M.J. Zhou, B.T. Liu, Y.H. Wang, The synthesis and afterglow luminescence properties of a novel red afterglow phosphor: SnO_2: Sm^{3+}, Z^{r4}+. Mater. Chem. Phys. **136**, 320–324 (2012)
108. T. Wang, Y. Hu, L. Chen, X. Wang, G. Ju, Luminescent properties of a reddish orange long afterglow phosphor $SrSnO_3$:Sm^{3+}. Radiat. Meas. **73**, 7–13 (2015)
109. S.W. Zhang, Z.S. Zheng, S.H. Yan, L.F. Huang, Effect of Mg^{2+}substitution on enhanced photoluminescence of red orange long persistent phosphor perovskite $Sr_{1-1.5x}Zr_{1-y}O_{3-y}$:$xSm^{3+}$,$yMg^{2+}$. Inorg. Chem. Commun. **148**, 110266 (2023)
110. W.M. Yen, S. Shionoya, H. Yamamoto, *Phosphor Handbook* (The CRC Press Laser and Optical Science and Technology Series, 2007), p. 1051
111. J. Hölsä, Persistent luminescence beats the afterglow: 400 years of persistent luminescence. Electrochem. Soc. Interface **18**, 42 (2009)

112. Q. le Masne, C. de Chermont, J. Chanéac, F. Seguin, S. Pellé, J.-P. Maîtrejean, D. Jolivet, M. Gourier, D.S. Bessodes, Nanoprobes with NIR persistent luminescence for in vivo imaging. Proc. Natl. Acad. Sci. U.S.A. **104**, 9266–9271 (2007)
113. A. Lecointre, A. Bessière, B. Viana, D. Gourier, Red persistent luminescent silicate nanoparticles. Radiat. Meas. **45**, 497–499 (2010)
114. W. Fan, B. Shen, W. Bu, F. Chen, Q. He, K. Zhao, S. Zhang, L. Zhou, W. Peng, Q. Xiao, D. Ni, J. Liu, J. Shi, A smart upconversion-based mesoporous silica nanotheranostic system for synergetic chemo-/radio-/photodynamic therapy and simultaneous MR/UCL imaging. Biomaterials **35**, 8992–9002 (2014)
115. R. Cooper, J.A. Capobianco, J. Seuntjens, Radioluminescence studies of colloidal oleate-capped β-Na (Gd, Lu) F_4:Ln^{3+} nanoparticles (Ln = Ce, Eu, Tb). Nanoscale **10**, 7821 (2018)
116. P. Pei, Y. Chen, C.X. Sun, Y. Fan, Y.M. Yang, X. Liu, L.F. Lu, M.Y. Zhao, H.X. Zhang, D.Y. Zhao, X.G. Liu, F. Zhang, X-ray-activated persistent luminescence nanomaterials for NIR-II imaging. Nat. Nanotechnol. **16**, 1011–1018 (2021)
117. H.M. Jiang, R.P. Wang, Q. Zhang, L. Song, X. Sun, J.P. Shi, Y. Zhang, A dual-functional nanoplatform based on NIR and green dual-emissive persistent luminescence nanoparticles for X-ray excited persistent luminescence imaging and photodynamic therapy. Nanoscale **14**, 15451–15461 (2022)
118. M.J. Sun, C. Yin, Z.C. Yan, Z.-J. Wei, Z.Y. Zhang, W. Wang, Z. Yuan, Energy recruitment via lanthanide-chelate to boost the persistent luminescence of nanophosphor for contrast-enhanced tumor navigation. Chem. Eng. J. **468**, 143814 (2023)
119. Y.X. Zhuang, J. Ueda, S. Tanabe, P. Dorenbos, Band-gap variation and a self-redox effect induced by compositional deviation in $Zn_xGa_2O_{3+x}$:Cr^{3+} persistent phosphors. J. Mater. Chem. C **2**, 5502–5509 (2014)
120. M. Allix, S. Chenu, E. Veron, T. Poumeyrol, E.A. Kouadri-Boudjelthia, S. Alahrache, F. Porcher, D. Massiot, F. Fayon, Considerable improvement of long-persistent luminescence in germanium and tin substituted $ZnGa_2O_4$. Chem. Mater. **25**, 1600–1606 (2013)
121. Z.W. Pan, Y.Y. Lu, F. Liu, Sunlight-activated long-persistent luminescence in the nearinfrared from Cr^{3+}-doped zinc gallogermanates. Nat. Mater. **11**, 58–63 (2012)
122. Y.L. Wu, Y. Li, X.X. Qin, R.C. Chen, D.K. Wu, S.J. Liu, J.R. Qiu, NIR longpersistent phosphor of $Zn_3Ga_2Ge_2O_{10}$:Cr^{3+} sintered in different atmosphere. Spectrochim. Acta, Part A **151**, 385–389 (2015)
123. Y. Li, S.F. Zhou, Y.Y. Li, K. Sharafudeen, Z.J. Ma, G.P. Dong, M.Y. Peng, J.R. Qiu, Long persistent and photo-stimulated luminescence in Cr^{3+}-doped Zn–Ga–Sn–O phosphors for deep and reproducible tissue imaging. J. Mater. Chem. C **2**, 2657–2663 (2014)
124. Z. Li, Y. Zhang, X. Wu, X. Wu, R. Maudgal, H. Zhang, G. Han, In vivo repeatedly charging near-infrared-emitting mesoporous SiO_2/$ZnGa_2O_4$:Cr^{3+} persistent luminescence nanocomposites. Adv. Sci. **2**, 1500001 (2015)
125. J. Wang, J. Li, J. Yu, H. Zhang, B. Zhang, Large hollow cavity luminous nanoparticles with nir persistent luminescence and tunable sizes for tumor afterglow imaging and chemo-/photodynamic therapies. ACS Nano **12**, 4246 (2018)
126. Z. Li, Y. Zhang, X. Wu, L. Huang, D. Li, W. Fan, G. Han, Direct aqueous-phase synthesis of sub-10 nm "Luminous Pearls" with enhanced in vivo renewable NIR persistent luminescence. J. Am. Chem. Soc. **137**, 5304 (2015)
127. J. Wang, Q. Ma, W. Zheng, H. Liu, C. Yin, F. Wang, X. Chen, Q. Yuan, W. Tan, Autofluorescence-free targeted tumor imaging based on luminous nanoparticles with composition-dependent size and persistent luminescence. ACS Nano **11**, 8185 (2017)
128. J. Wang, Q. Ma, X.-X. Hu, H. Liu, W. Zheng, X. Chen, Q. Yuan, W. Tan, ACS Nano **11**, 8010 (2017)
129. J. Wang, Q. Ma, H. Liu, Y. Wang, H. Shen, X. Hu, C. Ma, Q. Yuan, W. Tan, Anal. Chem. **89**, 12764 (2017)
130. Z. Li, Q. Wang, Y. Wang, Q. Ma, J. Wang, Z. Li, Y. Li, X. Lv, W. Wei, L. Chen, Q. Yuan, Nano Res. **11**, 6167 (2018)

131. Z.-Z. Chen, L.-C. Wang, D. Manoharan, C.-L. Lee, L.-C. Wu, W.-T. Huang, E.-Y. Huang, C.-H. Su, H.-S. Sheu, C.-S. Yeh, Adv. Mater. **31**, 1905087 (2019)
132. Y.-M. Yang, Z.-Y. Li, J.-Y. Zhang, Y. Lu, S.-Q. Guo, Q. Zhao, X. Wang, Z.-J. Yong, H. Li, J.-P. Ma, Y. Kuroiwa, C. Moriyoshi, L.-L. Hu, L.-Y. Zhang, L.-R. Zheng, H.-T. Sun, Light: Sci. Appl. **7**, 88 (2018)
133. Z. Xue, X. Li, Y. Li, M. Jiang, H. Liu, S. Zeng, J. Hao, Interfaces, X-ray-activated NIR persistent luminescent probe for deep-tissue and renewable in vivo bioimaging. ACS Appl. Mater. Interfaces **9**, 22132 (2017)
134. Z.Z. Chen, L.C. Wang, D. Manoharan, C.L. Lee, L.C. Wu, W.T. Huang, E.Y. Huang, C.H. Su, H.S. Sheu, C.S.J. Yeh, Low dose of X-ray-excited long-lasting luminescent concave nanocubes in highly passive targeting deep-seated hepatic tumors. Adv. Mater. **31**, 1905087 (2019)
135. B.M. Liu, R. Zou, S.Q. Lou, Y.F. Gao, L. Ma, K.L. Wong, J. Wang, Low-dose X-ray-stimulated LaGaO3: Sb, Cr NIR persistent luminescence nanoparticles for deep-tissue and renewable in vivo bioimaging. Chem. Eng. J. **404**, 127133 (2021)
136. Y.E. Serge-Correales, D. Neumeyer, S. Ullah, R. Mauricot, Q. Zou, S.J. Ribeiro, M.J. Verelst, Size control and improved aqueous colloidal stability of surface-functionalized $ZnGa_2O_4$: Cr^{3+} bright persistent luminescent nanoparticles. Langmuir **39**, 1495–1506 (2023)
137. W.Z. Yan, F. Liu, Y.Y. Lu, X.J. Wang, M. Yin, Z.W. Pan, Near infrared long-persistent phosphorescence in $La_3Ga_5GeO_{14}$:Cr^{3+} phosphor. Opt. Express **18**, 20215–20221 (2010)
138. X.S. Wang, W.S. Li, J.Q. Situ, X.Y. Ying, H. Chen, Y. Jin, Y.Z. Du, Multi-functional mesoporous β-Ga_2O_3: Cr^{3+} nanorod with long lasting near infrared luminescence for in vivo imaging and drug delivery. RSC Adv. **5**, 12886–12889 (2015)
139. W.Z. Liu, Y.J. Yan, Z.P. Chuang, J. Zhen, Z.W. Xie, Pan, Photostimulated NIR persistent luminescence as a new optical read-out from Cr^{3+}-doped $LiGa_5O_8$. Sci. Rep. **3**, 01554 (2013)
140. M. Allix, S. Chenu, E. Veron, T. Poumeyrol, E.A. Kouadri-Boudjelthia, S. Alahrache, F. Porcher, D. Massiot, F. Fayon, Considerable improvement of long-persistent lumines-cence in germanium and tin substituted $ZnGa_2O_4$. Chem. Mater. **25**, 1600–1606 (2013)
141. J. Xu, D.Q. Chen, Y.L. Yu, W.J. Zhu, J.C. Zhou, Y.S. Wang, Cr^{3+}:SrGa12O19: A broad-band NIR long-persistent phosphor. Chem.-Asian J. **9**, 1020–1025 (2014)
142. Y.J. Chuang, Z.P. Zhen, F. Zhang, F. Liu, J.P. Mishra, W. Tang, H.M. Chen, X.L. Huang, L.C. Wang, X.Y. Chen, J. Xie, Z.W. Pan, Photostimulable NIR persistent lumi-nescent nanoprobes for ultrasensitive and longitudinal deep-tissue bio-Imaging. Theranostics **4**, 1112–1122 (2014)
143. N. Basavaraju, K.R. Priolkar, D. Gourier, S.K. Sharma, A. Bessiere, B. Viana, The importance of inversion disorder in the visible light induced persistent luminescence in Cr^{3+} doped AB_2O_4 (A = Zn or Mg and B = Ga or Al). Phys. Chem. Chem. Phys. **17**, 1790–1799 (2015)
144. D.Q. Chen, Y. Chen, H.W. Lu, Z.G. Ji, A bifunctional Cr/Yb/Tm:$Ca_3Ga_2Ge_3O_{12}$ phos-phor with NIR long-lasting phosphorescence and upconversion luminescence. Inorg. Chem. **53**, 8638–8645 (2014)
145. Y. Li, Y.Y. Li, K. Sharafudeen, G.P. Dong, S.F. Zhou, Z.J. Ma, M.Y. Peng, J.R. Qiu, A strategy for developing near infrared long-persistent phosphors: taking $MAlO_3$:Mn^{4+}, Ge^{4+} (M = La, Gd) as an example. J. Mater. Chem. C. **2**, 2019–2027 (2014)
146. H. Xue, Y.H. Hu, L. Chen, H.Y. Wu, G.F. Ju, T. Wang, L. Yang, A novel rare-earth free red long-persistent phosphor: Mg_2GeO_4:Mn^{4+}. Ceram. Int. **43**, 15141–15145 (2017)
147. J.M. Liu, Y.Y. Liu, D.D. Zhang, G.Z. Fang, S. Wang, Synthesis of GdAlO3:Mn^{4+}, Ge^{4+}@Au core-shell nanoprobes with plasmon-enhanced NIR persistent luminescence for in vivo trimodality bioimaging. Acs. Appl. Mater. Inter. **8**, 29939–29949 (2016)
148. J.R. Du, D. Poelman, NIR persistent luminescence in Mn^{4+} doped perovskite type solid solutions. Ceram. Int. **45**, 8345–8353 (2019)
149. J. Du, O.Q. De Clercq, K. Korthout, D. Poelman, $LaAlO_3$:Mn^{4+} as NIR emitting persistent luminescence phosphor for medical imaging: a charge compensation study. Materials **10**, 1422–1436 (2017)
150. Y.J. Liang, F. Liu, Y.F. Chen, X.L. Wang, K.N. Sun, Z.W. Pan, Extending the applications for lanthanide ions: efficient emitters in short-wave infrared persistent luminescence. J. Mater. Chem. C **5**, 6488–6492 (2017)

151. N. Yu, F. Liu, X. Li, Z. Pan, Near infrared long-persistent phosphorescence in $SrAl_2O_4$:Eu^{2+}, Dy^{3+}, Er^{3+} phosphors based on persistent energy transfer. Appl. Phys. Lett. **95**, 231110 (2009)
152. P. F. Smet, Z. N. AVBi, D. Poelman, Red persistent luminescence in Ca_2SiS_4: Eu, Nd. J. Electrochem. Soc. **156**, H243 (2009)
153. Y. Lv, L. Wang, Y. Zhuang, T.L. Zhou, R.J. Xie, Discovery of the Yb^{2+}–Yb^{3+} couple as red-to-NIR persistent luminescence emitters in Yb-activated $(Ba_{1-x}Sr_x)$ $AlSi_5O_2N_7$ phosphors. J. Mater. Chem. **5**, 7095 (2017)
154. R. Li, L. Pang, D. Jiang, S. Li, H. Zhang, Zhang, Long persistent luminescence and photo-stimulated luminescence in Y_3GaO_6:Pr^{3+} phosphor. J. Solid State Chem. **325**, 124169 (2023)
155. J. Zhou, J.L. Leano Jr, Z. Liu, D. Jin, K.L. Wong, R.S. Liu, J.C.G. Bünzli, Impact of lanthanide nanomaterials on photonic devices and smart applications. Small **14**, 1801882 (2018)
156. W. Ge, Z. Li, Z. Lei, T. Chen, Z. Fu, R. Peng, M. Liu, Y. Lu, Synthesis of hexagonal phase $Gd_2O_2CO_3$: Yb^{3+}, Er^{3+} upconversion nanoparticles via SiO_2 coating and Nd^{3+} doping. CrystEngComm **17**, 5702–5709 (2015)
157. R. Jinnai, C. Kabe, Adachi, Wide-range tuning and enhancement of organic long-persistent luminescence using emitter dopants. Adv. Mater. **30**, 1800365 (2018)
158. Z. Li, Y. Zhao, K. Huang, L. Huang, Y. Zhang, H. Yang, G. Han, Enhancing rechargeable persistent luminescence via organic dye sensitization. Angew. Chem. **133**, 16022–16026 (2021)
159. H. Zhao, G. Shu, J. Zhu, Y. Fu, Z. Gu, D. Yang, Persistent luminescent metal-organic frameworks with long-lasting near infrared emission for tumor site activated imaging and drug delivery. Biomaterials **217**, 119332 (2019)
160. G. Shu, H. Zhao, X. Zhang, Persistent luminescent metal–organic framework nanocomposite enables autofluorescence-free dual modal imaging-guided drug delivery. Biomater. Sci. **11**, 1797–1809 (2023)
161. M. Palner, K.Y. Pu, S. Shao, J.H. Rao, Semiconducting polymer nanoparticles with persistent NIR luminescence for in vivo optical imaging. Angew. Chem. **127**, 11639–11642 (2015)
162. Q.Q. Miao, C. Xie, X. Zhen, Y. Lyu, H. Duan, X.G. Liu, J.V. Jokerst, K.Y. Pu, Molecular afterglow imaging with bright, biodegradable polymer nanoparticles. Nat. Biotechnol. **35**, 1102–1110 (2017)
163. X. Xie, Q.Q. Zhen, Y. Miao, K.Y. Lyu, Pu, Self-assembled semiconducting polymer nanoparticles for ultrasensitive NIR afterglow imaging of metastatic tumors. Adv. Mater. **30**, 1801331 (2018)
164. Y. Teng, J. Zhou, Z. Ma, M.M. Smedskjaer, J. Qiu, Persistent near infrared phosphorescence from rare earth ions co-doped strontium aluminate phosphors. J. Electrochem. Soc. **158**, K17–K19 (2011)
165. D. Xu, J. Murata, S. Ueda, Tanabe, NIR long persistent luminescence of Er^{3+} in garnet for the third bio-imaging window. J. Mater. Chem. C **4**, 11096–11103 (2016)
166. J. Xu, D. Murata, J. Ueda, B. Viana, S. Tanabe, Toward rechargeable persistent luminescence for the first and third biological windows via persistent energy transfer and electron trap redistribution. Inorg. Chem. **57**, 5194–5203 (2018)

X-Ray-Activated Persistent Luminescence

Bo Zhou, Zhengce An, and Chao Wang

Abstract X-ray-activated persistent luminescence (PersL) has emerged recently and is becoming a research hotspot with great promise in a set of frontier applications. Unlike traditional PersL materials, the use of high-energy X-rays as an excitation source can achieve multiple and full-spectral PersL emissions, opening a new door for the development of PersL materials and an extension of the library of luminescent materials. In this chapter, we focus on the latest progress of X-ray-activated PersL materials and their frontier applications. The luminescence mechanism and modulation strategy of X-ray-activated PersL are discussed. The multi-functional PersL nanoparticle implementation strategies based on both Stokes emissions and anti-Stokes emissions are summarized. The emerging frontier applications in biomedical diagnosis and treatment, information storage and encryption, advanced X-ray imaging, and other fields are also highlighted. The current scientific and technical issues together with future development perspectives are commented at the end.

Keywords X-ray-activated PersL · Tunable PersL properties · PersL models · Frontier applications

1 Introduction

Persistent Luminescence (PersL) is an interesting luminescence phenomenon, which is still able to generate luminescence after switching off the excitations, such as ultraviolet (UV) light and high-energy radiations (e.g., X-ray and electron beams) [1–5]. Despite an ancient discovery, PersL received systematic research until the late twentieth century due to the development of material preparation science [6–8]. For instance, in 1996 it was found that $SrAl_2O_4:Eu^{2+}/Dy^{3+}$ phosphor shows green PersL emission around 30 h after stopping UV light excitation [1]. At present, it is easy for PersL luminescent materials (such as $YPO_4:Pr^{3+}$, $CaAl_2O_4:Eu^{2+}/Nd^{3+}$, and

B. Zhou (✉) · Z. An · C. Wang
State Key Laboratory of Luminescent Materials and Devices, South China University of Technology, Guangzhou 510641, China
e-mail: zhoubo@scut.edu.cn

V. Kumar et al. (eds.), *Persistent Luminescence*, Progress in Optical Science and Photonics 30, https://doi.org/10.1007/978-981-97-4943-0_5

$ZnGa_2O_4:Cr^{3+}$) to produce emissions from UV, visible to near-infrared (NIR) regions [9–14]. The emission wavelengths of PersL depend on the doping ions with a stepped dense energy structure and host matrix which can adjust the crystal field environment of dopants for a micro-control of emission wavelength [15–20]. In addition, a tuning of lanthanide dopants can lead to rich emission wavelengths [21–24]. So far, PersL materials have shown great promise in various fields, including emergency signs, luminescence paint, national defense, and so on [25–29]. The recent emergence of uniformly dispersed PersL nanoparticles (PLNPs) further contributes to various biological applications [30–35].

Recently, X-ray light has become an effective charging source to activate the PersL emissions [36–43]. Compared with UV light excitation, X-ray has more advantages. For instance, it can effectively excite wide-band gap materials to achieve PersL emission in the UV region, which is difficult under UV excitation [44–46]. High-energy X-ray photons can cause structural changes inside the material matrix and consequently give rise to various defects as trapping centers [47–50]. Therefore, there is rapid progress in the research of X-ray-activated PersL materials. The doping of lanthanide and transition metal ions helps with the full-spectral PersL emissions, allowing for frontier applications in flexible panel X-ray imaging and information storage [51–56]. In addition, stable and biocompatible X-ray-activated PLNPs can improve cell targeting efficiency and find new opportunities in biological applications [57–59]. Moreover, the attenuation of near-infrared (NIR) light is significantly lower than that in the visible range, making the NIR PLNPs much more attractive for deep tissue imaging [60–62].

This chapter will focus on the recent advances of X-ray activated PersL materials and their frontier applications. The luminescent mechanism of X-ray-activated PersL together with its modulating strategies was discussed. The strategies for realizing multi-functional PLNPs based on both Stokes and anti-Stokes emissions were summarized. And their emerging applications in a diversity of fields from information storage and encryption, and biomedical diagnosis and therapy to advanced X-ray imaging were also highlighted (Fig. 1).

2 PersL Mechanism

Over the past years, extensive research on the PersL mechanism were carried out through thermo-luminescence (TL) spectroscopy and paramagnetic resonance techniques, and a series of PersL models were proposed [1, 63–65]. The mechanism of PersL associates closely with the process of carrier dynamics, including carrier capture and release. Energy storage in PersL materials can be achieved by trapping excited electrons and/or holes, and when they are released into the conduction band (CB) or valence band (VB), the carriers can move freely until they are captured by a recombination center or another trap. Here, we briefly discuss two commonly used PersL mechanism models, namely CB-VB theory and quantum tunneling theory, which are applicable to the luminescence mechanism of X-ray-activated PersL.

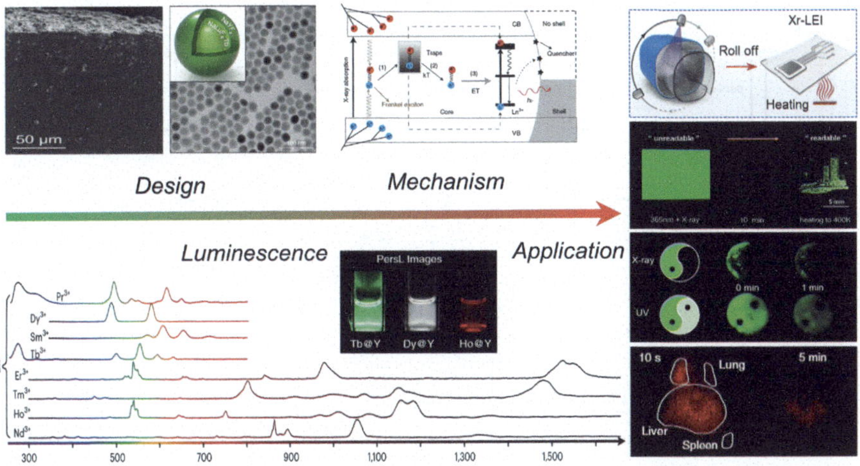

Fig. 1 Review of recent development of X-ray-activated PersL from material design to spectral tuning, PersL mechanism, and emerging frontier applications. (Refs. [7, 54, 57, 65, 79, 89, 94])

For the CB-VB theory, there are three models in detail. The hole transfer model is based on the CB/VB band participating in the excitation, capture, migration, and release of the carriers. It was first proposed in 1996 based on the research of $SrAl_2O_4$:Eu^{2+}/Dy^{3+} materials. Early in 1971, Abbruscato suggested that Eu^{2+} can produce a hole carrier after excitation according to the Hall effect [66]. On this basis, Yamamoto et al. [67] believed that there was also a hole trapping effect in $SrAl_2O_4$:Eu^{2+}/Dy^{3+} (Fig. 2a(i)). According to the photoconductivity experiment, Eu^{2+} can act as an electron trapping center to form Eu^{+}, while Dy^{3+} can capture a hole to form Dy^{4+}. This is a short-lived process. The hole can be captured when it migrates to the excited Eu^{+}. Consequently, the hole-electron recombination gives rise to the PersL. Afterwards, Dorenbos [68] reported an energy transfer model based on the research of Eu^{2+}. Dorenbos further developed various semi-empirical models and analyzed a series of experimental data, including chemical shift models of compounds, centroid shift models, redshift models, and charge transfer models (Fig. 2a(ii)). Based on these research, Zhuang et al. [69, 71] and Chen et al. [70] studied the binding energy of 4f electrons of rare earth ions in different substrates, and improved the host-referred binding energy (HRBE) and vacuum-referred binding energy (VRBE) schemes, which allows determining the energy level locations of bi- and tri-valent lanthanide ions in a matrix, as shown in Fig. 2b. Moreover, the PersL properties can be explained and predicted in various lanthanide-doped materials.

In addition, an electron trap model was proposed for $CaAl_2O_4$-based PersL materials [72, 73]. It was found that, in both Eu^{2+}-singly-doped and Eu^{2+}/RE^{3+} co-doped samples, the electrons were excited to the trap level and into the oxygen vacancy under UV excitation, while the free holes in the valence band migrated to the vacancy trap (V_{Ca}). After Eu^{2+} receives energy, the excited electron is excited from 4f level to 5d level, finally producing a PersL emission when the electron transitions down to the

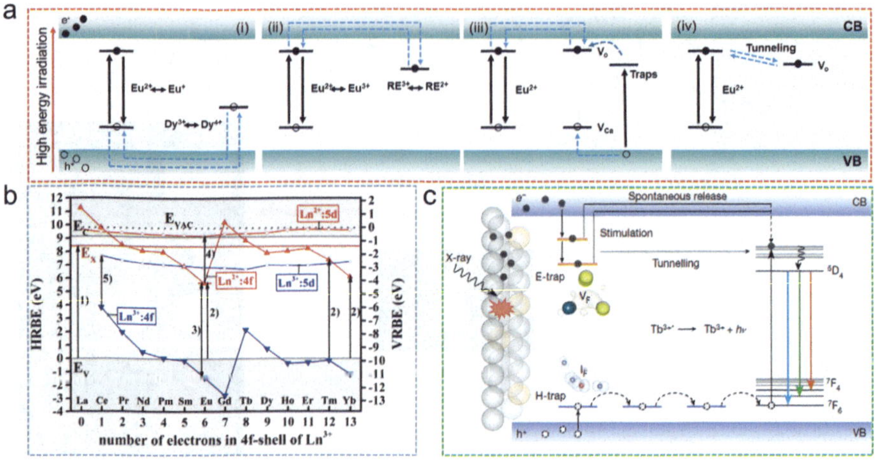

Fig. 2 Schematic illustration of models for X-ray-activated PersL. **a** Recombination accomplishing through several different processes after the cessation of excitation: (i) hole transfer model, (ii) energy transfer model, (iii) electron trap model, and (iv) quantum tunneling model. **b** Host-referred binding energy (HRBE) and vacuum-referred binding energy (VRBE) schemes of Ln^{2+}/Ln^{3+} dopants in YPO_4. **c** Proposed X-ray-activated PersL mechanism of Tb^{3+}-doped $NaLuF_4$ nanocrystals. (Refs. [69, 79])

ground state (Fig. 2a(iii)). The above models need the participation of CB and VB, and can be classified as CB-VB band theory [74].

In some material systems, the trap location is far away from CB and VB bands, and would have no contribution to the carrier migration. In this case, the PersL can be understood by the quantum tunneling model. Figure 2a (iv) shows the quantum tunneling process in which particles tunnel through a traditionally insurmountable barrier [75–77]. This process may take place between the trap in the host and the excited states of the activator. Compared with the CB-VB model, the electrons directly tunnel from the forbidden band to the luminescent center without passing through the CB band during the energy transfer process. Afterwards, Tanabe group [78] found that Eu^{2+} occupied the Sr site in $SrAl_2O_4$:Eu^{2+}/Dy^{3+}, and only the Eu^{2+} in the strong crystal field environment had a quantum tunneling with the electron traps to produce PersL emission according to photocurrent excitation spectra. In addition, there are some other models proposed in relevant experimental results. Although some theories have been widely recognized in certain fields, they still need further in-depth research for full confirmation.

Compared with UV light, X-ray can activate the PersL material systems with a wider band gap as mentioned above. Moreover, high-energy X-ray irradiation can cause collisions between the atoms/ions inside the host and high-energy electrons, and produce more defects. The generation of these defects is an indispensable prerequisite for the interpretation of X-ray-activated PersL. In 2021, Ou et al. [79] report an observation of 30-day-long PersL in lanthanide-doped $NaLuF_4$ nanoparticles under low-dose X-ray activation, and the PersL mechanism is shown in Fig. 2c. They believe that the X-ray energy is mainly absorbed by Lu atoms, which produce high-energy electrons due to the photoelectric effect. The elastic collision between large momentum X-ray photons and F^- ions leads to a deviation of F^- ions from their original sites in the lattice, forming Frankel defects including electron traps and hole traps which could trap high-energy electrons. Due to the difference in displacement distance of F^- ion, the trap depth varies from deep to shallow. When the external source of excitation is removed, the electrons in the shallow trap can spontaneously escape to the CB band and then return to the excited states of lanthanide ions with resultant PersL emission. While the electrons in the deep trap may be transferred to the excited states of the luminescence center by quantum tunneling, most of them need the help of external excitation conditions (such as light excitation) for release. The X-ray-activated PersL mechanism is similar to that with UV activation, except for producing new defects under X-ray irradiation. Similar PersL luminescence was observed in lanthanide-doped $NaYF_4$ nanoparticles [54], which can be ascribed to the transition of the trapped electrons from the CB band to the excited levels of lanthanide ions. In addition, due to the different HRBE and VRBE of lanthanide ions in $NaYF_4$, the emission intensity also varies for different lanthanide ions. In short, the current research on the mechanism of X-ray-activated PersL still concentrates on the generation, separation, and recombination process of electron–hole pairs. More general mechanisms still need further investigation in future.

3 Controlling X-Ray-Activated PersL Properties

Tunable X-PersL performance is very important for diversities of applications. Here, we mainly summarize the research progress in tuning PersL emission wavelengths, emission intensity, and duration times, together with the novel core–shell design for multiple optical responses.

3.1 Tuning X-Ray-Activated PersL Wavelengths

Lanthanide ions have abundant energy levels, providing a convenient method to produce rich emission bands for PersL materials [80–82]. The emission colors of early materials are mainly in the visible range, including blue, green, and red emissions (Fig. 3a). Pr^{3+}-doped $Ca_5Ga_6O_{14}$ phosphor is able to produce blue PersL with an emission peak of 485 nm (Pr^{3+}: $^3P_0 \rightarrow {}^3H_4$ transition) (Fig. 3b) [83]. Tb^{3+}-doped $NaYF_4$ nanoparticles can generate green PersL emission over more than 240 h (Fig. 3c) [84]. In addition, the introduction of transition metal elements (such as Mn^{2+}, Cr^{3+}, and Bi^{3+}) helps to expand the emission colors to cyan, yellow, and deep red gamut due to their d-d transition property. For instance, $LaGaO_3$:Sb^{3+}/Cr^{3+} phosphor is able to generate deep red PersL emission over 500 h under activation of low-dose X-ray, and successful realization of nano-sized material is much suitable for imaging in mouse tissues (Fig. 3d) [62]. By doping Mn^{4+} in $Sr_2Ca_{0.9}La_{0.1}WO_6$, its red PersL emission was observed. The TL spectra imply that X-ray irradiation results in a change in the trap depth, which is conducive to the enhancement of PersL emission [85]. With the increasing demand for ultraviolet light in killing bacteria, the design of ultraviolet PersL materials also gains attention. $Sr_2P_2O_7$:Gd^{3+} phosphor can realize a UVB PersL emission (peaking at 312 nm) over 24 h after being charged under an X-ray beam (Fig. 3e). In bright indoor environments, the UVB PersL light from charged $Sr_2P_2O_7$:Gd^{3+} phosphors can be clearly recorded by a UVB camera [86]. Zhou et al. [87] recently reported a heterovalent doping strategy to achieve X-ray-activated deep violet foreign PersL emission of Pr^{3+} in deficient Li_2CaGeO_4 matrix. In addition, the incident visible light can be converted into deep ultraviolet emission for sterilization, providing a new method for the development of UV light sources for sterilization (Fig. 3f).

Core–shell structure is also possible to design PersL materials with tunable colors. In contrast to traditional powder materials with possible cross-relaxation between different luminescent centers; the core–shell structure can spatially separate different types of emission centers to achieve their respective emission properties. Lei et al.

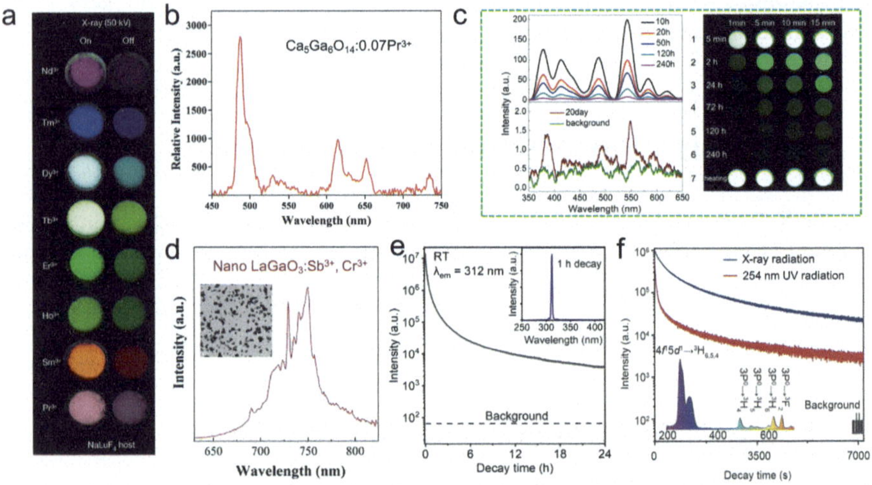

Fig. 3 Tuning PersL emission spectra of X-ray-activated materials. **a** Radioluminescence and PersL images of Ln-doped $NaLuF_4$ nanocrystals (Ln = Nd^{3+}, Tm^{3+}, Dy^{3+}, Tb^{3+}, Er^{3+}, Ho^{3+}, Sm^{3+} and Pr^{3+}). **b** X-ray-activated PersL emission spectra of $Ca_5Ga_6O_{14}$: $0.07Pr^{3+}$. **c** PersL spectra of the sample recorded at 10th hour, 20th hour, 50th hour, 120th hour, 240th hour, and 20th day of PersL attenuation (left) and emission images of four sample disks taken at different PersL times (5 min to 240 h) after X-ray irradiation for 1 to 15 min (right). **d** Emission spectrum of $LaGaO_3$:Sb^{3+},Cr^{3+} phosphor under X-ray irradiation (50 kV, 60 μA). Inset: TEM image of the sample (Scale bar, 200 nm). **e** UVB PersL decay curve of the $Sr_2P_2O_7$:Gd^{3+} phosphor monitored at 312 nm at room temperature after X-ray irradiation for 5 min. Inset: PersL spectrum of Gd^{3+} after switching off excitation for 1 h. **f** Room temperature PersL decay curves of Li_2CaGeO_4:$0.006Pr^{3+}$ after excitation by an X-ray beam (10 W, blue line) and a UV low-pressure mercury lamp for 5 min. Inset: PersL spectrum of Li_2CaGeO_4:$0.006Pr^{3+}$. (Refs. [62, 79, 83, 84, 86, 87])

[88] recently prepared a series of lanthanide-doped core–shell nanoparticles, such as $NaLuF_4$:15Gd/15 Tb@$NaLuF_4$:15Gd/0.5Sm. Because Tb^{3+} and Sm^{3+} have different PersL decay rates, the resultant PersL emission color changes from yellow to green over time. Recently, it was found that NIR PersL emission holds great advantages for in vivo imaging of biological tissues. Pei et al. [57] prepared a set of lanthanide-doped PersL nanoparticles with emission wavelengths in the second near infrared (NIR-II) window. The luminescence center, size, matrix phase, and core–shell structure of Ln-PLNPs were optimized to extend the PersL signal over 72 h, and multiple NIR PersL emissions were observed in such flexible core–shell structure samples.

3.2 *Modulating X-ray PersL Intensity and Duration Time*

It is important to enhance PersL emission intensity and increase duration for PersL materials toward frontier applications. Changing the number of luminescence centers is a direct way to increase the PersL intensity. In addition, tuning the energy level positions of defects by doping the auxiliary ions can also improve the intensity and duration of PersL. It was found that introducing Li^+ in Zn_2GeO_4:Mn^{2+} PLNPs results in enhanced PersL emission under X-ray activation, and the DFT calculation showed that Li^+ doping forms a defect energy level which is more suitable for forming excited states of Mn^{2+}. This makes the energy transfer from the defect to $^4T_1$ level of Mn^{2+} easier, thus increasing the emission intensity and duration, as shown in Fig. 4a [89]. Further, introducing Zr^{4+} ions into Mn^{2+}-doped perovskite phosphor can markedly enhance its red PersL performance. The TL spectra (Fig. 4b) showed that the introduction of Zr^{4+} is helpful in storing high-energy electrons by increasing the trap depth, and other impurity ions such as Ti^{4+}, Hf^{4+}, and Sn^{4+} have a similar result [65]. In addition, the PersL performance can be effectively improved by introducing defects indirectly via increasing the concentration of cations in a matrix. Recently, it has been reported that the addition of an excessive proportion of Na^+ during the synthesis process of $NaLuF_4$ nanoparticles can introduce appropriate interstitials in the $NaLuF_4$ matrix lattice. The presence of Na^+ interstitials promotes the formation of anionic Frankel defects and greatly enhances the X-ray-activated PersL of lanthanide ions (Tb^{3+}, Dy^{3+}, Pr^{3+}, and Sm^{3+}) [88]. Due to the small size and large surface area-to-volume ratio of nanoparticles, there are many surface defects, especially in lanthanide-doped fluoride nanoparticles. Excessive surface defects will cause serious quenching of the luminescence center, resulting in weak PersL intensity. Similar to upconversion, it was found that the $NaYF_4$:Tb^{3+}@$NaYF_4$ core–shell nanoparticles have much better PersL performance than the core only nanoparticles under X-ray irradiation; see Fig. 4c [54]. $NaLuF_4$:Tb^{3+}@$NaYF_4$ core–shell nanoparticles show a similar result, further demonstrating the effectiveness of inert layer protection [79]. It is worth noting that there is an optimal thickness for the inert shell to facilitate the strongest PersL emission, as indicated in cubic CaF_2:Dy^{3+}@$NaYF_4$ core–shell nanoparticles (Fig. 4d) [90]. In addition, a more stable phase composition

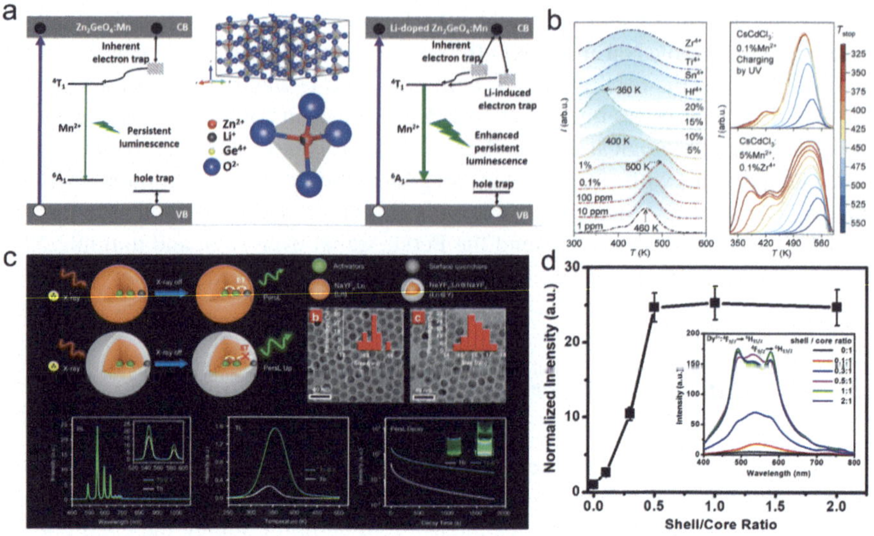

Fig. 4 Tuning X-ray-activated PersL intensity and duration times. **a** Engineering electron traps in Zn_2GeO_4:Mn^{2+} PersL nanocrystal through Li^+ doping. **b** TL curves of $CsCdCl_3$:$x Mn^{2+}$ (x = 1–100 ppm, 0.1–20%) and $CsCdCl_3$:5%Mn^{2+},0.1%R^{4+} (R = Hf, Zr, Sn, and Ti) (left) and results of the partial cleaning ($T_{max} - T_{stop}$) experiments (right). The TL curves of $CsCdCl_3$:0.1%Mn^{2+} and $CsCdCl_3$:5%Mn^{2+},0.1%Zr^{4+} recorded after thermal cleaning at elevating temperatures. **c** Schematic of the core–shell structure and possible PersL enhancement mechanism, including TEM images, RL spectra, TL curves, and PersL decay curves of $NaYF_4$:Tb^{3+} core and $NaYF_4$:Tb^{3+}@$NaYF_4$ core–shell nanoparticles monitored at the range of 400–750 nm. **d** Integrated PersL intensity of CaF_2:Dy^{3+}@$NaYF_4$ core–shell nanoparticles normalized to that of CaF_2:Dy^{3+} core. Inset: PersL spectra of CaF_2:Dy^{3+}@$NaYF_4$ core–shell nanoparticles with different shell thicknesses. (Refs. [54, 65, 89, 90])

is conducive to emission enhancement. Compared with the cubic phase, the hexagonal phase $NaYF_4$ has a higher synthesis temperature and a more stable structure, and results in better X-ray-activated PersL performance [57].

3.3 *Time-Dependent Color-Tuning of X-Ray-Activated PersL*

As one of the interesting characteristics, the emission intensity of PersL materials shows a gradual decline over time after removing the source of excitation. Different emission centers may result in a change in their decay rates. Hence, the emission color would be tuned by altering observation time for a PersL material system with multiple luminescent centers or evenly for a mixture of different kinds of PersL materials. However, simple mixing does not work well to realize good uniformity of the material system, so it is necessary to achieve this performance at the nanoscale.

It shows to be a good way to combine varieties of PersL nanoparticles by means of organic loading, such as metal–organic framework (MOF) materials combined with fluoride nanoparticles (MOF@NPs), which could modulate emission spectra greatly over a wide wavelength range. A variety of lanthanide nanoparticles containing fluorine can be connected to the charged and negatively charged MOF surface, as shown in Fig. 5a. By using the different decay curves of the PersL of different luminescence centers (Fig. 5b), an emission color change can be obviously observed with time by a proper combination (Fig. 5c). The MOF@NPs composites are used for the first time to realize full X-ray-activated optical luminescence toward multiple organic molecule sensing by using the spectral overlap of their absorption profiles and the X-ray-excited luminescence of specific nanoparticles that facilitates efficient energy transfer between them [91]. In addition, the core–shell nanostructure design can also be used to combine the luminescence center, especially in the $NaLnF_4$ matrix. For instance, by doping Tb^{3+} ions in $NaLuF_4$ core and other ions in shell, such a core–shell nanostructure is able to enable a better PersL performance of Tb^{3+} than other ions, resulting in a change of PersL color over time, as shown in Fig. 5d [88]. Interestingly, despite a continuous decrease with time for PersL luminescence, their emission intensity ratio is almost constant with time for some lanthanide ions, especially in NIR-II emission. Based on this phenomenon, Zhang's group found that the ratio of long afterglow signals of different wavelength channels in $NaYF_4$ core–shell nanoparticles was not affected by sample concentration, temperature, and time. By changing the thickness of the nanoparticle luminescent layer and inert layer, it is helpful to construct a ratio PersL signal coding library spanning two orders of magnitude. The stable ratio PersL signal solves the coding difficulties caused by the decay of the PersL signal over time, and the addition of dynamic information can further improve the encoding capacity of biological multi-detection and live biological information encryption [57].

3.4 *Multi-Mode Luminescent PLNPs*

Core–shell nanostructure has recently become a powerful tool in the research of luminescent materials because of its tunable luminescence properties. However, there are few reports on core–shell PLNPs that combine X-ray-activated PersL properties and other luminescence properties, due to the limit in the synthesis techniques of core–shell nanoparticles. The current reports mainly focus on lanthanide-doped rare earth/alkaline earth metal fluorides, such as lanthanide fluoride $NaLnF_4$ systems.

In general, UV light cannot directly activate the $NaLnF_4$ matrix, while X-rays can directly activate the CB band of the materials to produce electron motion. Therefore, reasonable material structure design can enrich the down-shifting emissions of the luminescent center while maintaining the PersL of X-ray activation. For example, $NaLuF_4$:15%Gd/15%Tb@$NaLuF_4$:15%Gd/10%Ce/Ln (Ln = Dy, Pr, and Sm) core–shell nanoparticles have both down-shifting and PersL emissions, which can promote the development of superior X-ray PLNPs with rich emission bands

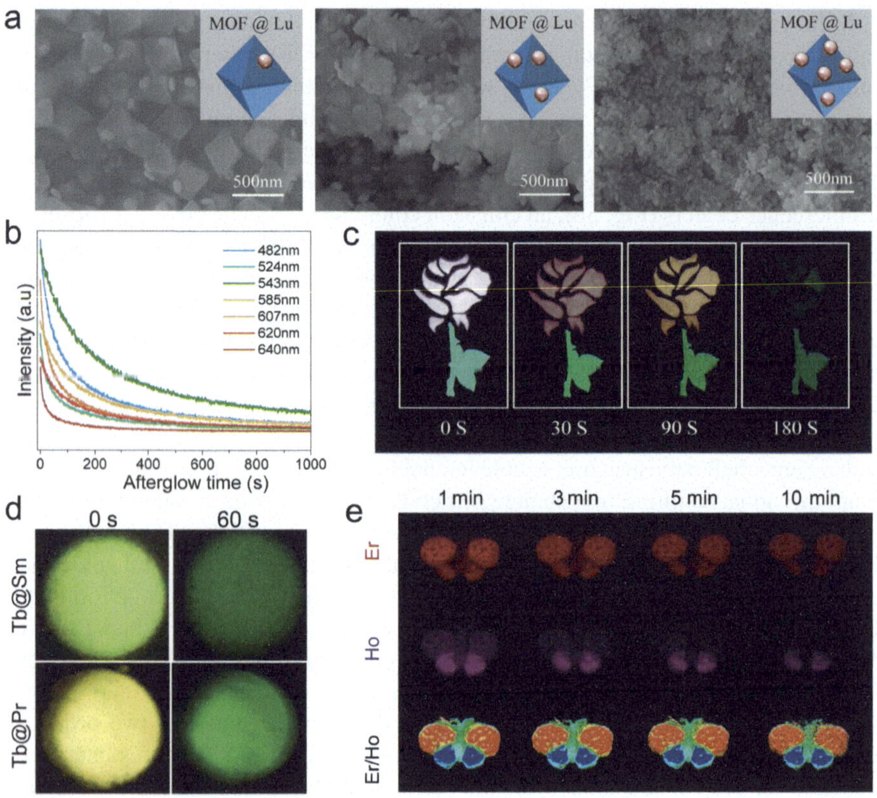

Fig. 5 Time-dependent color-change in X-ray-activated PersL. **a** TEM images of the MOF@$NaLuF_4$ composites with different $NaLuF_4$ contents of 0.04 mmol (left), 0.2 mmol (middle), and 0.4 mmol (right). Scale bar, 500 nm. **b** X-ray-activated PersL decay curves measured at different emission wavelengths. **c** X-ray-activated PersL photographs of the patterns taken at different time intervals (0–180 s). The patterns were pre-irradiated by an X-ray source (50 kV, 5 min). **d** X-ray-activated PersL photographs with different delay times for Tb@Sm (top) and Tb@Pr (down) core–shell nanoparticles. **e** NIR-II PL and ratiometric PL images of a PDMS butterfly pattern made of multilayer Ho/Er-doped PLNPs ($NaYF_4$:1%Ho@$NaGdF_4$@$NaYF_4$:3%Er@$NaGdF_4$). Scale bar, 5 mm. Imaging exposure time, 10 s. (Refs. [57, 88, 91])

[88]. Through appropriate selection of activator and sensitizer, it is easy to realize upconversion and down-shifting emissions in PersL phosphors. Recently, the combination of multiple luminescent functions on the nanoscale has been further realized in core–shell nanoparticles by spatially separating activator ions [92, 93]. However, to our best knowledge there is still no report on the combination of X-ray PersL and upconversion, which is an important field for the near future.

4 Emerging Applications of X-Ray-Activated PersL

The X-ray-activated PersL materials produce rich emission bands from visible to near-infrared ranges, greatly promoting their applications in many frontier fields of biomedical diagnosis and therapy, information encryption, and X-ray imaging.

4.1 Biomedical Diagnosis and Therapy

For biomedical diagnosis and treatment, the background fluorescence of biological tissue will greatly affect the signal-to-noise ratio and sensitivity of imaging. In contrast to upconversion or down-shifting nanoparticles, the PersL nanomaterials can achieve spontaneous radiation without activation, avoiding the interference of background fluorescence. In addition, compared with UV light as an excitation source, the X-ray penetration depth is much high, and thus is helpful in solving the problem of in vitro activation.

In 2011, Wu et al. [29] found that water-soluble Eu^{2+}- and Dy^{3+}-doped $Ca_{1.86}Mg_{0.14}ZnSi_2O_7$ PLNPs can be used for fluorescence resonance energy transfer inhibition of alpha-fetoprotein (AFP) which was secreted during cancer cell growth. Figure 6a shows the results of polythylenimine-coated PLNPs binding to gold nanoparticles coated with AFP antibodies to act as a sensitive and specific long-lasting luminescent probe for detecting AFP in serum samples and imaging AFP secreted during the cancer cell growth. In addition to X-ray-activated luminescence, the ZnS:Cu,Co nanoparticles conjugated with photosensitizer tetrabrodamine-123 (TBrRh123) also show a PersL emission that can be used as a light source for photodynamic therapy (PDT). By comparison to TBrRh123 alone or ZnS:Cu,Co nanoparticles, the ZnS:Cu,Co-TBrRh123 coupling showed low dark toxicity but still with high X-ray-induced toxicity to human prostate cancer cells [94]. Subsequently, it was found that $ZnGa_2O_4:Cr^{3+}$ PLNPs have effective X-ray-activated NIR emission and rechargeable activation, and both excitation and emission have high penetration in vivo (Fig. 6b). After stopping the X-ray excitation source, these PLNPs exhibit NIR emission over 6 h. Moreover, such PLNPs can be easily reactivated by a soft X-ray light source at a low excitation power (45 kVp, 0.5 mA) to recover bioimaging signals even at a depth of 20 mm in vivo (Fig. 6c). Renewable in vivo whole-body bioimaging by intravenous/oral PLNPs has also been successfully achieved after in situ X-ray activation. This result showed that NIR-emitted PLNPs can enable X-ray bioimaging of deep tissues in the body [60].

Recently, Pei et al. [57] investigated the X-ray-activated PersL property of $NaY(Gd)F_4$:X (X = Nd^{3+}, Ho^{3+}, Tm^{3+} or Er^{3+}) nanoparticles. In addition to visible and NIR-I PersL, the PersL in the NIR-II region (Nd^{3+} at 1064 nm, Ho^{3+} at 1180 nm, Tm^{3+} at 1475 nm and Er^{3+} at 1525 nm) was also available. Compared to NIR-I, NIR-II light is less affected by tissue, and thus provides a way for the ultra-high contrast sustained luminescence in deep tissue imaging. As a proof of concept, the

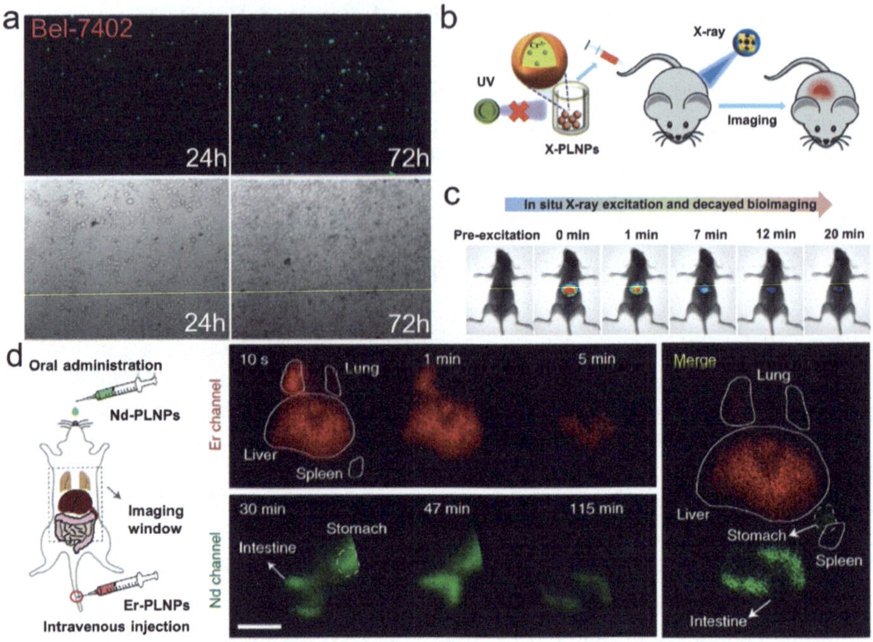

Fig. 6 NIR PersL for high-resolution animal imaging. **a** Fluorescence images of Bel-7402 cells stained with the FRET inhibition probe after the cells were cultured for 24 and 72 h (top) and corresponding bright-field images (down). **b** Schematic of in vivo whole-body PersL bioimaging of a mouse after intravenous injection of PEG-modified X-PLNPs under in situ X-ray excitation. **c** In vivo whole-body bioimaging of intravenously injected mouse after ceasing X-ray excitation (45 kVp, 5 min). **d** Dual-channel in vivo PL imaging of organs in a living mouse. Nd-PLNPs were used for the gastrointestinal tract through oral administration, and Er-PLNPs for the visceral organs (lung, liver, and spleen) by a tail vein injection. Scale bar, 1 cm. (Refs. [29, 57, 60])

authors pre-irradiated Er-doped PLNPs with X-rays at high power and then stored at -20 °C to minimize energy loss from the traps. After injection into mice, it was found that the sustained luminescence pattern of NIR-II shows a fluorescence pattern with an acuity of ~0.83 times and a signal-to-noise ratio of ~3.7 times on mouse capillary imaging, which highlights the sustained luminescence imaging ability of NIR-II (Fig. 6d).

4.2 Information Storage and Encryption

The generation of the X-PersL process depends closely on the trap centers in the host material. The electrons in shallow traps are gradually released over time, while for the electrons in deep traps, their release needs external stimulation such as heat, light, and force, resulting in thermal-induced, photo-induced, and force-induced PersL emission. Therefore, X-ray-activated PersL materials hold great promise in the field of data storage and information encryption.

In 2021, the Xie group reported that $NaMgF_3$:Tb^{3+}@$NaMgF_3$ core–shell nanoparticles show excellent optical storage ability to enter deep traps under X-ray irradiation and have controllable photon emission under thermal stimulation. Because of its excellent dispersion and stability in water, these nanoparticles were further used to prepare luminescent inks. According to the design pattern, the sapphire substrate was painted with ink and placed under X-ray irradiation for energy charging (5 min); see Fig. 7a. Under X-ray irradiation, the luminous pattern can be clearly observed, and once the excitation light is turned off, the luminous pattern quickly fades. After a few seconds of heat conduction, the stored pattern in the sapphire substrate is reproduced and recorded by a digital camera. This demonstrates its application in inkjet printing optical information storage and decryption [95]. The same group also reported a three-dimensional (3D) optical information storage scheme based on the multicolor PersL nanoparticles. As shown in Fig. 7b, three luminescent inks containing different $NaYF_4$:Ln^{3+}@$NaYF_4$ (Ln = Tb, Dy and Ho) core–shell nanoparticles were used to print overlapping patterns on the same piece of glass (information writing step). Under X-ray irradiation, the spectral information can be broken down into three independent sets of patterns, which can be interpreted into three readable quick-response code spectral information, allowing 3D information to be stored on a single-layer recording medium [54].

In addition, Han group [89] demonstrated a simple strategy that amplifies the sustained luminescence of light and X-ray charging in monodisperse Zn_2GeO_4. The nanoparticles can be stimulated by UV-light and X-ray excitations independently, so the spatiotemporal 2D PersL security of light/X-ray orthogonal encryption was realized by using these nanoparticles. As shown in Fig. 6c, they covered the top of the polydimethylsiloxane (PDMS) with a paper mask, made on a piece of black paper with a smiling face, and the center of the paper is a square picture with $PbCl_2$ solution. Black paper blocks UV rays while allowing X-ray to pass through, and the $PbCl_2$ layer can effectively attenuate and block X-rays. With this paper mask, two different PersL encrypted images can be clearly observed by matching orthogonal excitation resources (UV light and photons), which is difficult to duplicate showing promise for advanced anti-counterfeiting applications.

4.3 Advanced X-ray Imaging

Since the 1980s, X-ray storage phosphors with PersL characteristics have been widely developed as imaging boards for commercial computed radiography (CR) technology, which realizes the visualization of X-rays by optically or thermally stimulated luminescence. Recently, CR has been phasing out due to the presence of digital X-ray imaging which uses flat-panel X-ray detectors to convert X-rays into real-time electrical signals. However, these flat, non-flexible X-ray detectors are difficult to be applicable in high-resolution X-ray imaging of curved or irregular 3D objects,

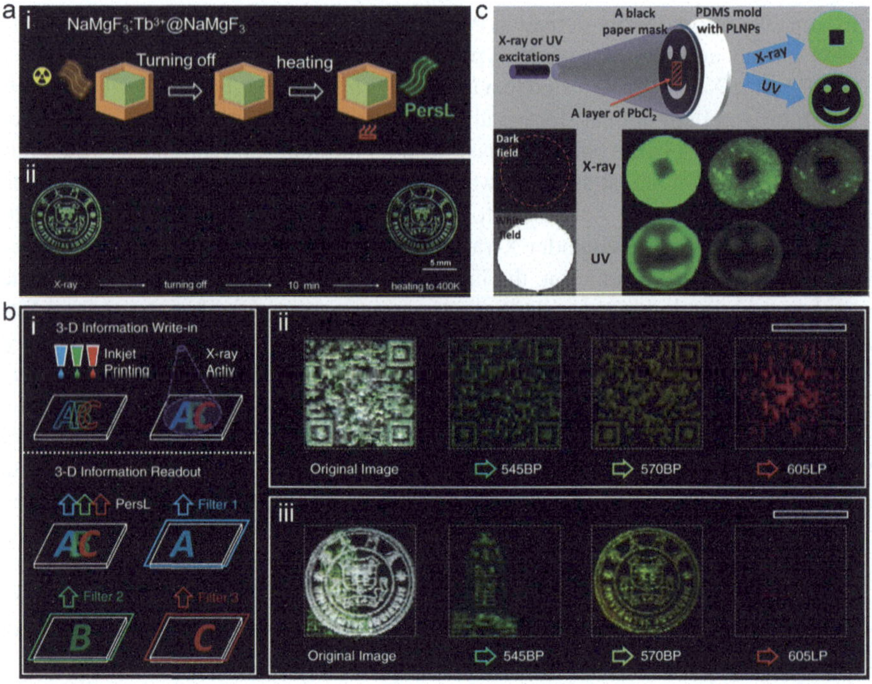

Fig. 7 PersL for anti-counterfeiting and information encryption. **a** Schematic of using core–shell structure to improve PersL intensity (i) and PersL emission photographs during X-ray irradiation at the moment of turning off and with a 10-min delay time and heat treatment (400 K) (ii). **b** Schematic of application in 3D information storage (i) and the original images and monochromatic images after passing through 545BP, 570BP, and 605LP filters (ii and iii). Scale bars, 100 μm. **c** Display of distinct patterns made of PDMS mold embedded with Li-doped Zn_2GeO_4:Mn^{2+} PLNPs after the X-ray or UV excitation through an X-ray/UV mask. (Refs. [54, 89, 95])

hindering their further development in security inspection, industrial inspection, and medical diagnosis.

In 2021, the Liu group reported the latest results of nanoscintillator research. They built a 3D radiographic flexible X-ray detector based on these materials, and developed a new technique for X-ray luminescent extended imaging (Xr-LEI); see Fig. 8a. By embedding nanocrystals into a highly stretchable silicone resin to prepare a flexible film, a high resolution of about 25 μm was obtained, much higher than that obtained using conventional flat-panel detectors. In particular, Fig. 8b shows the technique of performing radiography on highly curved 3D objects after terminating X-ray irradiation, which is not available for conventional flat-panel X-ray detectors or synchrotron-based X-ray microscopes [79]. Combined with other sensing properties, X-ray imaging can be further demonstrated in everyday life applications. Recently, Hou et al. [96] developed an ingestible X-ray dosimeter capsule based on $NaLuF_4$:Tb^{3+}@$NaLuF_4$ nanoparticles (Fig. 8c and d), which exhibit enhanced luminescence in the face of increased X-ray radiation when ingested and passed along the

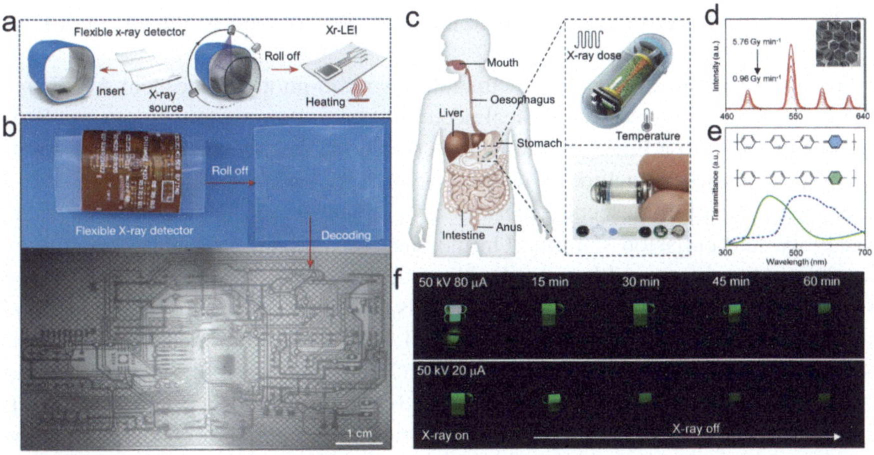

Fig. 8 X-ray-activated PersL nanoparticles for high-resolution X-ray luminescence extension imaging. **a** Schematic of 3D electronic imaging by a nanoscintillator-integrated, flexible detector and **b** Xr-LEI of a 3D electronic board using a prototype $NaLuF_4:Tb^{3+}$(15 mol%)@$NaYF_4$-based detector (50 kV, 80 °C). **c** Schematic and real view of the dosimeter for simultaneous in situ detection of absorbed radiation dose, pH, and temperature in the gastrointestinal tract. **d** Radioluminescence spectra of $NaLuF_4:Tb^{3+}$@$NaYF_4$ nanoscintillators under irradiation at room temperature at different dose rates (0.96–5.76 Gy min^{-1}). Inset: TEM image of the as-synthesized nanoscintillators. Scale bar, 100 nm. **e** Transmission spectra of the polyaniline film recorded in acidic and basic solutions. **f** Radioluminescence and PersL photographs of $NaLuF_4:Tb^{3+}$(15 mol%)@$NaYF_4$ nanocrystals embedded in an optical fiber under X-ray irradiation at 50 kV with tube currents of 80 μA (16.68 mGy/min) and 20 μA (4.17 mGy/min). (Refs. [79, 96])

gastrointestinal tract. The sensors inside the capsule measure the light emitted by the nanoscintillator to determine the radiation delivered to the target area. Meanwhile, the fluid module allows the collection of gastric fluid for pH detection through a film that shows a color change according to pH, as shown in Fig. 8e. The photoelectric signals from both sensors are processed by a microcontroller circuit board, which sends the information to a mobile phone application via Bluetooth technology and an antenna. The low cost of X-ray PersL nanoparticles and ease of implementation of X-ray imaging technology open up a promising path for non-invasive medical radiology and industrial flaw detection.

4.4 X-ray Radiation Dose Monitoring

Radiation dose verification in radiotherapy is a powerful technique to determine the dose delivered to irradiated tissue. While modern radiotherapy still produces side effects, including desquamation, acute gastrointestinal erythema, and necrosis, which are associated with overexposure and inaccurate X-ray dosing, X-ray-activated PersL nanomaterials can be used to monitor X-ray doses because their PersL intensity usually shows a linear relationship with the dose. Recent reports of such work have received widespread attention.

For instance, Liu et al. [97] documented a radio-photoluminescent MOF that allows for X-ray-induced free radical accumulation on an organic connector. This further gives rise to a linear relationship between the radiation dose and the intensity of the emission signature by using a metal center as an X-ray absorber. MOF-based radio-photoluminescent materials have advantages in dose detection range, reusability, and emission stability. On the basis of electronic structure and energy diagram studies, it is possible to predict the rational design and substantial expansion of radio-photoluminescent candidate materials. To further improve performance and find new candidates besides MOF materials, Luo et al. [98] reported a highly sensitive dosimetry technique based on $NaLuF_4$:Tb X PersL nanoparticles for monitoring X-ray dosimetry in clinical patients during radiotherapy. In this work, a set of PersL films consisting of lanthanide-doped nanoparticles were prepared by optimizing the nanoparticle concentration and spin coating speed. Such films emit uniform and continuous luminescence after stopping X-ray irradiation, making it possible to quantitatively report ionizing radiation. In addition, it also demonstrated that terrain-sustained luminescence dosimetry can record complex clinical dose distributions for dose validation and radiotherapy planning.

5 Conclusion

In summary, the progress of X-ray-activated PersL materials, including emission mechanism, emission modulation, structural design, and cutting-edge applications was reviewed in detail. Despite the impressive progress X-PLNP has made, there are still a number of issues that need to be addressed in the future:

(i) *Develop a new class of PersL materials*: The mainstream materials focus on the modification system of traditional PersL phosphors and a few fluoride nanocrystals, which greatly limits the performance and application of X-PersL materials. How to break the limitations of the material system has become a key problem, which requires the support of a large number of theoretical and experimental research.

(ii) *In-depth understanding of PersL mechanism*: At present, the relevant mechanism is still referring to the activation theory of visible light PersL, namely the energy transfer between trap level and luminescent center, and the difference lies in the generation process of defects. However, the underlying mechanism is not clear, especially since there is no more direct evidence to prove that the displacement of ions leads to the formation of Frenkel defects.

(iii) *Expand PersL emission ranges*: Recent studies on X-ray-activated PersL mainly focus on the materials with doping of lanthanide ions or transition metal ions. The bottleneck is that the emission of a single luminescence center cannot cover the full spectral range, and the co-doping of multiple ions will produce serious luminescence quenching. It is necessary to design reasonable nanostructures that can intelligently control PersL property.

(iv) *Increase emission intensity and duration time*: The methods to enhance the emission intensity and duration times are concentrated in the doping of heterogeneous ions; however, this method has no corresponding prediction method and needs a lot of experimental attempts. In addition, the doping method of heterogeneous ions is not desirable for nanoparticles, which will greatly destroy the uniform morphology of nanoparticles. Although PersL performance can be effectively improved by coating an inert shell for nanoparticles, the improvement is still limited.

(v) *Explore emerging applications*: The present applications of X-ray-activated PersL materials mainly include information security and anti-counterfeiting. X-ray-activated PLNPs can be further used for in vivo biological imaging and planar X-ray high-resolution imaging by taking advantage of their small particle size. Finding new application scenarios becomes a goal for future research.

Finally, we believe that the research on PersL will benefit much from the design and development of new nanomaterials, the demonstration of more accurate and general PersL theories, and the exploration of frontier applications, which need closer cooperations between material chemists, photo-physicists, and biologists.

References

1. T. Matsuzawa, Y. Aoki, N. Takeuchi, Y. Murayama, J. Electrochem. Soc. **143**(8), 2670–2673 (1996)
2. T. Maldiney, A. Bessière, J. Seguin, E. Teston, S.K. Sharma, B. Viana, A.J.J. Bos, P. Dorenbos, M. Bessodes, D. Gourier, D. Scherman, C. Richard, Nat. Mater. **13**(4), 418–426 (2014)
3. R.E. Rojas-Hernandez, F. Rubio-Marcos, M.Á. Rodriguez, J. Francisco Fernandez. Renew. Sust. Energ. Rev. **81**, 2759–2770 (2018)
4. R. Kabe, C. Adachi, Nature **550**, 384–387 (2017)
5. X. Qin, X. Liu, W. Huang, M. Bettinelli, X. Liu, Chem. Rev. **117**(5), 4488–4527 (2017)
6. W. Jia, H. Yuan, L. Lu, H. Liu, W. Yen, J. Lumin. **76**, 424–428 (1998)
7. L. Liang, J. Chen, K. Shao, X. Qin, Z. Pan, X. Liu, Nat. Mater. **22**, 289–304 (2023)
8. Y. Li, M. Gecevicius, J. Qiu, Chem. Soc. Rev. **45**(8), 2090–2136 (2016)
9. H. Chen, R. Lian, M. Yin, L. Lou, W. Zhang, S. Xia, J. Krupa, J. Phys. Condens. Matter **13**, 1151–1158 (2001)
10. Q. Zeng, Y. Wu, H. Cheng, K. Shao, T. Hu, Z. Pan, J. Inorg. Mater. **38**(8), 901–909 (2023)
11. C. Chang, J. Xu, L. Jiang, D. Mao, W. Ying, Mater. Chem. Phys. **98**(2–3), 509–513 (2006)
12. X. Teng, W. Zhuang, Y. Hu, C. Zhao, H. He, X. Huang, J. Alloys Compd. **458**(1–2), 446–449 (2008)
13. J.S. Kim, J.S. Kim, H.L. Park, Solid State Commun. **131**(12), 735–738 (2004)
14. W. Zhang, J. Zhang, Y. Li, Z. Chen, T. Wang, Appl. Surf. Sci. **256**(14), 4702–4707 (2010)
15. Y. Zhang, D. Chen, W. Wang, S. Yan, J. Liu, Y. Liang, Inorg. Chem. Front. **7**, 3063–3071 (2020)
16. H. Cai, Z. Song, Q. Liu, Mater. Chem. Front. **5**, 1468–1476 (2021)
17. S.K. Singh, RSC Adv. **4**(102), 58674–58698 (2014)
18. Y. Zhuang, Y. Katayama, J. Ueda, S. Tanabe, Opt. Mater. **36**(11), 1907–1912 (2014)
19. X. Fu, S. Zheng, J. Shi, Y. Li, H. Zhang, J. Lumin. **184**, 199–204 (2017)

20. S. Wang, X. Liu, B. Qu, Z. Song, Z. Wang, S. Zhang, F. Wang, W. Geng, Q. Liu, J. Mater. Chem. C **7**(40), 12544–12551 (2019)
21. Y. Liang, F. Liu, Y. Chen, X. Wang, K. Sun, Z. Pan, J. Mater. Chem. C **5**(26), 6488–6492 (2017)
22. X. Qin, J. Wang, Q. Yuan, Front. Chem. **8**(608578), 1–8 (2020)
23. M. Xu, J. Liu, X. Su, Q. Zhou, H. Yuan, Y. Wen, Y. Cheng, F. Li, Sci. China Chem. **64**, 2125–2133 (2021)
24. L. Li, T. Li, Y. Hu, C. Cai, Y. Li, X. Zhang, J. Qiu, Light: Sci. Appl. **11**(1), 51, 1–8 (2022)
25. J. Liu, T. Lécuyer, J. Seguin, N. Mignet, D. Scherman, B. Viana, C. Richard, Adv. Drug Delivery Rev. **138**, 193–210 (2019)
26. S. Ding, H. Guo, P. Feng, Q. Ye, Y. Wang, Adv. Opt. Mater. **8**(18), 2000097, 1–12 (2020)
27. H. Jiang, Y. Jia, T. Qu, Yao Pan, K. Yang, H. Luo, ACS Omega **4**(4), 6923–6930 (2019)
28. Q. Chermont, C. Chanéac, J. Seguin, F. Pellé, S. Maîtrejean, J. Jolivet, D. Gourier, M. Bessodes, D. Scherman, Proc. Natl. Acad. Sci. **104**(22), 9266–9271 (2007)
29. B. Wu, H. Wang, J. Chen, X. Yan, J. Am. Chem. Soc. **133**(4), 686–688 (2011)
30. A. Abdukayum, J. Chen, Q. Zhao, X. Yan, J. Am. Chem. Soc. **135**(38), 14125–14133 (2013)
31. J. Shi, X. Sun, J. Li, H. Man, J. Shen, Y. Yu, H. Zhang, Biomaterials **37**, 260–270 (2015)
32. Z. Li, Y. Zhang, X. Wu, X. Wu, R. Maudgal, H. Zhang, G. Han, Adv. Sci. **2**(3), 1500001, 1–6 (2015)
33. S. Sun, H. Wang, X. Yan, Acc. Chem. Res. **51**(5), 1131–1143 (2018)
34. T. Lécuyer, J. Seguin, A. Balfourier, M. Delagrange, P. Burckel, R. Lai-Kuen, V. Mignon, B. Ducos, M. Tharaud, B. Saubaméa, D. Scherman, N. Mignet, F. Gazeau, C. Richard, Nanoscale **14**, 15760–15771 (2022)
35. W. Algar, M. Massey, K. Rees, R. Higgins, K.D. Krause, G.H. Darwish, W.J. Peveler, Z. Xiao, H. Tsai, R. Gupta, K. Lix, M.V. Tran, H. Kim, Chem. Rev. **121**(15), 9243–9358 (2021)
36. Z. Yi, Z. Luo, X. Qin, Q. Chen, X. Liu, Acc. Chem. Res. **53**(11), 2692–2704 (2020)
37. S. Wu, Y. Li, W. Ding, L. Xu, Y. Ma, L. Zhang, Nano-Micro Lett. **12**, 1–26 (2020)
38. P. Lei, J. Feng, H. Zhang, Nano Today **35**(100952), 1–32 (2020)
39. B.J. MacGowan, L.B. Da Silva, D.J. Fields, C.J. Keane, J.A. Koch, R.A. London, D.L. Matthews, S. Maxon, S. Mrowka, A.L. Osterheld, J.H. Scofield, G. Shimkaveg, J.E. Trebes, R.S. Walling, Phys. Fluids B **4**(7), 2326–2337 (1992)
40. L. Ren, B. Zheng, H. Liu, J. X-Ray Sci. Technol. **26**(1), 1–28 (2018)
41. E. Coutino-Gonzalez, D. Grandjean, M. Roeffaers, K. Kvashnina, E. Fron, B. Dieu, G.D. Cremer, P. Lievens, B. Sels, J. Hofkens, Chem. Commun. **50**, 1350–1352 (2014)
42. X. Chen, J. Song, X. Chen, H. Yang, Chem. Soc. Rev. **48**(11), 3073–3101 (2019)
43. H. Chen, X. Sun, G.D. Wang, K. Nagata, Z. Hao, A. Wang, Z. Li, J. Xie, B. Shen, Mater. Horiz. **4**, 1092–1101 (2017)
44. N.R.J. Poolton, A.J.J. Bos, J. Wallinga, J.T.M. de Haas, P. Dorenbos, L. de Vries, R.H. Kars, G.O. Jones, W. Drozdowski, J. Lumin. **130**(8), 1404–1414 (2010)
45. N.S. Bajaj, S.K. Omanwar, J. Rare Earth. **30**(10), 1005–1008 (2012)
46. M. Ma, L. Li, C. Cai, Y. Han, Y. Yang, J. Lumin. **252**(119376), 1–6 (2022)
47. C. Richard, B. Viana, Light: Sci. Appl. **11**(1), 123, 1–3 (2022)
48. I. Villa, F. Moretti, M. Fasoli, A. Rossi, B. Hattendorf, C. Dujardin, M. Niederberger, A. Vedda, A. Lauria, Adv. Opt. Mater. **8**(1), 1901348, 1–10 (2020)
49. B. Liu, W. Gan, S. Lou, R. Zou, Q. Tang, C. Wang, J. Jiao, J. Wang, J. Appl. Phys. **129**(120901), 1–10 (2021)
50. M. Jain, Manju, R. Kumar, S.O. Won, K. H. Chae, A. Vij, A. Thakur, Sci. Rep. **10**(1), 385, 1–14 (2020)
51. Y. Yang, Z. Li, J. Zhang, Y. Lu, S. Guo, Q. Zhao, X. Wang, Z. Yong, H. Li, J. Ma, Y. Kuroiwa, C. Moriyoshi, L. Hu, L. Zhang, L. Zheng, H. Sun, Light: Sci. Appl. **7**(1), 88, 1–11 (2018)
52. Z. Li, H. Li, H. Sun, J. Rare Earth. **38**(2), 124–129 (2020)
53. P. Xiong, C. Zheng, M. Peng, Z. Zhou, F. Xu, K. Qin, Y. Hong, Z. Ma, J. Am. Ceram. Soc. **103**(12), 6922–6931 (2020)
54. Y. Zhuang, D. Chen, W. Chen, W. Zhang, X. Su, R. Deng, Z. An, H. Chen, R. Xie, Light: Sci. Appl. **10**(1), 132, 1–10 (2021)

55. D. Ding, S. Li, H. Xu, L. Zhu, S. Meng, J. Liu, Q. Lin, S.W. Leung, W. Sun, Y. Li, H. Chen, A.C.S. Appl, Mater. Inter. **13**(14), 16166–16172 (2021)
56. L. Song, X. Lin, X. Song, S. Chen, X. Chen, J. Li, H. Yang, Nanoscale **9**, 2718–2722 (2017)
57. P. Pei, Y. Chen, C. Sun, Y. Fan, Y. Yang, X. Liu, L. Lu, M. Zhao, H. Zhang, D. Zhao, X. Liu, F. Zhang, Nat. Nanotechnol. **16**, 1011–1018 (2021)
58. Z. Hong, Z. Chen, Q. Chen, H. Yang, Acc. Chem. Res. **56**(1), 37–51 (2022)
59. J. Li, J. Guo, H. Li, J. Qu, J. Song, Chem. Eng. J. **406**, 126008 (2021)
60. Z. Xue, X. Li, Y. Li, M. Jiang, H. Liu, S. Zeng, J. Hao, A.C.S. Appl, Mater. Inter. **9**(27), 22132–22142 (2017)
61. Z. Xue, M. Jiang, H. Liu, S. Zeng, J. Hao, Biomaterials **263**(120384), 1–10 (2020)
62. B. Liu, R. Zou, S. Lou, Y. Gao, L. Ma, K. Wong, J. Wang, Chem. Eng. J. **404**(127133), 1–10 (2021)
63. V. Boiko, J. Zeler, M. Markowska, Z. Dai, A. Gerus, P. Bolek, E. Zych, D. Hreniak, J. Rare Earth. **37**(11), 1200–1205 (2019)
64. G. Doke, A. Antuzevics, G. Krieke, A. Kalnina, M. Springis, A. Sarakovskis, J. Lumin. **234**(117995), 1–9 (2021)
65. X. Zhou, K. Han, Y. Wang, J. Jin, S. Jiang, Q. Zhang, Z. Xia, Adv. Mater. **35**(16), 2212022, 1–9 (2023)
66. V. Abbruscato, J. Electrochem. Soc. **118**(6), 930–933 (1971)
67. H. Yamamoto, T. Matsuzawa, J. Lumin. **72**, 287–289 (1997)
68. P. Dorenbos, J. Lumin. **104**(4), 239–260 (2003)
69. Y. Zhuang, Y. Lv, Y. Li, T. Zhou, J. Xu, J. Ueda, S. Tanabe, R. Xie, Inorg. Chem. **55**(22), 11890–11897 (2016)
70. C. Chen, Y. Zhuang, D. Tu, X. Wang, C. Pan, R. Xie, Nano Energy **68**(104329), 1–9 (2020)
71. Y. Zhuang, Y. Lv, L. Wang, W. Chen, T. Zhou, T. Takeda, N. Hirosaki, R. Xie, A.C.S. Appl, Mater. Inter. **10**, 1854–1864 (2018)
72. T. Aitasalo, A. Durygin, J. Hölsä, M. Lastusaari, J. Niittykoski, A. Suchocki, J. Alloys Compd. **380**(1–2), 4–8 (2004)
73. T. Aitasalo, J. Hölsä, H. Jungner, M. Lastusaari, J. Niittykoski, J. Phys. Chem. B **110**(10), 4589–4598 (2006)
74. H. Suo, X. Zhang, F. Wang, Trends Chem. **4**(8), 726–738 (2022)
75. W. Hahn, J.M. Lentali, P. Polovodov, N. Young, S. Nakamura, J. S. Speck, C. Weisbuch, M. Filoche, Y.-R. Wu, M. Piccardo, F. Maroun, L. Martinelli, Y. Lassailly, J. Peretti, Phys. Rev. B **98**(4), 045305, 1–5 (2018)
76. J. Zhang, Y. Long, X. Yan, X. Wang, F. Wang, Chem. Mater. **28**, 4052–4057 (2016)
77. K. Eeckhout, D. Poelman, P.F. Smet, Materials **6**(7), 2789–2818 (2013)
78. J. Ueda, T. Nakanishi, Y. Katayama, S. Tanabe, Phys. Status Solidi C **9**(12), 2322–2325 (2012)
79. X. Ou, X. Qin, B. Huang, J. Zan, Q. Wu, Z. Hong, L. Xie, H. Bian, Z. Yi, X. Chen, Yiming Wu, X. Song, J. Li, Q. Chen, H. Yang, X. Liu, Nature **590**, 410–415 (2021)
80. A. Nadort, J. Zhao, E.M. Goldys, Nanoscale **8**(27), 13099–13130 (2016)
81. T. Vitova, M. A. Denecke, J. Göttlicher, K. Jorissen, J. J. Kas, K. Kvashnina, T. Prüßmann, J. J. Rehr, J. Rothe, J. Phys.: Conf. Ser. **430**, 012117, 1–4 (2013)
82. H. Dong, L. Sun, C. Yan, Nanoscale **5**(13), 5703–5714 (2013)
83. N. Li, P. Zhang, Z. Wang, Z. Wei, Z. Jiang, Y. Shang, M. Zhang, Q. Qiang, L. Zhao, W. Chen, J. Alloys Compd. **858**(157719), 1–9 (2021)
84. Y. Hu, Y. Yang, X. Zhang, X. Wang, X. Li, Y. Li, T. Li, H. Zhang, J. Phys. Chem. C **124**(45), 24940–24948 (2020)
85. J. Du, K. Li, R.D. Van, D. Poelman, H. Lin, Adv. Opt. Mater. **10**(2), 2101714 (2022)
86. Y. Zhang, Y. Liang, X. Shan, D. Chen, S. Miao, R. Shi, F. Xie, W. Wang, Inorg. Chem. **61**(50), 20647–20656 (2022)
87. X. Zhou, J. Qiao, Y. Zhao, K. Han, Z. Xia, Sci. China Mater. **65**, 1103–1111 (2022)
88. L. Lei, Y. Wang, W. Xu, R. Ye, Y. Hua, D. Deng, L. Chen, P.N. Prasad, S. Xu, Nat. Commun. **13**(5739), 1–11 (2022)

89. K. Huang, X. Dou, Y. Zhang, X. Gao, J. Lin, J. Qu, Y. Li, P. Huang, G. Han, Adv. Funct. Mater. **31**(22), 2009920, 1–9 (2021)
90. K. Huang, Z. Li, Y. Li, N. Yu, X. Gao, L. Huang, S.F. Lim, G. Han, Nano Lett. **21**(12), 4903–4910 (2021)
91. W. Xu, L. Lei, Y. Wang, P.N. Prasad, L. Chen, S. Xu, Laser Photonics Rev. **17**(5), 2200997, 1–9 (2023)
92. M. Ding, B. Dong, Y. Lu, X. Yang, Y. Yuan, W. Bai, S. Wu, Z. Ji, C.Lu, K. Zhang, H. Zeng, Adv. Mater. **32**(45), 2002121, 1–10 (2020)
93. Z. An, J. Huang, L. Yan, L. He, B. Zhou, J. Phys. Chem. Lett. **13**(39), 9007–9013 (2022)
94. L. Ma, X. Zou, W. Chen, J. Biomed. Nanotechnol. **10**(8), 1501–1508 (2014)
95. Y. Wang, D. Chen, Y. Zhuang, W. Chen, H. Long, H. Chen, R. Xie, Adv. Opt. Mater. **9**(17), 2100624, 1–8 (2021)
96. B. Hou, L. Yi, D. Hu, Z. Luo, D. Gao, C. Li, B. Xing, J. Wang, C.N. Lee, R. Zhang, Z. Sheng, B. Zhou, X. Liu, Nat. Biomed. Eng. **7**, 1242–1251 (2023)
97. H. Liu, H. Qin, N. Shen, S. Yan, Y. Wang, X. Yin, X. Chen, C. Zhang, X. Dai, R. Zhou, X. Ouyang, Z. Chai, S. Wang, Angew. Chem. Int. Ed. **132**(35), 15321–15326 (2020)
98. Z. Luo, Y. Wu, Y. Wang, D. Hu, D. Gao, Y. Ge, Z. Sheng, X. Liu, H. Zheng, Nano Today **50**(101854), 1–10 (2023)

Different Synthesis Methods of Persistent Luminescence Nanoparticles

K. R. Nagabhushana and S. Satyanarayana Reddy

Abstract This chapter demonstrates the synthesis of various morphologies of persistent luminescence nanoparticles (PLNPs) and their composites using chemical and physical methods for a range of applications such as electroluminescent displays, solid-state lighting, lamp phosphors, thermoluminescent dosimeters, scintillation, etc. In contemporary material science research, the first stage of synthesizing various composites with different morphology of PLNPs is challenging. Particularly, this chapter concentrates on the synthesis of the different morphology of aluminate PLPNs. The divalent metal (M) and trivalent rare-earth (Ln) aluminates exist in different forms, of which MAl_2O_4 and $LnAlO_3$ are the most widely used and studied due to their persistent luminescence. While synthesizing the PLNPs, many factors need to be considered, including the desired phase, the desired site of incorporation of the activator, specific grain size, and morphology for the desired application. Further, it is also important to consider the manufacturing cost, scalability, repeatability, reproducibility, and so forth. Different synthesis techniques are available to encounter the above difficulties, which are classified into two types: bottom-up approach and top-down approach. The bottom-up approach is the chemical method, which includes solution combustion, sol–gel, hydro/salvo thermal, and coprecipitation synthesis methods. Similarly, the top-down approach is the physical method, which includes high-energy ball milling, pulsed laser ablation, spray pyrolysis, and solid-state diffusion strategies which are the most important. Hence, these strategies are detailed in-depth with their advantages and limitations in the chapter. The chemical methods of synthesis of these aluminates based PLNPs are more advantageous over the physical approach to tuning the luminescence properties by tuning the morphology of the nanoparticles, incorporation of favorable impurities and/or

K. R. Nagabhushana (✉)
Department of Physics, Amity School of Applied Sciences, Amity University, Bengaluru, Karnataka 562 110, India
e-mail: nagabhushanakr@blr.amity.edu

Federal University of São Paulo, Campus Baixada Santista, CEP: 11070-100, Santos, SP, Brazil

S. S. Reddy
Department of Physics, RV Institute of Technology and Management, Bengaluru, Karnataka 560076, India

V. Kumar et al. (eds.), *Persistent Luminescence*, Progress in Optical Science and Photonics 30, https://doi.org/10.1007/978-981-97-4943-0_6

defects for the PLNPs and, importantly, more homogenous chemical composition, where luminescence properties can be tuned to the desired extent.

Keywords Solution combustion · Sol–gel · Hydro/salvo thermal · Coprecipitation synthesis methods

1 Introduction

The ratio of particle surface area to volume in nanomaterials is substantially higher than that in the bulk material [1]. It is conceivable that surface conditions such as void bands, surface flaws, and surface disorders will affect a nanomaterial's luminescence characteristics [2]. Therefore, by altering the surface of nanomaterials, it is vital to reduce surface imperfections to boost luminescence efficiency. Like surface modification, the amorphous phase at the enclosed nanocrystalline phase's surface can efficiently reduce surface flaws and non-radioactive emissions [3]. In the nanoscale, the bulk of atoms are found on the particle surfaces. The material's intrinsic qualities will alter in such a situation.

Nanotechnology has sparked a revolution because it combines physics, chemistry, materials science and engineering, biology and medicine. Due to their special chemical and physical characteristics, nanomaterials are gaining more and more attention for possible technological uses [4]. The distribution and size of nanomaterial grains, the chemical makeup of the constituent phases, the existence of interfaces—more especially, heterophase interfaces, free surfaces, or grain boundaries—and the interactions between constituent ions and atoms determine a material's characteristics. Nanomaterials of various sizes and shapes, such as nanoparticles [5], nanowires [6, 7], nanotubes [8], nanorods [9–11], and nanocomposites [12], are currently the subject of intensive research due to their potential for revolutionary technological applications. These interesting materials are expected to be used in a variety of applications, including sensors [13–15], photonic band gap materials [16], single electron transistors [17], drug/gene delivery vehicles [18, 19], solar cells [20–22], plasmonic devices [23], cosmetics [24], coating materials [25, 26], catalysis [27–30], diagnostic tools [31], therapeutic agents [32], cell imaging [18, 33], fuel cells [34–36], and non-linear optics devices [37].

Persistent luminescence nanoparticles (PLNPs) generate bright visible light for many minutes after stimulation. Persistent luminescence (phosphorescence or long-lasting phosphorescence) in solids occurs when an inorganic host material is doped with small amounts of an activator metal, which changes the electronic structure and traps charge carriers in metastable states upon excitation. The gradual de-trapping caused by thermal activation results in light from electron–hole recombination. The fabrication and characterization of new materials (PLNPs in particular) for diverse luminescent applications is the first step in modern material science research. The creation of phosphors for many optical applications, including solid-state lighting

[38–40], lamp phosphors [41], thermoluminescent dosimeters [42–45], electroluminescent displays [46], scintillation [22, 47, 48], etc., is the subject of this study area. Aluminates are the most widely used and studied matrixes of persistence luminescence. The divalent aluminates exist in different forms such as MAl_2O_4 (MA), MAl_4O_7 (MA_2), $MAl_{12}O_{19}$ (MA_6), $M_3Al_2O_6$ (M_3A), $M_{12}Al_{14}O_{33}$ ($M_{12}A_7$), etc. [49], where M is the divalent metal ion such as Mg, Ca, Sr, Ba, Mn, Co, Ni, Cu, Zn etc. (M = MO, A = Al_2O_3). Similarly, the trivalent lanthanide aluminates exist in different forms, particularly $LnAlO_3$ is one of the promising materials in material science, where Ln is the trivalent lanthanides such as La [50], Gd, Lu [51], Ce [52], Tb [53], Tm, Sm [54], Nd [55] etc. Phosphors, notably MAl_2O_4 and $LnAlO_3$, are currently in high demand for various applications. However, the morphology, size, and optical characteristics of this pigment are significantly influenced by the synthesis method used to synthesize it. Though the synthesis method appears to be facile and straightforward, one must exercise caution when picking the process to be used for the synthesis of a suitable pigment. However, there are constraints during these material syntheses, which are mostly due to the physical and chemical properties of the basic materials utilized, as well as the limitations and ambient conditions of the process of synthesis and the nature of the application [56].

Dopants (activators) employed during the material synthesis are another crucial parameter. Selecting a suitable matrix activator combination for a certain application is an interesting study topic. The f-orbital electrons of rare earths are tightly shielded from external ligands; the position of the spectral lines varies merely with the surrounding crystal field, but their intensities strongly depend on the host where the rare earth is buried [57, 58]. It is an excellent moment to develop new materials synthesis processes for MAl_2O_4 [59] and $LnAlO_3$ that may lead to innovative ideas [60–62]. Researchers have already developed more fascinating approaches to synthesizing nanoscale materials [63–66]. However, synthesizing these materials with appealing features can be rather difficult. The difficulties arise due to numerous factors that must be considered, such as the desired phase [67], incorporation of the activator at desirable sites [68], the elimination of unwanted contaminants, specific grain size [69], and morphology [70–72] suitable for the application, manufacturing cost, batch homogeneity, and reproducibility, and so on.

2 Synthesis Methods of Aluminate PLNPs

Developing innovative methods for preparing any solid is an essential component of materials physics and chemistry. Chemical approaches for improving the characteristics of inorganic materials have become increasingly popular in recent years. Most wet chemical techniques have various disadvantages: they require a long processing time, expensive chemicals, a specific experimental setup, and a lengthy sintering or annealing process to produce the required output. In comparison, hydrothermal provides numerous benefits. High-throughput NPs with good and controlled quality are desirable for their commercialization in various fields of applications. The approaches to aluminate NP's preparation methods are commonly employed to prepare NPs.

2.1 *Top-Down Approach*

The top-down approach in which synthesis is started with the bulk counterpart, which leaches out little by bit, resulting in the formation of fine NPs. Top-down methods for mass manufacture of NPs include photolithography, electron beam lithography, milling processes, anodization, ion and plasma etching, and others.

2.2 *Bottom-Up Approach*

It involves the coalescence or assembly of atoms and molecules to produce various NPs. Self-assembly of monomer/polymer molecules, chemical or electrochemical nanostructural precipitation, sol–gel processing, laser pyrolysis, chemical vapor deposition (CVD), plasma or flame spraying synthesis, and bio-assisted synthesis are examples of bottom-up approaches. Several CD synthesis processes have been developed over the years. In particular, the bottom-up approach of hydrothermal, solvothermal, and microwave-aided synthesis approaches has recently piqued the interest of many researchers.

In general, NP synthesis methods can be divided into two groups—Chemical synthesis methods and physical synthesis methods.

3 Chemical Methods

Significant emphasis has been paid to creating MAl_2O_4 and $LnAlO_3$ nanophosphors for high efficiency, high quality, high stability, low cost, and large-scale productivity applications. Some of the unique techniques of aluminate PLNPs synthesis.

- Solution combustion synthesis
- Sol–gel synthesis
- Hydro/Salvo thermal synthesis
- Coprecipitation synthesis.

Synthesis methods of aluminate PLNP using different organic precursors: metal/lanthanide secondary butoxide, nitrate, isopropoxide etc. Some of these strategies are detailed in depth in the sections that follow.

3.1 Solution Combustion Synthesis Method

The solution combustion method is a simple, straightforward, quick, versatile, and cost-effective approach for synthesizing a wide range of very pure and homogeneous ceramics, oxides, and nano-sized phosphor materials with customized properties. This method of synthesis provides a high-temperature environment favorable for material synthesis by the self-heat generating process (SHGP) in an exothermic chemical reaction itself for the synthesis. Solution combustion is an exciting phenomenon that propagates self-sustained exothermic reactions along an aqueous or sol–gel media. This process allows for the synthesis of various nanoscale materials, including oxides, metals, alloys, and sulfides. The combustion process requires fuel (e.g., urea, glycine, hydrazides) and oxidizer (e.g., metal nitrates $M(NO_3)_{2/3}$ $0.6/9H_2O$); Metal nitrate acts as an oxidizing reactant and urea acts as a reducing reactant.

The coefficient of total oxidizing valency of metal nitrates $[M(NO_3)_{2/3}{\cdot}6/9H_2O]$ and fuel (urea) can be calculated by arithmetic summation of oxidizing and reducing valency of elements. The stoichiometric ratio (Φ_e) of an oxidizer and fuel mixture is given in Eq. (1) in terms of elemental coefficient [73].

$$\Phi_e = \frac{\sum(\text{coefficient of oxidising elements}) \times (\text{valancy})}{(-1)\sum(\text{coefficient of reducing elements}) \times (\text{valancy})} \quad (1)$$

The oxidizer to reducing ratio can be adjusted to vary the energy dissipation during the synthesis. The stoichiometric mixture maximizes energy dissipation when Φ_e=1, fuel lean when Φ_e >1, fuel rich when <Φ_e 1.

The flow chart for preparation of Al_2O_3 by the combustion method is shown in Fig. 1. The chemical reaction Eqs. (2) and (3) for the synthesis of MAl_2O_4 and $LnAlO_3$ must be balanced, respectively. To synthesize MAl_2O_4, the oxidizing (O) valencies of $M(NO_3)_2$, $2Al(NO_3)_3$, and reducing (F) valencies of fuel were computed. Similarly, $Ln(NO_3)_3$, $Al(NO_3)_3$ valencies, and fuel valencies were estimated to synthesize the $LnAlO_3$. Stoichiometrically balance the oxidizing and reducing valency coefficients. Exothermic reactions between oxidizer and reducer are exploited in the combustion synthesis. It is also known as propellant synthesis, thermally explosive synthesis, auto-ignition, etc. This method is both cost-effective and time-efficient, i.e., the water evaporation and combustion process is completed within 10 min and then results in very pure fine nanopowder. A stoichiometric amount of oxidizer and reducer dissolves by stirring in 20–50 ml of double distilled water in a cylindrical Pyrex dish of 250 ml capacity. Then, the solution was stirred well by a magnetic stirrer until we got a transparent homogeneous mixture called Redox mixture, then placed it in the preheated muffle furnace and kept it based on the required temperature ranging from 300 to 500 °C. During the combustion process, the Redox mixture undergoes boiling, and further ignition occurs under the rapid flame, causing nanopowder. Also, combustion gases such as CO_2, O_2, N_2, NO_2, H_2O, and NH_3 are produced during the reaction process, making it unfriendly to the environment. The final product will be ground using pestle-mortar. This method's control parameters include oxidizer flow rates, fuel and flame temperature.

$$M(NO_3)_2.6/9H_2O + 2Al(NO_3)_3.6/9H_2O + Fuel$$

$$\xrightarrow{573\text{ to }773\text{ K}} MAl_2O_4 + CO_2 + N_2 + NO_2 + O_2 + H_2O \quad (2)$$

$$Ln(NO_3)_3.6/9H_2O + Al(NO_3)_3.6/9H_2O + Fuel$$

$$\xrightarrow{573\text{ to }773\text{ K}} LnAlO_3 + CO_2 + N_2 + NO_2 + O_2 + H_2O \quad (3)$$

One example of an oxidiser and isoleucine ($C_6H_{13}NO_2$) as a reducer valence's calculation is given in Table 1. A stoichiometrically balanced equation of oxidizing (×33) and reducing valency (×40) coefficients is given in Eq. (4).

The ratio of valences of oxidizer and fuel should be unity.

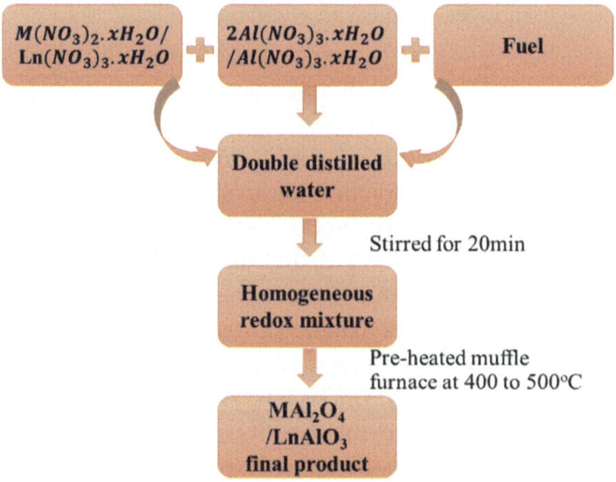

Fig. 1 The flow chart of solution combustion synthesis of MAl_2O_4/$LnAlO_3$ nanoparticles

Table 1 Calculation of the total oxidizers and fuel valences

Compound	Details of estimate of total valences	Total valance
$M(NO_3)_2.6H_2O$	(1*M) + (2*N) + (12*H) + (12*O) = [1*(+2)] + [2*0] + [12*(+1)] + [12*(−2)]	−10
$Al(NO_3)_3.9H_2O$	(1*Al) + (2*N) + (18*H) + (18*O) = [1*(+3)] + [2*0] + [18*(+1)] + [18*(−2)]	−15
Fuel $C_6H_{13}NO_2$	6*C + 13H + 1*N + 2*O = [6*(+4)] + [13*(+1)] + (1*0) + [2*(−2)]	33

$$\frac{O}{F} = \frac{-10 + 2 \times (-15)}{-1 \times 33}$$

$$= \frac{40}{33} \times \frac{33}{40} = 1$$

The stoichiometric equation for combustion synthesis is as follows:

$$33M(NO_3)_2.6H_2O + 66\,Al(NO_3)_3.9H_2O + 40C_6H_{13}NO_2$$

$$\xrightarrow{573\text{ to }773\text{ K}} 33MAl_2O_4 + 152\,N_2 + 240CO_2 + 1052H_2O \quad (4)$$

3.2 *Sol–gel Synthesis*

The sol–gel method is complicated and expensive. Still, it generates bulk metal oxides like ceramic, fibers, films, and glasses with increased features including high surface area, high purity, controlled chemical composition as well as nano-size particles. Nanoparticles are essential for the creation of nanocomposite or nanosized powder. The sol–gel synthesis at low-temperature hydrolysis and condensation processes enables the production of inorganic oxides and/or their initial precursors from simple alkoxides or chelates [74]. This approach obtains the reactants in a very fine form, which considerably reduces the reaction time and allows one to produce the products at lower reaction temperatures in a reasonable amount of time.

A chemical process transforms the initial material, such as organic precursors like metal/lanthanide secondary butoxide, nitrate, isopropoxide, etc., into a dispersible oxide, which, when added to diluted acid or water or both in a particular ratio, creates a colloidal dispersion (sol). Refluxing causes water and/or anions in the sol to slowly evaporate, resulting in a stiff gel that can take the shape of spheres, threads, pieces, or coatings. This transition is typically reversible. To create a powder, the gel was baked for an entire night at 373 K. After the salts broke down, calcining the gel in air produced an oxide product. Another method of sol–gel synthesis is the hydrolysis of metal–organic compounds, specifically metal alkoxides, to create ceramics [75]. This is a really practical and well-liked strategy that is becoming more and more common these days.

For gel preparation using nitrides, the stoichiometric of metal and aluminium nitrides (1:2 ratio) are dissolved in double distilled water for preparing MAl_2O_4 with an appropriate molar ratio. Metal nitrate solution was dropped wisely added to the pre-prepared aluminium nitrate solution with continuous stirring. The liquid "sol" was heated to 300 °C to eliminate the nitrates and gases. After the water evaporated and the solution became dense, nitrogen oxide emissions were discovered. With a continuous heating process, a white solid "gel" phase was generated, with odors

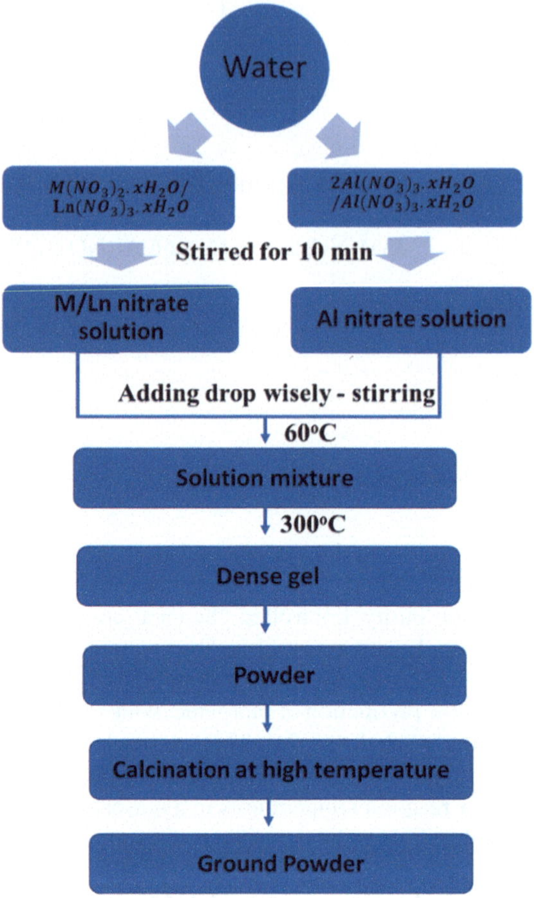

Fig. 2 Schematic diagram of the sol–gel synthesis procedure of metal and lanthanide aluminates nanoparticles

vanishing at the end of the heating operation. This technique is known as the aqueous sol–gel approach. Figure 2 depicts the synthesis scheme.

3.3 Hydro/Salvo Thermal Synthesis

One of the most popular techniques that attracted the interest of scientists and technicians working in the field of material science is the hydro/solvothermal synthesis process. As the name suggests, either water or any other suitable solvent or mixture of solvents is used in the process of synthesis and the word thermal refers to heat [44, 45]. Hydrothermal synthesis is a technique that uses a single or heterogeneous phase reaction at high pressures and temperatures to crystallize anhydrous ceramic or oxide materials straight from liquids. The reaction is carried out in water at high temperatures and pressures. Under these conditions, nucleation and particle growth can produce nanometer/sub-micron meter materials with controllable shape and size. This process is commonly used to synthesize a wide range of oxides, sulphides, carbonates, and tungsten-based nanomaterials with consistent size and shape on a large scale. The temperature, reaction duration, and solvent used are some of the governing factors in this reaction. The numerous benefits of hydrothermal/solvothermal synthesis, including its superior crystallinity, high purity, homogeneity, and low process temperature, make it a suitable technique. Additionally, this technique allows for control over the morphology [76]. Hence this Hydrothermal/Solvothermal synthesis [77, 78] route has been recognized as a suitable method.

The schematic representation of the synthesis process of the hydrothermal/solvothermal method is shown in Fig. 3. The aluminate PLNPs are synthesized using a two-step hydrothermal method. It involves the mixing of nitrate salts such as metal nitride ($M(NO_3)_2{\cdot}6H_2O$) and aluminum nitrate ($Al(NO_3)_3{\cdot}9H_2O$) acting as oxidizing agents to synthesize the MAl_2O_4. $Al(NO_3)_3{\cdot}9H_2O$ and $M(NO3)_2{\cdot}6H_2O$ are dissolved in 20 mL of deionized water/solvent at 0.4 and 0.2 mol concentrations, respectively, and stirred using a magnetic stirrer. The aluminium solution will be continuously stirred as the metal nitride solution is added dropwise. After that, the mixture had been stirred for around twenty minutes. Subsequently, a 25% ammonia solution was gradually added to the mixture to modify its morphology or pH, and it was swirled for 60 min until a white precipitate formed. The resultant solution was poured into a stainless-steel autoclave with a Teflon lining and tightly sealed. It was then placed in an oven set to around 180 to 220 °C for eight to twenty-four hours. The precipitate from the autoclave was collected then washed, multiple times for 10 min at 3000–6000 rpm using ethanol and deionized water at a 1:10 sample-to-solvent ratio, and dried for 6 h at around 60 to 80 °C. The resulting powder was dried, then annealed and manually milled to produce fine particles.

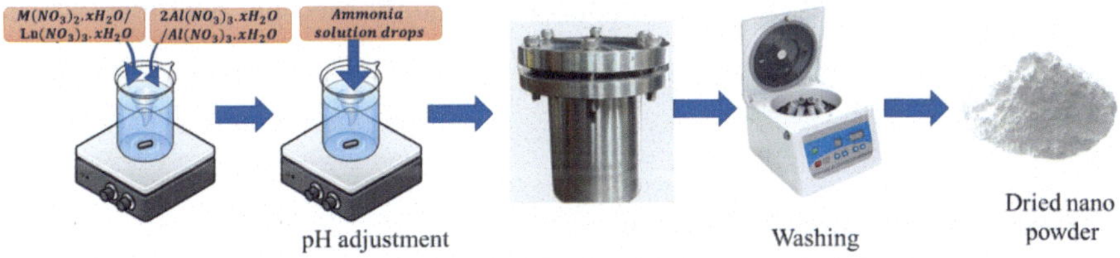

Fig. 3 Schematic diagram of the hydro/salvo thermal synthesis procedure of metal and lanthanide aluminates nanoparticles

3.4 Co-Precipitation Synthesis

Among the methods discussed above, co-precipitation with organic salts has been demonstrated to be feasible for the production of persistent luminescence aluminate nanoparticles with narrow size distribution and little agglomeration [79, 80]. When synthesizing materials, the co-precipitation technique works incredibly well. Furthermore, it is both convenient and cost-effective. Due to its superior stoichiometry control and higher purity, this technique is highly recommended by scientific researchers. It is a less complicated process that involves a chemical reaction to generate a metal/lanthanide nitrates precursor into a precipitate using ammonium bicarbonate as a precipitant. To create the desired product using this process, the sociometric amounts of chemical precursors are added and coupled with trace amounts of an activator/dopant as per the requirement. A more homogenous product requires thorough mixing, especially when co-precipitation is involved. After stirring for another hour, the resulting suspension is centrifuged. The precipitated precursor was rinsed multiple times with distilled water before being dried at 100 °C overnight. This approach can be used to synthesize phosphors of both micro and sub-micron-size particles [30, 81]. The obtained compound is sintered at a high temperature.

4 Physical Methods

Physical techniques use heat, electrical, mechanical, or high-energy radiation to abrade, melt, evaporate, or condense materials to produce nanoparticles (NPs). These techniques, which mostly employ a top-down approach, have the benefit of producing homogeneous, monodisperse NPs without contaminating the solvent. At the same time, physical processes are less efficient due to the copious amounts of waste generated during the synthesis. Some of the most commonly used physical methods to generate PLNPs are:

(a) High energy ball milling
(b) Pulsed laser ablation
(c) Spray pyrolysis
(d) Solid state diffusion synthesis, etc.

4.1 High Energy Ball Milling (HEBM)

The commercially available microcrystalline aluminate powder with high thermal stability is used as the starting material. The raw microcrystalline powder is milled nominally at room temperature using an oscillatory mill. The complex movements of the grinding vial in conjunction with back-and-forth oscillation and lateral movement of its bottom are used to optimise the collisions between the powder, ball, and inner wall of the vial to guarantee that the as-milled powder pattern is uniform. The monocrystalline aluminate persistence luminescence nanoparticles (MAPLNP's) produced from HPHT micro sized synthetic aluminates are commercially accessible with the smallest typical particle size available from a few tens of micrometers. The as-received powder was dry milled in a planetary mill (Pulverisette 5, Fritsch, Rheinland-Pfalz, Germany) using zirconia grinding media (10 mm YTZ balls, Tosoh, Tokyo, Japan). The powder was subjected to 30 cycles of milling at a speed of 200 rpm for 15 min, followed by a 5-min rest period; the entire milling process took 10 h [82]. The weight ratio of milling balls to the powder (BPR) was kept constant at 20:1. The samples generated from the as-received and milled powder using YTZ balls were labelled "As-received" and "Milled," respectively.

4.2 *Pulsed Laser Ablation*

Pulsed laser ablation (PLA) is one of the physical methods that has garnered a lot of interest because it makes it possible to produce highly pure NPs and their surfaces with a variety of compositions and morphology. The PLA technique generates NPs by applying laser pulses that are focused on the surface of a target material as shown in Fig. 4. High repeatability, scalability, and good universality can be obtained via laser approaches in the synthesis process. The production of environmentally benign and chemically pure PLA nanoparticles has developed into a subject of study that is drawing interest from both lab and industry users [22].

Pulsed laser ablation is the simplest technique and has become an attractive method for the synthesis of the liquid phase NP's through aluminate single crystals or micro-crystals. A single crystal of aluminates or about 0.5 mg per ml of micro crystals is disseminated in solvent (for example chloroform, toluene, isopropyl alcohol and distilled water) by sonicating the solution for an hour. The resultant graphene powder dispersed solution will be black in colour. Following that, the graphene dispersed sample underwent the Nd-YAG laser irradiation (wavelength 532 nm, power 2 W, pulse duration 5 ns, energy 80 mJ per pulse and pulse rate 10 Hz) or a nanosecond

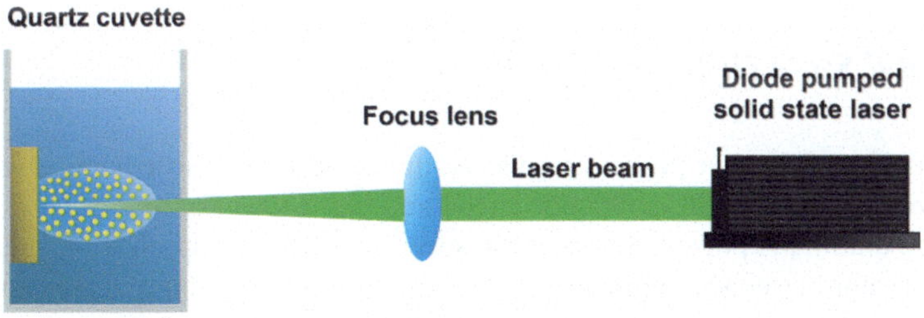

Fig. 4 Scheme of laser ablation synthesis of nanomaterials [22]

diode-pumped laser Alphalas PULSELAS P-355–100-HP (The following parameters: $\lambda \approx 1064$ nm; $\Delta t \approx 845$ ns; repetition rate ≈ 560 Hz, beam spot size $\approx$ 0.203 mm, pulse energy ≈ 212 μJ, laser fluence ≈ 1.31 J/cm^2), with vertical polarization focused by a lens with about 50 to 200 mm focal length for about 30 min on micro aluminates crystal that was placed in a quartz cuvette. The size of the laser spot on the target surface can be approximated to be between 0.1 and 0.5 mm. As a result, the fluence each pulse ranges between 40 and 1000 J cm^{-2}. In order to prevent gravitational settling at the solution's base during the laser irradiation, it is magnetically stirred throughout the laser irradiation process. The laser-ablated solution is then centrifuged or filtered to collect the precipitation of the Np's sample and dried overnight at 70 °C [22].

The solvent environment parameters for PLA and its purity, such as type, viscosity, temperature, compressibility, and so on, can have a significant impact on the composition, shape, and size of the produced NPs. Fluid purity, on the other hand, is an essential consideration, as high-pressure liquid chromatography or spectroscopic solvents, for example, may contain atmospheric gases such as N_2, O_2, H_2O, and CO_2, as well as traces of other substances. Under atmospheric circumstances, water also contains O_2 and N_2. In some circumstances, the degradation of the solvent over time may result in by-products. As a result, even if spectroscopic and high-pressure liquid chromatography solvents were utilised, numerous sorts of contaminants were present during PLA [83]. When using water as a PLA medium, colloidal solutions have no by-products, making it a very convenient environment for the PLA process. In comparison to organic solvents, colloidal solutions generated by PLA in distilled water are not appropriate for future applications in the form of thin films or combination with polymers. Organic solvents, on the other hand, can generate hydrogen, carbon, and acetyl group radicals and ions after laser absorption, which can form undesirable compounds in the solution. Amorphous carbon is typically formed as a result of the laser breakdown of organic solvent [23]. This effect was particularly noticeable in toluene, where black micro- and nanoparticles were found in substantial quantities. Taking into account the drawbacks of distilled water and toluene, chloroform emerged as the most convenient environment for PLA synthesis of PLNPs due to the very low number of by-products and the prospect of future practical uses.

4.3 Solid State Diffusion Synthesis

The solid-state diffusion process is one of the easiest and most popular ways to synthesize a wide range of lanthanide-metal and multi-metal oxides, such as metal/lanthanide aluminate phosphors for use in several fascinating applications including persistence luminescence applications. The primary disadvantage of this approach is that the reaction temperature is quite high and the nanostructure morphology is difficult to regulate. This synthesis technique frequently allows for control over particle size by altering the reaction temperature, reaction time and morphology in addition to using long and short chain surfactants [84, 85]. The precursors are made to react through a diffusion process, by ensuring that the temperature is just enough to have adequate diffusion and helps to complete the reaction without melting the constituents. However, aluminate PLNPs cannot be prepared by the solid-state diffusion method at temperatures up to 1400 °C. The sociometric ratio of starting oxides (Al_2O_3 and metal oxides or lanthanide oxides) are mixed with Ceria balls and ethanol, followed by milling at or grounded using pestle-mortar. Subsequently, the slurry was evaporated on a hot plate. Further, the prepared powder was made into pellets followed by sintering at different high temperatures and times [86].

4.4 Spray Pyrolysis Synthesis

High purity, sub-micron, spherical-shaped luminous materials can be produced in large quantities using the spray pyrolysis method. It can be of two types: (i) Aerosol spray pyrolysis (ASP) [87] and (ii) Flame spray pyrolysis (FSP). The latter type is generally preferred for the synthesis of nanomaterials.

For the synthesis of multi-component active redox water splitting materials, ASP has several built-in advantages, including the capacity to synthesize nanoparticles, ease of precisely controlled doping, the lack of a need for lengthy sintering at high temperatures, and scalability. Furthermore, the extremely quick synthesis durations, rapid cooling rates, and reactor's capacity to regulate oxygen partial pressure allow for the production of particles with high and "tunable" defect concentrations—specifically, oxygen vacancies—for the intended use. The precursors are dissolved in water in the necessary amounts of each nitrate for a particular intended product stoichiometry to create aqueous solutions. Three different types of these solutions with a concentration of 1 M total metal ions were used: "plain," as well as solutions with the addition of ammonia, citric acid, or glycine as solvents. An air-assisted atomizer that was created in-house is used to atomize the solution into fine droplets, which are then sprayed into a hot-wall tube reactor to become the appropriate powders.

Figure 5 shows a visualization of an actual FSP flame together with the synthesis parameters, including the pilot flame, precursor solution, dispersion, and sheath gas [88].

The preparation of a metal or metal–organic precursor solution, usually dissolved in a suitable solvent, is the first step in the process design of FSP [89]. This solution is atomized to create a thin spray of droplets, which is frequently assisted by a high-pressure nozzle [90]. The solvent is eventually burned and the metal precursors eventually decompose as a result of the subsequent burning of these droplets, frequently with the assistance of an auxiliary flame. The precursor breaks down in this high-temperature flame environment, allowing metal or metal-oxide nanoparticles to nucleate and develop [91]. The size, shape, crystallinity, and phase composition of the resulting nanoparticles can be regulated by carefully selecting the metal precursor, its concentration in the solution, the kind of solvent, the atomization technique, flame temperature, oxygen-to-fuel ratio, and other factors.

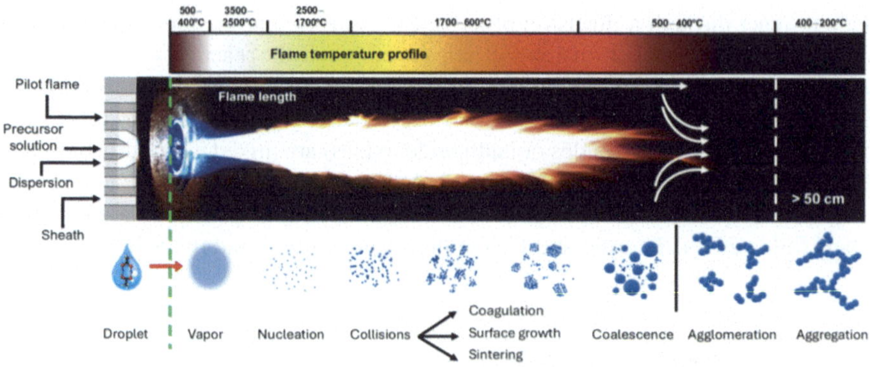

Fig. 5 Visualization of actual FSP flame, depicting the synthesis parameters (pilot flame, precursor solution, dispersion, sheath gas) [88]

5 MAl_2O_4 Based Persistent Luminescence Materials

The divalent aluminates exist in different forms in which MAl_2O_4 (where M = Mg, Ca, Sr, Ba, Mn, Co, Ni, Cu, Zn etc.) is one of the promising materials for a wide range of applications in material science. However, the luminescence properties if these pigments are highly influenced by its the morphology, size, and optical characteristics, which are significantly influenced by the synthesis processes.

Xi et al. [92] reported the germanium homogenous substitution in the original $ZnAl_2O_4:Cr^{3+}$ infrared long persistent phosphors $Zn_{1+x}Al_{2-2x}Ge_xO_4$: Cr (x = 0.1 to 1) was achieved using a high temperature solid-state process. The phase and material transformation of $Zn_{1+x}Al_{2-2x}Ge_xO_4$: Cr from the spinel structure of $ZnAl_2O_4$ to the tetrahedral structure of Zn_2GeO_4 with the increase of substitution of germanium appears $x \geq 0.5$ as evidenced by XRD data. The photoluminescence emission of infrared emissions of Cr^{3+} ions are reliably demonstrating how sensitive Cr^{3+} is to the host crystal field in this substance. An infrared emission of Cr^{3+} is an indication of a spinel structure of the host lattice. Nevertheless, the spectra become broader and less well-resolved with high substitution levels of germanium ions $x \geq 0.5$. Both R and N_2 lines decreased their intensity, with R lines losing intensity even more quickly than N_2 with increasing the substitution concentration. The thermoluminescence (TL) results, however, show that the substitution of germanium can not only alter the host structure but also produce flaws that enhance the materials' afterglow. TL peaks shift to lower temperatures, more intense, and broaden for $x = 0.4$, indicating an appropriate substitution of germanium, which are an important factor for room temperature afterglow phosphors. TL data shows how the traps dept being influenced by the germanium concentration. As a result, it might make an outstanding infrared long afterglow phosphor.

Jamalaiah et al. [93] reported the luminescence properties of Tb^{3+} and Bi^{3+} co-doped $SrAl_2O_4$ nanophosphors synthesized by solid state reaction for photonic applications. The monoclinic phase was observed in the XRD patterns of the examined materials. The SEM data show a grain-like morphology with a uniform dispersion of particles of varying sizes. The nanophosphors' improved luminosity may be attributed to their homogenous distribution and nanoscale crystallites. Bi^{3+} ions' effective sensitivity to excitation at 340 nm, which corresponds to the Bi^{3+}: $^1S_0 \rightarrow {}^1P_1$ transition and emit strong and enhanced green luminescence through Tb^{3+}: $^5D_4 \rightarrow {}^7F_J$ transitions. It is evident that there is energy transfer from Bi^{3+} to Tb^{3+} ions from the spectral overlap of the Tb^{3+}: $^7F_6 \rightarrow {}^5G_6$ and Bi^{3+}: $^3P_1 \rightarrow {}^1S_0$ transition. The colour perception and the luminescence quantum efficiency and luminescence decay time of Tb^{3+} and Bi^{3+} co-doped $SrAl_2O_4$ green emission suggest that the nanophosphors were greater promise for display applications. Chawla et al. [94] reported the synthesis of $SrAl_2O_4$:Pr^{3+} by high temperature solid state reaction for white LED application. The PL emission spectrum analysis of the nanophosphor exhibits the broad yellow-orange emission under blue light excitation. Stronger emission intensity, commercially available excitation blue LED, favourable broad band visible emission and faster decay of multi phonon relaxation makes $SrAl_2O_4$:Pr^{3+} is a potential material for solid state lighting.

Gurugubelli et al. [77] reported synthesis of Fe-doped $ZnAl_2O_4$ nanosheets by a co-precipitation assisted hydrothermal method (220 °C) for the bandgap engineering and room temperature ferromagnetic applications. TEM images of pure and Fe-doped $ZnAl_2O_4$ exhibit a micro-hexagonal structure and nanosheet-like structure respectively, which conforms that the dopant element also influence the morphology of the material even when synthesized by the same method. The variation of band gap of $ZnAl_2O_4$ with increasing Fe dopant concentration, which conforms that the band gap of the material can be engineered using a suitable dopant and its concentration.

Basavaraju et al. [95] studied an importance of inversion disorder in the visible light induced persistent luminescence in spinel Cr^{3+} doped AB_2O_4 (A = Zn or Mg and B = Ga or Al) nanophosphors compounds. All of these samples exhibited a long, near infrared persistent luminescence under UV or X-rays excitation. In addition, the persistent luminescence of AGa_2O_4 (A = Zn and Mg) are also be induced by visible light excitation of Cr^{3+}:$^4A_2 \rightarrow {}^4T_2$ transitions, which makes these compounds suitable as biomarkers for in vivo optical imaging of small animals. It appears that $ZnAl_2O_4$, which exhibits the largest crystal field splitting of Cr^{3+} and the smallest antisite disorder, does not show considerable persistent luminescence upon visible light excitation as compared to Zn/$MgGa_2O_4$.

Huang et al. [78] were reported the optical properties and theoretical study of spinal structured Mn doped $ZnAl_2O_4$ nanoparticles synthesized by the hydrothermal method. In this report it is concluded that the doping concentration greatly influence an average crystallite size, which is decreased with increasing doping concentration. As per TEM analysis the morphology of the samples shows spherical or ellipsoid particles with uniform particle size. EDX spectra of Mn doped $ZnAl_2O_4$ nanoparticles displayed that homogeneous distribution of the dopant and the host lattice elements without any enrichment phenomenon throughout the sample. Hence, Mn ions are successfully substituted in the intended Zn^{2+} site and uniformly entered in to the $ZnAl_2O_4$ host matrix. XPS spectra also conform the Mn^{2+} in the material and these dopant ions mainly occupy the tetrahedral sites by substituting the Zn^{2+} site. Band gap of the doped sample decreases with the increase of Mn content due to substitution of Mn which introduced the new energy levels conformed by the first principle calculations.

Large ionic radii of trivalent rare earth ion doping into the MAl_2O_4 form host material is one of the challenging task due to the imbalance of ionic size between the rare earth ion and divalent/trivalent (M/Al) site leading to a lattice distortion. It can be overcome by a new form of metal ion and lanthanide aluminate is one of the most important and a unique host material in the form of $MLnAlO_4$ by substitution of rare-earth trivalent and divalent ions into the Ln site. Devi et al. [96] reported the Tb^{3+} doped $SrLaAlO_4$ synthesized by solution combustion technique for RGB based white LED applications. It was studied that the influence of the Tb^{3+} doping on the structural and photoluminescence properties of $SrLaAlO_4$ nanophosphor. The SEM analysis depicts that the resulting sample is composed of large number of agglomerated particles with irregular morphology and TEM analysis conform the particle size lies in the range of 15–50 nm. Whereas, EDS spectral analysis confirmed the presence of the dopant along with the all-other existing elements in host matrix. The PL results reveal that an intense green emission was obtained at 545 nm due to the Tb^{3+} transition. Further, the radiative lifetime (1.17 ms), non-radiative rates (194.28 s^{-1}), quantum efficiencies (81%) were calculated using Auzel's model and Commission International de l'Eclairage (CIE) color coordinates confirm the emission in green region. Hence, it was concluded that Tb^{3+}-activated nanophosphors can be used as a green component in RGB based white LEDs.

6 $LnAlO_3$ Based Persistent Luminescence Materials

The trivalent lanthanide aluminates also existed in different forms, particularly $LnAlO_3$ is one of the most important and significant material for persistent luminescence phosphors (where Ln = La, Gd, Lu, Ce, Tb, Tm, Sm, Nd etc.). These materials luminescence properties are highly depending on their morphology, which is significantly influenced by the synthesis technique.

Siyuan et al. [97] Perovskite-type fine microsphere of $LaAlO_3$: xMn^{4+}, yGe^{4+} have been synthesized through thermal decomposition by the solvothermal process using ethylene glycol as the solvent in the presence of citric acid. These micro sized spheres emit broad and bright near infrared (NIR) emission ranging from 680 to 750 nm with the maximum at around 730 nm. These micro-sized spheres have a bright near-infrared (NIR) broad band emission ranging from 680 to 750 nm along with a small shoulder approximately at 730 nm. The highest intensity is found for $x = 0.1$ mol% of Mn^{4+} doping concentration in the $LaAlO_3$ host lattice. It also reported that, it exhibits an intense NIR afterglow for 10 min at room temperature in the absence of excitation light source, due to the formation of $[Mn_{Al}]$ defects. The band gap, thermal stability, morphology, and crystal structure were all unhampered by the addition of co-dopant Ge^{4+} ions. On the other hand, co-doping of Ge^{4+} resulted in the formation of more isolated Mn^{4+} ions and a stronger MnO_6 asymmetry, which enhanced the persistent luminescence NIR emission intensity about 260% time for y = 0.6% of Ge^{4+} ions. Positive defects of are formed when Ge^{4+} is substituted into the Al^{3+} site [Ge_{Al}]; this increases the concentration of the trap rather than its depth. This persistent luminescence material can be used in optical data storage because they show consistent afterglow when heat processing is applied correctly even in the absence of a light source.

Luo et al. [98] reported the role of Cr^{3+} and Ln (Nd, Dy, Ho, Er, Tm, Sm, Eu) co-doped $GdAlO_3$ and $LaAlO_3$ synthesized by solid state reaction at 1500 °C for 10 h in N_2:H_2 (93%:7%) atmosphere to prevent oxidation of Cr to 4+. In broad band gap compounds, trivalent Nd, Dy, Ho, Er, Tm, Sm, and Eu typically function as electron trapping centres, while trivalent Ce, Tb, and Pr function as hole trapping centres. Recombination luminescence at the electron trap is produced when a deep electron trap and a shallow hole trap are coupled. This occurs during the thermoluminescence glow. In contrast, the electron will be released to recombine at the hole trapping centre in the event of a shallow electron trap. In $GdAlO_3$, different trivalent lanthanides were co-doped with Cr^{3+}. A consistent interpretation of all the data was provided by merging the data from PL and TL. Different from all other lanthanides, Cr^{3+} can function as a deep hole trap as well as a deep electron trap. The placement of the Cr^{2+} and Cr^{3+} levels within the band gap and in relation to the vacuum level was inferred from the data. In addition to TL recombination through the conduction band, tunnelling recombination was verified. The outcomes on $GdAO_3$ are compared with those on $LaAlO_3$. An analogous system that predictably lowers the lanthanide and Cr trap depths has a higher lying valence band and a lower lying conduction band.

7 Summary

In this chapter, several synthesis methods of persistent luminescence nanoparticles in particular divalent and trivalent aluminate phosphors are demonstrated. The synthesis methods of PLNPs are classified viz bottom-up approach and top-down approach. The bottom-up approach is a chemical method wherein Solution combustion, Sol–gel, Hydro/Salvo thermal and Co-precipitation synthesis methods are discussed. The reaction kinetics, stoichiometric calculations and detailed experimental procedures along with necessary flow charts are presented. Further, a top-down approach is a physical method that includes High energy ball milling, Pulsed laser ablation, Spray pyrolysis and Solid-state diffusion synthesis. In conclusion, the bottom-up approach synthesis of PLNPs is more advantageous than the top-down approach because the former has a better chance of producing nanostructures with fewer defects, and more homogenous chemical composition, where luminescence properties can be tuned to the desired extent.

Acknowledgements The authors dedicate this chapter to **Dr. B.N. Lakshminarasappa**, a retired professor of Physics, Bangalore University, Bengaluru, Karnataka, India who passed away on January 02, 2024. He was a real fine experimentalist and a source of inspiration for many researchers.

Prof. BNL was a true physicist with sincerity and hard work, he was always eager to explore new horizons of physics in general and luminescence in particular.

References

1. I. Khan, K. Saeed, I. Khan, Nanoparticles: properties, applications and toxicities. Arab. J. Chem. **12**, 908–931 (2019). https://doi.org/10.1016/j.arabjc.2017.05.011
2. R. Singhal, F. Singh, A. Tripathi, D.K. Avasthi, A comparative study of ion-induced damages in in C60 and C70 fullerenes. Radiat. Eff. Defects Solids **164**, 38–48 (2009). https://doi.org/10.1080/10420150802479638
3. H.A. Alluhaybi, S.K. Ghoshal, B.O. Alsobhi, W.N. Wan Shamsuri, Visible photoluminescence from gold nanoparticles: a basic insight, Optik (Stuttg). **192**, 162936 (2019). https://doi.org/10.1016/j.ijleo.2019.162936
4. X. Liu, M.C. Patsavas, R.H. Byrne, Purification and characterization of meta-cresol purple for spectrophotometric seawater ph measurements. Environ. Sci. Technol. **45**, 4862–4868 (2011). https://doi.org/10.1021/es200665d
5. A.M. Volodin, A.F. Bedilo, V.O. Stoyanovskii, V.I. Zaikovskii, R.M. Kenzhin, I.V. Mishakov, A.A. Vedyagin, Nanocrystalline carbon coated alumina with enhanced phase stability at high temperatures. RSC Adv. **7**, 54852–54860 (2017). https://doi.org/10.1039/c7ra08841h
6. V. Veligura, G. Hlawacek, R. Van Gastel, H.J.W. Zandvliet, Investigation of ionoluminescence of semiconductor materials using helium ion microscopy. J. Lumin. **157**, 321–326 (2015). https://doi.org/10.1016/j.jlumin.2014.09.016
7. X.S. Fang, C.H. Ye, X.X. Xu, T. Xie, Y.C. Wu, L.D. Zhang, Synthesis and photoluminescence of alpha-Al_2O_3 nanowires. J. Phys.-Condensed Matter. (2004). https://doi.org/10.1088/0953-8984/16/23/030
8. Z.H. Khan, A. Kumar, S. Husain, M. Husain, Introduction to nanomaterials. Adv. Struct. Mater. **79**, 1–23 (2016). https://doi.org/10.1007/978-81-322-2668-0_1
9. Z.-L. Zhang, J.-F. Li, X.-L. Wang, J.-Q. Qin, W.-J. Shi, Y.-F. Liu, H.-P. Gao, Y.-L. Mao, Enhancement of perovskite solar cells efficiency using n-doped TiO_2 nanorod arrays as electron transfer layer. Nanoscale Res. Lett. **12**, 43 (2017). https://doi.org/10.1186/s11671-016-1811-0
10. B. Kumar, S. Saha, M. Basu, A.K. Ganguli, Enhanced hydrogen/oxygen evolution and stability of nanocrystalline (4–6 nm) copper particles. J. Mater. Chem. A. **1**, 4728–4735 (2013). https://doi.org/10.1039/c3ta01194a
11. F.H. Alkallas, A.B.G. Trabelsi, R. Nasser, S. Fernandez, J.M. Song, H. Elhouichet, Promising Cr-doped ZnO nanorods for photocatalytic degradation facing pollution. Appl. Sci. **12** (2022). https://doi.org/10.3390/app12010034
12. X. Li, F. Zhang, D. Zhao, Lab on upconversion nanoparticles: optical properties and applications engineering via designed nanostructure. Chem. Soc. Rev. **44**, 1346–1378 (2015). https://doi.org/10.1039/c4cs00163j
13. X.D. Wang, O.S. Wolfbeis, Optical methods for sensing and imaging oxygen: materials, spectroscopies and applications. Chem. Soc. Rev. **43**, 3666–3761 (2014). https://doi.org/10.1039/c4cs00039k
14. C. Pan, S.Y. Chen, P. Shen, Photoluminescence and transformation of dense Al_2O_3:Cr3+condensates synthesized by laser-ablation route. J. Cryst. Growth **310**, 699–705 (2008). https://doi.org/10.1016/j.jcrysgro.2007.11.089
15. Z. Zhu, Q. Sun, Z. Zhang, J. Dai, G. Xing, S. Li, X. Huang, W. Huang, Metal halide perovskites: stability and sensing-ability. J. Mater. Chem. C. **6**, 10121–10137 (2018). https://doi.org/10.1039/C8TC03164A
16. R.M. Almeida, A.C. Marques, Rare-earth photoluminescence in sol-gel derived confined glass structures. J. Non Cryst. Solids **352**, 475–482 (2006). https://doi.org/10.1016/j.jnoncrysol.2005.11.059

17. M.H. Devoret, C. Glattli, Single-electron transistors. Phys. World **11**, 29–34 (1998). https://doi.org/10.1088/2058-7058/11/9/26
18. D. Zupančič, P. Veranič, Nanodiamonds as possible tools for improved management of bladder cancer and bacterial cystitis. Int. J. Mol. Sci. **23**, 8183 (2022). https://doi.org/10.3390/ijms23158183
19. Q. Duan, L. Ma, B. Zhang, Y. Zhang, X. Li, T. Wang, W. Zhang, Y. Li, S. Sang, Construction and application of targeted drug delivery system based on hyaluronic acid and heparin functionalised carbon dots. Colloids Surfaces B Biointerfaces **188**, 110768 (2020). https://doi.org/10.1016/j.colsurfb.2019.110768
20. S. Ghosh, B. Pradhan, Lead-free metal halide perovskite nanocrystals: challenges, applications, and future aspects. ChemNanoMat. **5**, 300–312 (2019). https://doi.org/10.1002/cnma.201800645
21. E.A.R. Assirey, Perovskite synthesis, properties and their related biochemical and industrial application. Saudi Pharm. J. **27**, 817–829 (2019). https://doi.org/10.1016/j.jsps.2019.05.003
22. V. Vasylkovskyi, I. Bespalova, O. Gryshkov, M. Slipchenko, S. Tkachenko, P. Arhipov, I. Gerasymov, Y. Zholudov, Z. Zhao, A. Feldhoff, A. Sorokin, O. Slipchenko, B. Grynyov, B. Chichkov, Laser generation of CeAlO3 nanocrystals with perovskite structure. Appl. Phys. A **129**, 714 (2023). https://doi.org/10.1007/s00339-023-06977-4
23. A.H. Attallah, F.S. Abdulwahid, Y.A. Ali, A.J. Haider, Effect of liquid and laser parameters on fabrication of nanoparticles via pulsed laser ablation in liquid with their applications: a review. Plasmonics **18**, 1307–1323 (2023). https://doi.org/10.1007/s11468-023-01852-7
24. A. Verma, A. Verma, Synthesis, characterization, mechano-luminescence, thermoluminescence, and antibacterial properties of SrMgAl10O17: Eu phosphor. J. Alloys Compd. **802**, 394–408 (2019). https://doi.org/10.1016/j.jallcom.2019.06.209
25. R. Krishnan, R. Kesavamoorthy, S. Dash, A.K. Tyagi, B. Raj, Raman spectroscopic and photoluminescence investigations on laser surface modified α-Al_2O_3 coatings. Scr. Mater. **48**, 1099–1104 (2003). https://doi.org/10.1016/S1359-6462(02)00631-0
26. K.C. Patil, M.S. Hegde, T. Rattan, S.T. Aruna, Chemistry of nanocrystalline oxide materials combustion synthesis, properties and applications (2008). https://doi.org/10.1142/6754
27. G. Impellizzeri, V. Scuderi, L. Romano, P.M. Sberna, E. Arcadipane, R. Sanz, M. Scuderi, G. Nicotra, M. Bayle, R. Carles, F. Simone, V. Privitera, Fe ion-implanted TiO_2 thin film for efficient visible-light photocatalysis. J. Appl. Phys. **116** (2014). https://doi.org/10.1063/1.4901208
28. P.-C. Chen, T.-Y. Yeh, C.-M. Ou, C.-C. Shih, H.-T. Chang, Synthesis of aluminum oxide supported fluorescent gold nanodots for the detection of silver ions. Nanoscale **5**, 4691–4695 (2013). https://doi.org/10.1039/c3nr00713h
29. J.D. Benck, T.R. Hellstern, J. Kibsgaard, P. Chakthranont, T.F. Jaramillo, Catalyzing the hydrogen evolution reaction (HER) with molybdenum sulfi de nanomaterials. ACS Catal. **4**, 3957–3971 (2014)
30. S. Farhadi, S. Panahandehjoo, Spinel-type zinc aluminate (ZnAl2O4) nanoparticles prepared by the co-precipitation method: a novel, green and recyclable heterogeneous catalyst for the acetylation of amines, alcohols and phenols under solvent-free conditions. Appl. Catal. A Gen. **382**, 293–302 (2010). https://doi.org/10.1016/j.apcata.2010.05.005
31. N. Mercier, H. Valladas, J.L. Joron, J.L. Reyss, F. Lévêque, B. Vandermeersch, Thermoluminescence dating of the late Neanderthal remains from Saint-Césaire. Nature **351**, 737–739 (1991). https://doi.org/10.1038/351737a0
32. P.-C. Lin, S. Lin, P.C. Wang, R. Sridhar, Techniques for physicochemical characterization of nanomaterials. Biotechnol. Adv. **32**, 711–726 (2014). https://doi.org/10.1016/j.biotechadv.2013.11.006
33. T. Maldiney, A. Bessière, J. Seguin, E. Teston, S.K. Sharma, B. Viana, A.J.J. Bos, P. Dorenbos, M. Bessodes, D. Gourier, D. Scherman, C. Richard, The in vivo activation of persistent nanophosphors for optical imaging of vascularization, tumours and grafted cells. Nat. Mater. **13**, 418–426 (2014). https://doi.org/10.1038/nmat3908
34. C. Materials, S. Oxide, F. Cells, Ceramic materials for solid oxide fuel cells (2009)

35. A. Anastasopoulos, J. Blake, B.E. Hayden, Non-noble intertransition binary metal alloy electrocatalyst for hydrogen oxidation and hydrogen evolution. J. Phys. Chem. C **115**, 19226–19230 (2011). https://doi.org/10.1021/jp205287b
36. S. Sengodan, S. Choi, A. Jun, T.H. Shin, Y.W. Ju, H.Y. Jeong, J. Shin, J.T.S. Irvine, G. Kim, Layered oxygen-deficient double perovskite as an efficient and stable anode for direct hydrocarbon solid oxide fuel cells. Nat. Mater. **14**, 205–209 (2015). https://doi.org/10.1038/nmat4166
37. Y. Cui, J. Zhang, H. He, G. Qian, Photonic functional metal-organic frameworks. Chem. Soc. Rev. **47**, 5740–5785 (2018). https://doi.org/10.1039/c7cs00879a
38. S. Anupam, M. Aynyas, P. Pathak, A. Selot, Comparative thermoluminescence study of rare earth (Eu 3+, Dy 3+) and transition metal Mn 2+ with aluminate host. Adv. Phys. Theor. Appl. **19**, 3–8 (2013). 2224-719X
39. G. Li, Y. Tian, Y. Zhao, J. Lin, Recent progress in luminescence tuning of Ce3+and Eu2+-activated phosphors for pc-WLEDs. Chem. Soc. Rev. **44**, 8688–8713 (2015). https://doi.org/10.1039/c4cs00446a
40. C. Madhukar Reddy, B. Deva Prasad Raju, N. John Sushma, N.S. Dhoble, S.J. Dhoble, A review on optical and photoluminescence studies of RE3+ (RE=Sm, Dy, Eu, Tb and Nd) ions doped LCZSFB glasses. Renew. Sustain. Energy Rev. **51**, 566–584 (2015). https://doi.org/10.1016/j.rser.2015.06.025
41. N. Rakov, G.S. Maciel, Photoluminescence analysis of α-Al_2O_3 powders doped with Eu3+ and Eu2+ ions. J. Lumin. **127**, 703–706 (2007). https://doi.org/10.1016/j.jlumin.2007.04.001
42. P.R. González, E. Cruz-Zaragoza, C. Furetta, J. Azorín, B.C. Alcántara, Effect of thermal treatment on TL response of CaSO4: Dy obtained using a new preparation method. Appl. Radiat. Isot. **75**, 58–63 (2013). https://doi.org/10.1016/J.APRADISO.2013.02.005
43. I. Human, H. Series, Development of procedures for in vivo dosimetry in radiotherapy (2013). http://www-pub.iaea.org/MTCD/Publications/PDF/Pub1606_web.pdf
44. S.S. Reddy, N. Chauhan, K.R. Nagabhushana, F. Singh, TL/OSL properties of beta irradiated Al_2O_3:Tm3+ phosphor synthesized by microwave combustion method. Mater. Res. Bull. **104**, 236–243 (2018). https://doi.org/10.1016/j.materresbull.2018.04.019
45. S.S. Reddy, K.R. Nagabhushana, F. Singh, Mechanism of thermoluminescence in high energy carbon ion irradiated Tb 3+ doped Al_2O_3 phosphor for carbon ion beam dosimetry. Mater. Res. Express. **4**, 095023 (2017). https://doi.org/10.1088/2053-1591/aa8928
46. M. Leskelä, Rare earths in electroluminescent and field emission display phosphors. J. Alloys Compd. **275–277**, 702–708 (1998). https://doi.org/10.1016/S0925-8388(98)00421-6
47. L. Pidol, B. Viana, A. Galtayries, P. Dorenbos, Energy levels of lanthanide ions in a Lu2Si2O7 host. Phys. Rev. B - Condens. Matter Mater. Phys. **72**, 1–9 (2005). https://doi.org/10.1103/PhysRevB.72.125110
48. E. Zych, C. Brecher, A. Lempicki, Infrared spectroscopy of LuAlO3: Ce a useful tool to determine Ce concentration, Spectrochim. Acta Part A Mol. Biomol. Spectrosc. **54**, 1763–1769 (1998). https://doi.org/10.1016/S1386-1425(98)00109-7
49. A.A. Amer, H. El-Didamony, T.M. El-Sokkary, M.I. Wahdan, Synthesis and characterization of some calcium aluminate phases from nano-size starting materials, Boletín La Soc. Española Cerámica y Vidr. **61**, 98–106 (2022). https://doi.org/10.1016/j.bsecv.2020.07.006
50. M. Rizwan, S. Gul, T. Iqbal, U. Mushtaq, M.H. Farooq, M. Farman, R. Bibi, M. Ijaz, A review on perovskite lanthanum aluminate ($LaAlO_3$), its properties and applications. Mater. Res. Express. **6**, 112001 (2019). https://doi.org/10.1088/2053-1591/ab4629
51. C. Adelmann, J. Swerts, O. Richard, T. Conard, M. Popovici, J. Meersschaut, V.V. Afanas'ev, L. Breuil, A. Cacciato, K. Opsomer, B. Brijs, H. Tielens, G. Pourtois, H. Bender, M. Jurczak, J. Van Houdt, S. Van Elshocht, J.A. Kittl, Lanthanide aluminates as dielectrics for non-volatile memory applications: material aspects. J. Electrochem. Soc. **158**, H778–H784 (2011). https://doi.org/10.1149/1.3596557
52. M. Tanaka, T. Shishido, H. Horiuchi, N. Toyota, D. Shindo, T. Fukuda, Structure studies of $CeAlO_3$. J. Alloys Compd. **192**, 87–89 (1993). https://doi.org/10.1016/0925-8388(93)90194-R

53. Y. Onishi, T. Nakamura, H. Sone, S. Adachi, Luminescence properties of Eu 3+ -activated TbAlO 3 perovskite compound synthesized by metal organic decomposition. Jpn. J. Appl. Phys. **57**, 082601 (2018). https://doi.org/10.7567/JJAP.57.082601
54. B.B. de Oliveira Neto, C. da Costa Pinto, L. da Silva Marques, P. Chaudhuri, A. Ghosh, D.M. Trichês, S.M. de Souza, Experimental and theoretical characterization of SmAlO3 perovskite synthesized by mechanical alloying. J. Solid State Chem. **298**, 122097 (2021). https://doi.org/10.1016/j.jssc.2021.122097
55. H. Li, S. Ma, J. Ye, N. Li, X. Wang, The effects of pressure on the lattice of the rare-earth-based perovskite-type oxides $SmAlO_3$ and $NdAlO_3$. New J. Phys. **24**, 113008 (2022). https://doi.org/10.1088/1367-2630/ac9b82
56. M.A. Hernández-Rodríguez, V. Monteseguro, A.D. Lozano-Gorrín, F.J. Manjón, J. González-Platas, P. Rodríguez-Hernández, A. Muñoz, V. Lavín, I.R. Martín, U.R. Rodríguez-Mendoza, Structural, vibrational, and elastic properties of yttrium orthoaluminate nanoperovskite at high pressures. J. Phys. Chem. C **121**, 15353–15367 (2017). https://doi.org/10.1021/acs.jpcc.7b04245
57. E. van der Kolk, J.T.M. de Haas, a. J.J. Bos, C.W.E. van Eijk, P. Dorenbos, Luminescence quenching by photoionization and electron transport in a LaAlO3:Ce3+ crystal. J. Appl. Phys. **101**, 083703–083703–4 (2007). https://doi.org/10.1063/1.2715748
58. M.M. Smedskjaer, J. Wang, Y. Yue, Tunable photoluminescence induced by thermal reduction in rare earth doped glasses. J. Mater. Chem. **21**, 6614 (2011). https://doi.org/10.1039/c1jm10472a
59. T. Aitasalo, P. Dereń, J. Hölsä, H. Jungner, J.C. Krupa, M. Lastusaari, J. Legendziewicz, J. Niittykoski, W. Stręk, Persistent luminescence phenomena in materials doped with rare earth ions. J. Solid State Chem. 114–122 (2003). https://doi.org/10.1016/S0022-4596(02)00194-9
60. M. Puchalska, E. Zych, The effect of charge compensation by means of Na+ ions on the luminescence behavior of Sm3+-doped $CaAl_4O_7$ phosphor. J. Lumin. **132**, 826–831 (2012). https://doi.org/10.1016/J.JLUMIN.2011.11.015
61. J.B. Gruber, K.L. Nash, R.M. Yow, D.K. Sardar, U.V. Valiev, A.A. Uzokov, G.W. Burdick, Spectroscopic and magnetic susceptibility analyses of the 7FJ and 5D4 energy levels of Tb3+(4f8) in TbAlO3. J. Lumin. **128**, 1271–1284 (2008). https://doi.org/10.1016/j.jlumin.2007.12.041
62. D. Dutczak, T. Jüstel, C. Ronda, A. Meijerink, Eu 2+ luminescence in strontium aluminates. Phys. Chem. Chem. Phys. **17**, 15236–15249 (2015). https://doi.org/10.1039/C5CP01095K
63. G.L. Hornyak, H.F. Tibbals, J. Dutta, J.J. Moore, *Introduction to Nanoscience and Nanotechnology* (CRC Press, 2008)
64. M. Enhessari, A. Salehabadi, Perovskites-based nanomaterials for chemical sensors. Progresses Chem. Sens. (2016). https://doi.org/10.5772/62559
65. B.M. Tissue, Synthesis and luminescence of lanthanide ions in nanoscale insulating hosts. Chem. Mater. **10**, 2837–2845 (1998). https://doi.org/10.1021/cm9802245
66. T. Minami, K.C. Patil, Solution combustion synthesis of nanoscale oxides and their composites. Mater. Phys. Mech. **4**, 134–137 (2001). https://doi.org/10.1016/j.materresbull.2008.09.034
67. S.R. Chauruka, A. Hassanpour, R. Brydson, K.J. Roberts, M. Ghadiri, H. Stitt, Effect of mill type on the size reduction and phase transformation of gamma alumina. Chem. Eng. Sci. **134**, 774–783 (2015). https://doi.org/10.1016/j.ces.2015.06.004
68. T.A. Safeera, J. Johny, S. Shaji, Y.T. Nien, E.I. Anila, Impact of activator incorporation on red emitting rods of ZnGa2O4:Cr3+ phosphor. Mater. Sci. Eng. C **94**, 1037–1043 (2019). https://doi.org/10.1016/j.msec.2018.10.059
69. M. Ghasdi, H. Alamdari, S. Royer, A. Adnot, Electrical and CO gas sensing properties of nanostructured La 1-xCexCoO3 perovskite prepared by activated reactive synthesis. Sens. Actuators, B Chem. **156**, 147–155 (2011). https://doi.org/10.1016/j.snb.2011.04.003
70. N. Kamarudin, W.S.W. Abdullah, M.A.A. Hamid, M.T. Dollah, Thermoluminescence (TL) properties and x-ray diffraction (XRD) analysis of high purity CaSO4: Dy TL material. AIP Conf. Proc. **1614**, 94–98 (2014). https://doi.org/10.1063/1.4895178
71. Y.A.N. Xing-huang, Z. Song-sheng, Y.U. Rui-min, C.A.I. Jing, X.U. Zhi-wei, L.I.U. Chun-jia, L.U.O. Xue-tao, M. Science, Preparation of YAG Ce 3 + phosphor by sol-gel low temperature combustion method and its luminescent properties, 3–8 (2007)

72. J.R. Jayaramaiah, B.N. Lakshminarasappa, K.R. Nagabhushana, Luminescence performance of europium-doped yttrium oxide thin films. J. Lumin. **157**, 63–68 (2015). https://doi.org/10.1016/j.jlumin.2014.08.003
73. K.C. Patil, M.S. Hegde, T. Rattan, S.T. Aruna, *Nanocrystalline Oxide Materials* (2008)
74. L.C. Klein, Optical Materials | Sol-Gel Materials, Encycl. Mod. Opt., Elsevier, pp. 16–21 (2005). https://doi.org/10.1016/B0-12-369395-0/00873-3
75. A.R. Sanwaria, M. Nagar, R. Bohra, A. Chaudhary, S.M. Mobin, P. Mathur, B.L. Choudhary, Sol–gel synthesis of highly pure α-Al_2O_3 nano-rods from a new class of precursors of salicylaldehyde-modified aluminum(iii) isopropoxide. Crystal and molecular structure of [Al(OC 6 H 4 CHO) 3], RSC Adv. **4**, 30081–30089 (2014). https://doi.org/10.1039/C4RA03245D
76. S.V. Bhandare, R. Kumar, A.V. Anupama, H.K. Choudhary, V.M. Jali, B. Sahoo, Mechanistic insights into the sol-gel synthesis of complex (quaternary) Co–Mn–Zn-spinel ferrites: an annealing dependent study. Ceram. Int. **46**, 17400–17415 (2020). https://doi.org/10.1016/j.ceramint.2020.04.031
77. T.R. Gurugubelli, B. Babu, J. Kim, K. Yoo, Hydrothermal synthesis of Fe-doped ZnAl2O4 nanosheets: bandgap engineering and room temperature ferromagnetism. Chem. Pap. **75**, 6407–6416 (2021). https://doi.org/10.1007/s11696-021-01807-4
78. S. Huang, Z. Wei, X. Wu, J. Shi, Optical properties and theoretical study of Mn doped $ZnAl_2O_4$ nanoparticles with spinel structure. J. Alloys Compd. **825**, 154004 (2020). https://doi.org/10.1016/j.jallcom.2020.154004
79. P. Qi, Y. Cong, L. Yu, X. Fu, X. Ge, C. Hao, L. He, Effect of co-precipitation synthesis parameters on gadolinium aluminate nanoparticles. Mater. Lett. **341**, 134163 (2023). https://doi.org/10.1016/j.matlet.2023.134163
80. N. Salah, P.D.D. Sahare, A.A.A. Rupasov, Thermoluminescence of nanocrystalline LiF:Mg, Cu, P. J. Lumin. **124**, 357–364 (2007). https://doi.org/10.1016/j.jlumin.2006.04.004
81. B.W. Nuryadin, A. Sawitri, E.C.S. Mahen, A.Y. Nuryantini, Synthesis and morphology properties of Mn-doped zinc aluminate (ZnAl 2–x O_4: xMn 4+, x=1%) using co-precipitation methods with microwave thermal assisted. J. Phys. Conf. Ser. **812**, 012020 (2017). https://doi.org/10.1088/1742-6596/812/1/012020
82. A. Talimian, V. Pouchly, H.F. El-Maghraby, K. Maca, D. Galusek, Impact of high energy ball milling on densification behaviour of magnesium aluminate spinel evaluated by master sintering curve and constant rate of heating approach. Ceram. Int. **45**, 23467–23474 (2019). https://doi.org/10.1016/j.ceramint.2019.08.051
83. V. Amendola, M. Meneghetti, What controls the composition and the structure of nanomaterials generated by laser ablation in liquid solution? Phys. Chem. Chem. Phys. **15**, 3027–3046 (2013). https://doi.org/10.1039/C2CP42895D
84. S.-J. Cho, M.-J. Uddin, P. Alaboina, Review of nanotechnology for cathode materials in batteries. Emerg. Nanotechnologies Recharg. Energy Storage Syst. Elsevier, 2017, pp. 83–129. https://doi.org/10.1016/B978-0-323-42977-1.00003-0
85. A. Osman, N. Shahidzadeh, H. Stitt, N. Shokri, Morphological transformations during drying of surfactant-nanofluid droplets. J. Ind. Eng. Chem. **67**, 92–98 (2018). https://doi.org/10.1016/j.jiec.2018.06.019
86. A.-T. Tran, V.-T. Tran, N.T.M. Nguyet, A.T.-Q. Luong, T. Van Le, N.H.H. Phuc, Solid-state reaction synthesis of $MgAl_2O_4$ Spinel from MgO–Al_2O_3 composite particles prepared via electrostatic adsorption. ACS Omega **8**, 36253–36260 (2023). https://doi.org/10.1021/acsomega.3c04782
87. S. Lorentzou, C.C. Agrafiotis, A.G. Konstandopoulos, Aerosol spray pyrolysis synthesis of water-splitting ferrites for solar hydrogen production. Granul. Matter **10**, 113–122 (2008). https://doi.org/10.1007/s10035-007-0069-8
88. C. Dimitriou, P. Psathas, M. Solakidou, Y. Deligiannakis, Advanced flame spray pyrolysis (FSP) technologies for engineering multifunctional nanostructures and nanodevices. Nanomaterials **13**, 3006 (2023). https://doi.org/10.3390/nano13233006

89. A.B. Workie, H.S. Ningsih, S.-J. Shih, An comprehensive review on the spray pyrolysis technique: historical context, operational factors, classifications, and product applications. J. Anal. Appl. Pyrolysis **170**, 105915 (2023). https://doi.org/10.1016/j.jaap.2023.105915
90. W.Y. Teoh, R. Amal, L. Mädler, Flame spray pyrolysis: an enabling technology for nanoparticles design and fabrication. Nanoscale **2**, 1324 (2010). https://doi.org/10.1039/c0nr00017e
91. W. Teoh, A perspective on the flame spray synthesis of photocatalyst nanoparticles. Materials (Basel) **6**, 3194–3212 (2013). https://doi.org/10.3390/ma6083194
92. L. Xi, Y. Wang, L. Yin, P.D. Townsend, The influence of germanium ions on the infrared long persistent phosphor Zn1+xAl2-2xGexO4: Cr. J. Alloys Compd. **820**, 153094 (2020). https://doi.org/10.1016/j.jallcom.2019.153094
93. B.C. Jamalaiah, N. Madhu, Luminescence properties of $SrAl_2O_4$: Tb3+/Bi3+ nanophosphors for photonic applications. J. Mol. Struct. **1205**, 127599 (2020). https://doi.org/10.1016/j.molstruc.2019.127599
94. S. Chawla, N. Kumar, H. Chander, Broad yellow orange emission from $SrAl_2O_4$:Pr3+ phosphor with blue excitation for application to white LEDs. J. Lumin. **129**, 114–118 (2009). https://doi.org/10.1016/j.jlumin.2008.08.009
95. N. Basavaraju, K.R. Priolkar, D. Gourier, S.K. Sharma, A. Bessière, B. Viana, The importance of inversion disorder in the visible light induced persistent luminescence in Cr 3+ doped AB2O_4 (A = Zn or Mg and B = Ga or Al). Phys. Chem. Chem. Phys. **17**, 1790–1799 (2015). https://doi.org/10.1039/C4CP03866E
96. S. Devi, A. Khatkar, V.B. Taxak, A. Hooda, P. Sehrawat, S. Singh, S.P. Khatkar, Influence of Tb3+ doping on the structural and down-conversion luminescence behaviour of SrLaAlO4 nanophosphor. J. Lumin. **221**, 117064 (2020). https://doi.org/10.1016/j.jlumin.2020.117064
97. S. Li, Q. Zhu, X. Li, X. Sun, J.-G. Li, Near-infrared emitting microspheres of LaAlO3:Mn4+: Defects engineering via Ge4+ doping for greatly enhanced luminescence and improved afterglow. J. Alloys Compd. **827**, 154365 (2020). https://doi.org/10.1016/j.jallcom.2020.154365
98. H. Luo, P. Dorenbos, The dual role of Cr 3+ in trapping holes and electrons in lanthanide co-doped $GdAlO_3$ and $LaAlO_3$. J. Mater. Chem. C. **6**, 4977–4984 (2018). https://doi.org/10.1039/C8TC01100A

Persistent Luminescence and in Vivo Bioimaging

Soorya G. Nath and E. I. Anila

Abstract In vivo bioimaging encompasses the non-invasive visualization of internal cells or organs within the human body, facilitating the real-time monitoring of dynamic biological processes. This form of bioimaging holds paramount significance within clinical domains due to its capacity to facilitate early disease detection, notably in cases of life-threatening conditions such as cancer. Although established imaging modalities like ultrasound, magnetic resonance imaging, and fluorescence imaging exist, the quest for innovative imaging probes with enhanced visibility to provide high-resolution images remains an active area of research. The introduction of persistent luminescent nanoprobes, characterized by their unique afterglow luminescence property, presents an exciting avenue in the realm of bioimaging. These nanoprobes offer several advantages over traditional fluorescent probes, notably in their ability to surmount light attenuation challenges within biological tissues including issues such as autofluorescence and light scattering. The chapter offers a succinct introduction to two pivotal facets of fluorescence imaging, namely persistent luminescence and upconversion luminescence while elucidating the underlying mechanisms governing their luminescence behavior. Moreover, this chapter provides comprehensive insights into various fluorescence imaging techniques, encompassing near-infrared imaging, upconversion imaging, X-ray-activated luminescence imaging, persistent luminescence imaging, and various multimodal imaging techniques.

Keywords Persistent luminescence · In vivo imaging · Upconversion · NIR fluorescent probes · Multimodal imaging

S. G. Nath · E. I. Anila (✉)
Optoelectronic and Nanomaterials' Research Lab, Department of Physics, Union Christian College, Aluva, Kerala 683102, India
e-mail: anila.ei@christuniversity.in

E. I. Anila
Department of Physics and Electronics, CHRIST (Deemed to be University), Bengaluru, Karnataka 560029, India

V. Kumar et al. (eds.), *Persistent Luminescence*, Progress in Optical Science and Photonics 30, https://doi.org/10.1007/978-981-97-4943-0_7

1 Introduction

Persistent luminescent (PersL) materials possess unique properties that they have been applied in a variety of fields like lighting, energy and environmental, ornamental, biomedical, etc. The development of efficient, nontoxic PersL materials for biomedical applications is a hot research area for material scientists. PersL nanoparticles have wonderful applications in luminescence imaging of biological systems including diagnosis of diseases, lymphatic imaging, targeted drug delivery, understanding cellular processes, etc. Cells and cellular components are usually observed through an optical microscope after color staining with dyes, which is a direct technique providing minute details of cells and tissues. However excessive use of these organic dyes leads to toxicity issues. Background light and autofluorescence are also major concerns with phosphor materials when applied in luminescence bioimaging. So, the application of nontoxic phosphor materials giving PersL to the field of in vivo imaging is essential. This chapter includes an introduction to PersL, brief discussions about in vivo imaging, and various factors affecting luminescence during imaging. Also, brief descriptions of various fluorescent probes used for in vivo imaging techniques are covered in the upcoming sections.

2 Persistent Luminescence

Persistent luminescence (PersL) is an optical phenomenon where we obtain light emission even after the removal of the source of excitation. Here, the excitation source can be UV radiation, visible light, X rays, plasma beam, electron beam, etc. It is also referred to as long-lasting phosphorescence, though the luminescence mechanism is different in both cases. Phosphorescence can be understood using Jablonski diagram (Fig. 1a) in which both singlet and triplet states are involved. Consider an electron getting excited from singlet state S_0 to S_2 from where it decays to S_1 by radiation less decay known as internal conversion (IC) as S_2 is more unstable compared to S_1 with lifetimes ~10^{-11} s and 10^{-8} s respectively. From S_1, the electron can return to the singlet ground state through two paths; one is direct transition known as fluorescence. Another path is through intersystem crossing where the electron changes its spin multiplicity from excited singlet state to excited triplet state. This electron undergoes radiative transition to singlet ground state known as phosphorescence. This is a slow and forbidden transition and its timescale is 10^{-3}–100 s.

In the case of persistent phosphors there exist trap centers along with luminescence emission centers. The long decay time in persistent luminescence is due to the involvement of traps in the process. The trap centers are shallow lying at an energy gap of less than 1 eV below the conduction band or above the valance band in the case of electrons and holes respectively. The electron or hole trapping centers are generated by intrinsic defects such as vacancies, F centers, intentionally added co-dopants, etc. Excitation corresponds to the band gap energy, which means electrons

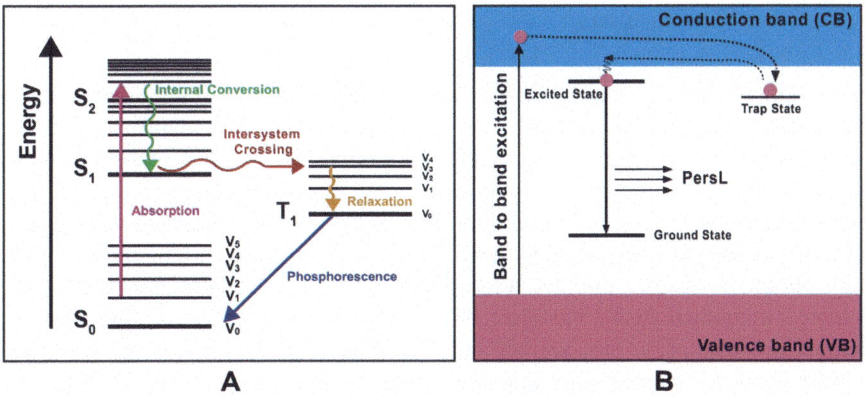

Fig. 1 **a** Partial Jablonski diagram for phosphorescence, **b** A schematic diagram for the illustration of the persistent luminescence (PersL) mechanism

are excited to the conduction band. These electrons fall into the shallow traps due to Coulombic force or by quantum tunneling process through the forbidden band, from where they will be detrapped again to the conduction band as the energy levels of trapping centers are equivalent to room temperature thermal energy. They are further migrated to the excited state of the luminescence center from where they undergo a radiative transition to the ground state. The whole process takes several minutes or hours leading to persistent luminescence (Fig. 1b). In the work reported by N Basavaraju et al. defects due to inversion sites act as trapping centers giving rise to red PersL in $MgGa_2O_4$: Cr^{3+}phosphor [1]. Luminescence lifetime in PersL is three orders of magnitude higher than that of phosphorescence.

PersL materials have a variety of applications. The safety signage in airplanes is persistent phosphors. They are used in toys, road markings, watches, wall decorations, etc. They can be used for other applications like solar energy storage, photocatalysis, temperature sensors, dosimeters, etc. depending on the depth of trap centers.

PersL materials are suitable for in vitro and in vivo bioimaging applications. Compared to normal fluorescent probes, better resolution can be obtained in imaging using PersL materials since imaging is done after the excitation source is switched off. Moreover, autofluorescence is inhibited due to the lack of an exciting source.

Before moving to the applications of PersL phosphors in in vivo imaging, brief discussions about in vivo imaging and various factors affecting luminescence during imaging are necessary. Also, brief descriptions of various fluorescent probes used for in vivo imaging techniques are covered in the upcoming sections.

3 In Vivo Bioimaging

The complex biological processes occurring inside the body of living organisms can be visualized after the introduction of biomedical imaging. The picturization of the internal structure of organs and cells plays a crucial role in the diagnosis of many diseases. Early detection of malignant cells is the most important and difficult part of cancer therapy. Conventional imaging techniques such as computed tomography (CT), magnetic resonance imaging (MRI), positron emission tomography (PET) and single photon emission computed tomography (SPECT) take significant roles in the biological imaging industry. However, these techniques usually require some contrast agents (CnA) to improve the accuracy and sensitivity of the diagnosis images and such contrast agents have inbuilt limitations like toxicity, less circulation period, etc. One of the major problems associated with in vivo imaging techniques is the absorption and scattering of radiations by the animal tissues, which are basically opaque in nature and these lead to the added requirements of contrast agents mentioned before. Usually, gadolinium (III) based contrast agents are widely employed in MRI but the toxicity concerns related to the unforeseen leakage of free Gd led to the use of other contrast agents such as highly paramagnetic iron oxide nanoparticles [2, 3]. The rapid developments in the bioimaging industry led to multimodal imaging techniques through the combination or merging of different imaging modes such as PET/CT, PET/MRI, etc.

Optical imaging is another technique used for in vitro and in vivo imaging using optical signals. Fluorescence imaging plays an important role here owing to its comparatively benign nature and this technique mainly requires fluorescent probes that generate optical signals upon activation. Organic fluorophores as well as luminescent proteins are extensively used for this purpose. The development of ultra-fine fluorescent nanoparticles for in vivo targeting seeks much attention today for its peculiarities such as increased specificity, high resolution, non-hazardous nature of optical radiations, less toxicity of the fluorophores, etc. The non-toxic nature and targeting efficiency of such probes can be further enhanced by surface functionalization using various proteins. Ultra-fine nanoprobes such as Quantum Dots (QD), Lanthanide activated nanoparticles, etc. are nowadays widely investigated for their clinical use in cancer targeting and therapy.

There are many works on using fluorescent nanoparticles for in vivo imaging which are excited at different wavelengths, i.e., UV (Ultraviolet), visible, NIR (Near-infrared), and X-ray wavelengths [4–7]. They include organic materials including dyes, inorganic semiconductor quantum dots, carbon dots, etc. UV and visible excitation suffer from a major limitation such as photon scattering, absorption, and tissue autofluorescence, which in turn affects the fluorescence output. Cells consist of certain molecules that give rise to color emission when interacting with UV/visible light, which is termed autofluorescence. This kind of intrinsic fluorescence of cells affects the overall performance of the fluorescent probes. More details regarding such effects can be found in previous reports [8]. This kind of autofluorescence and scattering effects are greatly avoided in NIR excitable fluorescent probes.

4 Factors Leading to Fluorescence Attenuation During in Vivo Imaging

The discovery of the microscope leads to huge advancements in the biomedical fields particularly in biological imaging. There are several methods developed for in vivo imaging such as MRI, CT, PET, X-ray imaging, optical imaging, etc. in which the more expensive and complicated radio-wave based (MRI) or radiation-based techniques (PET, X-ray) overpowered optical imaging. Biological tissues are usually opaque to visible range in the electromagnetic (EM) spectrum, which is the major drawback associated with optical imaging methods.

In vivo fluorescence imaging technique involves the excitation of an exogenous fluorophore introduced into an organ or tissue in an animal (or in some cases endogenous fluorophores) using photons usually in the UV, visible, NIR, or X-ray region of the EM spectrum. Fluorescent probes are usually distributed in animal tissues, which can be deep inside the body, which means that the excitation light needs to travel deep through the biological tissues to reach these probes thus making the interactions of light with tissues an important factor during imaging. The main processes occurring during in vivo fluorescence imaging in the UV/visible region due to such interactions are interface reflection/refraction, scattering, absorption as well as intrinsic fluorescence or autofluorescence. Before moving to in vivo fluorescence imaging using nanoparticles, a brief discussion about these effects can help to understand more about the concept.

4.1 Interface Reflection

When the excitation light touches the surface of the biological tissues, there is a change in the refractive index of the medium, which causes reflection of the incident light at the interface. The amount of reflected light depends on the angle of incidence and the difference in refractive index. This kind of interface reflection usually does not cause any signal loss or image distortion as long as the roughness and curvature of the surface are small compared to the imaged object and also the excitation is close to normal. When the surface is rough, diffuse reflection occurs at the air/tissue interface leading to signal loss.

4.2 Scattering

The un-reflected excitation light can face scattering effects by different elements inside the tissue also due to the non-homogenous nature of refractive indices. Different components inside the tissue such as lipid membranes, water, etc. can lead to scattering effects. Scattering of light can be greatly minimized by using higher wavelength excitation, especially in the NIR range i.e., scattering is inversely proportional to the fourth power of the excitation wavelength.

4.3 Absorption

The light entered inside the tissues faces another problem other than scattering, which is the absorption of light by various components in the tissue. Many biological molecules such as chromophores inside animal tissues absorb incoming light and convert it into heat. An example of this is hemoglobin (both oxygenated and non oxygenated), which absorbs light in the wavelength range below 600 nm and thus gives red color to blood. Many such molecules show strong absorption of light in the UV/visible wavelength range such as aromatic amino acid deposits in various proteins such as tyrosine, phenylalanine, tryptophan, various kinds of melanin, flavins, reduced nicotinamide adenine dinucleotide (NADH) [9], etc. These kinds of absorption effects can also be largely avoided by using NIR radiation.

4.4 Autofluorescence

Another factor restricting deep tissue fluorescence imaging is the intrinsic fluorescence or autofluorescence effect. The light absorbed by certain biological molecules can be re-emitted to give fluorescence emission. That is, certain components inside the tissue can be excited using light in the UV or visible range and they deexcite to the ground state by emission of radiation usually in the visible range. This effect is termed autofluorescence to differentiate this from the fluorescence observed from externally supplied fluorescent probes. The amount of autofluorescence emission depends on the concentration as well as the distribution of such molecules inside the tissue.

There are many biological components inside the tissue as mentioned in Sect. 4.3 which act as endogenous fluorescence centers. The emission of flavins lies in the range 500–600 nm and of NADH is centered at 460 nm [9]. There are many other such components leading to the autofluorescence effect and the excitation as well as emission ranges of such molecules can be found in published reports [8, 10]. Figure 2 depicts a simple representation of all light attenuation effects.

All these kinds of effects limiting deep tissue imaging occur usually in the UV/visible region but there are some elements such as water showing such kinds of attenuation effects in NIR region also. But moving to far NIR region i.e., beyond 1500 nm, these effects are completely removed. Therefore, the development of fluorescent probes in the NIR II region can bring significant developments in the fluorescence imaging techniques.

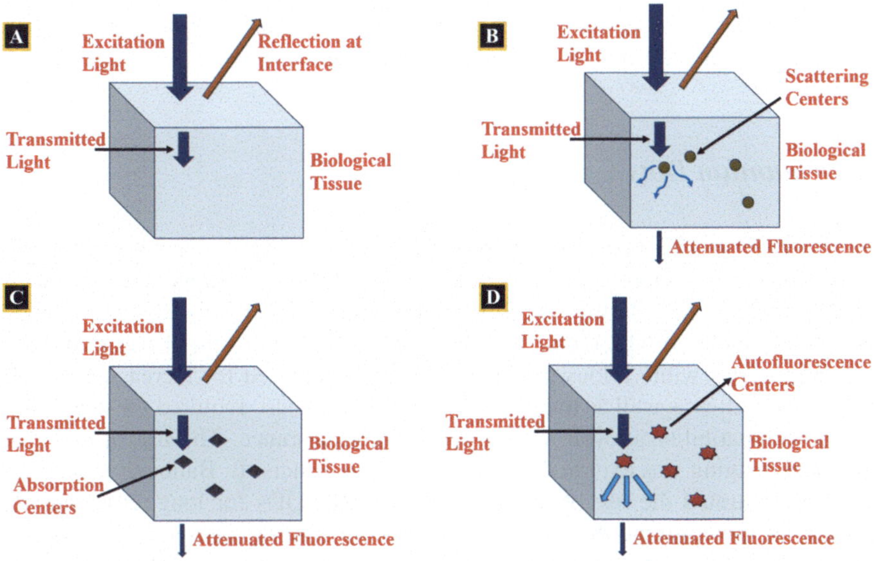

Fig. 2 Light attenuation effects inside biological tissues: **a** Interface reflection, **b** scattering of light, **c** absorption of light by various absorption centers, and **d** autofluorescence emission from various biological molecules

5 Fluorescent Probes for in Vivo Imaging

Fluorescence imaging is a widely investigated area in molecular imaging techniques and requires an exogenous or endogenous fluorescent probe which can emit detectable electromagnetic radiation. In the following section, let us discuss the various types of fluorescent probes used for in vivo imaging.

5.1 Organic Fluorescent Probes

Many organic fluorescent probes have been developed so far, and the commonly used ones include cyanines or carbocyanine dyes. This is the most common probe used for in vivo applications and one of its derivatives namely ICG or indocyanine green is the first ever NIR based fluorescent dye approved by FDA (Food and Drug Administration) for clinical trials. Other organic probes include methylene blue (MB) (also approved by FDA after ICG), BIODPY (difluoroboron dipyrromethene), nicotinamide adenine dinucleotide (NADH), 5-aminolevulinic acid (5-ALA), rhodamine, flavin adenine dinucleotide (FAD), porphyrins, etc.

Conventional imaging techniques use organic dyes and the major drawback suffered by those dyes are low photostability as well as wide emission range. These led to the development of many other fluorescent probes such as quantum dots with high photostability and size-dependent emission characteristics.

5.2 *Quantum Dots*

Fluorescent probes based on semiconductor quantum dots (QDs) are now extensively studied for in vivo fluorescence imaging in view of their excellent size dependent properties such as long-lasting stability and brightness. Their emission range can be tuned from visible to NIR region and they exhibit high quantum yields. Functionalization of QDs with various proteins are usually required for increasing its cell binding properties as well as tumor targeting and this is also required for increasing their biocompatibility as well as successful body clearance. The use of QDs for biological imaging was explored well by many researchers. B. Ballou et al. in their work demonstrated the use of polymer encapsulated QDs for long-term imaging in live mice thus proving the stability of QDs long after the injection [11]. Water-soluble QDs were used in an in vivo imaging experiment using two-photon imaging in which the blood vessels in live mice were visible [12]. Another work used bio-conjugated and polymer-coated CdSe-ZnS QDs as fluorescent probes for in vivo imaging of human prostate cancer cells developing in mice [13]. The preparation of self-illuminating QDs and its use for in vivo imaging were explored in a work by Sun et al. [14]. Rare-earth doped semiconductor QDs provide high quantum efficiency with a narrow emission range. Details of similar studies performed using QDs can be found in the review article written by Michalet et al. [15].

Even though there are many advantages for quantum dots like narrow emission range which can be tuned by altering particle size, high photostability, enhanced fluorescence emission, higher quantum yields, etc. QDs suffer from a major disadvantage of cytotoxicity, especially those containing the toxic Cd ions. Accidental leakage of Cd ions can lead to adverse effects when used for in vivo imaging. Yet, the material has greater potential due to the development of self-illuminating as well as NIR-emitting QDs. So, the development of non-toxic quantum dots with PersL can provide remarkable contributions to the fluorescence imaging industry.

5.3 *Noble Metal Nanoparticles*

Among the various metal nanoparticles, silver (Ag) and gold (Au) are the most explored materials for in vivo applications. There are many works exploring the in vivo applications of Ag nanoparticles but the major concern related to this material is their toxic nature of the material. The toxic nature of Ag nanoparticles as well as ionic silver was reported in many works [16, 17]. In this context, Au nanoparticles seek more attention as they show much less toxicity compared to Ag. Au nanoparticles have properties such as high photostability, tunable PL emission from UV to

NIR region, large lifetime, etc. thus making it an important nanomaterial for imaging applications [18]. In a recent work by Yang et al., Au nanoclusters were prepared using the protein alpha-lactalbumin (α-LA) for the detection of breast cancer in mice. The material acts as a fluorescent probe with multiple functions by providing tumor imaging through fluorescence detection, and breast cancer detection through CT and MRI and is also utilized for anticancer therapy [19]. Other reports using metal nanoparticles for in vivo applications can be found in published review articles [20, 21]. Noble metal nanoparticles can also be used for fluorescence imaging applications by reducing their toxicity.

5.4 Silica Nanoparticles

Silica-based nanomaterials used for many biological applications are non-luminescent nanoparticles of silica, which is non-toxic. This material can be easily doped with rare-earth elements, organic fluorophores, etc. for fluorescence imaging applications. Chen et al., developed dye-doped and PEG-coated silica nanoparticles for targeted fluorescence imaging of melanoma cells [22]. In vivo applications of silica-based nanomaterials can be further found in published review articles [20, 23]. The biological distribution as well as the long-term toxic effects of silica nanoparticles need to be studied more for their applications in more in vivo imaging experiments.

5.5 Carbon Dots

Carbon dots (CD) are zero-dimensional nanostructures of carbon with unique optical properties and they were first synthesized in 2004 [24]. They are fluorescent nanoparticles with tunable emission characteristics. Their emission wavelength can be changed by changing the synthesis method and usually, the emission range covers UV to yellow region. Their fluorescence is usually observed in the blue region but excitation with long wavelength gives green as well as yellow emission. Their unique properties such as easy synthesis, tunable fluorescence, low cytotoxicity, increased photostability, etc. make this material ideal for imaging applications.

In vivo imaging using CDs was first reported by Yang et al. in 2009 in which CD solution was injected into mice to perform imaging and this report opened the door for using CDs as an efficient fluorescent probe for in vivo imaging [25]. In another study by Liu et al., CDs were prepared from *F. nucleatum* (an anaerobe), and they were used for in vivo imaging in mice. The fluorescence from the material was observed from different parts of the body through blood circulation and the material was cleared from the body within 24 h [26]. Recent works also focus on obtaining persistent luminescence from carbon dots which can give more opportunities for in vivo imaging applications [27]. More applications using CD materials could be obtained from published reports [23, 28]. One of the major drawbacks of the material is its emission range, which normally falls in the blue-green region, which is also the region where autofluorescence occurs.

6 Developments in in Vivo Fluorescence Imaging Techniques Using NIR, Upconversion, and X-ray Activated Fluorescent Probes

Fluorescent probes for in vivo imaging applications such as NIR-activated materials, upconversion luminescent materials as well as X-ray-activated fluorescent probes having high tissue penetration depth can avoid autofluorescence thereby realizing high-resolution images. In the following section, we discuss the developments in in vivo imaging experiments using these materials emphasizing PersL.

6.1 NIR Fluorescent Probes for in Vivo Imaging and the Concept of Biological Windows

One of the major limitations associated with in vivo fluorescence imaging is lower penetration depth. This is because most animal tissues are usually opaque to the electromagnetic spectrum in the visible region so that light cannot penetrate deep into the tissues due to high photon scattering/absorption in the tissues which leads to poor-resolution images. Autofluorescence is another reason for low-contrast images. To get the best imaging results, the signal intensity must be high compared to the background noises, or in other words, the signal-to-background ratio should be high. As mentioned earlier, fluorescent probes in the NIR region can overcome most of the above-mentioned limitations.

6.1.1 Biological Windows

NIR wavelength range of the electromagnetic spectrum can be divided into three regions and they are usually called the first, second, and third biological windows. NIR-I region with a wavelength range of 700–950 nm is called the first biological window, the NIR-II region with a wavelength range of 1000–1350 nm is known as the second biological window, and NIR-III region with wavelength in the range 1550–1870 nm is called the third biological window [29]. In these two regions, the NIR-I region is usually called the 'biological transparent window' in which most of the current research is focused. The development of material science and chemistry over the last few years led to the development of new probes in the NIR-II and NIR-III regions for in vivo imaging.

6.1.2 In Vivo Imaging Using NIR Fluorescent Probes

The first ever NIR fluorophore that came into clinical use is indocyanine green (ICG) and it has an emission wavelength of ~800 nm [30]. This was followed by the use of methylene blue (MB) with emission at ~700 nm. But both these dyes are charged and this causes non-specific binding to body proteins which in turn results in increased noise. Thus, the research focused on the development of new probes in the NIR region with low noise levels that can specifically bind to the target tissues. Many such NIR-emitting fluorescent probes have been developed so far and this includes QDs, lanthanide-doped nanoparticles, carbon nanotubes, semiconducting polymers, etc. [9]. NIR fluorophores which do not require external excitation can greatly avoid the autofluorescence effect by the tissues. Such self-illuminating fluorophores based on NIR-I emitting QDs usually exhibit in vivo fluorescence emission through energy transfer mechanisms such as bioluminescence, chemiluminescence, and fluorescence resonance energy transfer [9]. For example, on the introduction of persistent luminescence (PersL) agents along with NIR fluorophores can transfer energy through fluorescence resonance energy transfer mechanism thus enabling self-illumination during in vivo imaging. PersL is a phenomenon in which the optical energy is released slowly and it can last from minutes to hours even after cutting off the excitation. The application of PersL for in vivo imaging was tested in a work by Maldiney et al. by preparing nanoparticles of $Ca_2Si_5N_8$:Eu^{2+}, Tm^{3+} from bulk material [31]. Mikael Palner and his co-workers report a semiconducting conjugating polymer nanoparticle that exhibits both NIR emission as well as PersL which lasts for about an hour and was also utilized for in vivo imaging in mice [32]. In another work, the authors developed a nanoprobe based on chromium-activated zinc gallate, which exhibited PersL upon excitation with red light. The material was also used for in vivo tumor targeting in mice after PEG capping [33]. Mn^{2+} doped nanophosphor showing long-lasting luminescence was used for in vivo imaging in yet another work [34]. The biodistribution and toxicity effects of zinc gallium oxide doped with Cr^{3+} ions were tested in mice in a work by Xia Sun and his co-workers and the work demonstrates the risk effect of the PEG functionalized $Zn_{1.1}Ga_{1.8}Sn_{0.1}O_4$:$Cr^{3+}$ nanoparticles when used for long term PersL bioimaging [35]. Figure 3 shows the in vivo and ex-vivo persistence luminescence imaging performed in Balb/c mice. For in vivo imaging, the mouse was injected with the sample (10 mg/kg) and in situ irradiated with an LED with wavelength 655 nm for 5 min. PersL signal was obtained one day after injection from most parts of the body and predominantly it occurred in the reticuloendothelial system (RES). The signal intensity slowly decreased with time but it existed even 60 days after injection which shows that the sample was distributed almost in all organs through blood circulation and it stays in the body for a long time (Fig. 3a). To study the biodistribution of the nanoparticles in the long term, the major organs from the body were collected at different times post-injection and they were irradiated by LED for 3 min and then imaged. One day after injection, strong PersL was observed from the liver, spleen, and lungs and no signals were obtained from the heart and kidney (Fig. 3b). The signals from the liver and lungs decreased with time

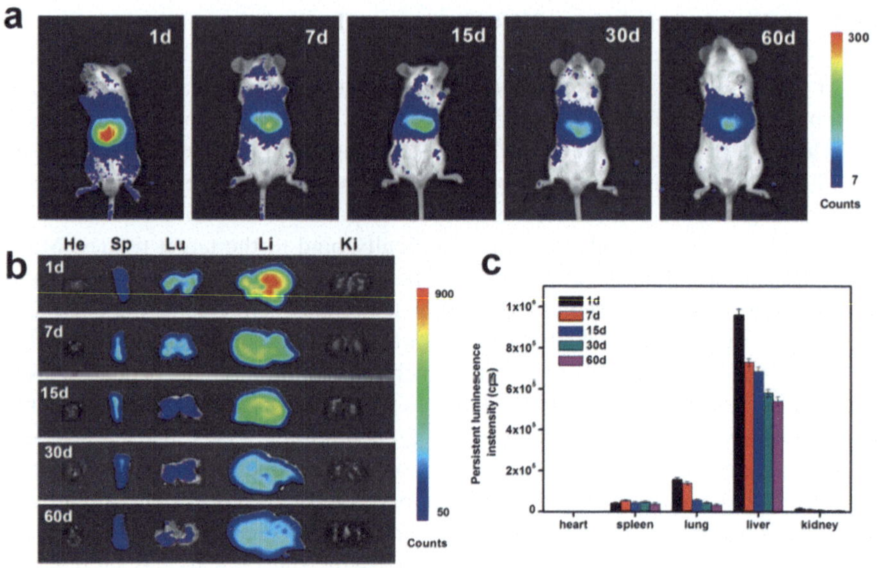

Fig. 3 In vivo and ex vivo persistent luminescence imaging. **a** In vivo persistent luminescence images of Balb/c mice with the intravenous injection of ZGO (10 mg/kg) at different time points, **b** ex vivo persistent luminescence imaging of the heart (He), liver (Li), spleen (Sp), lungs (Lu) and kidneys (Ki) collected from Balb/c mice injected with ZGO at different time points, and **(c)** persistent luminescence intensities of ZGO from isolated organs at different time points. The data represent the mean ± standard deviation (n = 5). Reprinted from Ref. [35]

(Fig. 3c) but existed up to 60 days after injection proving the long term biodistribution of the material. More details about NIR fluorescent probes and PersL can be found in published review articles [9, 36]. NIR fluorescent probes with the highest quantum yields are listed in Table 1 and the other NIR probes and their details are available from the published review article [9]. In vivo imaging in the third biological window (NIR-III region) became more prominent after it was first reported in 2011 by Kamimura et al. in which they used 1550 nm NIR radiation from rare-earth doped nanophosphors for in vivo imaging in mouse [37]. More details about such works in this window can be found from the review article by Hemmer et al. [29].

6.2 Upconversion Nanophosphors for in Vivo Imaging

Most of the materials used in the bioimaging industry are excited at higher energy and emit lower energy photons i.e., they exhibit Stokes shift. Certain materials possess the capability to exhibit anti-Stokes luminescence processes such as two-photon luminescence, second-harmonic generation, and upconversion luminescence (UCL). Among these, the two-photon absorption process as well as second-harmonic generation requires coherent excitation sources like lasers. Two photon process requires very

Table 1 List of NIR fluorescent probes with the highest quantum yield

NIR fluorescent probes	Quantum yield (%)	References
Boron dipyrromethene (BODIPY)	~100	Quan et al. [38], Watanabe et al. [39]
Rare-earth nanoparticles (e.g.: SrF_2: Nd, LaF_3: Nd)	50–90	Villa et al. [40], Rocah et al. [41]
Non-sulfonated cyanine dyes (CyX Series)	28	Hadrup et al. [17], Phillips et al. [42]
Quantum dots	~8–17 and higher	Xie et al. [43], Zhang et al. [44], Hong et al. [45]
Zwitterionic cyanine dyes	15.10	Choi et al. [46]

expensive pulsed lasers normally femtosecond lasers with high power for exciting the materials whereas UCL nanophosphors require only low-power excitation sources such as normal xenon lamps, continuous-wave lasers, etc.

Before moving to the in vivo imaging applications using UCL materials, a brief introduction about the UCL mechanisms needs to be discussed. UCL nanophosphors convert lower energy photons to higher energy photons or in other words these materials can convert NIR wavelengths to UV/visible wavelengths thus exhibiting the anti-Stokes process. Upconversion process can occur via various mechanisms namely excited state absorption (ESA), energy transfer upconversion (ETU) and photon avalanche (PA). These are illustrated in the Fig. 4 in which Fig. 4a represents the ESA process where the electron moves to an excited metastable state E_I through the absorption of a pump photon ($h\nu_1$) and then to an excited state E_{II} using another pump photon ($h\nu_2$). The upconversion emission occurs when the electron returns to the ground state (G) with maximum energy equal to $h\nu_1 + h\nu_2$. This kind of ESA process occurs in the case of singly doped material.

The ETU mechanism is the most efficient upconversion process which occurs in doubly doped materials (i.e., it occurs in 2 neighboring lanthanide ions) in which one ion act as sensitizer (S) and the other act as activator (A). There are five different types of ETU process namely ET followed by ESA, successive ET, cross relaxation (CR), cooperative sensitization, and cooperative luminescence of which ET following ESA is depicted in Fig. 4b. Here, the S ion absorbs the pump photon to get excited to metastable level E_I and this energy is then transferred to A ion which is used for exciting the A ion to its metastable level E_I and this is followed by ESA process to excite the A ion to E_{II}. More ETU schemes can be found from previous reports [47]. PA process involves the combined effects of ESA and ETU processes and in this process, the pump energy is resonant with the metastable state and excited state only and a simple energy level diagram representing PA mechanism is shown in Fig. 4c. Suppose ion 1 is in its ground state and ion 2 absorbs a photon through non-resonant absorption and get excited to its metastable level E_I and it is further excited to E_{II} through ESA process (through absorption of pump photon). Now, ion 2 can lose its energy through many processes including fluorescence and this de-excites the second ion to E_I and during which this energy is transferred to first ion (through cross

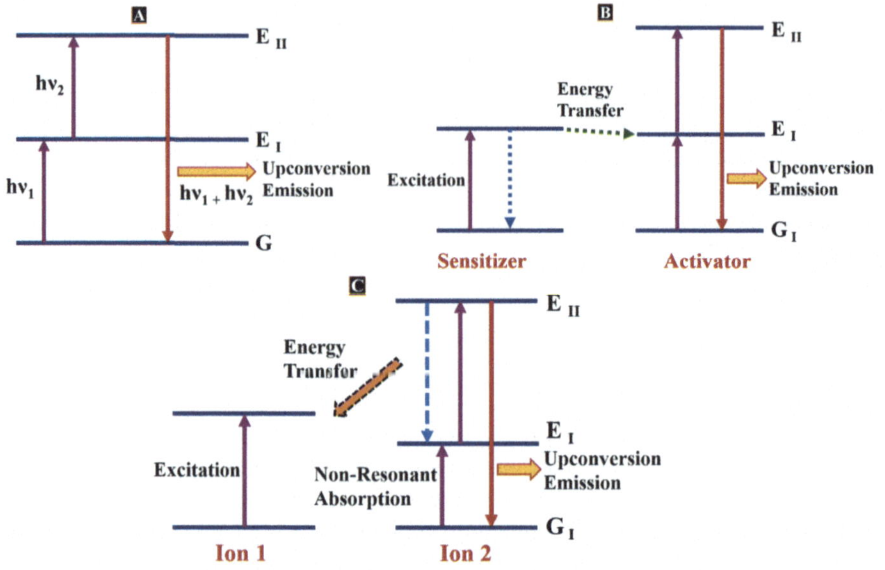

Fig. 4 UCL Mechanisms. **a** ESA process, **b** energy transfer followed by ESA process in ETU mechanism, and **c** PA process

relaxation) thus exciting ion 1 to its metastable level (E_I), thus making the two ions in their metastable levels. Thus, both these ions can now act as sensitizer ions, which can again produce four ions in metastable state and the process can continue, thus forming the avalanche effect leading to upconversion luminescence. This process occurs only when the excitation intensity is above a threshold value and hence it is the least occurring upconversion process. The detailed processes can be found from published reports [47–49].

Recent studies demonstrate on the use of UCL materials for in vivo applications and the predominant UCL materials include lanthanide-doped materials. One of the widely studied materials in this context is $NaRF_4$, where R represents rare earth elements and Yttrium (Y). $NaYF_4$ upconversion nanophosphor doped with Tm and Yb was used for in vitro imaging of pancreatic cancer cells and in vivo imaging in mouse [50]. Another work reported the synthesis of amine functionalized UCL material ($NaYF_4$: Yb, Er) denoted by UCNPs-NH_2 and the same conjugated with folic acid (FA) denoted as UCNPs-FA through simple hydrothermal microemulsion method [51]. The presence of amine group enabled the conjugation of the material with folic acid (FA) and both the materials (with and without FA conjugation) were injected to a nude mouse bearing HeLa tumor (in the right hind leg) for demonstrating the capability of the material for targeted in vivo imaging (Fig. 5). This was followed by the imaging experiment in which excitation was at 980 nm and the UC luminescence was observed at 600–700 nm range. UC luminescence signals were obtained from the tumor when injected with FA conjugated material and no signals were obtained for the UCNPs-NH_2 injected tumor (Fig. 5a, b). Liqin Xiong et al.

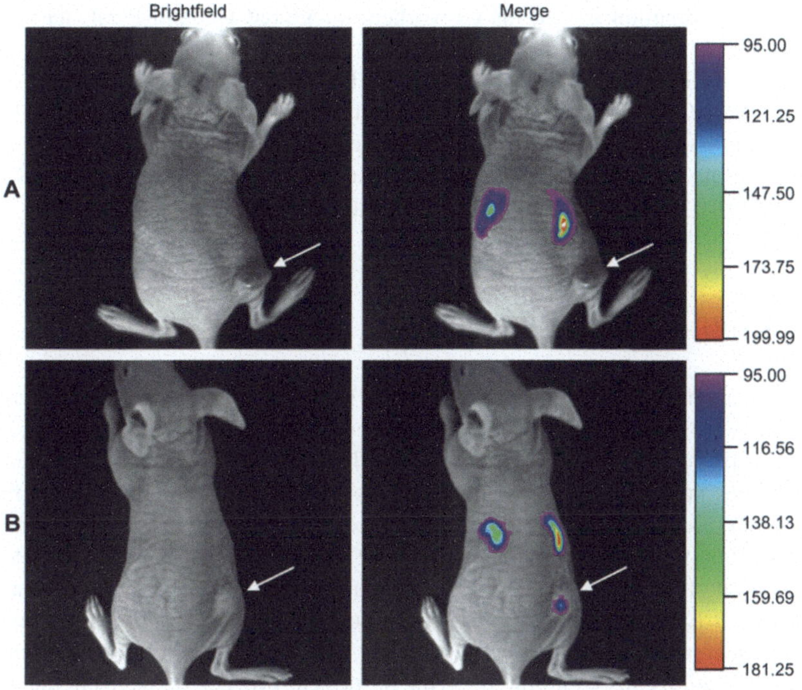

Fig. 5 In vivo up-conversion luminescence imaging of subcutaneous HeLa tumor-bearing athymic nude mice (right hind leg, pointed by white arrows) after intravenous injection of amine functionalized UC nanoparticles (**a**) and FA conjugated UC nanoparticles (**b**). All images were acquired under the same instrumental conditions (power density z 120 mW/cm^2 on the surface of mouse). Reprinted with permission from Ref. [51]. Copyright 2009 Elsevier Ltd.

reported the use of peptide modified UCL nanoparticles for tumor targeting in small animals [52].

Qian Liu and his co-workers successfully performed in vivo imaging using UCL material in a black mouse (which usually exhibit high absorption) and they reported a penetration depth of ~2 nm [53]. UC nanophosphors were used by several researchers for imaging the blood vessels or vascular imaging in small animals [54, 55]. There are many other imaging studies using upconversion nanophosphors, which can be found from published review articles [47, 56].

UC nanophosphors exhibiting the PersL phenomenon are now a day widely investigated due to the peculiar properties of UC materials. Xiaochen Qiu et al. in their work reported the synthesis of a hybrid nanocluster comprising both upconverting nanoparticles and PersL nanomaterial and they also reported the use of the material for in vivo lymphatic imaging in nude mice [57]. UC and PersL nanoparticles were developed in a work by P. Zhao et al. for in vivo imaging-guided photothermal therapy [58]. Zhanjun Li and his co-workers integrated UC and PersL nanoparticles to develop NIR-activated upconversion persistent luminescent (UCPersL) material [59]. $Zn_3Ga_2GeO_8$: Cr^{3+}, Yb^{3+}, Er^{3+} UCPersL nanoparticles exhibiting long-lasting NIR-UC PersL were developed in yet another work and the material was used for in vivo imaging in mice [60]. Also, in this work both excitation and emission fall in the NIR region (biological transparent region) and the emission signal lasted up to 10 h during the in vivo experiment.

6.3 X-Ray Activated Fluorophores

The mechanism behind the PersL phenomena was explained before, and it involves slow release of electrons from the traps to generate long-lasting luminescence. The activation of PersL by visible/NIR wavelength was also discussed. Usually, materials such as sulfides, oxides, etc. with bandgap energy in the intermediate range i.e., 3–7 eV require UV/ visible light as an excitation source to generate luminescence. Certain wide band gap materials can be charged by X-rays to produce PersL. The creation of defects in lanthanides (Ln^{3+}) upon charging by X-ray radiation was discussed by Li et al. and the work also proposed the mechanism behind PersL in large bandgap materials doped with Ln^{3+} ions [61]. The use of low wavelength UV light as excitation can be phototoxic to the tissues causing cell damage but the materials that are excited at low energy X-rays and emit deep UV persistent radiation (high energy can find several applications such as disinfection, sterilization, etc. [62].

The applications of NIR emitting nanomaterials for in vivo imaging applications were discussed in the previous sections but all of them were excited in the visible/ NIR regions. Few works are reporting the X-ray-activated PersL materials for in vivo imaging. In such a work, the authors developed X-ray activated Ln doped nanoparticles which exhibit NIR persistent luminescence and they studied the material for its in vivo imaging applications [7]. They performed in vivo imaging studies in several areas of a living mouse such as blood vessels, tumors, ureters, etc. In another work, Cr-doped zinc gallate nanophosphors (X-PLNPs) with persistent emission in the NIR region and activated by X-rays were used for in vivo imaging after modification with polyethylene glycol (PEG). The in vivo imaging was performed through oral administration as well as intravenous injection (Figs. 6 and 7) of nanoparticles in a mouse [63]. After intravenous injection of the nanoparticles and in situ X-ray irradiation, whole-body imaging was performed. A significant NIR PersL signal was obtained in the liver after excitation (Fig. 6) and the signal decayed slowly but it was again obtained after recharging using an x-ray. This shows that the material can be successfully recharged constantly.

The in vivo imaging experiment was also performed after oral administration of the nanoparticles followed by x-ray excitation. In this case, the NIR PersL signals were obtained from the stomach and intestines (Fig. 7), and it existed up to 50 min. Here

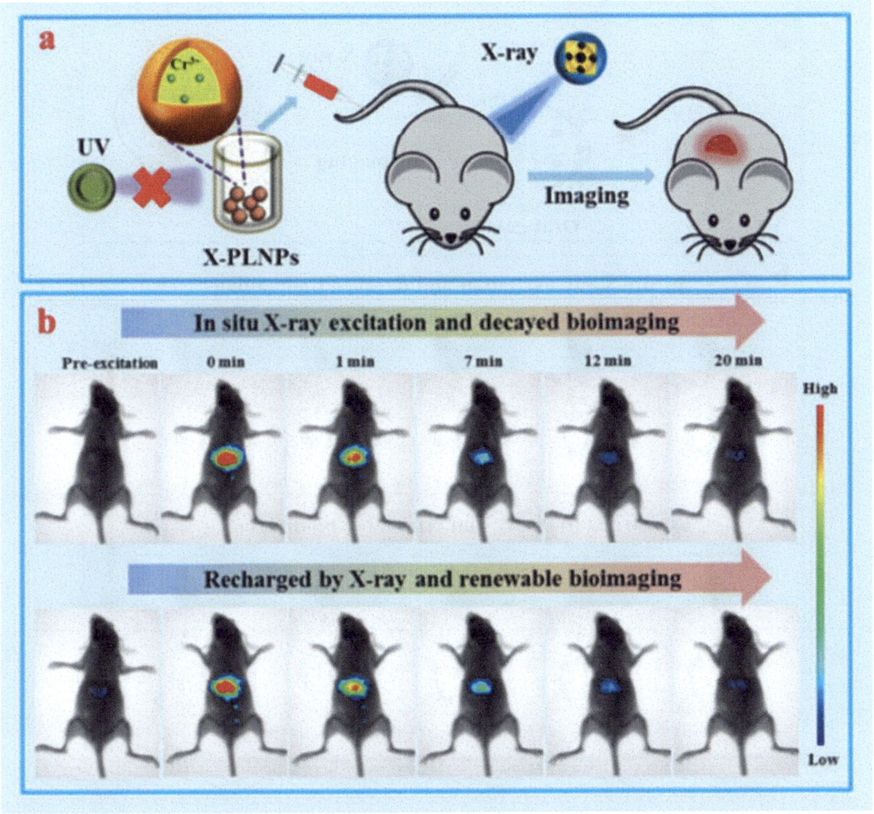

Fig. 6 **a** A schematic representation of in vivo whole-body persistent bioimaging of mouse after intravenous injection with PEG-modified X-PLNPs by in situ X-ray excitation, **b** in vivo renewable and decayed whole-body bioimaging of intravenously injected mouse after ceasing X-ray excitation (5 min, 45 kVp). Reprinted with permission from Ref. [63] Copyright © 2017, American Chemical Society

also, the luminescence was obtained after repeated excitation. This work demonstrated the use of X-ray activated NIR PersL material for in vivo deep penetrating and renewable imaging.

7 Molecular Imaging Techniques and Multimodal Imaging

Molecular imaging techniques utilize a combination of both molecular biology and in vivo imaging for explaining the biological processes inside the body (both molecular and cellular levels) using various procedures. These kinds of imaging techniques are used for the identification of various disorders or abnormalities in the cellular

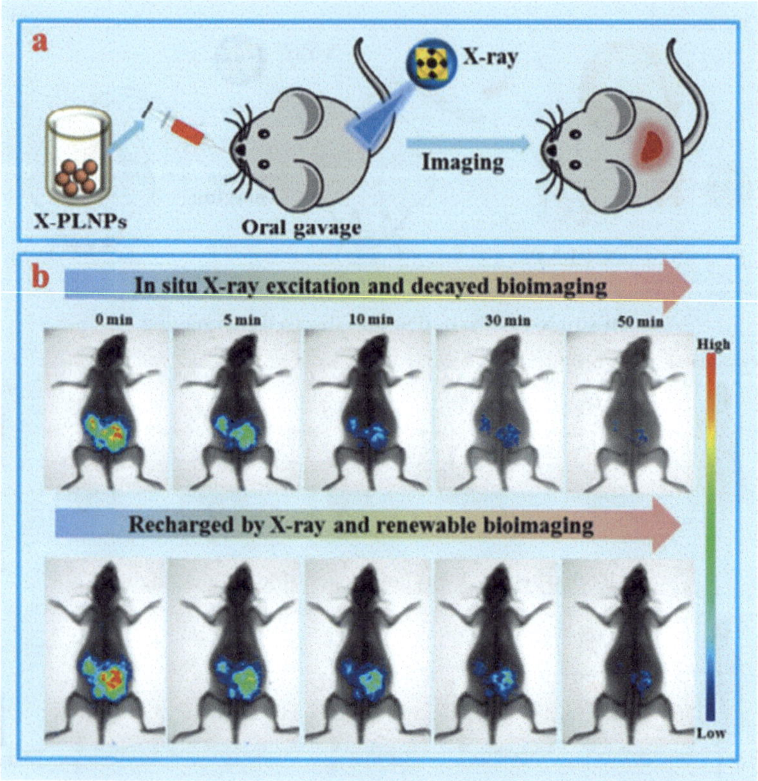

Fig. 7 **a** A schematic diagram of in vivo bioimaging of mouse with oral administration of PEG-modified X-PLNPs after in situ X-ray excitation, **b** rechargeable in vivo NIR persistent bioimaging of mouse at different time via oral gavage of PEG-modified X-PLNPs after stoppage of X-ray excitation (5 min, 45 kVp). Reprinted with permission from Ref. [63] Copyright © 2017, American Chemical Society

levels as well as molecular levels for diagnosis of many diseases. Various kinds of imaging techniques are used for the same and they include ultrasound, CT, MRI, PET, SPECT, and optical imaging techniques.

Ultrasound imaging system uses transducers to produce sound waves with frequencies in the ultrasound region and these waves sent to the body are reflected back to the transducers by the boundaries between the tissues. In CT imaging, different images of the part of the body are collected by passing X-rays at different angles and they are processed for creating the final image. MRI uses powerful magnets to produce strong magnetic fields for aligning the protons in the body with the field. These protons get stimulated when radio-waves are passed through the body and the protons are re-aligned by cutting off the radio current. The energy released during re-alignment is detected by the MRI sensors. The time of re-alignment and the released energy are the main factors influencing MRI imaging. Sometimes, CnA are used for MRI imaging for getting high contrast images and they are magnetic materials which can change the longitudinal (T_1) and transverse (T_2) relaxation rates of the water protons. PET scan utilizes small amounts of radioactive isotopes, which emits positrons during their decay process. SPECT uses radioactive molecules injected to the patient's body for creating three-dimensional images and both PET and SPECT differ by the type of radioactive tracers used. Optical imaging techniques uses bioluminescent materials or exogenous fluorescent probes which produce optical signals. In bioluminescence, light signals are created through chemical reactions in which conversion of chemical energy to light occurs (e.g.: Firefly). In fluorescence imaging, fluorophores which exist endogenously inside the tissues or which are supplied externally are triggered so that they emit light in the UV, visible, or NIR region.

7.1 Multimodal Imaging

In multimodal imaging, two or more imaging techniques are combined to create a new imaging mode for improving the imaging efficiency and creating more accurate images. For example, PET and CT methods are combined to produce PET/CT instrument, which takes the advantages of both the imaging modalities. Similarly, the combination of MRI, SPECT, and fluorescence imaging can create efficient MRI/SPECT/Fluorescence imaging mode thus giving more clear images. PET/MRI imaging system can reduce the scan time compared to using single MRI mode for imaging. Thus, multimodal imaging helps to create high-resolution images by combining the properties of all the combined techniques.

8 Role of Up-Conversion Nanoparticles and PersL Nanoparticles in Multimodal Imaging

8.1 Upconversion Nanophosphors for Multimodal Imaging Applications

As mentioned earlier, CnA is usually used for conventional imaging techniques like MRI to create the best images and also to improve the contrast of structures thus helping differentiate the diseased tissue from normal tissue. Usually, Gd^{3+}-based CnA is used as T_1 CnA, and superparamagnetic iron oxide is used as T_2 CnA in MRI. Gd^{3+} containing UC nanoparticles can be thus utilized as T_1 weighted CnA in MRI. For example, in a work by Park et al., $NaGdF_4$: Yb, Er UC nanoparticles were used as a new imaging probe for both MRI and optical imaging [64]. $NaGdF_4$ UC nanophosphor co-doped with Tm^{3+}, Yb^{3+}, and Er^{3+} was used for dual-modal in vivo MRI and UC imaging in living mice [65]. Liu et al. report the synthesis of upconversion nanocomposite material by coating mesoporous silica shell on Gd^{3+} co-doped $NaYF_4$ crystals and used the same for in vivo NIR imaging as well as for T_1 weighted MRI imaging [66]. In a work by Xia et al., core–shell nanocrystals of $NaYF_4$:Yb^{3+}, Tm^{3+}@Fe_xO_y were synthesized, and the material was applied for UC NIR luminescence imaging and T_2 weighted MRI imaging in animal lymphatic system [67]. In this work, the material was coated with dopamine for ensuring water solubility, and the material was used for UC luminescence imaging of lymph nodes in a mouse. The material was injected to the forepaw of the animal, and the luminescence at 800 nm was observed at the injection site and the UC luminescence was observed at 0, 10, and 20 min after injection. The signal was obtained at three different lymphatic sites at 10 min and at 20 min, the signal was obtained from the forepaw and lymph node (Fig. 8). Ex-vivo UC luminescence was also observed after cutting off the area and strong emission was observed in this case also.

The material was also tested for its in vivo application as T_2 weighted contrast agent in MRI imaging and the test was performed in the lymph node of the armpit

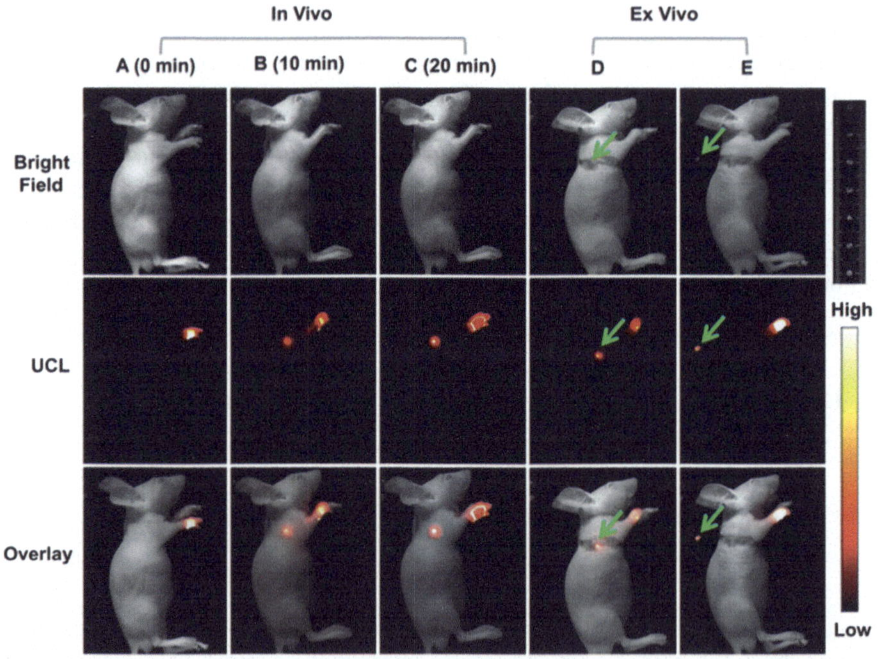

Fig. 8 In vivo UCL imaging of lymphatic system after injection with $NaYF_4$:Yb^{3+}, Tm^{3+}@Fe_xO_y nanocrystals into a nude mouse at various time **a** 0 min, **b** 10 min, and **c** 20 min. All images were acquired under the same instrumental conditions (power density w 150 mW/cm^2 on the surface of nude mouse). **d, e** UCL ex vivo images of the lymphatic system of a nude mouse sacrificed after injection with $NaYF_4$:Yb^{3+}, Tm^{3+}@Fe_xO_y nanocrystals for 40 min. Green arrow marked the lymphatic nodes. All images were acquired under the same instrumental conditions (power density w 100 mW /cm^2). Reprinted with permission from Ref. [67]. Copyright 2011 Elsevier Ltd.

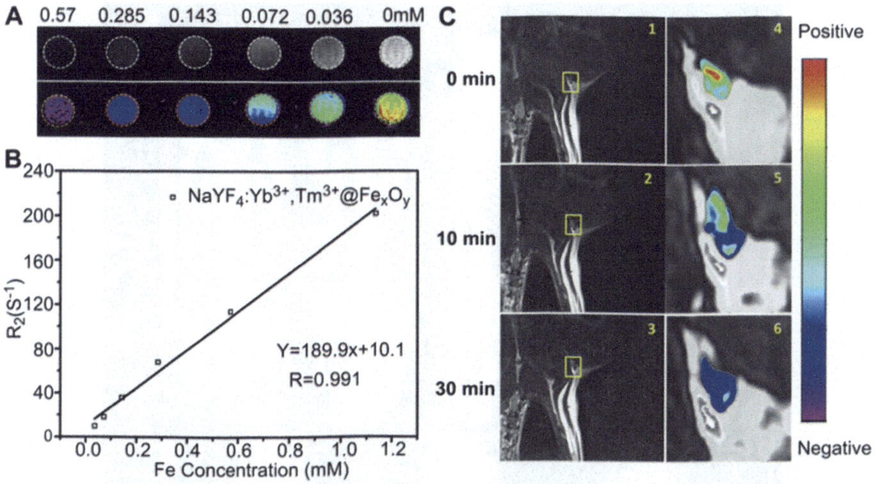

Fig. 9 **a** T2-weighted and color-mapped MR images of various molar concentrations of $NaYF_4$:Yb^{3+}, Tm^{3+}@Fe_xO_y nanocrystals. Deionized water (0 mg/mL) was the reference. **b** Relaxation rate R2 (1/T2) versus various molar concentrations of hydrophilic $NaYF_4$:Yb^{3+}, Tm^{3+}@Fe_xO_y nanocrystals at room temperature using a 3 T MRI scanner. **c** MR images of the armpit region after injection with $NaYF_4$:Yb^{3+}, Tm^{3+}@Fe_xO_y nanocrystals and Color-mapped coronal images of lymph node at various time. The dose of $NaYF_4$:Yb^{3+}, Tm^{3+}@Fe_xO_y nanocrystals is 1.5 mg/kg. Reprinted with permission from Ref. [67]. Copyright 2011 Elsevier Ltd.

of a rat. The images were taken at different times and the change in contrast at the lymph node was observed (Fig. 9).

UC nanomaterials were also used as contrast agents for CT imaging method and lanthanide based materials with high value of atomic number were reported to work as efficient CT contrast agents with high X-ray attenuation property [68, 69]. In one such work, Wu and co-researchers synthesized PEG capped EuOF: Sm^{3+} nanocrystals in which ^{153}Sm is a rare-earth radio isotope metal which emits beta particles with intermediate energy and also gamma rays and thus it is useful for long term SPECT imaging in animals [68]. In this work, both SPECT and CT imaging were combined to produce highly sensitive and high-resolution imaging mode. Through in vivo SPECT imaging in mice, the biodistribution of the nanoparticles was estimated and thus proved the higher blood retention time of the small particles. The SPECT imaging was performed in a Kunming mouse after injecting 0.2 ml of the material into its tail vein and the images were recorded at different times post injection (Fig. 10). The presence of Eu with high atomic number was utilized for CT imaging in the lymph nodes in mice due to the high x-ray absorption property of the element. CT imaging was performed by injecting 150 μL of PEG-EuOF nanoparticles into the left paw of the mouse and the images showing the contrast enhancement in the lymph nodes were recorded 20 min after injection (Fig. 11). Thus, the work demonstrated the application of the material in multimodal imaging.

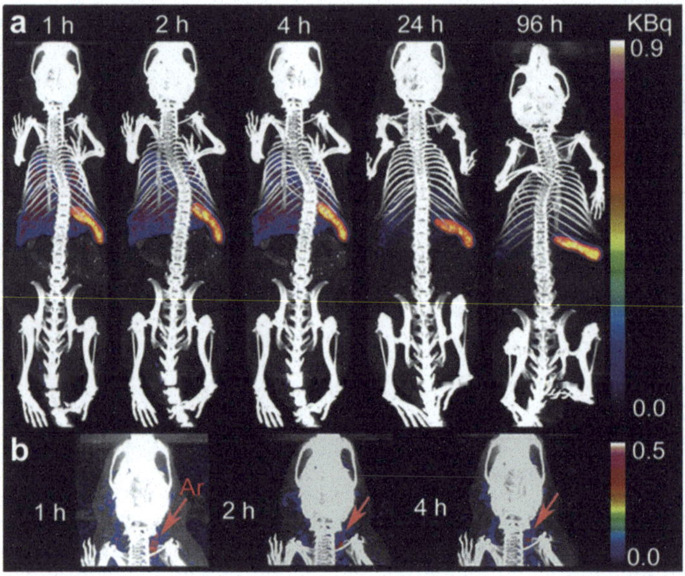

Fig. 10 SPECT/CT imaging of Kunming mice after injection with PEG-EuOF:153Sm at 1 h, 2 h, 4 h, 24 h, and 96 h (**a**). **b** SPECT/CT imaging of Kunming mice after injection with PEG-EuOF:153Sm at 1 h, 2 h, and 4 h with adjusting the intensity scale, Ar: artery. Reprinted with permission from Ref. [68]. Copyright 2014 Elsevier Ltd.

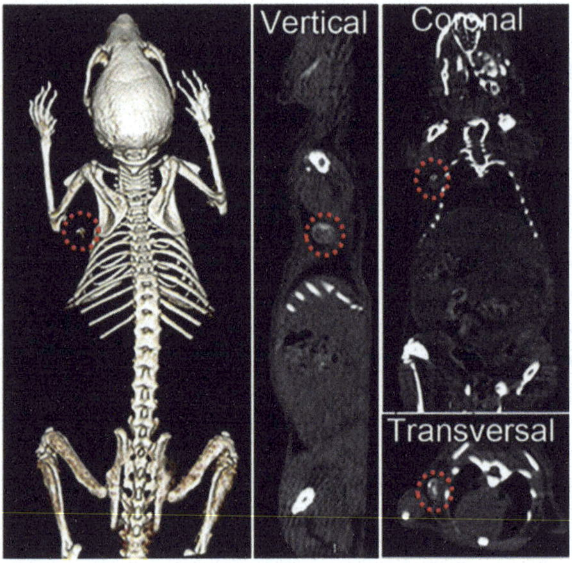

Fig. 11 In vivo CT volume-rendered and intensity of vertical, coronal, transversal images of 20 min after injection of 4.0 mg/mL PEG-EuOF. The position of lymph node was marked by circles. Reprinted with permission from Ref. [68]. Copyright 2014 Elsevier Ltd

Other than these applications, UC nanomaterials were also tested for their applications in PET and SPECT imaging modes [70, 71]. $NaYF_4$:Yb/Tm UC nanoparticles were used in another work for trimodal imaging combining UC luminescence, MRI and CT [72]. Four modes of imaging namely UC luminescence, MRI, CT, and SPECT were combined in another work for studying tumor angiogenesis [73].

8.2 Persistent Luminescence Nanophosphors for Multimodal Imaging Applications

Multimodal imaging techniques combining different imaging modes were developed to compensate the limitations of individual imaging techniques. Such examples combining different imaging modes and UC nanomaterials were discussed in the previous section. The advantages of materials showing PersL were discussed already and now let us focus on combining the PersL technique with other imaging modes, thus showing the applications of such materials in the multimodal imaging industry.

PersL nanoprobes show major properties such as deep tissue penetration, very low noise levels and easy mode of operation which make them an ideal candidate for linking with other imaging modes such as MRI, CT, etc. so that the problems such as low resolution, sensitivity, etc. can be eliminated. There are works reported on such multimodal imaging techniques using PersL materials. Combination of PersL nanomaterial along with MRI imaging was first tried by Abdukayum et al. in which they have synthesized $Zn_{2.94}Ga_{1.96}Ge_2O_{10}$: Cr^{3+}, Pr^{3+} PersL nanophosphor and surface modified it using 3-amino propyltriethoxysilane (APTES) and diethylenetriamine pentaacetic acid (DTPA) and this was followed by the introduction of Gd^{3+} ions to the material [74]. In the work, the material was used for in vivo NIR PersL imaging of a nude mouse after injecting 300 μL of the material solution prepared in PBS. 10 min before the injection, the sample solution was excited using a UV lamp (254 nm) and the luminescence images for more than six hours were taken without any excitation after injection (Fig. 12). The luminescence signals were mainly obtained from liver and this demonstrated an example of long-term in vivo imaging.

The material was also used for in vivo T_1-weighted MRI imaging in mouse after injecting 300 μL of the sample solution. Before injection, the liver was not clearly seen in the MRI image but it was seen 15 min after the injection of the PersL nanoparticles (Fig. 13). There are many other reported works using PersL nanomaterials for multimodal imaging applications—some of which are listed in Table 2.

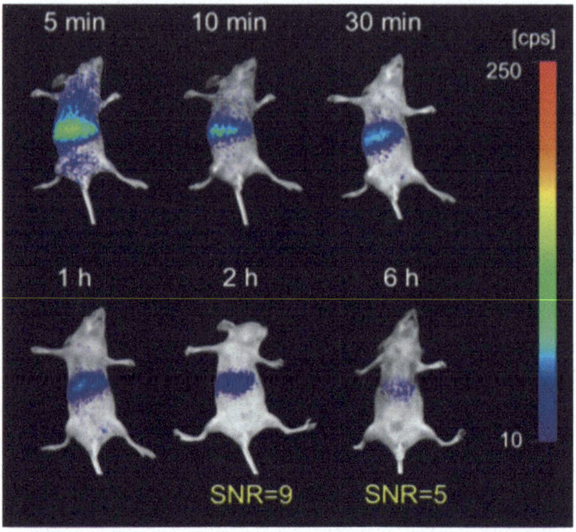

Fig. 12 In vivo NIR luminescence images of a normal mouse after intravenous injection of Gd (III)-PLNPs (0.3 mg, 10 min irradiation with a 254 nm UV lamp before injection but without irradiation during imaging). Reprinted with permission from Ref. [74]. Copyright © 2014, American Chemical Society

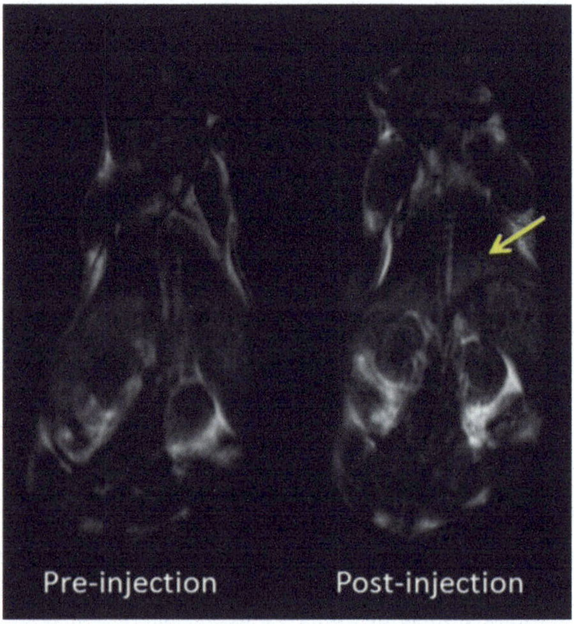

Fig. 13 In vivo T1-weighted MR images of the mouse before and after intravenous injection of Gd (III)-PLNPs (0.3 mg); the yellow arrow indicates the liver. Reprinted with permission from Ref. [74]. Copyright © 2014, American Chemical Society

9 Conclusions

To sum up, the chapter includes the basics of persistent luminescence (PersL) materials and their applications in in vivo imaging applications. In order to get a clear idea about the use of luminescent materials used for in vivo imaging applications, the

Table 2 Some works reporting use of PersL nanomaterials for multimodal imaging

Imaging modes combined	Material	Refs.
PersL/MRI	Gd (III)-$Zn_{2.94}Ga_{1.96}Ge_2O_{10}$: Cr^{3+}, Pr^{3+}	Abdukayum et al. [74]
PersL/MRI	Gd_2O_3@$mSiO_2$@$CaTiO_3$: Pr^{3+}	Teston et al. [75]
PersL/MRI	$ZnGa_2O_4$: Cr^{3+}, Gd^{3+}	Maldiney et al. [76]
PersL/MRI	Gd_2O_3-$ZnGa_2O_4$:Cr^{3+}	Wang et al. [77]
PersL/CT	$Zn_{2.94}Ga_{1.96}Ge_2O_{10}$: Cr^{3+}, Pr^{3+}@TaO_x@SiO_2	Lu et al. [78]
PersL/MRI/CT	Silica Coated $GdAlO_3$:Mn^{4+}, Ge^{4+}@Au	Liu et al. [79]

chapter investigated about various types of fluorescent probes, their drawbacks and also studied recent works focusing their use in medical imaging. With the increasing research interest on the use of inorganic fluorescent nanoparticles for clinical applications especially in in vivo imaging experiments, the development of materials such as upconversion nanomaterials, NIR fluorescent probes and persistent luminescent materials, etc. become extremely important in biomedical industry. So, the chapter discussed the recent progress of these materials in in vivo imaging as well as multimodal imaging applications. In vivo imaging applications mainly uses organic dyes as fluorescence markers and they suffer from a major drawback of less photostability. Here comes the importance of developing fluorescent probes with enhanced photostability and high quantum yield. Inorganic nanoparticles such as semiconductor quantum dots and rare-earth doped nanomaterials possess very high quantum yield, and they are extremely photostable. The major concern regarding such materials is their toxicity while using for in vivo applications. So, the development of inorganic nanomaterials or quantum dots which does not pose any toxicity concern as well as exhibiting PersL mechanism can bring drastic developments in the medical imaging field since they can be used for long-lasting imaging experiments without any harmful side effects.

References

1. N. Basavaraju, S. Sharma, A. Bessière, B. Viana, D. Gourier, K.R. Priolkar, Red persistent luminescence in MgGa2O4 : Cr 3+; a new phosphor for in vivo imaging. J. Phys. D. Appl. Phys. **46** (2013)
2. Z. Zhou, Z.R. Lu, Gadolinium-Based Contrast Agents for Magnetic Resonance Cancer Imaging, Wiley Interdisciplinary Reviews: Nanomedicine and Nanobiotechnology
3. H. Yang, et al., Effects of iron oxide nanoparticles as t 2-mri contrast agents on reproductive system in male mice. J. Nanobiotechnol. **20** (2022)
4. J. Gao, K. Chen, R. Xie, J. Xie, S. Lee, Z. Cheng, X. Peng, X. Chen, Ultrasmall near-infrared non-cadmium quantum dots for in vivo tumor imaging. Small **6** (2010)

5. H. Chen, B. Zheng, C. Liang, L. Zhao, Y. Zhang, H. Pan, W. Ji, X. Gong, H. Wang, J. Chang, Near-infrared persistent luminescence phosphors znga2o4:cr3 + as an accurately tracker to photothermal therapy in vivo for visual treatment. Mater. Sci. Eng. C **79** (2017)
6. Y. Zhong, et al., In vivo molecular imaging for immunotherapy using ultra-bright near-infrared-IIb rare-earth nanoparticles. Nat. Biotechnol. **37** (2019)
7. P. Pei, et al., X-ray-activated persistent luminescence nanomaterials for NIR-II imaging. Nat. Nanotechnol. **16** (2021)
8. J. Lifante, Y. Shen, E. Ximendes, E. Martín Rodríguez, D.H. Ortgies, The role of tissue fluorescence in in vivo optical bioimaging. J. Appl. Phys. **128** (2020)
9. G. Hong, A.L. Antaris, H. Dai, Near-infrared fluorophores for biomedical imaging. Nat. Biomed. Eng.
10. A.C. Croce, G. Bottiroli, Autofluorescence spectroscopy and imaging: a tool for biomedical research and diagnosis. Eur. J. Histochem. **58** (2014)
11. B. Ballou, B.C. Lagerholm, L.A. Ernst, M.P. Bruchez, A.S. Waggoner, Noninvasive imaging of quantum dots in mice. Bioconjug. Chem. **15** (2004)
12. D.R. Larson, W.R. Zipfel, R.M. Williams, S.W. Clark, M.P. Bruchez, F.W. Wise, W.W. Webb, Water-soluble quantum dots for multiphoton fluorescence imaging in vivo. Science **300**(80) (2003)
13. X. Gao, Y. Cui, R.M. Levenson, L.W.K. Chung, S. Nie, In vivo cancer targeting and imaging with semiconductor quantum dots. Nat. Biotechnol. **22** (2004)
14. X. Sun, et al., Self-illuminating 64Cu-doped cdse/zns nanocrystals for in vivo tumor imaging. J. Am. Chem. Soc. **136** (2014)
15. X. Michalet, F.F. Pinaud, L.A. Bentolila, J.M. Tsay, S. Doose, J.J. Li, G. Sundaresan, A.M. Wu, S.S. Gambhir, S. Weiss, Quantum dots for live cells, in vivo imaging, and diagnostics. Science
16. B. Dąbrowska-Bouta, M. Zięba, J. Orzelska-Górka, J. Skalska, G. Sulkowski, M. Frontczak-Baniewicz, S. Talarek, J. Listos, L. Strużyńska, Influence of a low dose of silver nanoparticles on cerebral myelin and behavior of adult rats. Toxicology **363–364** (2016)
17. N. Hadrup, K. Loeschner, A. Mortensen, A.K. Sharma, K. Qvortrup, E.H. Larsen, H.R. Lam, The similar neurotoxic effects of nanoparticulate and ionic silver in vivo and in vitro. Neurotoxicology **33** (2012)
18. E. Porret, X. Le Guével, J.L. Coll, Gold nanoclusters for biomedical applications: toward: in vivo studies. J. Mater. Chem. B
19. J. Yang, et al., Gold/alpha-lactalbumin nanoprobes for the imaging and treatment of breast cancer. Nat. Biomed. Eng. **4** (2020)
20. S. Li, J. Wei, Q. Yao, X. Song, J. Xie, H. Yang, Emerging ultrasmall luminescent nanoprobes for in vivo bioimaging. Chem. Soc. Rev.
21. D.F. Báez, E. Gallardo-Toledo, M.P. Oyarzún, E. Araya, M.J. Kogan, The influence of size and chemical composition of silver and gold nanoparticles on in vivo toxicity with potential applications to central nervous system diseases. Int. J. Nanomed. **16** (2021)
22. F. Chen, et al., Molecular phenotyping and image-guided surgical treatment of melanoma using spectrally distinct ultrasmall core-shell silica nanoparticles. Sci. Adv. **5** (2019)
23. O.S. Wolfbeis, An overview of nanoparticles commonly used in fluorescent bioimaging. Chem. Soc. Rev.
24. X. Xu, R. Ray, Y. Gu, H.J. Ploehn, L. Gearheart, K. Raker, W.A. Scrivens, Electrophoretic analysis and purification of fluorescent single-walled carbon nanotube fragments. J. Am. Chem. Soc. **126** (2004)
25. S.T. Yang, L. Cao, P.G. Luo, F. Lu, X. Wang, H. Wang, M.J. Meziani, Y. Liu, G. Qi, Y.P. Sun, Carbon dots for optical imaging in vivo. J. Am. Chem. Soc. **131** (2009)
26. L. Liu, S. Zhang, X. Zheng, H. Li, Q. Chen, K. Qin, Y. Ding, Y. Wei, Carbon dots derived from: fusobacterium nucleatum for intracellular determination of Fe3+and bioimaging both in vitro and in vivo. Anal. Methods **13** (2021)
27. S. Yoo, Y. Song, S. Hahn, Ultralong persistent luminescence from carbon dots. Light: Sci. Appl.

28. C. He, X. Lin, Y. Mei, Y. Luo, M. Yang, Y. Kuang, X. Yi, W. Zeng, Q. Huang, B. Zhong, Recent advances in carbon dots for in vitro/vivo fluorescent bioimaging: a mini-review. Front. Chem.
29. E. Hemmer, A. Benayas, F. Légaré, F. Vetrone, Exploiting the biological windows: current perspectives on fluorescent bioprobes emitting above 1000 nm. Nanoscale Horizons.
30. J.V. Frangioni, In vivo near-infrared fluorescence imaging. Curr. Opin. Chem. Biol.
31. T. Maldiney, G. Sraiki, B. Viana, D. Gourier, C. Richard, D. Scherman, M. Bessodes, K. Van den Eeckhout, D. Poelman, P.F. Smet, In vivo optical imaging with rare earth doped Ca_2Si_5N_8 persistent luminescence nanoparticles. Opt. Mater. Express **2** (2012)
32. M. Palner, K. Pu, S. Shao, J. Rao, Semiconducting polymer nanoparticles with persistent near-infrared luminescence for in vivo optical imaging. Angew. Chemie - Int. Ed. **54** (2015)
33. T. Maldiney, et al., The in vivo activation of persistent nanophosphors for optical imaging of vascularization, tumours and grafted cells. Nat. Mater. **13** (2014)
34. Q.L.M. De Chermont, C. Chanéac, J. Seguin, F. Pellé, S. Maîtrejean, J.P. Jolivet, D. Gourier, M. Bessodes, D. Scherman, Nanoprobes with near-infrared persistent luminescence for in vivo imaging. Proc. Natl. Acad. Sci. U. S. A. **104** (2007)
35. X. Sun, J. Shi, X. Fu, Y. Yang, H. Zhang, Long-term in vivo biodistribution and toxicity study of functionalized near-infrared persistent luminescence nanoparticles. Sci. Rep. **8** (2018)
36. Y. Chen, S. Wang, F. Zhang, Near-infrared luminescence high-contrast in vivo biomedical imaging. Nat. Rev. Bioeng. **1** (2023)
37. M. Kamimura, N. Kanayama, K. Tokuzen, K. Soga, Y. Nagasaki, Near-infrared (1550 Nm) in vivo bioimaging based on rare-earth doped ceramic nanophosphors modified with PEG-b-Poly(4-Vinylbenzylphosphonate). Nanoscale **3** (2011)
38. L. Quan, S. Liu, T. Sun, X. Guan, W. Lin, Z. Xie, Y. Huang, Y. Wang, X. Jing, Near-infrared emitting fluorescent BODIPY nanovesicles for in vivo molecular imaging and drug delivery. ACS Appl. Mater. Interfaces **6** (2014)
39. H. Watanabe, M. Ono, K. Matsumura, M. Yoshimura, H. Kimura, H. Saji, Molecular imaging of β-amyloid plaques with near-infrared boron dipyrromethane (BODIPY)-based fluorescent probes. Mol. Imaging **12** (2013)
40. I. Villa, et al., 1.3 μm emitting SrF2:Nd3+ nanoparticles for high contrast in vivo imaging in the second biological window. Nano Res. **8** (2015)
41. U. Rocha, et al., Neodymium-doped LaF3 nanoparticles for fluorescence bioimaging in the second biological window. Small **10** (2014)
42. E. Phillips, et al., Clinical translation of an ultrasmall inorganic optical-PET imaging nanoparticle probe. Sci. Transl. Med. **6** (2014)
43. R. Xie, K. Chen, X. Chen, X. Peng, InAs/InP/ZnSe core/shell/shell quantum dots as near-infrared emitters: bright, narrow-band, non-cadmium containing, and biocompatible. Nano Res. **1** (2008)
44. Y. Zhang, et al., Biodistribution, pharmacokinetics and toxicology of Ag2S near-infrared quantum dots in mice. Biomaterials **34** (2013)
45. G. Hong, J.T. Robinson, Y. Zhang, S. Diao, A.L. Antaris, Q. Wang, H. Dai, In vivo fluorescence imaging with Ag2S quantum dots in the second near-infrared region. Angew. Chemie - Int. Ed. **51** (2012)
46. H.S. Choi, et al., Synthesis and in vivo fate of zwitterionic near-infrared fluorophores. Angew. Chemie - Int. Ed. **50** (2011)
47. J. Zhou, Q. Liu, W. Feng, Y. Sun, F. Li, Upconversion luminescent materials: advances and applications. Chem. Rev.
48. D. Song, S. Zhao, Z. Xu, Upconversion luminescent materials: properties and luminescence mechanisms, in *Principles and Applications of Up-Converting Phosphor Technology* (2019)
49. K.K. Markose, R. Anjana, M.K. Jayaraj, Upconversion nanophosphors: an overview, in *Materials Horizons: From Nature to Nanomaterials* (2020)
50. M. Nyk, R. Kumar, T.Y. Ohulchanskyy, E.J. Bergey, P.N. Prasad, High contrast in vitro and in vivo photoluminescence bioimaging using near infrared to near infrared up-conversion in Tm 3+ and Yb 3+ doped fluoride nanophosphors. Nano Lett. **8** (2008)

51. L.Q. Xiong, Z.G. Chen, M.X. Yu, F.Y. Li, C. Liu, C.H. Huang, synthesis, characterization, and in vivo targeted imaging of amine-functionalized rare-earth up-converting nanophosphors. Biomaterials **30** (2009)
52. L. Xiong, Z. Chen, Q. Tian, T. Cao, C. Xu, F. Li, High contrast upconversion luminescence targeted imaging in vivo using peptide-labeled nanophosphors. Anal. Chem. **81** (2009)
53. Q. Liu, Y. Sun, T. Yang, W. Feng, C. Li, F. Li, Sub-10 nm hexagonal lanthanide-doped NaLuF 4 upconversion nanocrystals for sensitive bioimaging in vivo. J. Am. Chem. Soc. **133** (2011)
54. S.A. Hilderbrand, F. Shao, C. Salthouse, U. Mahmood, R. Weissleder, Upconverting luminescent nanomaterials: application to in vivo bioimaging. Chem. Commun. (2009)
55. J. Pichaandi, J.C. Boyer, K.R. Delaney, F.C.J.M. Van Veggel, Two-photon upconversion laser (scanning and wide-field) microscopy using Ln3+-doped NaYF4 upconverting nanocrystals: a critical evaluation of their performance and potential in bioimaging. J. Phys. Chem. C **115** (2011)
56. W. Feng, X. Zhu, F. Li, Recent advances in the optimization and functionalization of upconversion nanomaterials for in vivo bioapplications. NPG Asia Mater.
57. X. Qiu, X. Zhu, M. Xu, W. Yuan, W. Feng, F. Li, Hybrid nanoclusters for near-infrared to near-infrared upconverted persistent luminescence bioimaging. ACS Appl. Mater. Interfaces **9** (2017)
58. P. Zhao, W. Ji, S. Zhou, L. Qiu, L. Li, Z. Qian, X. Liu, H. Zhang, X. Cao, Upconverting and persistent luminescent nanocarriers for accurately imaging-guided photothermal therapy. Mater. Sci. Eng. C **79** (2017)
59. Z. Li, L. Huang, Y. Zhang, Y. Zhao, H. Yang, G. Han, Near-infrared light activated persistent luminescence nanoparticles via upconversion. Nano Res. **10** (2017)
60. Z. Xue, X. Li, Y. Li, M. Jiang, G. Ren, H. Liu, S. Zeng, J. Hao, A 980 nm laser-activated upconverted persistent probe for nir-to-nir rechargeable in vivo bioimaging. Nanoscale **9** (2017)
61. L. Li, T. Li, Y. Hu, C. Cai, Y. Li, X. Zhang, B. Liang, Y. Yang, J. Qiu, Mechanism of the trivalent lanthanides' persistent luminescence in wide bandgap materials. Light Sci. Appl. **11** (2022)
62. Y.M. Yang, et al., X-ray-activated long persistent phosphors featuring strong UVC afterglow emissions. Light Sci. Appl. **7** (2018)
63. Z. Xue, X. Li, Y. Li, M. Jiang, H. Liu, S. Zeng, J. Hao, X-ray-activated near-infrared persistent luminescent probe for deep-tissue and renewable in vivo bioimaging. ACS Appl. Mater. Interfaces **9** (2017)
64. Y. Il Park, et al., Nonblinking and nonbleaching upconverting nanoparticles as an optical imaging nanoprobe and T1 magnetic resonance imaging contrast agent. Adv. Mater. **21** (2009)
65. J. Zhou, Y. Sun, X. Du, L. Xiong, H. Hu, F. Li, Dual-modality in vivo imaging using rare-earth nanocrystals with near-infrared to near-infrared (NIR-to-NIR) upconversion luminescence and magnetic resonance properties. Biomaterials **31** (2010)
66. J. Liu, W. Bu, S. Zhang, F. Chen, H. Xing, L. Pan, L. Zhou, W. Peng, J. Shi, Controlled synthesis of uniform and monodisperse upconversion core/mesoporous silica shell nanocomposites for bimodal imaging. Chem. - A Eur. J. **18** (2012)
67. A. Xia, Y. Gao, J. Zhou, C. Li, T. Yang, D. Wu, L. Wu, F. Li, Core-shell NaYF4:Yb3+, Tm3+@FexOy nanocrystals for dual-modality T2-enhanced magnetic resonance and NIR-to-NIR upconversion luminescent imaging of small-animal lymphatic node. Biomaterials **32**, 7200 (2011)
68. Y. Wu, Y. Sun, X. Zhu, Q. Liu, T. Cao, J. Peng, Y. Yang, W. Feng, F. Li, Lanthanide-based nanocrystals as dual-modal probes for SPECT and X-ray CT imaging. Biomaterials **35** (2014)
69. J. Ma, P. Huang, M. He, L. Pan, Z. Zhou, L. Feng, G. Gao, D. Cui, Folic acid-conjugated LaF3:Yb,Tm@SiO2 nanoprobes for targeting dual-modality imaging of upconversion luminescence and X-ray computed tomography. J. Phys. Chem. B **116** (2012)
70. Y. Sun, et al., Fluorine-18 labeled rare-earth nanoparticles for positron emission tomography (PET) imaging of sentinel lymph node. Biomaterials **32** (2011)
71. J. Peng, Y. Sun, L. Zhao, Y. Wu, W. Feng, Y. Gao, F. Li, Polyphosphoric acid capping radioactive/upconverting NaLuF4: Yb,Tm,153Sm nanoparticles for blood pool imaging in vivo. Biomaterials **34** (2013)

72. J.W. Shen, C.X. Yang, L.X. Dong, H.R. Sun, K. Gao, X.P. Yan, Incorporation of computed tomography and magnetic resonance imaging function into NaYF4:Yb/Tm upconversion nanoparticles for in vivo trimodal bioimaging. Anal. Chem. **85** (2013)
73. Y. Sun, X. Zhu, J. Peng, F. Li, Core-shell lanthanide upconversion nanophosphors as four-modal probes for tumor angiogenesis imaging. ACS Nano **7** (2013)
74. A. Abdukayum, C.X. Yang, Q. Zhao, J.T. Chen, L.X. Dong, X.P. Yan, Gadolinium complexes functionalized persistent luminescent nanoparticles as a multimodal probe for near-infrared luminescence and magnetic resonance imaging in vivo. Anal. Chem. **86** (2014)
75. E. Teston, Y. Lalatonne, D. Elgrabli, G. Autret, L. Motte, F. Gazeau, D. Scherman, O. Clément, C. Richard, T. Maldiney, Design, properties, and in vivo behavior of superparamagnetic persistent luminescence nanohybrids. Small **11** (2015)
76. T. Maldiney, B.T. Doan, D. Alloyeau, M. Bessodes, D. Scherman, C. Richard, Gadolinium-doped persistent nanophosphors as versatile tool for multimodal in vivo imaging. Adv. Funct. Mater. **25** (2015)
77. Y. Wang, C.X. Yang, X.P. Yan, Hydrothermal and biomineralization synthesis of a dual-modal nanoprobe for targeted near-infrared persistent luminescence and magnetic resonance imaging. Nanoscale **9** (2017)
78. Y.C. Lu, C.X. Yang, X.P. Yan, Radiopaque tantalum oxide coated persistent luminescent nanoparticles as multimodal probes for in vivo near-infrared luminescence and computed tomography bioimaging. Nanoscale **7** (2015)
79. J.M. Liu, Y.Y. Liu, D.D. Zhang, G.Z. Fang, S. Wang, Synthesis of GdAlO3:Mn4+,Ge4+@Au core-shell nanoprobes with plasmon-enhanced near-infrared persistent luminescence for in vivo trimodality bioimaging. ACS Appl. Mater. Interfaces **8** (2016)

Theranostic Applications of Persistent Luminescence Nanoparticles

Joydip Dutta and Mitesh Chakraborty

Abstract In the present book chapter, we have discussed all fundamental aspects related to synthesis techniques, different types of characterizations, mechanisms of afterglow and applications in the medical fields of persistent luminescence nanoparticles (PLNPs). Different types of applications in the context of luminescence imaging, photodynamic therapy, different forms of tumor diagnosis, and means of treatments are reported. Recent studies reported by researchers based on modification and design transfer of biological surfaces of PLNPs are specified in our content. Exploration of the afterglow property to upgrade the sensitivity of nanoprobes in viz. biosensing, bioimaging, theranostics, etc. are reviewed in the present study. We have also debated about various techniques for tumor-targeted imaging, methods to improve the accuracy of stem cell tracking, different therapies based on stem cells, highly efficient multimodal nanoprobes for imaging, and nano-agents for theranostic applications. Limitations and scope for improvement of PLNPs for applications in biomedical systems are also debated in the current study.

Keywords Persistent luminescence · Theranostic · Tumor · Therapy · Biomedical

1 Introduction

The phenomenon by which materials absorb energy and emit light is known as luminescence. This is also known as the cold body radiation phenomenon. Among these materials, those that convert absorbed energies like ultraviolet, beta rays, gamma rays, X-rays, etc. into emission wavelength in visible regions are known as phosphors. In general, phosphors emit this radiation for some milli-seconds, once the external source is removed. On the other hand, after glow phosphors continue to emit light

J. Dutta (✉)
Central Research Facility, Indian Institute of Technology (Indian School of Mines), Dhanbad, Jharkhand 826004, India
e-mail: joydip.duttaism@gmail.com

M. Chakraborty
Department of Physics, St. Xavier's College, Ranchi, Jharkhand 834001, India

V. Kumar et al. (eds.), *Persistent Luminescence*, Progress in Optical Science and Photonics 30, https://doi.org/10.1007/978-981-97-4943-0_8

say for some minutes to hours. This phenomenon can be explored in displays, safety signals, night vision surveillance, and nowadays in medical imaging.

The first scientific report of persistent luminescence (PL) was discovered in the famous Bologna stone in the year 1603 by shoemaker and alchemist Vincenzo Casciarolo. In the year 1640, the unexpected glow of this stone was discovered by his "Fortunius Licetus" in his ".Litheosphorus Sive De Lapide Bononiensi.". The strange glow in the rock may be due to the Barium sulfide present in the rock [1]. The long-lasting glow may be due to natural ingredients present in the rock material [2]. Active research related to PL is important in the areas of spectroscopy, photochemistry, photonics, and material science. In recent times, research has focused on changing the morphology and time duration of novel PL materials. Due to this, it is possible to design bio labels, photocatalysts, biosensors, detectors, and photonic devices.

Persistent luminescence nanoparticles (PLNPs) are unique optical materials that emit long-lasting luminescence after the external excitation source is withdrawn [3]. Persistent luminescence or long-term persistence is also said to be persistent luminescence (PL) in abbreviated form. PLNPs can store some excitation energy in different metastable states for a substantial amount of time and emit photons corresponding to different wavelengths. These wavelengths can be in the ultraviolet (UV), visible, or in the near-infrared (NIR) regions. This continuous emission also said to be afterglow luminescence is observed due to defects and/or specific dopants in the host. The physical phenomenon of electron capture or electron release may be responsible for this luminescent behavior in these types of materials [4, 5]. From the literature survey, it has been observed that a wide range of emissions is possible from ultraviolet (UV) to near infra-red (NIR) regions by suitable choice of hosts and dopants. From optical studies, it is known that emission due to PL is dependent upon four physical phenomena: (a) Charge carrier excitation: in this process, electrons and/or holes are created by suitable gamma ray radiation, and migration of charge carriers takes place; (b) Accumulation of charge carriers: the excited charge carriers are trapped at defect sites and the optical transition is dependent upon the storage capacity of charge carriers; (c) Release of charge carriers which is dependent upon mechanical, thermal, and optical disturbance in the system which ultimately governs the duration of PL (d) charge carriers recombination. Various models have been proposed by material scientists related to these mechanisms viz. hole model, electron model, energy band gap model, and quantum tunneling model to explain long PL in novel materials. It is also important that before any investigation about defects in synthesized materials, measurements about structure, microstructure, and spectroscopic parameters can provide information about qualitative (nature of hosts and dopants) and quantitative information (number of trap centers, energy levels) about hosts, activators, and defects. This optical property can be applied for luminescence detection without any external optical source to stray light from bodily fluids and tissues [4]. The application of NIR-emitting PLNPs in long-term bioimaging is independent of PL lifetime, and the near-infrared PLNPs have the advantage of penetrating deep into biological cells and tissues by self-activation of PL by high wavelength source viz. red or NIR light [6]. To explore PLNPs for biological systems, researchers have developed PLNP-based nanoprobes to intertwine the features of PLNPs and biological systems [7]. Precision medicine's environment has been rapidly changing as a result of recent developments in the synthesis of nanomaterials with diagnostic and therapeutic capabilities. The development and manufacturing of novel theranostic nanomaterials to treat a range of diseases has advanced impressively, but their widespread clinical application is currently hampered by factors such as low bioavailability, poor biocompatibility, and unfavorable pharmacokinetics.

Early bulk persistent phosphors were produced by solid-state reactions [8] and were mainly used for civilian applications, viz., safety signs, escape route signs, etc. [4]. However, bulky persistent phosphors prepared by conventional high-temperature solid-state reactions are unsuitable for biomedical applications due to their poor biocompatibility and irregular size. Therefore, new synthetic approaches to fabricate PL nanoparticles (PLNPs), such as sol-gel, solvothermal, and template methods, have been proposed to control the morphology and manage the emission properties at the nanolevel [9]. Many new phosphors have been reported by researchers primarily due to the development of new host matrices and dopants. For example, $Ca_{0.2}Mg_{0.9}Zn_{0.9}Si_2O_6$: Eu, Dy, and Mn are the first persistent phosphors studied in sustained emission imaging [10]. Cr-doped gallium oxide-related phosphors represent highly efficient PLNPs [11]. Novel Cr/Ga-free persistent phosphors are being explored such as $Zn_3Ga_2Ge_2O_{10}$: Ni^{2+} [12], $LaAlO_3$: Mn^{4+}, Ge^{4+} [13], $BaZrSi_3O_9$: Eu^{2+}, Pr^{3+} [14], $KGaGeO_4$: Bi^{3+} [15], semiconducting polymer [16], and metal-organic framework [17]. Till date, $Sr_2MgSi_2O_7$: Eu^{2+}, Dy^{3+} is one of the most efficient PLNPs used for commercial purposes [18].

Case 1: Absorption of a photon by a Eu^{2+} impurity transfers an electron from the 4f to the 5d orbital. After returning to the ground state within microseconds results in the emission of a photon, this phenomenon is said to be photoluminescence.

Case 2: The excited electron can be captured at a defect center, said to be a trap. A certain trap length features the associated energy levels. For a specific range of trap lengths, sufficient thermal energy is available at room temperature (RT), which releases the trapped charge carrier within seconds up to hours, after which recombination and delayed emission occur at the Eu^{2+} center. Persistent phosphors are thus photoluminescent materials with a high density of traps where trapping and de-trapping of excited electrons are responsible for the optical phenomenon.

2 Synthesis of Nanotheranostics

New applications demand new approaches to synthesize PL materials. Different synthesis methods throw light on the defects distribution, afterglow mechanism, quantum efficiency of fluorescent emission, and features of structure and microstructure. Based on a literature survey of different raw ingredients related to PL, it is evident that other synthesis techniques are applied to these materials. In general, the primary approach to synthesizing optical materials is high-temperature solid-state reactions. This conventional approach though produces long-duration PL but due to high calcination temperature, the prepared particles become agglomerated. Hence, wet chemical techniques, such as sol–gel, hydrothermal, or precipitation reactions are also applied due to the homogenous distribution of constituent ions in these preparation techniques. Combustion is another method to synthesize materials but the distribution of grains is non-uniform. The ultimate objective of synthesizing materials by any of the chosen methods described above is to obtain single-phase resultant compounds, otherwise, if secondary phases are formed luminescence of the material will decrease due to cross-relaxation and non-radiative decay. Hence, a thorough microstructural characterization viz. X-ray diffraction (XRD) analysis, scanning electron microscopy (SEM) analysis, and Tunneling electron microscopy (TEM) are important at the start of the investigation. The wide range of synthesis techniques indicates the importance of precise synthesis conditions on the PL. It has been observed through literature survey that sometimes it is not possible to obtain proper crystallographic phase by solid-state or wet chemical methods. This is because the valence state of the dopant ions, ionic radii of the dopants and lattice sites, the purity of starting compounds, and lattice defects are dominant factors responsible for the purity/phase of the desired compound.

Researchers have been investigating and refining PLNP synthesis techniques for over ten years to produce nanoscale PLNPs. At present, the synthesis pathways for PLNPs widely used in bio-based applications are mainly as follows: Sol–gel, template, hydrothermal/solvothermal, and co-synthesis. In the sol–gel pathway, the precursor is firstly hydrolyzed into the solvent and then the solvent is transformed into a gel before subsequent calculation (800–1100 °C) resulting in high-yield, small-size PLNPs. Scherman et al. synthesized a series of silicate PLNPs by sol–gel method, such as $Ca_{0.2}Zn_{0.9}Mg_{0.9}Si_2O_6$: Eu^{2+}, Dy^{3+}, Mn^{2+} [19], $CaMgSi_2O_6$: Eu^{2+}, Mn^{2+}, Pr^{3+} [20], $Sr_{1.6}Mg_{0.3}Zn_{1.1}Si_2O_7$: Eu^{2+}, Dy^{3+} [21], $Ca_{1.86}Mg_{0.14}ZnSi_2O_7$: Eu^{2+}, Dy^{3+} [22], $Sr_2MgSi_2O_7$: Eu^{2+}, and Dy^{3+} [6]. In addition, gallate PLNPs, such as $LiGa_5O_8$:Cr^{3+} [23], $Zn_{2.94}Ga_{1.96}Ge_2O_{10}$:$Cr^{3+}$, Pr^{3+}, and $Zn_{1.1}Ga_{1.8}Ge_{0.1}O_4$: Cr^{3+} [24], were similarly created using the sol–gel technique. One of the most popular techniques for creating PLNPs is the sol–gel approach. Nevertheless, this approach presents many drawbacks, including non-uniform shapes, unpredictable dimensions, and inadequate dispersion.

Zhang and associates have made significant progress in the template-based synthesis of PLNPs. They controlled generated PLNPs by using carbon nanospheres, hollow silica, or mesoporous silica as a template, such as SiO_2@Zn_2SiO_4: Mn [25], SiO_2@$SrMgSi_2O_6$: $Eu_{0.01}$, $Dy_{0.02}$ [26], SiO_2@$CaMgSi_2O_6$: Eu^{2+}, Pr^{3+}, Mn^{2+} [27], SiO_2@ $CaTiO_3$: Pr [28], and Gd_2O_3@$mSiO_2$@$CaTiO_3$: Pr [29]. The particle size range of these PLNPs is between 50 and 500 nm, and they exhibit a consistent spherical shape with good monodispersity. NIR photoluminescence is highly appropriate for use in biomedical applications. This technique is thought to be an accurate means of creating nanocarriers for PL imaging-guided medication distribution, notwithstanding several difficulties in the preparation of ultra-tiny PLNPs. The co-synthesis method was described as a means of synthesizing small-size and brilliant PLNPs while taking into account the benefits and drawbacks of each of these techniques. The synthesis of $ZnGa_2O_4$:Cr^{3+} by co-synthesis method was reported by Richard et al.

Table 1 Comparison of synthesis methods of nano-PLNPs

Synthesis methods	Size (nm)	Morphology	Uniformity	PL	References
Sol–gel	50–100	Poor	Poor	Long	Shi et al. [33]
Template	50–100	Fine	Fine	Medium	Maldiney et al. [20]
Hydrothermal/ solvothermal	5–20	Fine	Fine	Short	Wang et al. [34]
Co-synthesis	20–60	Average	Average	Long	Zou et al. [35]

[30]. This process involved first synthesizing the PLNPs precursor by hydrothermal method and then calcining at 750 °C to improve the PL characteristics, by introducing cetyltrimethylammonium bromide (CTAB) into the hydrothermal system as a precursor, Yan et al. [31].

However, from the literature survey, it has been observed that many of the synthesized PLNPs after injecting into living organisms (small animals) are functionalized in the liver. To overcome this problem, special surface coatings are done in the prepared PLNPs, which makes nanoparticle circulation in the bloodstream for sufficient time. Liu et al. have developed poly (N-2-hydroxypropyl) methacrylamide (pHPMA). It is a hydrophilic polymer usually used in the preparation of nanomedicine. Investigation of in vivo imaging using pHPMA as coating layer in $ZnGa_2O_4$: Cr^{3+} improved the blood circulation time [32] (Table 1).

It is clear from studies of investigations performed by researchers worldwide that the main interest is in preparing PLNPs from rare earth-doped inorganic oxides and other materials. This process of PLNP synthesis is costly and tedious. Hence, the need of the hour is on preparing cost-effective, long-emission PLNPs, good biocompatibility, economically friendly, and rare earth-free ions [36, 37]. A new type of luminescent material of the form metal–organic framework with a long decay time was reported in the studies done by Yang et al. [37]. To date, this material is not used in bio-imaging investigations due to the high microstructural diameter of the synthesized material and emissions in the UV region. Polymers with semiconducting properties prepared at the nanolevel have recently attracted bio-molecular scientists for in vivo imaging studies. These polymers emit wavelengths in the NIR region with long emission times about 100 times greater than normal PLNPs [38, 39]. These polymers have been used for lymph nodes and imaging of tumors with high signal-to-noise ratio. The persistent afterglow has made these polymers suitable for specific drug release on passive tumors [40].

3 Diagnosis and Therapy

(a) UV pre-excitation

In 2007, Scherman et al., for the first time applied $Ca_{0.2}Zn_{0.9}Mg_{0.9}Si_2O_6{:}Eu^{2+}$, Dy^{3+}, Mn^{2+} for in vivo (process which takes place in living organisms) PL imaging of tumor bearing mice [41]. After pre-excitation of PLNPs by UV lamp, it was observed that the intravenously injected PLNPs accumulated in the tumor region in 2 minutes. This was detected by the PL signal. Subsequently, Yan et al. synthesized Cr^{3+}, Pr^{3+} codoped $Zn_{2.94}Ga_{1.96}Ge_2O_{10}$ with almost 15 days of NIR PL [42]. The details of tumor regions marked by PLs are shown in Fig. 2.

(b) LED in situ excitation

Richard et al. investigated an orange/red LED as in situ (at the original place of the tumor) excitation source for in vivo PL imaging of tumors using $ZnGa_2O_4{:}Cr^{3+}$ nanoparticles [43]. The UV pre-excited PLNPs were intravenously injected for in vivo tumor PL imaging at two hours. After the PL signal of the tumor site is decayed, the red LED was used for in-situ re-excitation, which can restore the PL signal of the tumor site and realize in vivo re-excitation imaging of the tumor as shown in Fig. 5. Later, Pan et al. intravenously injected peptide conjugated $LiGa_5O_8{:}Cr^{3+}$ in 4T1 tumor model and used white LED to in situ stimulate tumor accumulated $LiGa_5O_8{:}Cr^{3+}$ for PL imaging up to 24 hrs [44]. Besides, Yan et al. reported FA-modified $Zn_{1.25}Ga_{1.5}Ge_{0.25}O_4{:}Cr^{3+}$, Yb^{3+}, Er^{3+} for diligently targeting MCF-7 tumors orally [24]. With the excitation of 650 nm LED, the tumor regions achieve a long time about 160 minutes and high sensitivity (SNR > 20) PL imaging. Though, this results are satisfactory but work on deep tissue imaging is limited due to availability of LEDs in the visible region only.

(c) NIR laser excitation

The light in the biological window has better tissue penetration and is suitable for deep tissue reactivation [45]. Therefore, rare-earth doped PLNPs have been developed for NIR laser excitation. Zhang et al. studied that the $ZnGa_2O_4$: Cr, Eu and Zn_2SnO_4:

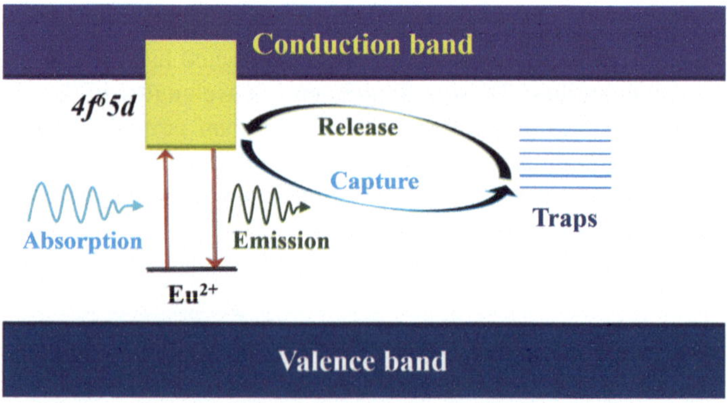

Fig. 1 Schematic view of PL in a Eu^{2+} doped persistent phosphor. (Reproduced from Elsevier, 2015; 48: 1–108 with permission from Elsevier Publishing)

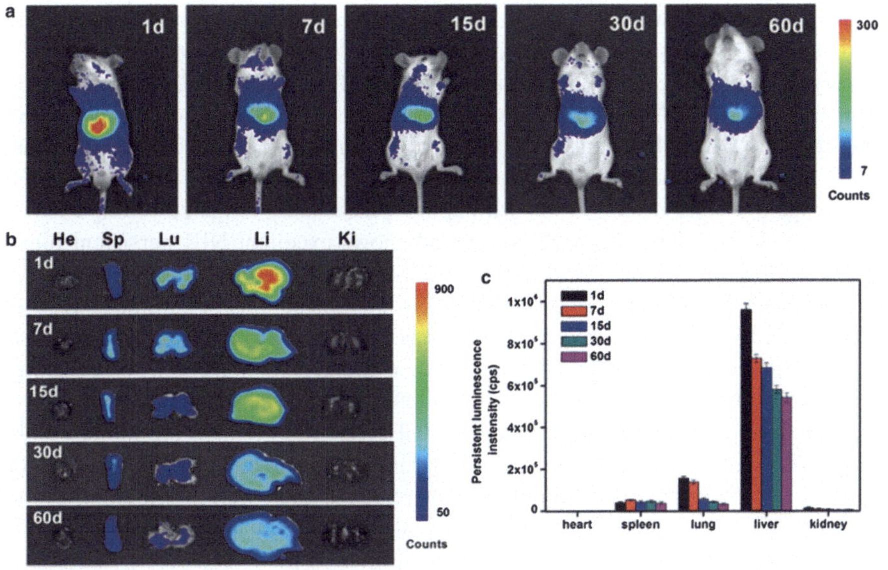

Fig. 2 **a** In vivo biodistribution; **b** distribution outside the body regions (ex-vivo) of the mice under observation; and **c** PL intensities of $Ca_{0.2}Zn_{0.9}Mg_{0.9}Si_2O_6$: Eu^{2+}, Dy^{3+}, Mn^{2+} in vital organs at different time points. Reproduced from Scientific Reports, 2018; 8: 1–11 with permission from the Nature Publishing

Cr, Eu can be easily modified with folic acid molecules through a simple condensation reaction [46]. Ultraviolet PLNPs were used for targeted imaging in mice. After the signal of the tumor site decayed, it was re-excited by using a light source of 808 nm with more substantial tissue penetration, which can achieve highly sensitive tumor imaging. Some of the investigations pursued by researchers are presented in this section. Hao et al. reported a novel 980 nm laser-activated upconverted PLNPs ($Zn_3Ga_2GeO_8$: Yb/Er/Cr) for in vivo PL imaging. Due to efficient energy transfer (Er^{3+}-Cr^{3+}), the produced NIR PL remained up to 15 hrs. Meanwhile, these upconverted PLNPs also can be effectively recharged in vivo under 980 nm laser excitation [47]. Chang et al. investigated [Zn_2SiO_4:Mn] Y^{3+}, Yb^{3+}, and Tm^{3+} upconverting PLNPs for deep tumor imaging under a 980 nm laser [48]. An arrangement for in-vivo PL based on LED Zn Ga_2O_4: Cr^{3+} and PLNPs in the NIR region prepared by the sol-gel method on mice is illustrated in Figs. 3 and 4, respectively. In another study by Sun et al. about one of the most novel materials, i.e., $Zn_{1.1}Ga_{1.8}Sn_{0.1}O_4$: Cr^{3+} (ZGSO) for near-infrared persistent luminescence, the potential risk assessment for bioimaging was investigated. A study in vitro shows that ZGSO is low cytotoxic, also the results of in vivo biodistribution after administering intravenous injection reveal that ZGSO accumulated in the reticuloendothelial system gets easily drained from the digestive system. No toxicity was found in the living organism, i.e., mice over two months. Also, the toxic content is almost negligible both in vitro and in vivo [49]. Sun et al. first synthesized ZGSO using a solvothermal method at 200 ^{0}C. Then the prepared sample was calcined at 700 ^0C for some time. It was observed that the original ZGSO shows poor biocompatibility and solubility, which restrict the applicability of bioimaging. As discussed in the section of synthesis, coating improves the features of PLNPs. Hence, in the present study, ZGSO was grafted with polyethylene glycol (PEG), which improves the solubility and biocompatibility of the host matrix. The TEM images reveal monodispersion of the prepared compound with an average diameter of 20–30 nm with spinel phase as investigated through X-ray diffraction (XRD). The prepared compound shows excellent NIR persistent luminescent properties. It was observed that NIR persistent luminescence shows optical emission for 3 hours after excitation with a 254 nm light source for 5 minutes. Hence, it can be said that the prepared compound has a slow decay rate. The potential risk of the prepared ZGSO under NIR light was evaluated for long-term bioimaging. Cytotoxic studies on carcinoma cells in the lungs (A549), hepatocellular carcinoma cells (HepG2) and umbilical vein endothelial cells (HUVEC) in the human body using NIR external source did not induce major changes in the cell viabilities, minor injury to cell membranes, negligible oxidative stress and small assimilation in the different variants of human lung cells. It was also observed that after injecting PEG-fabricated ZGSO in mice, was mainly concentrated in the liver, spleen, and lungs. Also, the histopathological investigation shows non-significant toxic concentration in mice for 60 days. Studies by Sun et al. encourage further research on different molar concentration of ZGSO for potential risk calculation.

(d) X-ray excitation

Since X-rays are high-energy collimated sources that may maximize penetration depth and reduce scattering in biological tissues and living cells, they have emerged as an inventive excitation source for in vivo optical imaging in recent years [50]. Non-invasive, fast, and extremely sensitive, optical imaging works by triggering the luminous centers of certain phosphors with X-ray photons. Since PLNPs mitigate the effects of tissue auto-fluorescence, bioimaging research on them has accelerated. Despite this, the technique's short decay period is a restriction. Song et al. introduced a novel method for characterizing materials for biological imaging. They have imaged tumors in vivo using soft-x radiation light sources ex vivo. During their examination, they noticed that X-rays used as an external optical excitation source to activate $SrAl_2O_4$:Eu^{2+} PLNPs for imaging deep tissues can penetrate up to 2.5 cm [51]. Hao et al. studied X-rays to stimulate $ZnGa_2O_4$: Cr PLNPs for renewable NIR PL imaging of deep tissues [52]. To synthesize and control the morphology of PLNPs, Yang et al. prepared kiwifruit-like structures of SiO_2@$ZnGa_2O_4$: Cr@SiO_2 with the help of a silica template. Then the X-ray irradiated PLNPs showcased excellent PL performance and long-term imaging from deep tissue [53]. Zhang et al. synthesized $MgGeO_3$: Mn^{2+}, Yb^{3+}, Li^+ (MGO/MYL) PLNPs with NIR-I and NIR-II emission. Under the influence of soft X-rays in the synthesized material (MGO/MYL), it was observed that PL imaging is possible from deep tissues [54]. In another investigation by Yeh et al. it was inferred that polyethylene-gylated $ZnGa_2O_4$: Cr^{3+} concave nanocubes using highly passive target and X-ray excitation can be used for PL imaging of deep-seated orthotopic hepatic tumors [55].

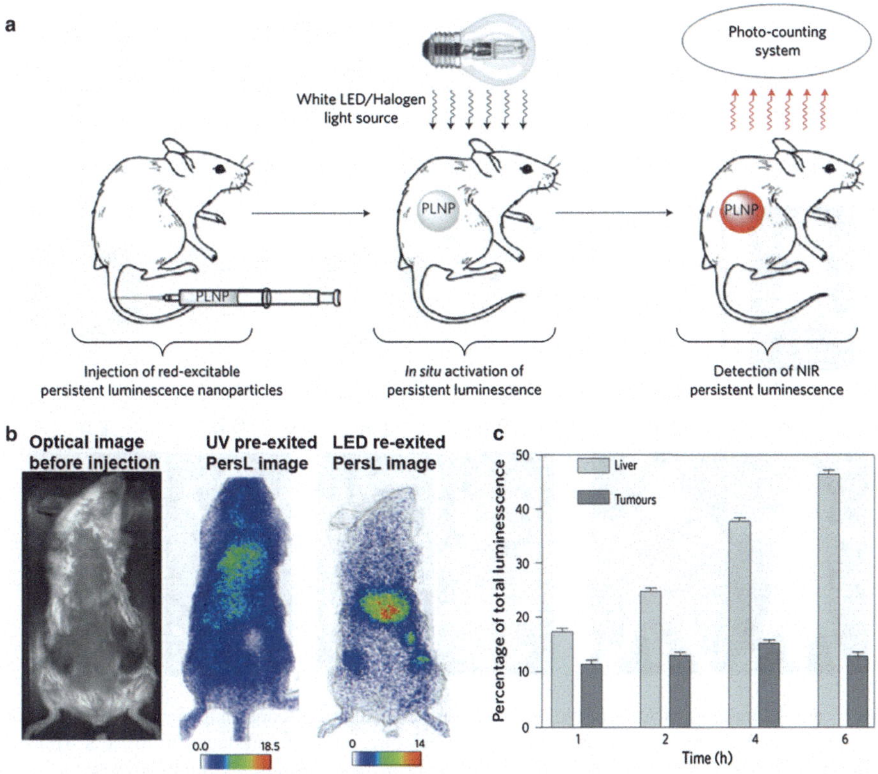

Fig. 3 **a** An arrangement for in vivo PL imaging based on LED Zn Ga_2O_4: Cr^{3+}; **b** optical image of a tumor-bearing mouse; and **c** Bar graph representation of PL intensities of liver and tumors measured at different time points. (Reproduced from Scientific Reports, 2014; 13: 418–26 with permission from the Nature Publishing)

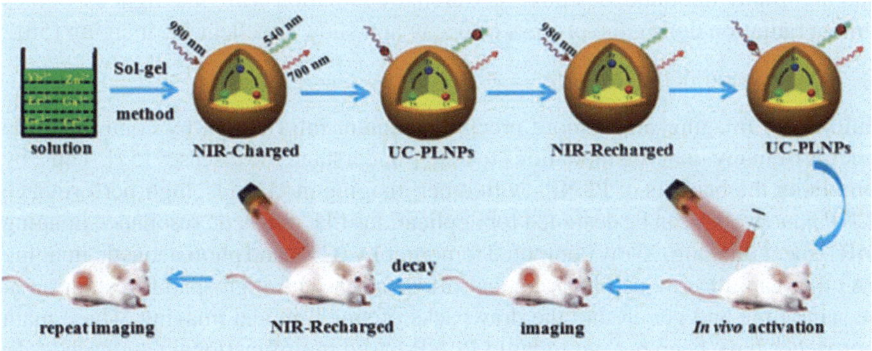

Fig. 4 Schematic representation of the in vivo bioimaging based on 980 nm laser upconversion PLNPs in the NIR region prepared by sol–gel method. (Reproduced from Nanoscale, 2017; 9: 7276–728 with permission from the RSC Publishing)

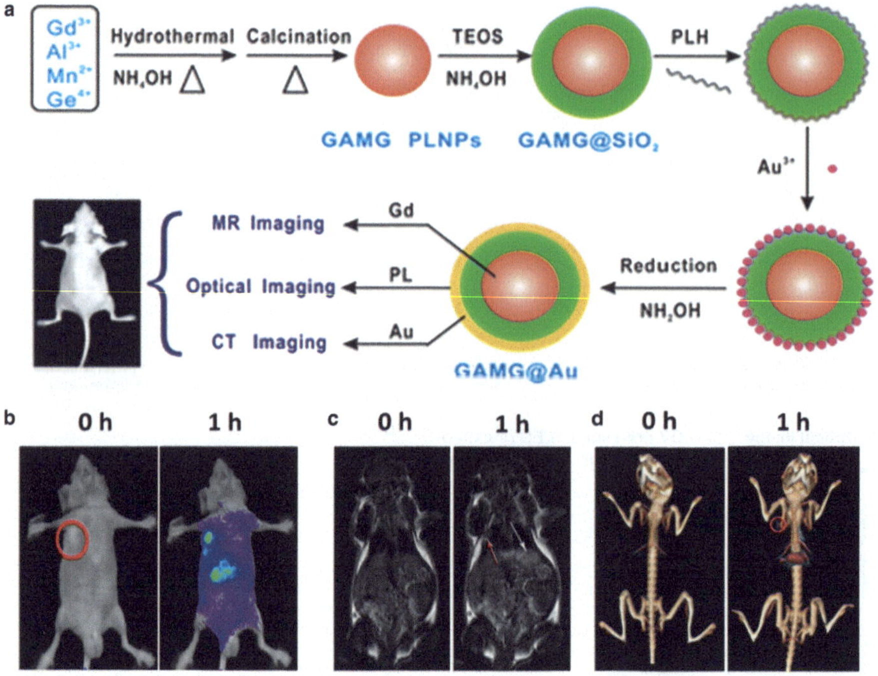

Fig. 5 **a** Schematic illustration of the preparation of GAMG@ Au-core shell nanostructure. **b–d** In vivo PL/MR/CT images before and after intravenous injections. (Reproduced from ACS Applied Materials and Interfaces, 2016; 8: 29,939–49 with permission from the ACS Publishing)

(e) Radiopharmaceutical excitation

Radiopharmaceuticals can be regarded as an internal excitation light for deep tissue imaging. Many radionuclides possess Cerenkov radiation with blue light as well as gamma radiation during the process of decay and show excellent PL imaging [56].

(f) In vivo multimodal imaging

Multimodal imaging offers more precise treatment information by combining the benefits of many imaging modalities to better understand the diseases [57]. Thus, by combining the benefits of PLNPs with other imaging modalities, high-performance PLNP nanoprobes can be designed for applications like magnetic resonance imaging (MR), spect imaging, X-ray computed tomography (CT), and photoacoustic imaging (PA) that can all provide physiological information with high spatial resolution at the same time and get around the drawbacks of single-modal imaging. Three main approaches have been put out to build PLNP-based multifunctional nanoprobes. To test alternative imaging modalities, the first technique involves altering the surface using chemical reagents. Gao et al., for instance, created c(RGDyK) peptide and radioisotope 99mTc labeled paraffin leaf patches (PLNPs) for oral delivery and targeted PL/SPECT imaging of orthotopic breast cancer [58]. Gd-DTPA-modified

PLNPs, or {Gd^{3+}-PLNPs}, were described by Yan et al. [59] for use in T1 weighted MRI imaging and in vivo NIR persistent luminescence. Furthermore, Yan et al. conjugated Gd_2O_3 (HA-Ga_2O_3) functionalized with hyaluronic acid on polylactic acid nanoparticles [60]. The conjugation demonstrated strong MR and PL signals in tumor locations and has the potential to actively target tumors. The second technique involves coating or sequential growth to introduce the core–shell structured PLNPs. Multifunctional core–shell nanostructures such as ($Zn_{2.94}Ga_{1.96}Ge_2O_{10}$: Cr^{3+}, Pr^{3+}@TaOx@SiO_2) were reported by Yan et al. for in vivo PL/CT imaging of tumors [61]. Polypyrrole-coated PLNPs were produced by Zhang et al. and may be investigated for dual-modal PL/PA imaging of tumors [62]. Wang et al. also investigated the host matrix $GdAlO_3$:Mn^{4+}, Ge^{4+}@Au (GAMG@Au) core–shell nanoprobes with MR/CT/PL third modals imaging properties [63]. The nanoparticles may actively collect in the tumor locations that are visible by MR/CT/PL imaging after being treated with folic acid PEG-SH. This core–shell approach, however, inevitably results in a size increase and a PL fall because of photon absorption or reflection from the shell structure. These investigations lead us to the conclusion that the doping strategy is far better due to its straightforward preparation and consistent physicochemical characteristic. Figure 5 shows a schematic illustration of the creation of the GAMG@ Au-core shell nanostructure as well as in vivo PL/MR/CT images taken both before and after intravenous injections.

4 PLNPs Based Surgery

Fluorescence-based surgery (FBS) uses real-time fluorescence images of disease to guide surgical operation, which is a cost-effective and more accessible way for minute resection of tumors compared to conventional imaging technologies [64]. In one of the studies by Tian et al., it was observed that $ZnGa_2O_4$: Cr^{3+} synthesized can be used for long-term image-guided surgery of hepatocellular carcinoma (HCC) due to the excellent advantages of PLNPs. It shows NIR emission, long PL, and a high signal-to-background ratio [35]. These $ZnGa_2O_4$:Cr^{3+} were prepared following a reported method [65]. It was observed that there was no intake of $ZnGa_2O_4$: Cr^{3+} by HCC tumor tissue, but a substantial amount was assimilated in the normal liver tissue, which made the precise mapping of the HCC tumor area with more radical excision. Endomicroscopic laser images of liver and peri-tumoral area after 6 hrs post injection of $ZnGa_2O_4$: Cr^{3+}, PL images of the exposed tumor lesion administered PLNPs injection, LEDs irradiation and fluorescence images of the 3 exposed tumor lesions illuminated with white light, stains of the resected tissue, and graphical illustration of the survival rates of mice are shown in Fig. 6.

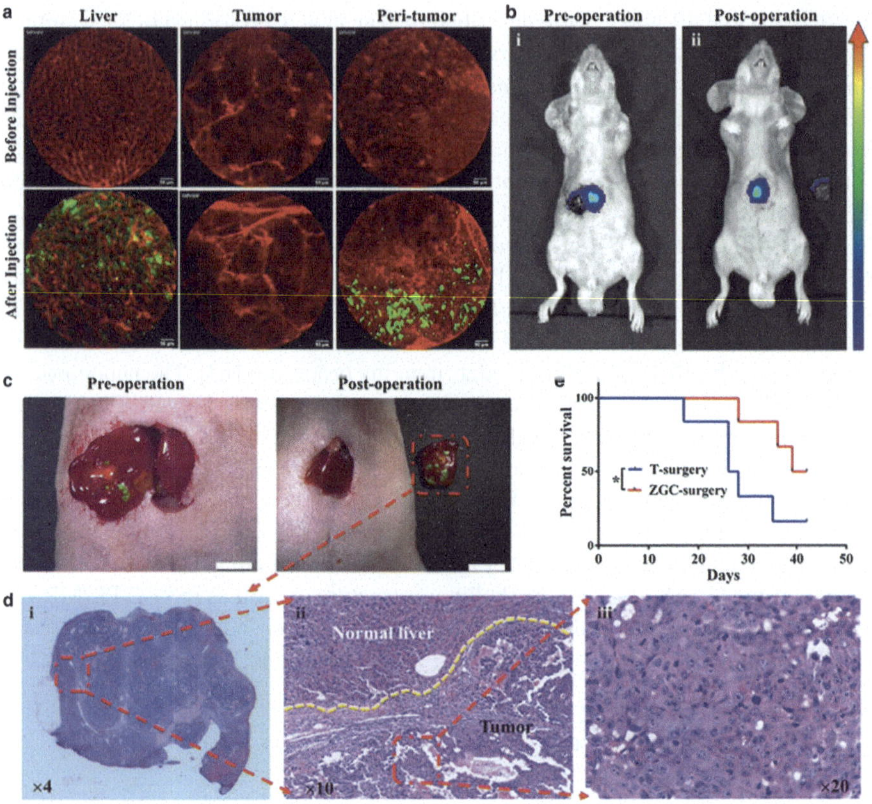

Fig. 6 **a** Endomicroscopic laser images of liver and peri-tumoral area after 6 h post injection of $ZnGa_2O_4$: Cr^{3+}; **b** PL images of the exposed tumor lesion administering PLNPs injection and LEDs irradiation; **c** Fluorescence images of the 3 exposed tumor lesions were illuminated with white light images; **d** Staining of the resected tissue; and **e** graphical representation of the survival rates of mice from different treatments. (Reproduced from Biomaterials 2018; 167: 216–25 with permission from the Elsevier Publishing)

5 DOX-Loaded Nanotheranostics

Mesoporous silica ($mSiO_2$) consisting of silica nanoparticles containing many pores in its structure is often used as a carrier to deliver drugs [66]. But this approach is limited as the biodistribution of mesoporous silica could not be traced by any method. In 2014, Maldiney et al. proposed a solution using PL as a potent property to follow in vivo the bio distribution of $mSiO_2$ loaded with doxorubicin (DOX) [67]. In 2016, Zhang and co-workers developed a variant of the previous system in which mesoporous silica can be made not only with PLNPs (ZGO) as before but also with an MRI contrast agent (Gd_2O_3) [68]. This nanotheranostic was loaded with DOX to monitor its bio distribution in healthy and in tumor-bearing mice. The details of the schematic illustration are represented in Fig. 7.

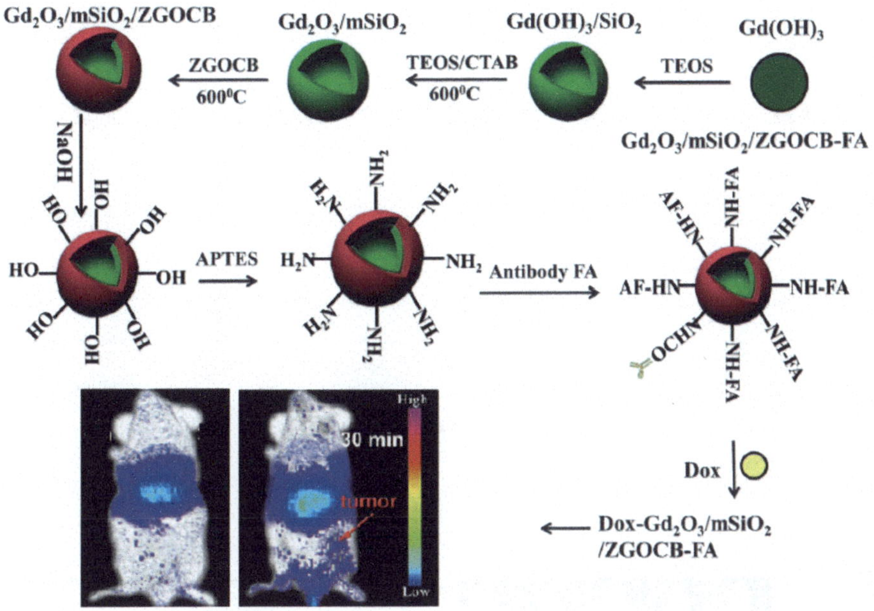

Fig. 7 In vivo application using DOX in the synthesis of Gd_2O_3/ZGO nanostructures. (Reproduced from Journal of Materials Chemistry B, 2016, 4, 1842–1852 with permission from the RSC Publishing)

6 PLNPs-Based Photothermal Therapy

Activatable optical imaging plays a crucial role in sensitive tumor-targeted imaging. Sensitivity can be further enhanced by incorporating PLNPs into activatable nanoprobes. For tumor therapy, photothermal therapy (PTT), which uses PTT drugs to generate heat from light for tumor ablation, offers the unique advantages of high selectivity and minimal invasiveness [69]. Furthermore, it has been observed that PTT agents with strong NIR absorption are good emission quenchers for activatable luminescent nanoprobes. Therefore, it is essential to fabricate PLNP-based activation imaging PTT nanoprobes for theranostics. Recently, Chen et al. developed activatable multifunctional PLNPs-CuS nanoprobes for in vivo PL imaging-guided PTT as shown in Fig. 8. They have used CuS nanoparticles as biocompatible PTT agents with excellent luminescence quenching ability due to strong NIR absorption and used matrix metalloproteinase (MMP)-specific peptide substrates to conjugate PLNPs and CuS resulting in MMPs. They have developed an activatable theranostic system MMP-2 activated PLNP-CuS nanopharmaceuticals with good biocompatibility for sensitive PL imaging of MMP-2 secreted by SCC-7 tumor cells [69]. Furthermore, in their investigation, they have also demonstrated robust PTT in tumors at the cellular level and in vivo. The successful combination of PLNP-based activatable imaging

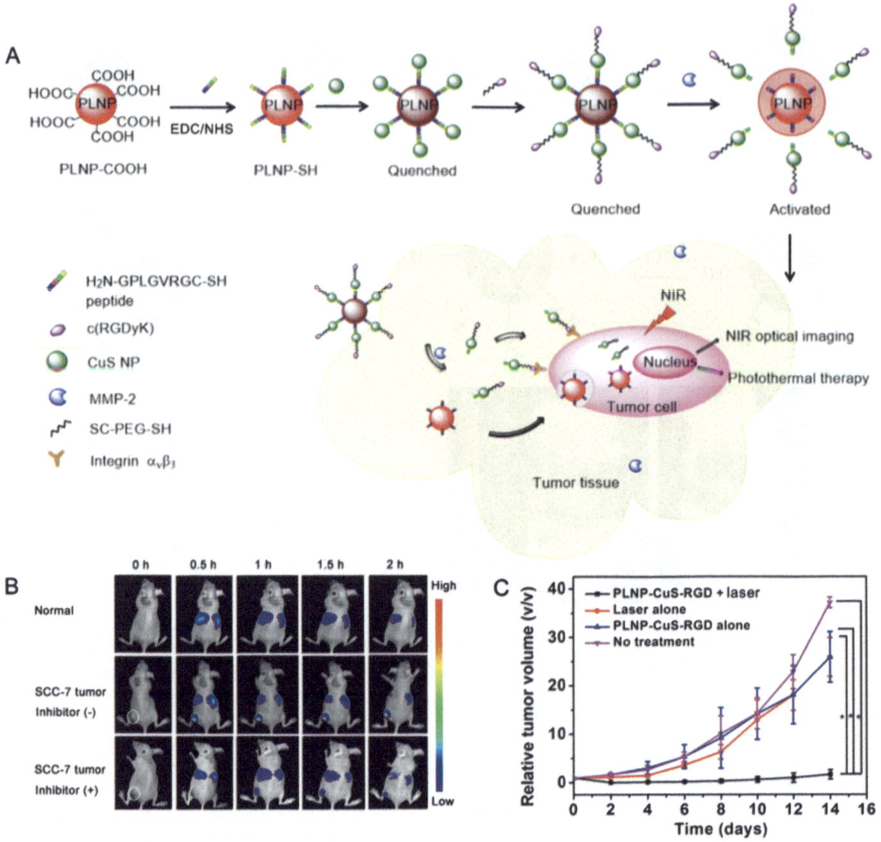

Fig. 8 **a** Synthesis of PLNP-CuS-RGD for in vivo imaging-guided PTT; **b** PL images of tumor-bearing mice administered with intravenous injection with or without inhibitor; and **c** graphical representation of the tumor growth curves by different treatments. (Reproduced from ACS Applied Materials and Interfaces, 2016, 8, 32,667–32,674 with permission from the ACS Publishing)

and PTT provides new ideas for developing highly sensitive and efficient therapeutic nanoprobes for tumor diagnosis and therapy in vivo [69].

7 PLNPs-Based Photodynamic Therapy

The extended afterglow properties of PLNPs also provide an internal light source for photodynamic therapy (PDT) [70]. PDT uses photosensitizers (pss) to generate singlet oxygen ($^1O_2$) or reactive oxygen species under the action of light to eliminate tumors and is clinically approved for the treatment of various tumors [71]. This technique is known for its efficient approach to treat some types of cancers or precancerous lesions [70]. The pss when exposed to a light source of some wavelength related to its absorption peaks, the photosensitizer changes the chemical form of oxygen from molecular oxygen to oxygen species which have high cytotoxic characteristics that result in tumor necrosis. In general, photosensitizers are non-toxic and hence, when exposed to external optical sources, are highly efficient in selecting cancerous tissues [72]. Under light illumination, the photosensitizer (ps) is excited to a higher energy state say $^{1}ps^{*}$. The excited $^{1}ps^{*}$ can return to the ground state through two possible transitions. It can directly move to the ground state by emission of fluorescence or through an intersystem crossing mechanism (where the state of the electron spin inverts to a triplet state, which remains excited for a long time) to $^{3}ps^{*}$. In the type I reaction approach, the long-lived triplet state transfers its energy through a proton or exchange of electrons which results in anions or cations. This free radical reacts with oxygen resulting in hydroxyl radicals, hydrogen peroxides, or superoxide anions. In another approach, say type II reactions, the energy of the $^{3}ps^{*}$ is transformed to molecular oxygen with the release of $^{1}O_2$. In both these types of reactions, the resultant is reactive oxygen, which may damage mitochondria, the golgi body, endoplasmic reticulum, and plasma membranes in its reaction time. It all depends on the location of the photosensitizer. In research investigation by Kanofsky, the lifetime of $^{1}O_2$ was found to be 0.4 ± 0.2 μs near living cell membranes [73] and in another study by Jarvi, it was observed to be 1.2 ± 0.3 μs for in vivo blood vessels [74]. With long-lasting and photo refresh ability, PLNPs show great potential as an internal light source for PDT without any external illumination. The technique of PDT is not specific to any type of living cells, and it applies to all types of microorganisms viz. bacteria, fungi, and viruses for the treatment of local infections [75]. PDT, which is used to cure pathogenic organisms, is often termed as photodynamic inactivation or anti-microbial PDT [76].

However, due to the limited tissue penetration depth of external light sources, PDT fails to apply tumor therapy to deep tissues. Furthermore, continuous laser irradiation risks tissue damage due to overheating, restricting the widespread use of PDT. One alternative to overcome this problem is introducing PDT mediators which can excite a ps present in some in-depth tissue region with the help of an external light source. Mediators used in PDT techniques are usually small-sized objects ranging from atomic to nanometer scale. There are two types of mediators, i.e., energy transducers and self-illuminating probes. Energy transducers work on the principle of energy conversion. It converts an external light beam of high energy into penetrating radiation to excite the ps. On the other hand, self-illuminating probes are based on chemiluminescence, bioluminescence, or Cerenkov radiation [77]. In this method, activation of pss is through resonance energy transfer using molecules such as luminol (H_2O_2) or oxalate ester (H_2O_2) and luciferase enzymes can generate intrinsic light [78]. Cerenkov radiations (high-energy radiations which travel with speed greater than light), usually emitted from β-emissions of radionuclides like ^{18}F, ^{64}Cu, and ^{68}Ga, are applied in some small animals by some researchers through Cerenkov radiation energy transfer (CeRET) from these emissions to pss. Initially, ^{18}F, titanocene-charged TiO_2 and apo transferrin were co-injected into the animal.

At later stages of research, radionuclides and pss are localized using a nano-agent [79].

8 PLNPs-Based Chemotherapy

Cancer is a complex disease because it encompasses a variety of disorders that are all characterized by unchecked and rapid cell multiplication. There can be many possible reasons for this complex disease, and one of them is a genetic abnormality. Another reason can be epigenetic. There are various modes of cancer treatments viz. chemotherapy, radiation, immunotherapy surgery etc. Among them, chemotherapy is a common mode of cancer treatment but it is limited by several side effects on the human body, low intensity on the target tumor area, and difficult in administering many hydrophobic agents [80]. As the probability of medicinal molecules reaching the tumor area is low, chemotherapy remains an ineffective means of cancer treatment. It also has widespread side effects on the patient's body. As per the literature survey, it has been observed that the concept of nanoscience is useful in the treatment of cancer. Here, the medicines are transformed at the nanolevel through various laboratory processes. In comparison with conventional pharmacological therapies, nanoparticles can effectively transport a wide variety of compounds while using minimum chemical reagents. Due to its high surface-to-volume ratio, nanomedicines are less degraded by physiological processes. The concept of nano-imaging enables targeted delivery of hydrophilic and hydrophobic drugs in tumor region; it is also useful for real-time monitoring of therapeutic response to administered medicines, which overcomes the limitations of conventional cancer treatment approaches. The investigation of active or passive usage of nanomedicine at tumor sites has been shown to increase the therapeutic index in cancer treatment. It is seen that novel PLNP has developed as a star material for highly sensitive luminescence image-guided drug delivery [80]. Novel PLNP-based nanocarriers are prepared with simple synthetic methods, good biocompatibility, and high drug-loading efficiency which are urgently needed for drug delivery by sensitive imaging. Liposomes are classical and commonly used nanocarriers with excellent biocompatibility, biodegradability, and low toxicity [80]. Combined treatment of PLNPs and liposomes offers excellent opportunities for creating intelligent PLNP-based drug delivery systems [81]. It has been observed that polymeric micelles, polymer-drug conjugates, and liposomes are a few examples of nanomedicines that have substantial clinical value [81]. Nanomedicines have longer half-lives than conventional small molecules; hence, medications are swiftly removed from the bloodstream. They also have enhanced tumor target delivery and biological availability is also high. The addition of imaging concepts into nanomedicine will improve the diagnosis of theronastics. We could advance the use of theranostics in the treatment of oncogenic diseases by using this approach, which has several benefits. A simple thin-film evaporation method was used to coat {chromium (Cr)} Cr-PLNPs with liposomes. The prepared PLNPs-liposomes exhibited PL, emitting red LED light, highly biocompatible, and high drug encapsulation efficiency. PLNPs-liposomes enabled cellular PL imaging in vivo while paclitaxel-encapsulated liposome-PLNPs (PLNPs-PTX) achieved significant improvement in therapeutic index. PLNPs-PTX has been used for passively targeted tumor imaging, monitoring drug delivery, and dramatic tumor growth inhibition without autofluorescence interference. The integration of liposomes and PLNPs provides a promising route to create susceptible PL imaging drug carriers with high drug loading and therapeutic efficacy.

9 PLNPs-Based Cell Tracking Gene Therapy

The exceptional capacity to detect stem cells makes it easier to design stem cell-based treatments to cure tumors by employing PLNPs as extremely sensitive tracking nanoprobes. Glioblastoma can be effectively treated with mesenchymal stem cell (MSC)-based gene therapy [82]. However, monitoring the movement, destiny, and distribution of therapeutic stem cells in brain tissue necessitates the use of a dependable, durable, non-invasive, and sensitive imaging method. As seen in Fig. 9a [83], we have recently developed bifunctional MSCs engineered with PLNPs for acute PL-directed gene therapy of glioblastoma. For the EGFP-TRAIL plasmid DNA (enhanced green fluorescent protein-human tumor necrosis factor-related apoptosis-inducing ligand), sequential conjugation of PEI, PEG, and TAT penetrating peptides to PLNP enabled internalization through its high water dispersibility and facilitation. PLNPs-PPT (positively charged PEG-TAT-PLNPs-PEI) nanoprobes were further attached to the His EGFP-TRAIL plasmid DNA by electrostatic adsorption. The tagged stem cells of the PLNPs-PPT and PLNPs-PPT/TRAIL hybrids showed little negative impacts on the potential for MSC development, and both hybrids demonstrated acceptable biocompatibility. Studies on in vitro gene transfection have shown that PLNPs-PPT can follow stem cells over an extended period and deliver therapeutic genes. Additionally, we discovered that after being administered to animals carrying U87-MG cancers, PLNPs-PPT-tagged MSCs moved through the corpus callosum into tumors but not into the brain tissue of normal mice. These findings show that PLNPs-PPT-tagged MSCs can specifically target tumors and allow for sensitive in vivo stem cell movement tracking. Furthermore, both in vitro and in vivo, PLNPs-PPT/TRAIL-generated MSCs demonstrated notable anti-tumor activity. The effective creation of MSCs with intriguing therapeutic capabilities that are PLNPs-PPT/TRAIL-engineered opens the door for the creation of multifunctional stem cell-based therapeutic nanoprobes that employ PLNPs.

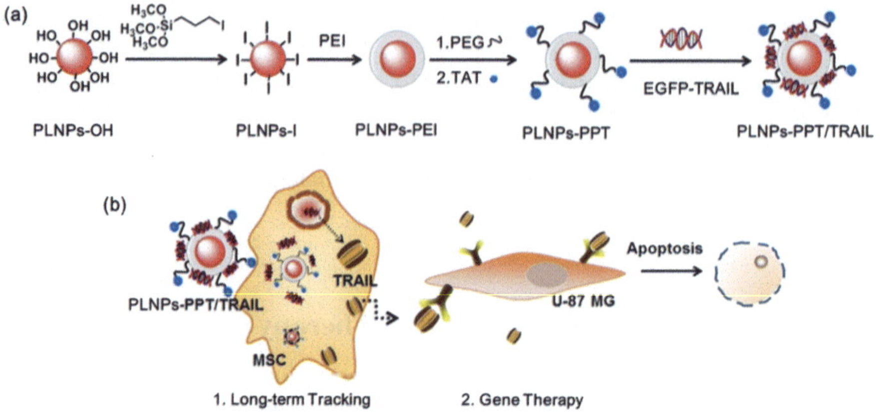

Fig. 9 Dual-functional PLNPs-engineered MSC. A **a** synthesis of PLNPs-PPT, A **b** 1. Long-term tracking 2. Gene therapy. (Reproduced from Advanced Functional Materials, 2017, 27, 1,604,992 with permission from the Wiley Online Library Publishing)

10 PLNPs-Based Tracking and Combined Therapy

Treatment for cancer patients must include accurate detection and efficient treatment of metastasis because it is well-recognized that metastasis can cause death in patients either directly or indirectly [84]. However, because of their tiny size, dispersed location, and inadequate anatomical vascularity, metastases are challenging to identify and treat in vivo. Chen et al. recently reported the biomimetic development of PLNPs for 808 nm laser-guided chemical PDT and long-term auto fluorescence-free metastatic tracking [85]. A PLNPs core modified with Si-PC photosensitizers (SPLNPs), a hollow silica ($hSiO_2$) interlayer filled with doxorubicin (DOX), and a cancer cell membrane (CCM) base acting as a protective sheath make up the nanoplatforms (DSPLNPs@$hSiO_2$@CCM). Figure 10b illustrates the tracking of metastatic tumors and cascading therapy. Drug leakage is avoided and targeted metastases can grow on the nanoplatform because of the CCM coating. Furthermore, PDT caused by PL has the potential to interfere with CCM and release medications into silica to treat metastases. The nanoplatform based on PLNP light retractable allows for long-term tracking of metastasis without the need for autofluorescence. It makes metastasis-related 808 nm laser cascade chemo photodynamic treatment possible in vivo and permits a precise diagnosis and manageable therapy. This study clearly indicates how surface engineering and energy transfer engineering may completely utilize the special PL features of PLNPs, allowing nanoplatforms to be used to address important scientific issues in the biomedical area.

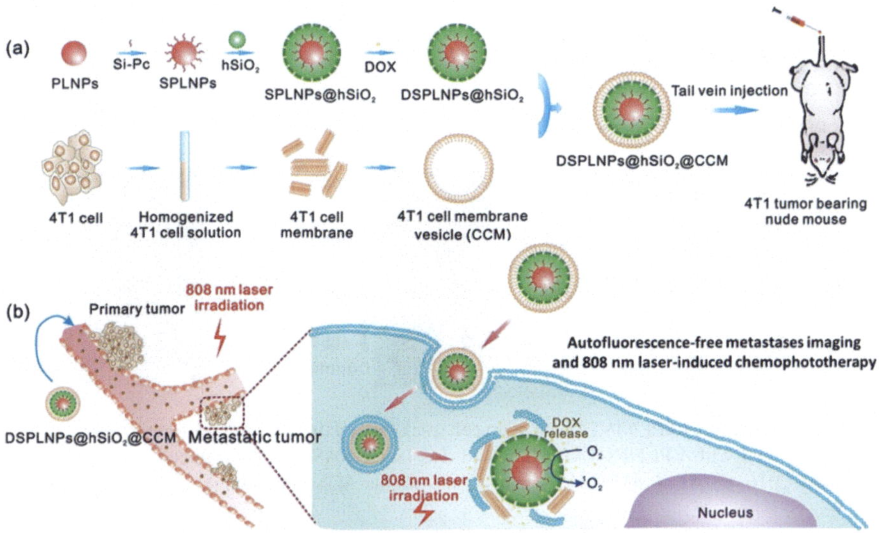

Fig. 10 DSPLNPs@hSiO$_2$@CCM for in vivo autofluorescence-free metastases imaging and chemophototherapy: **a** synthesis technique and **b** in vivo auto fluorescence-free imaging and chemophototherapy of metastases. (Reproduced from Analytical Chemistry 2018, 90, 6, 4188–4195 with permission from the ACS Publishing)

11 PLNPs for Stem Cell Tracking and Therapy

Stem cells not only serve as drug and gene carriers for tumor-targeted treatment due to their natural high tumor affinity but also play a vital role in regeneration medicine because of good self-renewability and low immunogenicity [86]. As stem-cells based therapy usually takes a long time, noninvasive and sensitive nanoprobes are important for tracking their long-term fate and migration [86]. Various imaging agents have been employed to detect in vivo stem cells through direct labeling to overcome the drawbacks of genetic transfection of reporter genes such as tedious manufacture and clinical safety concerns [86]. Nevertheless, traditional down conversion fluorescent nanoprobes suffer from the interferences of auto fluorescence and scattering light of tissues, while upconversion nanoparticles with low quantum yield exhibit the risk of tissue damage caused by constant high-energy laser excitation. Therefore, novel sensitive and safe labeling techniques are highly desired for stem cell tracking and therapy. High SNR PLNPs show great potential for reliable and sensitive stem cell tracking. Recently, we developed PLNPs equipped with a TAT (a transcriptional transactivator) penetrating peptide for long-term tracking of adipose-derived stem cells (ASCs) without the need for constant external stimuli as illustrated in Fig. 11 [87]. The TAT peptide bound to the surface of PLNP and improved the PLNP's biocompatibility and cellular uptake. In vitro, labeling, and imaging studies demonstrate efficient labeling of ASCs by TAT-PLNP for at least 3 days with no apparent leakage from labeled cells. From investigation, it has been observed that

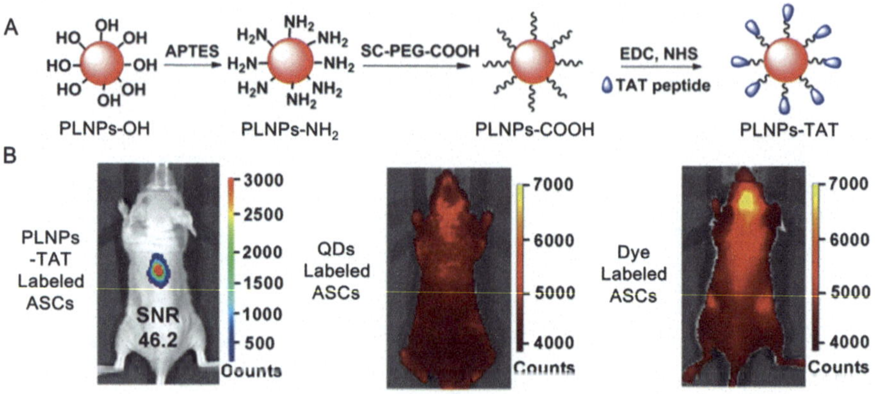

Fig. 11 **a** Fabrication of PLNP-TAT for stem cell tracking and therapy, and **b** Comparison of in vivo images obtained with LPLNPs-TAT, QDs, and dye-labeled ASCs. (Reproduced from Analytical Chemistry 2016, 88, 4114–4121 with permission from the ACS Publishing)

in vivo stem cell tracking shows successful detection of as few as 10 PLNP-tagged ASCs with SNR = 2.9 due to low background and red light regeneration ability. In contrast, both the quantum dots (QDs) and the dye-labeled ASCs had a lower SNR due to strong autofluorescence. To further demonstrate the long-term tracking ability of PLNP in stem cell-based therapy in vivo, we used TAT PLNP to track ASCs during wound healing and tumor-homing processes. PLNP-labeled ASCs effectively promoted wound healing as free stem cells without affecting stem cell proliferation or differentiation. A clear PL signal from PLNP could be observed until scar exfoliation after 7 days and residual PLNP-labeled stem cells could be detected even after 21 days. In tumor models, PLNP-tagged stem cells not only showed a negligible negative effect on the tumor-homing ability of stem cells, but also sensitively tracked stem cell migration during the tumor-homing process.

12 Analysis of Biosafety Applications of PLNPs

The biodistribution and biosafety of PLNPs synthesis procedures and characterization approaches have improved with the increased applicability of in vivo biomedical imaging evaluation methods. Numerous investigators have assessed the impact of crystal size, microstructure, surface variation, and modifications in the biodistribution of polymeric live nanoparticles (PLNPs) following in vivo interventions such as intravenous injections in small animals such as mice [88–90]. A scan of the literature has shown that surface changes have an impact on tissue dispersion. One well-known technique for changing the surface layer is PEG grafting. It has been discovered that by decreasing the amount of protein adsorption on the surface of PLNPs, an increase in PEG density will enhance the effect on tumor cells [91]. Due to these findings, more optical scientists are interested in delving deeper into the various PLNP dimensions in the field of bio-imaging. Evaluating the proportion of hazardous components in bio-applications is also crucial [92]. After incubating animals with various PLNP variants, there are several ways to assess the level of toxicity: cell proliferation, cell differentiation, physical and behavioral changes, histology or hematology changes, blood circulation biochemistry analysis, or a straightforward cell viability test following intravenous PLNP injection. In a study by Wu et al., materials like as adipogenesis, chondrogens, and ostogens are added to mesenchymal human cells to enhance the differentiation of the stem cells. It was noted that in all forms of differentiation, there were no differences between the supplemented and unlabeled stem cells. Therefore, we deduce that in the differentiation studies of tagged stem cells, PLNPs are independent [93, 94]. Fascinatingly, in a different study on in vivo toxicity, Abdurahman et al. found no differences in the behavioral features and appearance of the experimental and control groups [70]. Zhao et al. also conducted studies to determine the impact of PLNPs on toxicity levels by histologically examining the liver, heart, spleen, kidney, and other organs in mice before and after injection. It was found that PLNPs had no negative effects on lesions or tissues. In actuality, no inflammation was discovered in the mice's bodies [92]. It is challenging to draw any inferences from these investigations, though. Since these techniques' dependability needs to be increased. More thorough research is required to rule out the potential that the introduction of PLNPs will raise hazardous levels. More precise and accurate in vivo examinations are also needed. For a better understanding of the side effects of PLNPs in in-vivo biotherapies, changes in toxic levels must be connected with changes in crystallites, surfaces, and microstructure.

13 Conclusion and Outlook

This chapter provides an analysis in detail of PLNPs, covering everything from their basic properties to all of their potential scientific uses. Presenting new developments in synthesis methods, materials, characterizations, afterglow mechanisms, and applications is the primary contribution of this chapter. The PLNPs demonstrate a broad range of potential uses in photodynamic therapy and luminescence imaging to help understand biological processes and diagnose or treat illnesses. These materials shine in vivo because of a critical characteristic that allows extended emission for hours after halted excitation, which improves detection because of the background noise ratio. While background interference is eliminated by persistent luminescence, the requirement for the produced light to bypass the biological processes still poses a barrier. Recent developments in the surface modification and manipulation of energy transfer design principles used to create PLNPs for biomedical applications. The manufacturing, surface modification, and biological applications of polymeric living nanoparticles (PLNPs) in bioimaging and cancer therapy have advanced significantly. However, several topics still require more research. Studies conducted by several research groups enhance the sensitivity and selectivity of nanoprobes in biosensing, bioimaging, and theranostics by utilizing the intriguing properties of PLNPs, such as their enigmatic extended afterglow and red light renewability. High-performance multimodal imaging nanoprobes, intelligent theranostic nano-agents, tumor-targeted imaging, stem cell-based therapy, accurate and dependable stem cell tracking, and highly sensitive and selective biosensors are just a few of the crucial problems that have been addressed by a variety of PLNP engineering strategies. It is reasonable to acknowledge that the development of PLNPs in biological applications is still in its infancy, despite the significant advancements made by research teams. First, there is a severe shortage of high-quality PLNPs with small, uniform sizes, bright initial PL, and long PL duration. To manage the shape, morphology, homogeneity, and photoluminescence (PL) performance of PLNPs and other irradiation with no constraint on tissue penetration depth, new universal synthetic approaches are desperately needed. Despite significant progress in PLNP synthesis, there are still certain issues with PLNP morphological control. More sophisticated synthetic techniques are required when PLNPs find use in biological applications to accurately control PLNP shape, particle size, surface characteristics, intensity, and time. The excitation light source has a significant impact on how PLNPs are used in biomedicine. X-rays, radionuclides, and NIR light will be the light sources for PLNP excitation in future research since the tumor buildup of PLNPs cannot be supported by the persistent luminescence time for pre-excitation by UV lamps. Better tissue transmittance can be attained in the interim by emitting PLNPs in the NIR I or NIR-II range. Since NIR-II emitting PLNPs have not been frequently reported for bioimaging, the development of PLNPs with excellent NIR-II persistence luminescence will be one of the most demanded areas of research in the near future. PLNP nanoprobes may be able to do radiation and CT imaging in addition to deep tissue tumor imaging using PLNPs. Further research into PLNPs' luminescence process is necessary since it will have a significant impact on the production of PLNPs with more intriguing features in the long run. Although they are unable to offer all the information required for both disease diagnosis and treatment, PLNPs can offer persistent luminescence with high sensitivity for disease diagnosis. Consequently, PLNPs with multimodal imaging capabilities are being studied for the diagnosis of cancer. PLNPs can be used as smart drug delivery devices for gene therapy, PTT, PDT, chemotherapy, and combination therapy. Developing a PLNP nanoplatform with target-specific delivery, sufficient cellular internalization, effective loading, and appropriate protection of therapeutic payloads during circulation remains the primary challenge to date. PLNPs are a novel class of fluorescent nanomaterial that finds application in tumor detection and therapy. In contrast to other fluorescent nanoprobes like quantum dots and upconversion nanoparticles, they have comparatively fewer biomedical application possibilities. More PLNP uses in the biomedical industry, including in vitro diagnostics, cell imaging, and antibacterial disinfection, will be developed in the future. As a novel bioimaging material, researchers have also focused on the biosafety investigations of PLNPs. While studies on the biotoxicity of PLNPs have been conducted at several levels, including cellular and animal, and have achieved many research results, the biosafety studies of PLNPs are still in their infancy. More focus will be placed on in-depth biotoxicity research in the future, including molecular biology-level changes in proteins and genes, the toxicity of chronic exposure, migratory patterns, and animal-level PLNP transformation.

Research on the fundamentals and applications of numerous elements of the luminescence phenomenon is rapidly progressing, using both traditional and cutting-edge experimental techniques. Numerous long-lasting phosphors with colors ranging from blue to green to yellow to orange to red have been created thus far. There may be uses for long-lasting phosphors in some niche markets. This decade has seen a notable rise in the commercial interest in long-lasting phosphors that emit electromagnetic radiation in the visible spectrum. This is mostly because these materials are used in illumination engineering, dark vision display systems, and a variety of other home and industrial applications. There are now created phosphors that emit blue, green, and yellow-green light and have a lengthy decay time. Researchers are still having difficulty producing consistent red emitting long persistent phosphor. To obtain lengthy afterglow phosphors, it will be worthwhile to investigate novel host lattices such as titanates, silicates, oxysulphides, etc. for example. Other metal ions, such as transition ions, can be employed as dopants in addition to rare earth ions. It may be possible to generate phosphors with numerous emission colors and lower production costs by reaching the LP phosphors with transition ions because of d $\rightarrow$ d electronic transition. These long persistent phosphors could lead to new developments in the field of phosphor chemistry. The significance of cation and anion defects in enhancing the delayed emission process needs to be thoroughly investigated by analyzing thermoluminescence glow curves in conjunction with other rare-earth ions. Commercial manufacturing of long-persistent phosphors utilizing domestic materials and technology is mandated. Finally, the primary applications of this particular PLNP-based nanoprobe are in tumor diagnosis and treatment. Considerable effort will be made to investigate the use of PLNPs in the diagnosis and treatment of various critical conditions, including gastrointestinal, orthopedic, cerebrovascular, and cardiovascular disorders.

References

1. V. Shanker, D. Haranath, G. Swati, Defect and diffusion. Forum **361**, 69–94 (2015)
2. W.M. Yen, S. Shionoya, H. Yamamoto, *Phosphor Handbook*, 2nd edn. (CRC Press/Taylor and Francis, Boca Raton, FL, USA, 2007)
3. S.K. Sun, H.F. Wang, X.P. Yan, Engineering persistent luminescence nanoparticles for biological applications: from biosensing/bioimaging to theranostics. acc. Chem. Res. **51**, 1131–1143 (2018)
4. J. Xu, S. Tanabe, Persistent luminescence instead of phosphorescence: history, mechanism, and perspective. J. Lumin. **205**, 581–620 (2019)
5. Y. Li, M. Gecevicius, J.R. Qiu, Long persistent phosphors-from fundamentals to applications. Chem. Soc. Rev. **45**, 2090–2136 (2016)
6. B.Y. Wu, H.F. Wang, J.T. Chen, X.P. Yan, Fluorescence resonance energy transfer inhibition assay for α-fetoprotein excreted during cancer cell growth using functionalized persistent luminescence nanoparticles. J. Am. Chem. Soc. **133**, 686–688 (2011)

7. A. Abdukayum, J.T. Chen, Q. Zhao, X.P. Yan, Functional near infrared emitting Cr^{3+}/Pr^{3+} co-doped zinc gallogermanate persistent luminescent nanoparticles with superlong afterglow for in vivo targeted bioimaging. J. Am. Chem. Soc. **135**, 14125–14133 (2013)
8. Z. Pan, Y.Y. Lu, F. Liu, Sunlight-activated long-persistent luminescence in the near-infrared from Cr^{3+}-doped zinc gallogermanates. Nat. Mater. **11**, 58–63 (2012)
9. S.K. Singh, Red and near infrared persistent luminescence nanoprobes for bioimaging and targeting applications. RSC Adv. **4**, 58674–58698 (2014)
10. L. Zhang, J. Lei, J. Liu, F. Ma, H. Ju, Persistent luminescence nanoprobe for biosensing and lifetime imaging of cell apoptosis via time-resolved fluorescence resonance energy transfer. Biomaterials **67**, 323–334 (2015)
11. Z. Xue, X. Li, Y. Li, M. Jiang, H. Liu, S. Zeng, J. Hao, X-ray activated near infrared persistent luminescent probe for deep-tissue and renewable in vivo bioimaging. ACS Appl. Mater. Interfaces **9**, 22132–22142 (2017)
12. F. Liu, Y.J. Liang, Y.F. Chen, Z.W. Pan, Divalent nickel activated gallate- based persistent phosphors in the short-wave infrared. Adv. Opt. Mater. **4**, 562–566 (2016)
13. Y. Li, Y.Y. Li, K. Sharafudeen, G.P. Dong, S.F. Zhou, Z.J. Ma, M.Y. Peng, J.R. Qiu, A strategy for developing near infrared long persistent phosphors: taking $MAlO_3$: Mn^{4+}, Ge4+ (M=La, Gd) as an example. J. Mater. Chem. C **2**, 2019–2027 (2014)
14. H.J. Guo, Y.H. Wang, G. Li, J. Liu, P. Feng, D.W. Liu, Cyan emissive super-persistent luminescence and thermoluminescence in $BaZrSi_3O_9$: Eu^{2+}, Pr^{3+} phosphors. J. Mater. Chem. C **5**, 2844–2851 (2017)
15. W.Z. Sun, R. Pang, H.M. Li, D. Li, L.H. Jiang, S. Zhang, J.P. Fu, C.Y. Li, Investigation of a novel color tunable long afterglow phosphor $KGaGeO_4$: Bi^{3+}: luminescence properties and mechanism. J. Mater. Chem. C **5**, 1346–1355 (2017)
16. M. Palner, K.Y. Pu, S. Shao, J.H. Rao, Semiconductor polymer nanoparticles with persistent near infrared luminescence for in vivo optical imaging. Angew. Chem. Int. Ed. **54**, 11477–11480 (2015)
17. X.G. Yang, D.P. Yan, Long afterglow metal-organic frameworks: reversible guest-induced phosphorescence tenability. Chem. Sci. **7**, 4519–4526 (2016)
18. K. Van den Eeckhout, P.F. Smet, D. Poelman, Persistent luminescence in Eu^{2+}-doped compounds: a review. Materials **3**, 2536–2566 (2010)
19. Q.L.M. De Chermont, C. Chanéac, J. Seguin, F. Pellé, S. Maîtrejean, J.-P. Jolivet, D. Gourier, M. Bessodes, D. Scherman, Nanoprobes with nearinfrared persistent luminescence for in vivo imaging. Proc. Natl. Acad. Sci. **104**, 9266–9271 (2007)
20. T. Maldiney, A. Lecointre, B. Viana, A. Bessiere, M. Bessodes, D. Gourier, C. Richard, D. Scherman, Controlling electron trap depth to enhance optical properties of persistent luminescence nanoparticles for in vivo imaging. J. Am. Chem. Soc. **133**, 11810–11815 (2011)
21. L. Zhang, J.P. Lei, J.T. Liu, F.J. Ma, H.X. Ju, Persistent luminescence nanoprobe for biosensing and lifetime imaging of cell apoptosis via time-resolved fluorescence resonance energy transfer. Biomaterials **67**, 323–334 (2015)
22. N. Li, Y.H. Li, Y.Y. Han, W. Pan, T.T. Zhang, B. Tang, A highly selective and instantaneous nanoprobe for detection and imaging of ascorbic acid in living cells and in vivo. Anal. Chem. **86**, 3924–3930 (2014)
23. Y.J. Chuang, Z.P. Zhen, F. Zhang, F. Liu, J.P. Mishra, W. Tang, H.M. Chen, X.L. Huang, L.C. Wang, X.Y. Chen et al., Photostimulable near-infrared persistent luminescent nanoprobes for ultrasensitive and longitudinal deep-tissue bio-imaging. Theranostics **4**, 1112–1122 (2014)
24. X. Sun, J.P. Shi, S.H. Zheng, J.L. Li, S.G. Wang, H.W. Zhang, Visualization of inflammation in a mouse model based on near-infrared persistent luminescence nanoparticles. J. Lumin. **204**, 520–527 (2018)
25. Z. Li, H. Zhang, H. Fu, Facile synthesis and morphology control of Zn_2SiO_4: Mn nanophosphors using mesoporous silica nanoparticles as templates. J. Lumin. **135**, 79–83 (2013)
26. L. Zhan-Jun, Z. Hong-Wu, S. Meng, S. Jiang-Shan, F. Hai-Xia, A facile and effective method to prepare long-persistent phosphorescent nanospheres and its potential application for in vivo imaging. J. Mater. Chem. **22**, 24713–24720 (2012)

27. Z.J. Li, J.P. Shi, H.W. Zhang, M. Sun, Highly controllable synthesis of nearinfrared persistent luminescence SiO2/$CaMgSi_2O_6$ composite nanospheres for imaging in vivo. Opt Expr. **22**, 10509–10518 (2014)
28. Z.J. Li, Y.J. Zhang, H.W. Zhang, H.X. Fu, Long-lasting phosphorescence functionalization of mesoporous silica nanospheres by $CaTiO_3:Pr^{3+}$ for drug delivery. Microporous Mesoporous Mater. **176**, 48–54 (2013)
29. J.P. Shi, H.X. Fu, X. Sun, J.S. Shen, H.W. Zhang, Magnetic, long persistent luminescent and mesoporous nanoparticles as trackable transport drug carriers. J. Mater. Chem. B. **3**, 635–641 (2015)
30. T. Maldiney, A. Bessière, J. Seguin, E. Teston, S.K. Sharma, B. Viana, A.J. Bos, P. Dorenbos, M. Bessodes, D. Gourier, The in vivo activation of persistent nanophosphors for optical imaging of vascularization, tumours and grafted cells. Nat. Mater. **13**, 418–426 (2014)
31. Y.J. Li, X.P. Yan, Synthesis of functionalized triple-doped zinc gallogermanate nanoparticles with superlong near-infrared persistent luminescence for long-term orally administrated bioimaging. Nanoscale **8**, 14965–14970 (2016)
32. J. Liu, L. Kotrchová, T. Lécuyer, Y. Corvis, J. Seguin, N. Mignet, T. Etrych, D. Scherman, E. Randárová, C. Richard, Coating persistent luminescence nanoparticles with hydrophilic polymers for in vivo imaging. Front. Chem. **8**(584114), 1–10 (2020)
33. J. Shi, X. Sun, S. Zheng, J. Li, X. Fu, H. Zhang, A new near-infrared persistent luminescence nanoparticle as a multifunctional nanoplatform for multimodal imaging and cancer therapy. Biomaterials **152**, 15–23 (2018)
34. J. Wang, Q. Ma, X.-X. Hu, H. Liu, W. Zheng, X. Chen, Q. Yuan, W. Tan, Autofluorescence-free targeted tumor imaging based on luminous nanoparticles with composition-dependent size and persistent luminescence. ACS Nano **11**, 8010–8017 (2017)
35. R. Zou, J.J. Huang, J.P. Shi, L. Huang, X.J. Zhang, K.L. Wong, H.W. Zhang, D.Y. Jin, J. Wang, Q. Su, Silica shell-assisted synthetic route for monodisperse persistent nanophosphors with enhanced in vivo recharged near-infrared persistent luminescence. Nano Res. **10**, 2070–2082 (2017)
36. L. Wang, J. Zhang, B. Qu, Q. Wu, R. Zhou et al., Mechanistic insights into tunable luminescence and persistent luminescence of the full-color-emitting bcno phosphors. Carbon **122**, 176–184 (2017)
37. X. Yang, D. Yan, Long-afterglow metal–organic frameworks: reversible guest-induced phosphorescence tunability. Chem. Sci. **7**(7), 4519–4526 (2016)
38. P. Xue, P. Wang, P. Chen, B. Yao, P. Gong, J. Sun, Z. Zhang, R. Lu, Bright persistent luminescence from pure organic molecules through a moderate intermolecular heavy atom effect. Chem. Sci. **8**(9), 6060–6065 (2015)
39. M. Palner, K. Pu, S. Shao, J. Rao, Semiconducting polymer nanoparticles with persistent near-infrared luminescence for in vivo optical imaging. Angew. Chem. Int. Ed. **127**(39), 11639–11642 (2015)
40. J. Li, J. Rao, K. Pu, Recent progress on semiconducting polymer nanoparticles for molecular imaging and cancer phototherapy. Biomaterials **155**, 217–235 (2018)
41. J.B. Marion, 21-radiation effects and uses, in *Physics in the Modern World*, 2nd edn., ed. by J.B. Marion (Academic Press, New York, 1981), pp.603–631
42. A. Abdukayum, J.T. Chen, Q. Zhao, X.P. Yan, Functional near infrared-emitting Cr^{3+}/Pr^{3+} co-doped zinc gallogermanate persistent luminescent nanoparticles with superlong afterglow for in vivo targeted bioimaging. J. Am. Chem. Soc. **135**, 14125–14133 (2013)
43. T. Maldiney, A. Bessiere, J. Seguin, E. Teston, S.K. Sharma, B. Viana, A.J. Bos, P. Dorenbos, M. Bessodes, D. Gourier et al., The in vivo activation of persistent nanophosphors for optical imaging of vascularization, tumours and grafted cells. Nat. Mater. **13**, 418–426 (2014)
44. Y.J. Li, X.P. Yan, Synthesis of functionalized triple-doped zinc gallogermanate nanoparticles with superlong near infrared persistent luminescence for long-term orally administrated bioimaging. Nanoscale **8**, 14965–14970 (2016)
45. R. Weissleder, A clearer vision for in vivo imaging. Nat. Biotechnol. **19**, 316–317 (2001)

46. J.L. Li, J.P. Shi, C.C. Wang, P.H. Li, Z.F. Yu, H.W. Zhang, Five-nanometer $ZnSn_2O_4$: Cr, Eu ultra-small nanoparticles as new near infrared-emitting persistent luminescent nanoprobes for cellular and deep tissue imaging at 800 nm. Nanoscale **9**, 8631–8638 (2017)
47. Z.L. Xue, X.L. Li, Y.B. Li, M.Y. Jiang, G.Z. Ren, H.R. Liu, S.J. Zeng, J.H. Hao, A 980 nm laser-activated upconverted persistent probe for NIR to NIR rechargeable in vivo bioimaging. Nanoscale **9**, 7276–7283 (2017)
48. B. Zheng, Y. Bai, H. Chen, H. Pan, W. Ji, X. Gong, X. Wu, H. Wang, J. Chang, Near-infrared light-excited upconverting persistent nanophosphors in vivo for imaging-guided cell therapy. ACS Appl. Mater. Interfaces **10**, 19514–19522 (2018)
49. X. Sun, J. Shi, X. Fu, Y. Yang, H. Zhang, Long-term in vivo biodistribution and toxicity study of functionalized near-infrared persistent luminescence nanoparticles. Sci. Rep. **8**, 10595–10606 (2018)
50. X. Chen, J. Song, X. Chen, H. Yang, X-ray-activated nanosystems for theranostic applications. Chem. Soc. Rev. **48**, 3073–3101 (2019)
51. L. Song, X.H. Lin, X.R. Song, S. Chen, X.F. Chen, J. Li, H.H. Yang, Repeatable deep-tissue activation of persistent luminescent nanoparticles by soft X-ray for high sensitivity long-term in vivo bioimaging. Nanoscale **9**, 2718–2722 (2017)
52. Z. Xue, X. Li, Y. Li, M. Jiang, H. Liu, S. Zeng, J. Hao, X-ray-activated nearinfrared persistent luminescent probe for deep-tissue and renewable in vivo bioimaging. ACS Appl. Mater. Interfaces **9**, 22132–22142 (2017)
53. X.H. Lin, L. Song, S. Chen, X.F. Chen, J.J. Wei, J. Li, G. Huang, H.H. Yang, Kiwifruit-like persistent luminescent nanoparticles with high-performance and in situ activable near-infrared persistent luminescence for longterm in vivo bioimaging. ACS Appl. Mater. Interfaces **9**, 41181–41187 (2017)
54. S. Zheng, J. Shi, X. Fu, C. Wang, X. Sun, C. Chen, Y. Zhuang, X. Zou, Y. Li, H. Zhang, X-ray recharged long afterglow luminescent nanoparticles $MgGeO_3$:Mn^{2+}, Yb^{3+}, Li^{+} in the first and second biological windows for long term bioimaging. Nanoscale **12**, 14037–14046 (2020)
55. Z.Z. Chen, L.C. Wang, D. Manoharan, C.L. Lee, L.C. Wu, W.T. Huang, E.Y. Huang, C.H. Su, H.S. Sheu, C.S. Yeh, Low dose of X-Ray-excited long-lasting luminescent concave nanocubes in highly passive targeting deep-seated hepatic tumors. Adv. Mater. **31**, 1905087 (2019)
56. X. Sun, X. Huang, J. Guo, W. Zhu, Y. Ding, G. Niu, A. Wang, D.O. Kiesewetter, Z.L. Wang, S. Sun, X. Chen, Self-illuminating 64Cu-doped CdSe/ ZnS nanocrystals for in vivo tumor imaging. J. Am. Chem. Soc. **136**, 1706–1709 (2014)
57. M. Wu, J. Shu, Multimodal molecular imaging: current status and future directions. Contrast Media. Mol. Imaging **2018**, 1382183 (2018)
58. H. Liu, F. Ren, H. Zhang, Y. Han, H. Qin, J. Zeng, Y. Wang, Q. Sun, Z. Li, M. Gao, Oral administration of highly bright Cr3+ doped ZnGa2O4 nanocrystals for in vivo targeted imaging of orthotopic breast cancer. J. Mater. Chem. B. **6**, 1508–1518 (2018)
59. A. Abdukayum, C.X. Yang, Q. Zhao, J.T. Chen, L.X. Dong, X.P. Yan, Gadolinium complexes functionalized persistent luminescent nanoparticles as a multimodal probe for near-infrared luminescence and magnetic resonance imaging in vivo. Anal. Chem. **86**, 4096–4101 (2014)
60. Y. Wang, C.X. Yang, X.P. Yan, Hydrothermal and biomineralization synthesis of a dual-modal nanoprobe for targeted near-infrared persistent luminescence and magnetic resonance imaging. Nanoscale **9**, 9049–9055 (2017)
61. Y.C. Lu, C.X. Yang, X.P. Yan, Radiopaque tantalum oxide coated persistent luminescent nanoparticles as multimodal probes for in vivo near-infrared luminescence and computed tomography bioimaging. Nanoscale **7**, 17929–17937 (2015)
62. S. Wu, Y. Li, R. Zhang, K. Fan, W. Ding, L. Xu, L. Zhang, Persistent luminescence-polypyrrole nanocomposite for dual-modal imaging and photothermal therapy of mammary cancer. Talanta **221**, 121435 (2021)
63. J.M. Liu, Y.Y. Liu, D.D. Zhang, G.Z. Fang, S. Wang, Synthesis of $GdAlO_3$:Mn^{4+}, Ge^{4+}@Au core-shell nanoprobes with plasmon enhanced near-infrared persistent luminescence for in vivo trimodality bioimaging. ACS Appl. Mater. Interfaces **8**, 29939–29949 (2016)

64. T. Ai, W. Shang, H. Yan, C. Zeng, K. Wang, Y. Gao, T. Guan, C. Fang, J. Tian, Near infrared-emitting persistent luminescent nanoparticles for Hepatocellular Carcinoma imaging and luminescence guided surgery. Biomaterials **167**, 216–225 (2018)
65. Z.J. Li, Y.W. Zhang, X. Wu, L. Huang, D.S. Li, W. Fan, G. Han, Direct aqueousphase synthesis of sub-10 nm "luminous pearls" with enhanced in vivo renewable near-infrared persistent luminescence. J. Am. Chem. Soc. **137**, 5304–5307 (2015)
66. T. Asefa, Z. Tao, Biocompatibility of mesoporous silica nanoparticles. Chem. Res. Toxicol. **25**, 2265–2284 (2012)
67. T. Maldiney, B. Ballet, M. Bessodes et al., Mesoporous persistent nanophosphors for in vivo optical bioimaging and drug delivery. Nanoscale **6**, 13970–13976 (2014)
68. W.B. Dai, Y.F. Lei, S. Ye, E.H. Song, Z. Chena, Q.Y. Zhang, Mesoporous nanoparticles Gd_2O_3@$mSiO_2$/$ZnGa_2O_4$:Cr^{3+}, Bi^{3+} as multifunctional probes for bioimaging. J. Mater. Chem. B **4**, 18421852 (2016)
69. L. Cheng, C. Wang, L. Feng, K. Yang, Z. Liu, Functional nanomaterials for phototherapies of cancer. Chem. Rev. **114**, 10869–10939 (2014)
70. R. Abdurahman, C.X. Yang, X.P. Yan, Conjugation of a photosensitizer to near infrared light renewable persistent luminescence nanoparticles for photodynamic therapy. Chem. Commun. **52**, 13303–13306 (2016)
71. D.V. Straten, V. Mashayekhi, H.S. Bruijn, S. Oliveira, D.J. Robinson, Oncologic photodynamic therapy: basic principles, current clinical status and future directions. Cancers **9**(2) (2017)
72. J.P. Celli, B.Q. Spring, I. Rizvi et al., Imaging and photodynamic therapy: mechanisms, monitoring, and optimization. Chem. Rev. **2010**(110), 2795–2838 (2010)
73. J.R. Kanofsky, Measurement of singlet-oxygen in vivo: progress and pitfalls. Photochem. Photobiol. **87**, 14–17 (2011)
74. M.T. Jarvi, M.J. Niedre, M.S. Patterson, B.C. Wilson, The influence of oxygen depletion and photosensitizer triplet-state dynamics during photodynamic therapy on accurate singlet oxygen luminescence monitoring and analysis of treatment dose response. Photochem. Photobiol. **87**, 223–234 (2011)
75. G. Jori, C. Fabris, M. Soncin et al., Photodynamic therapy in the treatment of microbial infections: basic principles and perspective applications. Laser Surg. Med. **38**, 468–481 (2006)
76. T. Dai, Y.Y. Huang, M.R. Hamblin, Photodynamic therapy for localized infections-state of the art. Photodiagnosis Photodyn. Ther. **6**, 170–188 (2009)
77. N.T. Blum, Y. Zhang, J. Qu, J. Lin, P. Huang, Recent advances in self-exciting photodynamic therapy. Front. Bioeng. Biotechnol. **8**(1136), 2020 (2020)
78. L. Jiang, L. Liu, F. Lv, S. Wang, X. Ren, Integration of self- luminesce and oxygen self-supply: a potential photodynamic therapy strategy for deep tumor treatment. ChemPlusChem **85**, 510–518 (2020)
79. A. Kamkaew, L. Cheng, S. Goel et al., Cerenkov radiation induced photodynamic therapy using chlorin e6-loaded hollow mesoporous silica nanoparticles. ACS Appl. Mater. Interfaces **8**, 26630–26637 (2016)
80. Y. Zhang, H.F. Chan, K.W. Leong, Advanced materials and processing for drug delivery: the past and the future. Adv. Drug Delivery Rev. **65**, 104–120 (2013)
81. L.J. Chen, C.-X. Yang, X.P. Yan, Liposome-coated persistent luminescence nanoparticles as luminescence trackable drug carrier for chemotherapy. Anal. Chem. **89**, 6936–6939 (2017)
82. M. Cihova, V. Altanerova, C. Altaner, Stem cell based cancer gene therapy. Mol. Pharmaceutics **8**, 1480–1487 (2011)
83. S.-Q. Wu, C.-X. Yang, X.-P. Yan, Dual-functional persistent luminescence nanocomposite engineered mesenchymal stem cells for homing and gene therapy of glioblastoma. Adv. Funct. Mater. **27**, 1604992 (2017)
84. C. Liang, L. Xu, G. Song, Z. Liu, Emerging nanomedicine approaches fighting tumor metastasis: animal models, metastasistargeted drug delivery, phototherapy, and immunotherapy. Chem. Soc. Rev. **45**, 6250–6269 (2016)
85. Y.-J. Li, C.-X. Yang, X.-P. Yan, Biomimetic persistent luminescent nanoplatform for autofluorescence-free metastasis tracking and chemophotodynamic therapy. Anal. Chem. **90**, 4188–4195 (2018)

86. M. Janowski, J.W.M. Bulte, P. Walczak, Personalized nanomedicine advancements for stem cell tracking. Adv. Drug Delivery Rev. **64**, 1488–1507 (2012)
87. S.Q. Wu, C.W. Chi, C.X. Yang, X.P. Yan, Penetrating peptide-bioconjugated persistent nanophosphors for long-term tracking of adipose-derived stem cells with superior signal-to-noise ratio. Anal. Chem. **88**, 4114–4121 (2016)
88. T. Maldiney, B. Ballet, M. Bessodes, D. Scherman, C. Richard, Mesoporous persistent nanophosphors for in vivo optical bioimaging and drug-delivery. Nanoscale **6**(22), 13970–13976 (2014)
89. T. Maldiney, C. Richard, J. Seguin, N. Wattier, M. Bessodes, D. Scherman, Effect of core diameter, surface coating, and peg chain length on the biodistribution of persistent luminescence nanoparticles in mice. ACS Nano **5**(2), 854–862 (2011)
90. T. Maldiney, G. Byk, N. Wattier, J. Seguin, R. Khandadash, M. Bessodes, C. Richard, D. Scherman, Synthesis and functionalization of persistent luminescence nanoparticles with small molecules and evaluation of their targeting ability. Int. J. Pharmaceut. **423**(1), 102–107 (2012)
91. T. Maldiney, M. Rémond, M. Bessodes, D. Scherman, C. Richard, Controlling aminosilane layer thickness to extend the plasma half-life of stealth persistent luminescence nanoparticles in vivo. J. Mater. Chem. B **3**(19), 4009–4016 (2015)
92. Y. Jiang, Y. Li, C. Richard, D. Scherman, Y. Liu, Hemocompatibility investigation and improvement of near-infrared persistent luminescent nanoparticle $ZnGa_2O_4:Cr^{3+}$ by surface pegylation. J. Mater. Chem. B **7**(24), 3796–3803 (2019)
93. G. Ramírez-García, S. Gutiérrez-Granados, M.A. Gallegos-Corona, L. Palma-Tirado, F. D'Orlyé, et al., Long-term toxicological effects of persistent luminescence nanoparticles after intravenous injection in mice. Int. J. Pharmaceut. **532**(2), 686–695 (2017)
94. S.Q. Wu, C.W. Chi, C.X. Yang, X.P. Yan, Penetrating peptide-bioconjugated persistent nanophosphors for longterm tracking of adipose-derived stem cells with superior signal-to-noise ratio. Anal. Chem. **88**(7), 4114–4121 (2016)

Engineering of Persistent Luminescence Nanoparticles for Biosensing

Khouloud Hamraoui, Mounir Ferhi, and Karima Horchani-Naifer

Abstract Because of their unique optical features, persistent luminescence nanoparticles (PLNPs) have emerged as a viable tool for biosensing applications. After being activated by an external light source, these nanoparticles may emit light for a sustained period of time, making them ideal for long-term imaging and detection in biological systems. The PLNLs have attracted considerable attention for biosensing due to their unique advantages such as high signal-to-noise ratio, low background interference, and multiplexing capability. To engineer PLNPs for biosensing, several key factors need to be considered, including the choice of host material, dopant ions, and surface modifications. The design and development of nanoparticles with persistent luminescence for use in biosensing applications are covered in this chapter. This includes a discussion of different types of persistent luminescence nanoparticles, their synthesis and surface modification, and the key parameters that affect their performance as biosensors. The chapter also highlights the various applications of persistent luminescence nanoparticles in biosensing, including in vitro and in vivo sensing, and discusses the advantages and limitations of this technology.

Keywords Persistent photoluminescence · Biosensing · Long-lasting emission · Background free · Biological analytes

1 Introduction

Persistent luminescence nanoparticles (PLNPs) have arisen as an intriguing and potential biosensing technique [1]. The energy storage mechanism in PLNPs involves the formation of trapped charge carriers, typically holes and/or electrons, within the nanoparticle's crystal lattice or defect sites. These trapped carriers are created when the PLNPs are excited by various energy sources. Because of the barrier inhibiting recombination or relaxation to the ground state, trapped carriers in PLNPs can have

K. Hamraoui · M. Ferhi · K. Horchani-Naifer (✉)
Laboratory of Physical Chemistry of Mineral Materials and Their Applications, National Research Center in Materials Science, Technopole Borj Cedria, Soliman, Tunisia
e-mail: karima.horchani@inrst.rnrt.tn

V. Kumar et al. (eds.), *Persistent Luminescence*, Progress in Optical Science and Photonics 30, https://doi.org/10.1007/978-981-97-4943-0_9

long lifetimes, which leads to a persistence effect, where the PLNPs can emit light long after the excitation source is ceasing [2]. Depending on the precise nanoparticle composition and structure, the duration of the produced light might range from minutes to hours or even days. The capacity to develop persistent luminescence materials with customizable persistent luminescence and decay characteristics is critical for addressing detection demands in a variety of applications [3]. To meet these diverse demands, it is critical to establish methodologies that enable the regulated preparation of PLNPs. This entails careful designing and engineering of the material composition, crystal structure, and activator ions. Researchers can modify the luminescence and decay properties of PLNPs by changing these parameters. It has been demonstrated that by altering the doping level of Tb^{3+}, the color of persistent luminescence may be changed from blue to green [4]. The involvement of the 5d level is essential in the energy transfer mechanism and the accumulation of electrons in the traps. This has been established through thermoluminescence and fading time measurements, which validate the significance of the depth of the traps related to Tb^{3+} doping level in the persistence luminescence process. Trivalent terbium Tb^{3+} doped powders with different concentration were obtained using the sol–gel technique. These powders exhibited the longest and most intense persistent luminescence [4]. Using crystal field control, Nie et al. [5] successfully tuned afterglow in the NIR biological window. The local crystal field surrounding the activator ions can be adjusted by replacing Zn and Sn for Ga in the crystal structure, resulting in changes in the emission spectra. This allows for precise control and tuning of the emission properties of the phosphors. The discovery of multi-band afterglow phosphors with great resolution and little light disturbance opens up new possibilities for enhanced optical imaging. Guchowski et al. [6] investigated the effect of synthesis method on the shape and spectroscopic properties of powders of $Gd_{3-x}Ce_xGa_3Al_2O_{12}$. This statement implies that the powder created utilizing the combustion process with a doping concentration of 0.5 mol% Ce^{3+} ions produced the most intense luminescence. Similarly, the powder with a higher doping concentration of Ce^{3+} ions (1 mol %) demonstrated the most intense luminescence during the sol–gel synthesis. When it came to persistent luminescence, however, the powders doped with lower concentrations of Ce^{3+} ions exhibited efficient long afterglow luminescence for both synthesis processes [6]. Overall, PLNPs enable long-lasting luminescence behavior with potential applications in a variety of domains by storing and releasing excitation energy in the form of delayed photonic emission [7].

With their unique properties, including long-lasting emission, background-free emission, and high sensitivity, PLNPs have garnered significant attention for their potential applications in the detection of biological and chemical analytes without ongoing additional excitation sources [1–8]. The ability of these nanoparticles to emit light long after excitation opens up new possibilities for ultrasensitive and real-time biosensing, revolutionizing the field of biomedical diagnostics and environmental monitoring.

This chapter explores the intriguing field of PLNPs and how they can be engineered for biosensing applications. It provides a comprehensive overview of the key principles, synthesis methods, and optical properties of PLNPs, highlighting their significance in biosensing applications. Moreover, it explores the challenges associated with the fabrication and characterization of PLNPs, emphasizing the need for precise control over various parameters to optimize their biosensing performance.

The versatility of PLNPs in biosensing is showcased through a range of applications, including immunoassays [7], enzymatic assays [9], and DNA sensing [7]. These nanoparticles have also found utility as bioimaging agents, bio-labels, and carriers for targeted drug delivery [10], further expanding their potential impact in the biomedical field. Integration with advanced detection techniques, such as fluorescence microscopy [11], flow cytometry [12], and multiplexed assays [3], enhances the sensitivity and multiplexing capabilities of PLNPs, enabling comprehensive and efficient biosensing platforms. While PLNPs hold immense promise, there are still challenges to overcome. The chapter addresses the current limitations and discusses the prospects in the field of PLNPs for biosensing. Enhanced synthesis techniques, improved biocompatibility, and increased sensitivity are identified as key areas for further development. By overcoming these challenges, PLNPs have the potential to transform the landscape of biosensing, enabling ultrasensitive, multiplexed, and real-time detection of biomarkers in various biomedical and environmental applications. In summary, this chapter provides a comprehensive exploration of the engineering of PLNPs for biosensing. It not only highlights the unique properties and synthesis methods but also showcases their versatility and potential applications in the field. By addressing current challenges and discussing prospects, this chapter sets the stage for further advancements in PLNP-based biosensing, offering exciting opportunities for researchers and practitioners to revolutionize the field of biomedical diagnostics and environmental monitoring.

2 In Vitro Biosensing

In vitro biosensing refers to a type of biosensing that is performed outside of a living organism or cell [13]. In vitro, biosensors are vital in the therapeutic context for early cancer detection. These biosensors are analytical devices that are used to detect and measure specific biomarkers present in biological samples such as blood, urine, or tumor tissues. These biomarkers can include proteins, DNA/RNA, enzymes, small molecules, and cells. They provide a rapid, sensitive, and accurate detection method, allowing for early identification and subsequent treatment of cancer [14]. Additionally, in vitro biosensors eliminate the need for complex sample preparation and reduce the risk of interference from other components present in a living organism or cell [15]. These benefits make in vitro biosensing an attractive option for various fields, including biomedical research and clinical diagnostics. Immunosensors are one of the most widely utilized in vitro biosensors for clinical diagnosis. These biosensors use antibodies or antigens to detect and analyze specific analytes or compounds in biological samples [16]. Molazemhosseini et al. [17] developed an in vitro electrochemical immunosensor based on a three-electrode configuration consisting of 10-nm-thin gold film working and counter electrodes and a thick-film printed Ag/AgCl reference electrode was fabricated on a polyethylene terephthalate (PET) substrate. The detection assay of HbA1c in undiluted human serum was performed using differential pulse voltammetry (DPV). This biosensor found a potential application for in vitro measurement of HbA1c for diabetics.

Another type of in vitro biosensors commonly used in the detection of molecular biomarkers is one that uses DNA molecules as sensing elements to detect and quantify analytes in a biological sample. A DNA probe and a signal transduction system are generally used. DNA-based in vitro biosensors have several advantages including high specificity and sensitivity, as the DNA probe can be designed to specifically bind to the target analyte. They are also versatile in application both in vitro and in vivo, as different DNA probes can be easily synthesized to detect different analytes. Additionally, they are highly biocompatible relatively simple, and inexpensive to manufacture [18, 19].

In vitro biosensing approach holds great promise for various applications, including medical diagnostics, environmental monitoring, and drug discovery.

When compared to traditional fluorescence sensors based on dyes or quantum dots (QDs), PLNPs have various benefits. To begin, PLNPs can provide in vitro noninvasive sensing by detecting and monitoring numerous analytes within biological systems such as cells, tissues, or living organisms. Second, by removing autofluorescence interferences, PLNPs can enhance the detection limit. Traditional fluorescent sensors may have difficulty distinguishing between the fluorescence emitted by the analyte of interest and background autofluorescence. PLNPs, on the other hand, have long-lasting luminescence that continues even after the excitation source is removed. Because of this persistence, short-lived autofluorescence signals can be eliminated, considerably enhancing the detection limit and sensitivity of the sensors [20].

Near-infrared-emitting PLNPs, in particular, have demonstrated good penetration depth into biological tissues, avoiding tissue damage and living-tissue autofluorescence, resulting in minimal toxicity and unprecedented detection sensitivity, making them particularly useful in chem/biosensing processes [20].

Liu et al. [9] developed PLNPs $ZnGa_2O_4:Cr^{3+}$ (ZGO) for in vitro application to detect and quantity glucose. They employed glucose oxidase (GOD) to see if in-situ H_2O_2 generation could boost ZGO signal strength. Glucose oxidase is an enzyme that catalyzes the breakdown of glucose in the presence of oxygen, producing (H_2O_2) as a byproduct. The amount of H_2O_2 produced is directly proportional to the amount of glucose present, meaning that the more glucose there is the more H_2O_2 will be generated. For this reason, GOD was incubated with glucose and ZGO in a microplate at 37 °C. Following incubation, the plate was illuminated with a UV light and photographed using an ICCD camera. When ZGO, glucose, and GOD are combined in the same well, the luminescence signal increases significantly in comparison with control trials containing simply ZGO, ZGO with GOD, or ZGO with glucose, where no enhancement occurs. The LOD of glucose was determined which is equivalent to 35.8 μM. Another experiment was carried out in the same manner as with glucose, but with other saccharides that are not GOD substrates, such as sucrose, maltose, fructose, galactose, mannose, xylose, and lactose. The

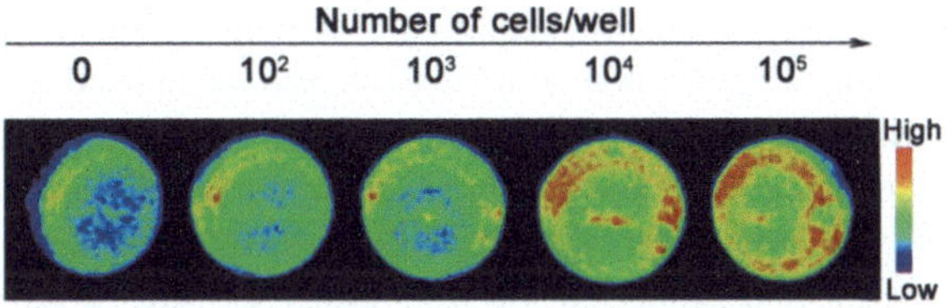

Fig. 1 Images showing NIR persistent luminescence of PLNP–CuS–RGD incubated with SCC-7 cell supernatant with having varying cell density. Reprinted with permission from [22]. Copyright (2016) ACS Applied Materials and Interfaces

researchers demonstrate that substituting saccharides for glucose has no effect on signal amplification. This observation indicates that the glucose sensor based on ZGO exhibits high selectivity specifically towards glucose.

Shi et al. [21] used TAT peptide (derived from the transactivator of transcription TAT) and folate (FA) to modify the surface of $ZnGa_2O_4$: Cr, Eu NPs (ZGO) for in vitro and in vivo imaging experiments. One of the objectives of the study is to perform cell imaging in a controlled in vitro using 808 nm NIR in situ re-excitation without harming the cells for afterglow luminescent in vitro imaging. This study attributed the excellent photo-stimulated afterglow properties of ZGO as the reason behind the high SNR cell imaging which can recover and persist for more than 7 min after excitation at 808 nm NIR light. The obtained results indicate that ZGO has unique features that make it appropriate for long-term in vitro imaging after NIR excitation.

A multifunctional PLNP/copper sulfide (CuS)-based nanoprobe for imaging-guided photothermal treatment in vivo was successfully developed [22]. In vitro assay was realized to study how the PLNP–CuS–Arginylglycylaspartic acid (RGD) material behaves and whether it retains its luminescence properties when placed in external environment. The incubation of SCC-7 cell supernatant with PLNP–CuS–RGD was performed at 37 °C for 2 h. It was found a noticeable luminescence recovery due to the secreted enzyme (matrix metalloproteinase-2 "MMP-2"), however the researchers observed "unapparent recovery.“ This means that the luminescence did not increase or was not significantly restored when the inhibitor was added. Figure 1 shows that as the cell density increases, the luminescence intensity also increases. The experimental results have shown that PLNP–CuS–RGD responds to the presence of MMP-2 in vitro by undergoing a specific activation process.

3 In Vivo Biosensing

In vivo biosensing refers to a type of biosensing that is performed inside a living organism or cell [23]. It has been used to detect various analytes such as glucose, lactate, and oxygen. In vivo biosensors can be implanted or injected into the body and can be used to monitor physiological parameters in real-time [24]. For example, glucose monitoring is important for people with diabetes and can be done using an implanted glucose sensor [25]. Lactate biosensors have been used to monitor lactate levels in athletes during exercise. Xuan et al. [26] developed a versatile epidermal patch as a biosensor capable of monitoring lactate levels, pH, and temperature on the skin during physical activity. It can also be used for post-stimulation analysis following controlled sweat induction via iontophoresis. Such a device could have applications in sports science, healthcare, and research for assessing athletic performance and physiological responses to various stimuli. Moreover, oxygen sensors are other types of sensors that interestingly used in biomedical and clinical research to measure and monitor oxygen levels within tissues. This information can aid in understanding tissue function, oxygen transport, and the impact of oxygen availability on various physiological processes [26]. In vivo biosensors can be implanted or injected into the body. They can be made of biocompatible materials such as polyethylene glycol (PEG) [27], poly (lactic-co-glycolic acid) (PLGA) [28], and chitosan [29] and can be designed to be biodegradable. Other applications of in vivo biosensors include drug delivery and gene therapy. In vivo biosensor drug delivery suggests a scenario in which a biosensor is employed within a living organism to monitor specific biological parameters or conditions with the capability to trigger drug release [30]. BioMEMS (Biological Micro-Electro-Mechanical Systems) have been widely utilized as electrochemical sensors that combine mechanical and electrical components with biological components. Indeed, BioMEMS offer several advantages such as lower response time, high scalability and reproducible production and high sensitivity. They are designed to interact with biological systems at a microscale. In this context, BioMEMS act as intelligent drug delivery systems capable of monitoring various physiological parameters within the body, responding to the unique needs of each patient [30].

PLNPs-based autofluorescence-free sensors have significant promise for in vivo biosensing by providing enhanced sensitivity, specificity, long-term monitoring of physiological parameters and biomarkers within the body, and real-time monitoring capabilities [31].

Li et al. [32] synthesized a nanoprobe that is a potent instrument for detecting and visualizing GSH in living cells and in vivo. Its unique design leveraging the quenching and recovery of PLNPs' luminescence by MnO_2 in the presence of GSH enables sensitive and specific GSH detection.

Liu et al. [33] reported a flexible NIR PLNPs-based on Cr^{3+}/Nd^{3+} codoped $ZnGa_2O_4$ (A-ZGCN) nanoparticle with a more than 15-day afterglow time for detection and suppression of breast cancer cell metastasis, as well as treatment (Fig. 2). The utilization of long-persistent luminescence and the background-free feature of (PLNPs) are extremely valuable for long-term in vivo tracking of breast cancer cells and imaging of lymphatic metastases (>25 days). The study shows that NIR-PL imaging is more sensitive than traditional bioluminescence imaging for detecting cancer cell metastasis.

Ongoing research in this field continues to advance the development and application of PLNP-based biosensors in various biomedical and clinical contexts.

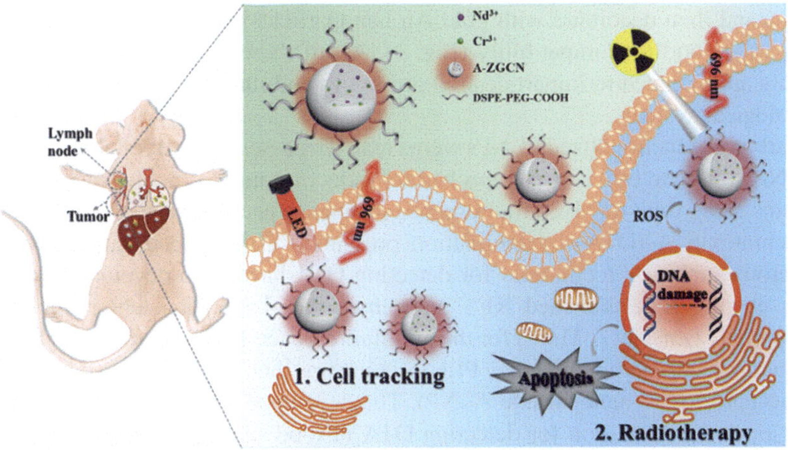

Fig. 2 Schematic illustration showing in vivo tracking of PLNPs (Cr^{3+}/Nd^{3+} codoped $ZnGa_2O_4$) for the detection and inhibition of breast cancer cell metastasis. Reprinted with permission from [33]. Copyright (2019) ACS Analytical Chemistry

4 Detection and Biosensing Based on Resonance Energy Transfer

Resonance energy transfer (RET) also called Förster resonance energy transfer (FRET) is a process that occurs when the energy from an excited donor molecule is transferred to an acceptor molecule through nonradiative dipole–dipole interactions. RET typically occurs at distances of 1–10 nm (nm), and this range can vary based on several factors, including the combination of donor and acceptor molecules and the relative orientation of their transition dipole moments [34].

FRET is indeed highly sensitive to the distance between a donor and an acceptor molecule, and it has proven to be a powerful tool in the study and sensing of molecular interactions and various applications in molecular biology, biochemistry, and biophysics [35]. Furthermore, the efficiency of this energy transfer depends on the quantum yield of the donor, spectral overlap, background noise and the quenching efficiency of the receptor [35]. Therefore, the choice of appropriate donor–acceptor pairs with favorable optical properties, along with careful consideration of experimental conditions and sample preparation, is crucial for the success of FRET-based sensors and measurements. RET-based persistent luminescence nanoparticles sensors can maintain emissive and keep detectable without need for in situ excitation. PLNPs are promising RET-biosensors due to their remarkable advantages in avoiding background signal and scattered light from real samples [36, 37].

Feng et al. [36] established PLNPs as afterglow RET sensors for quantitative determination and imaging of fibroblast activation protein-alpha (FAPα) in live cells. Functionalization plays a crucial role in engineering PLNPs for biological applications. Feng et al. [36] modified the surface of PLNPs with 3-mercaptopropionic acid

(MPA) and then decorated with gold Au nanoparticles, which bind with mercapto groups for good biocompatibility (Fig. 3). All of these designs and surface engineering are intended to impart desired properties and functionalities for application in biological systems.

Carboxyl-functionalized PLNPs were created by assembling dye-labeled peptides or DNA to create efficient nanoprobes for detecting nucleic acids, caspase-3, and proteins using time-resolved fluorescence resonance energy transfer [12].

Ochratoxin A (OTA) contamination presents a critical food safety issue that numerous analytical techniques for detecting OTA have been developed. Recently Jiang et al. [38] established RET aptasensor based on persistent luminescence nanoparticle (Zn_2GeO_4:Mn^{2+}) for the autofluorescence-free detection of OTA. OTA aptamer was used to functionalize PLNP, which was then modified by a donor (DNA) and acceptor (gold nanoparticle AuNP). The developed aptasensor based on PLNP offer a powerful solution for detecting OTA in food samples with high sensitivity and selectivity while minimizing interference from autofluorescence and background signals. Furthermore, the detecting OTA was tested in beer samples with no autofluorescence interference. This could be particularly important in food safety where the presence of OTA needs to be accurately and reliably determined [38].

Recently, Wang et al. [37] developed an efficient PLNPs ($ZnGa_2O_4$: Cr^{3+}) probe based on molecular beacon (MB) for miR21 sensing with a quantitative detection limit of 0.1 nM. The experiment of miR21 detection in human serum was performed in diluted serum and cell extraction, yielding acceptable values that matched well with those produced by the commercial NIR-MB probe, confirming the PLNPs-MB

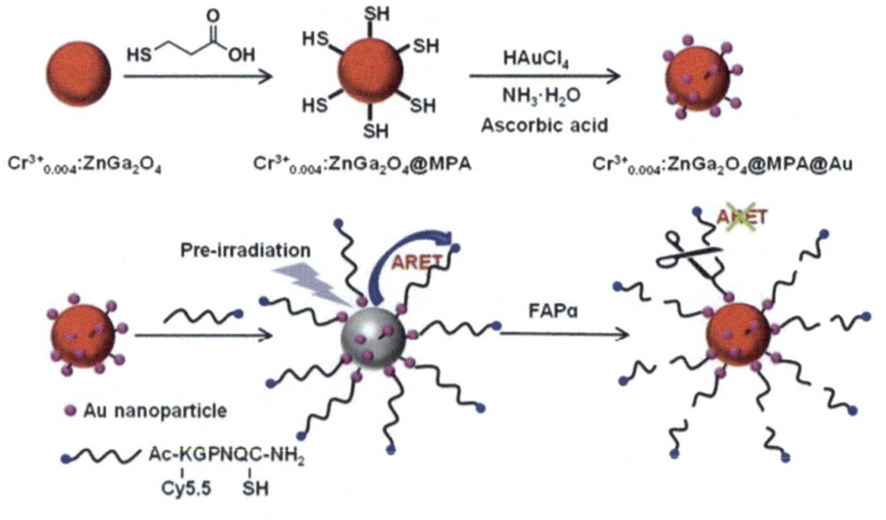

Fig. 3 Schematic illustration of synthesis steps and surface modification of an afterglow RET ($Cr^{3+}{}_{0.004}$:$ZnGa_2O_4$@MPA@Au) for fibroblast activation protein-α. Reprinted with permission from [36]. Copyright (2018) ACS Sensors

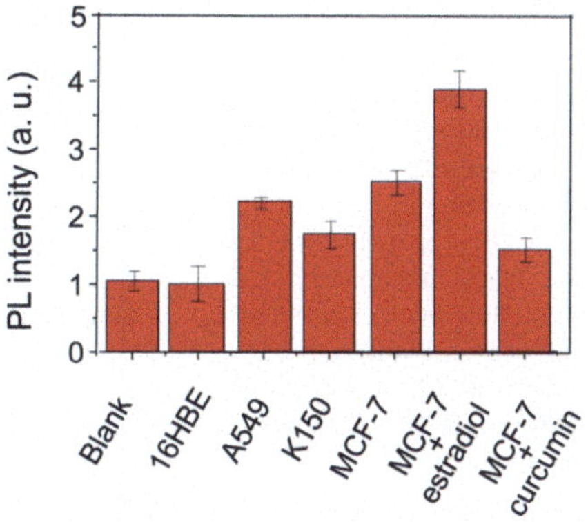

Fig. 4 The PLNPs-MB probe used to measure the levels of miR21 in various human cells. Reprinted with permission from [37]. Copyright (2022) Biosensors and Bioelectronics

probe's good testing accuracy. Furthermore, the suggested PLNPs-MB probe was used to monitor miR21 levels in three different cancer cells (MCF-7, A549, and K-151) as well as human bronchial epithelial cells (16HBE). Figure 4 demonstrates that when compared to the 16HBE group, the three selected cancer cells had higher PL intensities, confirming the over-expression of miR21 in all selected cancer cells. The developed probe has the potential to be used in miR21-related cytobiology research and illness diagnosis. However, this probe presents some limitation associated with the use of PLNPs in intracellular imaging due to their inherent ultraviolet (UV) excitation. The solution being pursued involves hybridizing PLNPs with NIR triggered up-conversion phosphors.

5 Detection and Biosensing of Tumor Biomarkers

Tumor biomarkers, also called biological markers, are substances, often molecules that can indicate the presence of cancer or provide information about the state and progression of a tumor. Genes, gene products, DNA, RNA, proteins, enzymes, hormones, and specialized cells are examples of biomolecular components that are commonly released through genetic modification and cell division [39]. These biomarkers can be produced either directly by cancer cells or by the body in response to the presence of cancer. Tumor biomarkers can be detected in bodily fluids (such as blood, urine, or saliva) or through tissue analysis [40].

In fact, cancer biomarkers can be categorized based on the principle of transduction, and different types of biosensors are designed to detect these biomarkers. These biosensors rely on various mechanisms for transducing the signal associated with the presence or concentration of specific biomolecules. The common categories of cancer biomarkers and the corresponding types of biosensors are: electrochemical, mass-sensitive, and optical [41].

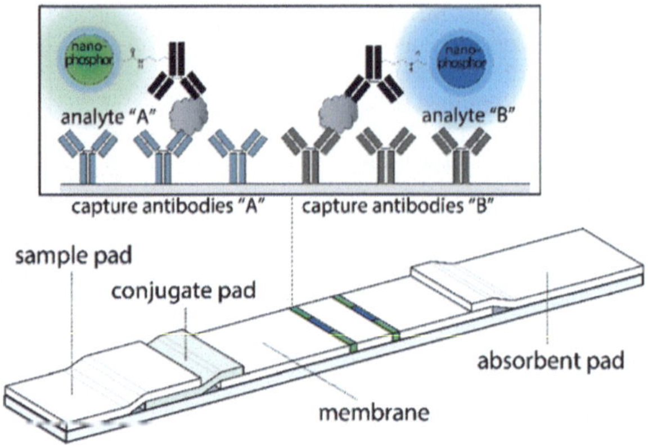

Fig. 5 Schematic representation showing the reporters based on PLNPs (green-emitting $SrAl_2O_4{:}Eu^{2+}$, Dy^{3+} and blue-emitting $(Sr_{0.625}Ba_{0.375})_2MgSi_2O_7{:}Eu^{2+}$, Dy^{3+}) used in a duplex lateral flow assay. Reprinted with permission from [42]. Copyright (2020) Analytical methods: advancing methods and applications

Several biosensors have been developed for the detection of tumor biomarkers such as prostate-specific antigen (PSA), carcinoembryonic antigen (CEA), and alpha-fetoprotein (AFP).

To name a few, prostate-specific antigen (PSA) is a well-known biomarker for prostate cancer, and it can be detected in both semen and blood samples [39]. Danthanarayana's group [42] employed a single lateral flow assay (LFA) containing two distinct PLNPs in a single sample, one for green-emitting $SrAl_2O_4{:}Eu^{2+}$, Dy^{3+} (SAO) and the other for blue-emitting $(Sr_{0.625}Ba_{0.375})_2MgSi_2O_7{:}Eu^{2+}$, Dy^{3+} (SBMSO) for the simultaneous detection of PSA and human chorionic gonadotropin (hCG) (Fig. 5).

The detection limits of both PSA and hCG were determined using SAO and SBMSO PLNPs respectively. The concept is based on gradually increasing each antigen concentration from 0.02 to 10 ng mL^{-1}, thereby calculating the ratio of test line intensity (TL) to control line intensity (CL). The detection limits values of PSA with SAO and hCG with SBMSO are 0.1 ng mL^{-1} and 1 ng mL^{-1}, respectively (Fig. 6), which are comparable to commercially available LFAs.

Carcinoembryonic antigen (CEA) is a tumor marker or biomarker that is found in elevated levels in the blood of some individuals with certain types of cancer, most commonly colorectal cancer [43]. Shi et al. [43] reported ratiometric aptasensor based on two PLNPs: NIR-emitted $ZnGa_2O_4{:}Cr^{3+}$ (ZGC NPs) coated with polydopamine (ZGC@PDA NPs) and Bleu-emitted $Zn_2GeO_4{:}Sr^{3+}$ persistent luminescence nanorods (ZGO:Sr NRs) for carcinoembryonic antigen detection. CEA-Apta modified ZGO:Sr NRs were coupled with ZGC@PDA NPs via the π-π stacking effect between aptamers and PDA, inducing luminescence quenching of ZGO:Sr NRs.

The detection signal's luminescence intensity at 490 nm grew significantly with increasing CEA concentration, despite the fact that the fluorescence intensity at 696 nm barely altered. To detect carcinoembryonic antigen (CEA), a ratiometric persistent luminescence sensor directed by an aptamer with a LOD of 0.46 pg mL^{-1} and good linearity ($R^2 = 0.993$) was developed. By inserting the reference signal,

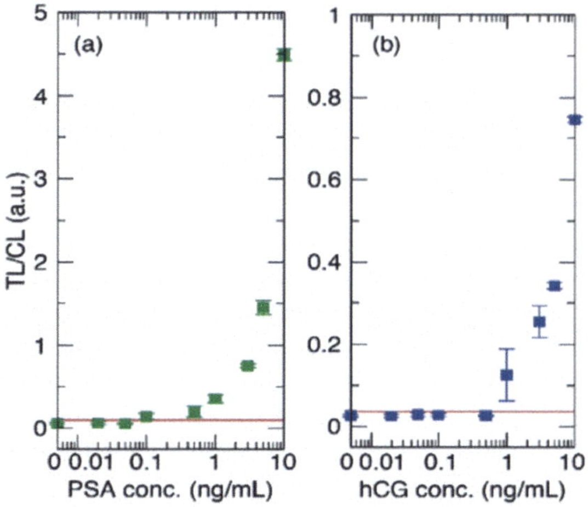

Fig. 6 Serial dilution of **a** Prostate-specific antigen with $SrAl_2O_4:Eu^{2+}, Dy^{3+}$ and **b** human chorionic gonadotropin with $(Sr_{0.625}Ba_{0.375})_2MgSi_2O_7:Eu^{2+}, Dy^{3+}$ was detected by the iPhone 5S. The detection limit cutoff of the no-analyte control tests is represented by the red line, which is computed as the mean plus three times the standard deviation ($\mu + 3\,\sigma$). Reprinted with permission from [42]. Copyright (2020) Analytical methods: advancing methods and applications

the proposed ratiometric aptasensor can eliminate interference detection environment effect on luminescence intensity in addition to preventing autofluorescence interference induced by the afterglow features of PLNPs. As a result, this property is appropriate for tumor marker screening to ensure the best possible outcomes in early cancer detection [43].

PLNPs are a promising platform for detecting biomolecules like alpha-fetoprotein (AFP). Alpha-fetoprotein is a protein that can be used as a tumor marker, particularly for detecting hepatocellular carcinoma (liver cancer) and certain other cancers [44].

PLNPs need to be functionalized or coated with specific molecules (ligands) that can selectively bind to AFP. This functionalization can be achieved by attaching antibodies, aptamers, or other affinity ligands that have high specificity for AFP [20].

For the first time, Yan et al. [45] developed an innovative approach to cancer biomarker detection (AFP) and imaging using PEI-coated PLNPs and time-resolved FRET. The combination of these techniques offers the advantage of autofluorescence-free detection of AFP in living cancer cells, making it a promising tool for cancer research, diagnostics.

6 Detection and Biosensing of Bioactive Molecules

Bioactive molecules are molecules that have a direct and measurable effect on living organisms, including humans. These molecules can interact with biological systems, such as cells, tissues, or organisms, and elicit specific biological responses or activities [46].

Advanced scientific and technological approaches, such as the use of bioactive compounds, sensors, and analytical techniques, could assist the control and early research of many diseases for diverse applications such as drug discovery and environmental monitoring [47]. Indeed, numerous biosensors have been developed for the detection of bioactive molecules like glucose, lactate, and oxygen. Biosensors are analytical devices that combine a biological recognition element (such as enzymes or antibodies) with a transducer to convert a biological response into a measurable signal [48, 49].

The controllable synthesis of PLNPs with uniform morphologies and strong persistent luminescence has been a significant challenge in the field of detection and biosensing. In this context, researchers have been making progress in addressing these issues to advance the applications of PLNPs. Recently, Yang et al. [50]'s research efforts aimed at precisely controlling the synthesis of zinc gallogermanate ZGGO PLNPs by modifying their electronic structures through nonstoichiometric reactions which represent an important step forward in the development of highly uniform nanocubes with controllable size and persistent luminescence (Fig. 7). This approach involves adjusting the ratios of the constituent elements (Ge/Ga) in the reaction to achieve the desired electronic and structural properties. The developed PLNPs have the potential for autofluorescence-free serum ROS detection and act as an efficient biosensor for glucose oxidase (GOx) detection.

The quenching effect of persistent luminescence issued from the produced H_2O_2 suggests that ZGGO PLNPs can detect glucose/GOx (Fig. 8a).

Figure 8c, d show a decrease in the persistent luminescence signal as the GOx concentration increases, indicating that the increased amount of GOx increases H_2O_2 productivity. This work encourages the use of zinc ZGGO Persistent Luminescence

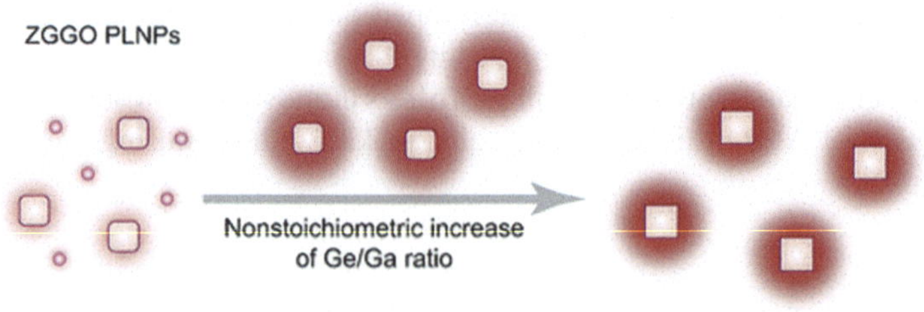

Fig. 7 Schematic Illustration of ZGGO PLNPs modified with nonstoichiometric reactions for uniform morphology and strong Persistent Luminescence. Reprinted with permission from [50]. Copyright (2023) ACS Applied Materials

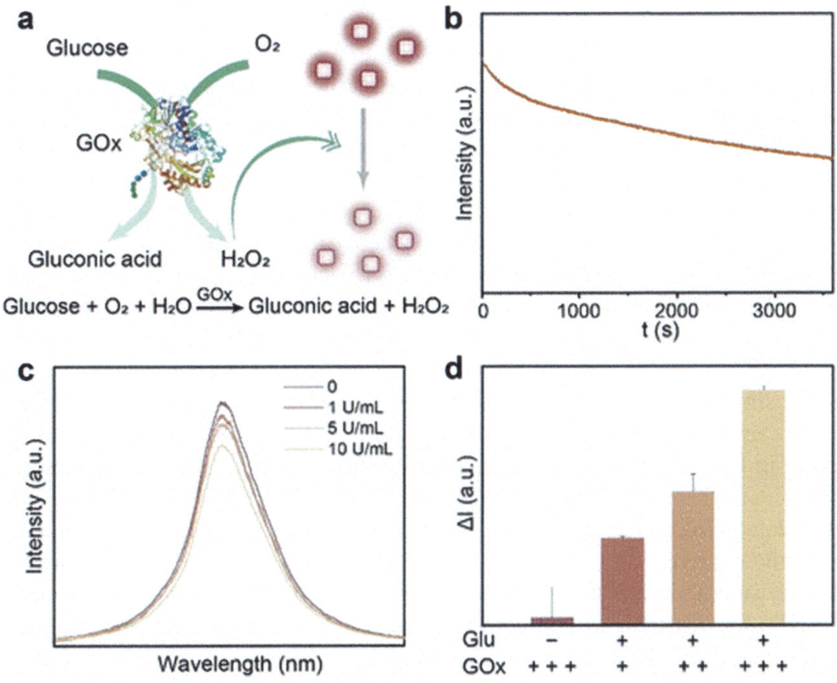

Fig. 8 **a** Diagram of glucose oxidase detection based on the response of ZGGO PLNPs to H_2O_2. **b** The intensity of the ZGGO-glucose- glucose oxidase reaction system's persistent luminescence with time. **c** Spectra of persistent luminescence and **d** Changes in the intensity of the ZGGO-glucose- glucose oxidase reaction system's persistent luminescence with varying glucose oxidase concentrations. Reprinted with permission from [50]. Copyright (2023) ACS Applied Materials

Nanoparticles for background-free serum (GOx) detection, which offers substantial promise in the context of identifying glucose metabolic disorder-related disorders such as diabetes mellitus.

Lactate is a major bioactive molecule in sports medicine, food industry, clinical diagnostics, metabolic research, and bioprocess monitoring [51].

Lactate is a key metabolite produced during anaerobic respiration or when there is an oxygen deficit in cells. Its concentration in biological samples, such as blood or cell culture media, can provide valuable information about metabolic processes and disease states. Lactate detection typically involves enzymatic reactions, where lactate is converted into a product that can be quantified [52]. Lactate biosensors are significant in this setting because to their multiple advantages, such as low cost, ease of use, rapid response, and minimal or no sample pre-treatment. These properties make them useful in a variety of sectors, such as healthcare, sports science, and environmental monitoring [51].

Because of their capacity to efficiently monitor the metabolic events related with lactate measurement, electrochemical and optical-based sensors are the most commonly investigated approaches for assessing lactate levels [51].

The majority of lactate biosensors described in the literature use an electrochemical approach [53–55]. Here's a recent study that used an electrochemical sensor to detect lactate [56]. The electrochemical properties of polyaniline screen-printed electrode (PANI-/SPCE) sensor to lactate are studied by Zhu et al. [56]. The results reveal that the PANI-/SPCE sensor has a sensitivity of 63.72 mV/pH and 67.53 mV/pH in the ranges of 1–12 pH and 4–8 pH, respectively, and that it is an efficient sensor with good repeatability, selectivity, and long-term stability toward lactate detection. In contrast to other reported sensors based on lactate oxidase (LOX) for sensing materials, the PANI-/SPCE sensor in this work [56] was not loaded with LOX, which has the advantage of inexpensive sensor production cost even if the sensitivity suffers.

Optical biosensors for lactate detection have received less attention than electrochemical biosensors. Optical lactate biosensors assess lactate concentrations in a sample using light-based approaches. These sensors frequently employ specialized receptors or indicators that interact with lactate molecules to produce measurable changes in light characteristics such as absorption, fluorescence, or luminescence [51].

The detection of lactate using PLNPs is not much well reported in the literature despite their interesting proprieties. Biosensors based on PLNPs is an innovative and promising approach that takes advantage of the unique luminescent properties of PLNPs for sensitive and selective lactate detection. This type of luminescent material can emit light for an extended period after exposure to an excitation source. In the context of lactate detection, PLNPs can be functionalized with molecules or ligands that can selectively bind to lactate or lactate-producing enzymes (e.g., lactate oxidase). This functionalization imparts lactate specificity to the PLNPs. The interaction between lactate and the functionalized PLNPs can result in luminescence quenching or enhancement, depending on the design of the biosensor. For example, binding to lactate may quench the persistent luminescence of the PLNPs due to energy transfer processes.

Sensors for oxygen detection, often referred to as oxygen sensors or oxygen analyzers, are devices designed to measure the concentration of oxygen (O_2) in a gas or liquid sample. These sensors are used in various industries, including healthcare, environmental monitoring, industrial processes, and scientific research [57].

Electrochemical sensors are widely used for oxygen detection because they can realize in situ online measurements of which the most common types include Clark cells and galvanic cells.

They consist of an electrolyte, working electrode, and reference electrode [57]. Oxygen molecules diffuse through a gas-permeable membrane and interact with the working electrode, causing an electrochemical reaction [57]. This generates a current or voltage signal proportional to the oxygen concentration. Indeed, optical oxygen sensors have seen rapid development and adoption in recent years. Optical oxygen sensors rely on the measurement of luminescence properties, such as fluorescence or phosphorescence, of certain indicator molecules that are sensitive to oxygen [58]. The indicator molecules emit light when excited, and the rate of emission is inversely proportional to the oxygen concentration. The change in luminescence is used to determine oxygen levels [59]. These sensors are often used in applications where non-invasiveness is critical, such as in biomedical and environmental monitoring. Zhu et al. [60] developed long persistent luminescence (LPL) metal–organic framework (MOF) from LIFMZCY-1 ([Cd (TzPCz)(DMA) $(H_2O)_3]_n$, DMA = N, N-Dimethylacetamide) for multi-level oxygen detection. By applying a high temperature to LIFMZCY-1, its structural properties change which results in the new sample known as LIFM-ZCY-1'. When a light source is turned off anywhere in the vacuum or N_2 environment, an orange extended afterglow luminescence discernible by sight is produced. As shown in Fig. 9a, after UV light excitation, the emission intensity (around 620 nm) drastically suffers from quenching effect while increasing the oxygen concentration. Furthermore, in an O_2 environment, the CIE coordinates change from (0.41, 0.25) in vacuum to (0.32, 0.20) (Fig. 9b). The phosphorescence lifetime of LIFM-ZCY-1 was discovered to be 30 ms longer than that of pure oxygen, demonstrating a considerable sensitivity to oxygen below 200 mbar (Fig. 9c). Interestingly, the visible light LPL of LIFM-ZCY-1' can last for more than two seconds, but when the oxygen pressure is 2 mbar, the LPL persists less than 0.5 s (Fig. 9e). Researchers continue to develop and refine biosensors for bioactive molecules detection and measurements, enhancing their sensitivity, accuracy, and usability in various fields.

7 Detection and Biosensing of Pathogenic Microorganism

Pathogenic microorganisms, often referred to simply as pathogens, are microorganisms (microscopic organisms) that can cause diseases in humans, animals, or plants. These pathogens have evolved various mechanisms to infect their hosts and evade the host's immune system, leading to illness or disease [61, 62].

The detection and biosensing of pathogenic microorganisms are critical for various applications, including medical diagnostics, food safety, environmental monitoring, and biodefense [63]. Several methods and technologies are used for this purpose, each with its advantages and limitations.

Indeed, various biosensors for the detection of harmful bacteria such as *E. coli, Salmonella*, and *Listeria monocytogenes* have been created. These biosensors are designed to identify microorganisms quickly, accurately, and with low-cost, making them useful tools in for the detection of pathogenic bacteria [63].

Detection of *Escherichia coli (E. coli)* using a microfluidic device is a promising approach for rapid, on-site, and sensitive detection of this pathogenic bacterium. Microfluidic devices, also known as lab-on-a-chip systems, enable precise control and manipulation of small volumes of fluids within microchannels. They can be designed to incorporate various detection methods and are well suited for point-of-care diagnostics and field applications [64]. Several research works have utilized microfluidic devices for the detection of *Escherichia coli* [65–67].

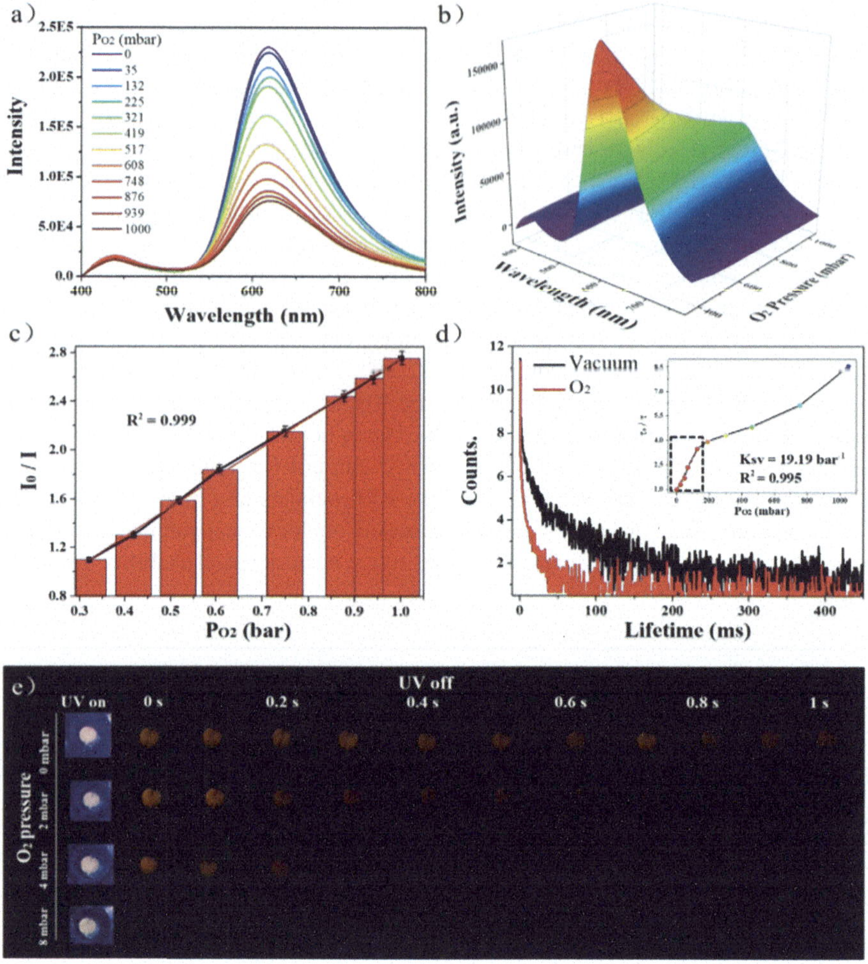

Fig. 9 **a** Photoluminescence spectra of LIFM-ZCY-1' subjected to different oxygen concentration. **b** 3D graph depicting the link between luminescent intensity and oxygen pressure. **c** LIFM-ZCY-1' Stern–Volmer quenching curve concerning the emission intensity at 620 nm. **d** The phosphorescence lifetimes decay curves of LIFM-ZCY-1' whether or not oxygen exists, as well as the Stern–Volmer quenching diagram as a function of oxygen fraction (insert). **e** LIFM-ZCY-1' LPL images at various oxygen pressures. Reprinted with permission from [60]. Copyright (2020) Journal of materials chemistry. C, Materials for optical and electronic devices

Alonzo et al. [68] developed a portable and scalable microfluidic technology capable of executing a quick and highly sensitive bacteriophage-based assay to identify *E. coli* cells in a fraction of the time required by existing approaches. The microfluidic system uses membrane filtration and T7-NanoLuc-CBM fusion to detect 4.1 *E. coli* CFU in water samples in less than 5.5 h. The researchers' goals were to boost test throughput and automate the instrumentation to enable for timely surveillance and convenience of usage. Another method for detecting *E. coli* O157:H7 was to employ a colorimetric signal from a membrane filter-integrated finger-powered microfluidic device. The method is based on immunomagnetic separation, which separates the target microorganisms from food using antibody-conjugated magnetic nanoparticles (MNPs). After optimizing the pore size of the membrane filter and applying a gold amplifier solution, the detection of *E. coli* O157:H7 was determined to be approximately 10^2 CFU/mL [64].

Fluorescent dyes, both organic and inorganic, have gained significant attention and are widely used in various bio-applications due to their ability to facilitate easy and sensitive detection under a fluorescence microscope [69]. These sensors use fluorescent tags attached to *E. coli*-specific probes (such as antibodies or DNA) that emit light when bound to *E. coli*. Changes in fluorescence intensity or emission spectra are then measured. Ananda et al. [69] designed Fluorescein isothiocyanate incorporated silica nanoparticles via microemulsion method for the detection of *E. coli*. To enable for antibody conjugation with *Escherichia coli*, the nanoparticle surface was chemically modified with amine and carboxyl functionalities. The work demonstrates how single *E. coli* cells could be clearly imaged (Fluorescence microscopy) following antibody conjugation onto the nanoparticle. This reveals the nanoprobe's sensitivity to bacterial cell detection. Other interesting biosensors for *E. coli* detection include optical fibers, which can be functionalized with *E. coli*-specific receptors and used to detect changes in light propagation through the fiber when *E. coli* binds to the surface [70, 71].

Biosensors for the detection of *Salmonella* are valuable tools in food safety and public health; they offer the advantage of rapid response, specificity, low cost and potential for real-time monitoring [72]. Among the current sensors for *Salmonella* detection are gold nanoparticles (GNPs), which can encapsulate antigen or antibody onto the surface of an electrode to create an electrochemical immunosensor [73]. Yang et al. [74] developed a GNP-based signal amplified lateral flow immunoassay (LFIA) for the quick detection of *Salmonella enteritidis* (*S. enteritidis*) in milk utilizing $HAuCl_4$ and $NH_2OH{\cdot}HCl$ as the enhanced solution. The increased signal at the test and control lines was plainly visible with the naked eye, and this finding was discovered to be dependent on GNPs growth and accumulation. With the addition of the improved solution, the detection range for *S. enteritidis* expanded, with a lower detection limit of 10^4 CFU/mL that was enhanced 100 times over the traditional LFIA for qualitative detection. It was also revealed that the improved method's test time was significantly shorter than many earlier findings, showing that the signal amplified immunochromatographic assay was both time efficient and easily readable. Another study used gold nanoparticles in a microfluidic device along with loop-mediated isothermal amplification to detect *Salmonella spp.* in food samples [75]. The successful coupling of microfluidic isothermal DNA amplification with GNPs enables naked-eye differentiation and reliable detection of *Salmonella* even at low concentrations (10 cfu/25 g), making it a promising strategy for usage in the food sector.

Listeria monocytogenes is a pathogenic bacterium that can cause serious foodborne illnesses, making it important to detect its presence quickly and accurately. Biosensors are analytical devices that can be used for the detection of *Listeria monocytogenes* in various samples, including food, water, and clinical specimens [76]. Electrochemical biosensors for the detection of *Listeria monocytogenes* utilize electrochemical principles to detect changes in electrical properties or signals that occur when the bacterium is present. These biosensors often employ specific antibodies or aptamers to capture *Listeria monocytogenes*, leading to measurable electrochemical responses [77]. Li et al. [78] developed an electrochemical transduction technique to boost the sensitivity of CRISPR/Cas12a for the detection of *Listeria monocytogenes*.

The methodology is based on combining a CRISPR/Cas12a -electrochemical biosensor (E-CRISPR) with recombinase-assisted amplification (RAA) to create a cost-effective, selective, and ultrasensitive tool. In comparison with prior studies on Cas12a-mediated E-CRISPR nucleic acid detection, this work [78] supports a high limit-of-detection (LOD) that is 10^{8-} fold lower. In terms of sensitivity, E-CRISPR has been shown to be superior to previously reported CRISPR/Cas-based detection methods such as fluorescence detection methods, surface-enhanced Raman scattering, and lateral flow test. These findings support the use of RAA-based E-CRISPR as a highly sensitive and specific biosensor for the detection of *L. monocytogenes* [78].

Optical biosensors for the detection of *Listeria monocytogenes* are also promising approaches that rely on changes in light properties, such as absorbance, fluorescence, or surface plasmon resonance, when the bacterium is present. These biosensors are often highly sensitive and can provide real-time or near-real-time detection [79]. Ariani et al. [80] created a fast optical biosensor for the detection of *L. monocytogenes* by combining the molecular recognition capability of antimicrobial peptides known as Leucocin A (LeuA) with the fluorescence labeling ability of gold nanoclusters (AuNCs).

Figure 10 depicts the biosensor experiment for detecting *L. monocytogenes* after the bacteria bind to the LeuA peptide on a solid surface and are labeled with the complex MPA-AuNCs, which uses 3-mercaptopropionic acid (MPA) to stabilize gold nanoclusters to label bacteria. The developed biosensor detects *Listeria* in a short period of time (45–50 min) from a small sample volume (10 L), with a detection limit of 2 105 cfu/mL. This technology paves the way for the developed fluorescence biosensor to be employed as a screening tool for *L. monocytogenes* in food and pharmaceutical items.

In another study, Chen et al. [79] created an optical biosensor for the rapid and sensitive detection of *Listeria monocytogenes* by combining immunomagnetic separation, urease catalysis, and pH indication. To synthesize magnetically tagged bacteria, a couplage of magnetic nanobeads (MNBs) with monoclonal antibodies (MAbs) via streptavidin–biotin binding was used to separate and identify *Listeria* cells. The gold nanoparticles (AuNPs) electrostatically modified with urease and polyclonal antibodies (PAbs) were then used to generate MNB-MAb-*Listeria*-PAb-AuNP-urease complexes with the magnetically tagged bacteria. Urease is primarily used in complexes to catalyze the hydrolysis of urea into ammonium carbonate,

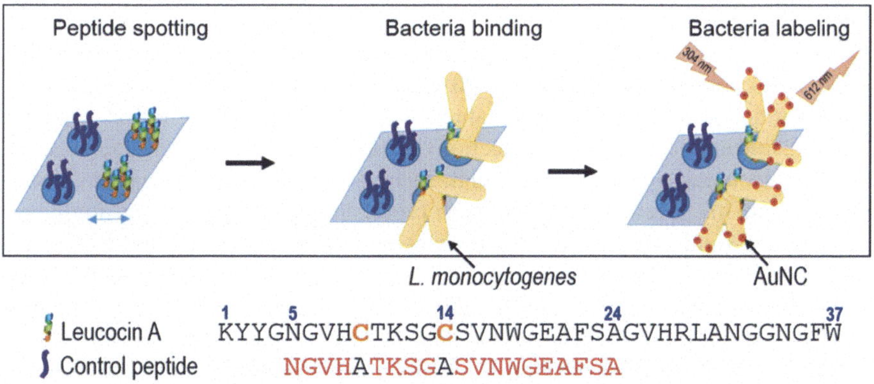

Fig. 10 Diagram illustrating the steps for detecting *Listeria monocytogenes* using leucocin A (LeuA) and gold nanoclusters (AuNCs). Reprinted with permission from [80]. Copyright (2018) ACS Applied Nano Materials

raising the pH of the urea solution. The best pH indicator, bromocresol purple, could monitor the color change, which could be detected at the characteristic wavelength of 588 nm for quantifiable detection of *L. monocytogenes.* The developed biosensor detected *Listeria monocytogenes* at concentrations as low as 1.0×10^2 CFU/mL, showing its suitability for practical applications of foodborne bacterium detection.

The importance of PLNPs results from long-lasting after being exposed to an external light source, and from their high sensitivity and real-time monitoring which allow the detection of different bacteria cells in complex sample matrixes (serum, water, or food). PLNPs can be functionalized with specific biomolecules, such as antibodies or DNA aptamers that have a high affinity for example to *E. coli* or its surface components. This allows for selective binding of PLNPs to the target bacteria [81]. It is important to note that while PLNPs offer several advantages for bacteria detection, there are also challenges, including the need for specialized equipment for luminescence measurement and the optimization of PLNP functionalization for specific bacterial strains or targets. In spite of vast development of PLNPs as sensors in various fields, still it is a crucial challenge for formulation scientists to develop appropriate PLNP-based detection methods to improve sensitivity, specificity, and practicality for bacterial detection.

8 Detection and Biosensing of Pollutants

The detection and biosensing of pollutants are critical for monitoring and mitigating environmental contamination, ensuring public health, and complying with regulatory standards. Biosensors are analytical devices that utilize biological components, such as enzymes, antibodies, or microorganisms, in combination with transducers to detect and quantify pollutants [82]. Pollution detection and biosensing can be utilized for a variety of purposes, including environmental monitoring and water quality control. Several biosensors for the detection of pollutants such as heavy metals, pesticides, and organic chemicals have been created [82].

Carbon nanotubes (CNTs) have gained significant attention in recent years for their potential applications in the detection of heavy metals due to their unique structural and electrical properties. Their high surface area, excellent electrical conductivity, and chemical reactivity make them promising candidates for sensing heavy metal ions. Recently, Gupta et al. [83] developed for the first time an electrochemical sensor using carbon nanotube microelectrodes for the detection of heavy metals in as-is tap water with parts per trillion (PPT) limits of detection (LOD). The microsensor is built on a single platform of six highly densified carbon nanotube fiber (HD-CNTf) (rod shape). This microsensor has an excellent limit of detection towards heavy metals Cu^{2+}, Pb^{2+} and Cd^{2+} of 6.0 nM (376 ppt), 0.45 nM (92 ppt), and 0.24 nM (27 ppt) in tap water and 0.32 nM (20 ppt), 0.26 nM (55 ppt), and 0.25 nM (28 ppt) in simulated drinking water, respectively.

Persistent luminescence nanoparticles also known as long-lasting phosphorescent nanoparticles, have emerged as a promising tool for the detection of heavy metals.

PLNPs can be engineered to selectively bind with heavy metal ions, and their luminescence properties can be used for sensitive and specific detection. Rare-earth-doped compounds or other phosphorescent materials are commonly used to make PLNPs. They are frequently functionalized with ligands or receptors that have high affinity for specific heavy metal ions to make them useful for heavy metal detection. These ligands are intended to bind preferentially to the target metal ions [84].

The detection of metal ions in real samples with no background interference was accomplished using a persistent luminescent nanobeacon. The nanobeacon is made up of metal-specific DNAzymes and persistent luminescence nanomaterials (Mn doped Zn_2GeO_4) that have nanorods shape (PLNRs). As a proof of concept, PLNRs were synthesized using hydrothermal techniques and subsequently surface functionalized using 17E DNAzyme by crosslinker Sulfo-SMCC for lead ion (Pb^{2+}) detection. Figure 11 shows how the background interference in the luminescent technique was reduced, allowing just the persistent fluorescence of nanorods to be detected. As a result of using persistent luminescence detection, the nanobeacon demonstrated improved signal-to-background ratios for sensitive and selective detection of metal ion in real samples.

A quenching group was added to the complementary strand of DNAzyme. In the absence of Pb^{2+}, the quenching group reduced the fluorescence of nanorods. After the Pb^{2+} was induced, the quenching group was released by cutting the substrate strand. The recovery of nanorods luminescence results from the dissociation of PLNRs from the quenching group (Fig. 12a and b). The fluorescence signal was disturbed by background fluorescence in real water samples, resulting in a low SBR (0.657), significantly lower than that in buffer (2.68) (Fig. 12c). When the background interference was removed in the luminescent method, only the persistent fluorescence of nanorods could be measured. Under the same conditions, the calculated SBR of persistent fluorescence is substantially higher; 6.62 for target ion detection in buffer and 6.95 for actual water sample detection (Fig. 12c). These findings demonstrate that

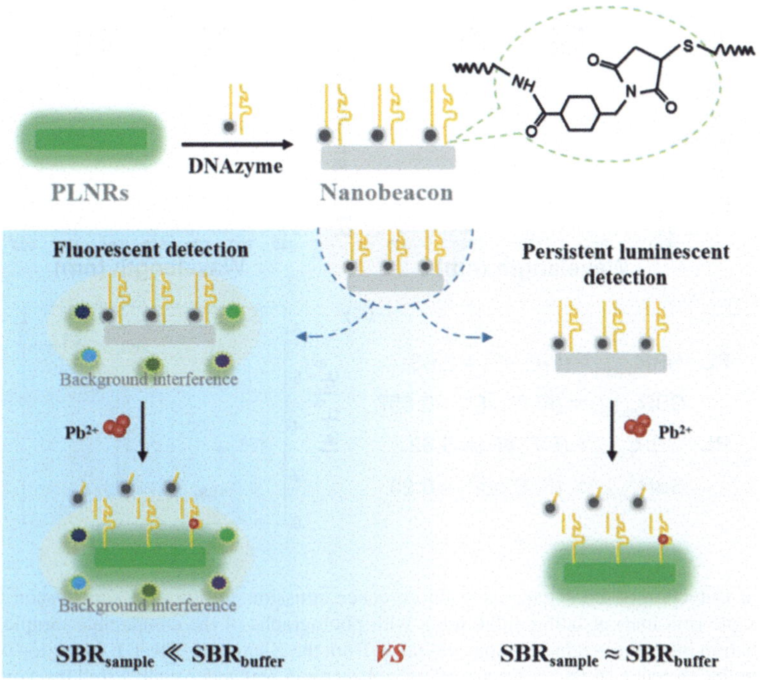

Fig. 11 Diagram of a persistent luminescent nanobeacon for precise metal ion measurement by eliminating background interference. Reprinted with permission from [84]. Copyright (2022) Biosensors and Bioelectronics

Pb^{2+} may be detected accurately utilizing PLNRs and DNAzyme. The new persistent luminescence method effectively eliminates autofluorescence background interference, resulting in greater signal-to-background ratios and more sensitive detection of metal ions in real water samples.

In another study, Li et al. [85] synthesized NIR fluorescent based probe (PLNPs-PEI/DTT-AuNPs) to detect arsenic (III) (As(III)). ZGGO (PLNPs) and gold nanoparticles (AuNPs) were created using a hydrothermal technique coupled with calcination and a seed-mediated growth mechanism. To increase the water solubility of PLNPs and supply positive surface charges, polyethyleneimine (PEI) was used. Furthermore, the chelating sulfur-containing ligand dithiothreitol (DTT) ensures the link to the surface of AuNPs via Au–S bonds, allowing each As (III) to bind. When compared to conventional instrumental analysis methods for As (III) detection (such as atomic spectrometric method and electrochemical detection method and method), the developed PLNPs-PEI/DTT-AuNPs promotes interesting detection properties such as high sensitivity (LOD = 55 nmoL.L^{-1}) and appropriate sensing linear range (0.067–13.35 μmoL.L^{-1}). Furthermore, these detection properties outperform fluorescence methods in regard to carbon dots and other carbon-based quantum and quantum dots [85–87]. Biosensors are valuable tools for the detection of pesticides

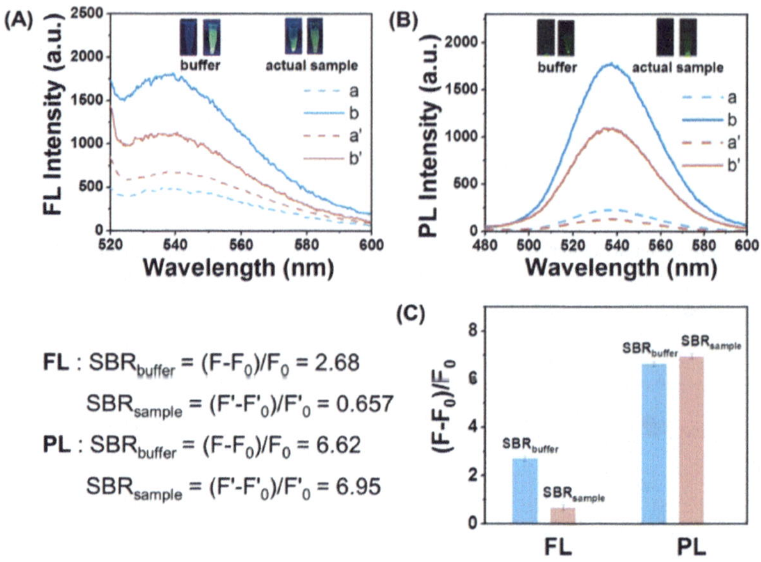

Fig. 12 **a** Detection of Pb^{2+} ions using fluorescence emission spectra of a nanobeacon in real water sample (red line) or buffer (blue line), with photographs of the comparable sample under UV irradiation inset; the original sample was taken from the Xiangjiang River. **b** Detection of Pb^{2+} ions using fluorescence emission spectra of a nanobeacon in real water sample (red line) or buffer (blue line), with photographs of the comparable sample in the dark inset. **c** The signal-to-background ratios (SBR) in buffer or actual sample for both fluorescence (FL) and phosphorescence (PL) modes, respectively. Reprinted with permission from [84]. Copyright (2022) Biosensors and Bioelectronics

in various environmental samples, including soil, water, and food. They rely on biological components such as enzymes, antibodies, or microorganisms to interact with the target pesticide and generate a measurable signal [88]. For pesticide analysis, traditional methods such as gas chromatography, high-performance liquid chromatography, and chromatography-mass spectrometry have been used [88]. In addition to these techniques, electrochemical biosensors are widely used for the detection of pesticides due to their high sensitivity, rapid response, and ease of miniaturization. They work by measuring changes in electrical properties, such as current or voltage, that result from the interaction between the pesticide and a biologically active component [88]. Tsounidi et al. [89] fabricated acetylcholinesterase (AChE)-based biosensor for the detection of pesticides in different types of vegetable oils. Pesticide synergies like carbofuran and fatty acids had no negative impact on sensor sensitivity or the linear dynamic range over which different pesticide doses can be detected. More research into the complex interactions that occur between different fatty acids and AChE is needed, however, to acquire a better understanding of the mechanisms of enzyme inhibition in the presence of multiple interferents. Caratelli et al. [90] developed a flower-like origami paper-based electrochemical device for precision agriculture that detects pesticides in aerosol phase at the same time. The scientific development of pesticides analysis can revolutionize the development of biosensors based on PLNPs, which is a promising approach due to the sensitivity and selectivity that PLNPs can offer. A suitable PLNP selection type with the requisite persistent luminescence proprieties is interestingly needed. PLNPs should also be functionalized with ligands or receptors that have a high affinity for the target pesticide compounds. These ligands could be antibodies, aptamers, or molecularly imprinted polymers (MIPs) that bind to the insecticide of interest preferentially.

Detecting organic compounds is also essential in various fields, including chemistry, environmental science, healthcare, and food safety. There are several techniques and methods available for the detection of organic compounds, each with its own advantages and limitations [91]. Surface enhanced Raman scattering (SERS) technology has been regarded as a feasible material strategy for organic pollutants sensing due to its unique qualities such as ultra-sensitivity that provides molecular level information on surface adsorbed molecules and the capacity to detect even a single molecule [92]. When compared to undoped TiO_2 nanofibers, the generated tantalum (5% Ta)-doped electrospun TiO_2 nanofibers have better photocatalytic activity and enhanced SERS signals of adsorbed methylene blue (MB) molecules. This was explained by increased visible light absorption, smaller crystalline size, and somewhat higher surface area [92]. Hurot et al. [93] developed sensitive olfactory biosensors that detect volatile organic compounds (VOCs) employing odorant binding proteins (OBPs) as sensing materials via surface plasmon resonance imaging. The obtained olfactory biosensors have very low LOD, strong selectivity, particularly at low VOC concentrations, and good repeatability, making them appropriate for VOC detection in solution. In literature, persistent luminescence nanoparticles for detection of organic compounds have not been reported yet. PLNP-based sensors can provide a powerful platform for the detection of organic compounds, offering a unique combination of features such as prolonged luminescence, high sensitivity, and the potential for real-time monitoring that make them valuable in various analytical and monitoring applications.

9 Detection and Biosensing Based on Time-Resolved Fluorescence Resonance Energy Transfer

Detection and biosensing based on time-resolved fluorescence resonance energy transfer (TR-FRET) is a powerful and versatile technique used to study molecular interactions, detect specific molecules, and measure binding events in various biological and chemical applications. TR-FRET combines the principles of fluorescence resonance energy transfer (FRET) with time-resolved fluorescence measurements to improve sensitivity, reduce background noise, and enhance the specificity of detection [34]. TR-FRET assays have been widely used in the detection and sensing of biomolecules by utilizing various types and structures of biosensors such as quantum dots [94–96] or lanthanides doped nanoparticles [97–99] and most importantly persistent luminescence nanoparticles [12]. The combination of PLNPs and TR-FRET holds great promise for advancing the field of biosensing, particularly in applications where long-term signal stability and high sensitivity are critical [12]. Li et al. [100] reported a dual-signaling nanoplatform consisting of PLNPs and gold nanoparticles (AuNPs) to detect L-cysteine (L-Cys) and insulin (Ins) via a radiometric absorption (Ratio-Abs) and TR-FRET. The AuNPs were prepared using the traditional sodium citrate synthesis process, whereas the PLNPs were synthesized using a hydrothermal approach and functionalized with amino groups. Figure 13 depicts the development of a dual-signaling nanoplatform based on Ratio-Abs and TR-FRET for the simultaneous detection of L-cysteine and insulin. AuNPs and PLNPs were used as signal elements due to their intriguing chemical and physical properties, such as AuNPs' large molar extinction coefficient and tunable optical properties, and PLNPs' long afterglow nature. When L-Cys is detected, a Ratio-Abs signal appears, which is explained by a substantial red shift in the absorption peak of gold nanoparticles with complementary sequence modifications (AuNP-CS) caused by L-Cys aggregation. Depending on the assay design, PLNPs can act as donors or acceptors in a TR-FRET biosensing system. In this study [100], PLNP-IBA (insulin binding aptamer) was utilized as the donor, PLNPs may be activated by an external light source and emit persistent luminescence, providing a long-lasting signal, while aggregated AuNP-CS was used as the acceptor. The TR-FRET between PLNP-IBA and aggregated AuNP-CS causes a quenching effect on the PL of PLNPs due to two occurrences: (i) the emission spectra of PLNPs overlaps nicely with the absorption spectrum of aggregated AuNPs, and (ii) IBA-CS ssDNA interaction that is complimentary. The presence of Ins inhibits TR-FRET and activates the persistent luminescence of the PLNPs due to the significant interaction between Ins and IBA. These findings explain the dual-signaling nanoplatform to detect selectively L-cysteine and insulin because of the intriguing optical properties of AuNPs and PLNPs.

Wang et al. [37] developed a molecular beacon (MB) probe based on persistent luminescence resonance energy transfer (LRET) for very sensitive detection of microRNA in biological samples. $ZnGa_2O_4:Cr^{3+}$ NIR-PLNPs were used as the energy donors and bioconjugated with poly-(acrylic acid) (PAA) and functionalized MB oligonucleotide (5′-NH_2 and 3′-BHQ3). LRET and PL quenching were generated by the presence of PLNPs and BHQ3. The complex probe (PLNPs-MB) was hybridized in the presence of miR21, directing BHQ3 away from PLNPs and restricting the LRET process. Consequently, the PLNPs' PL was restored, paving the way for miR21 measurement. The suggested probe detects miR21 with a linear response range of 0.1 to 10 nM, a lower quantification limit than previous nucleic acid MB probes, and efficient detection of miR21 in biological fluids such as human serum and cell extraction. The prolonged decay time, PLNP-free autofluorescence, and time-resolved fluorescence tool allow for the use of persistent luminescence biosensor-based aptasensors in order to identify carcinoembryonic antigen in pleural effusion [101]. This biosensor-based aptasensor has a low detection limit of 0.0851 pg mL^{-1} (for detection of carcinoembryonic antigen in both cancer and benign illness groups) and excellent selectivity from coexisting biomarkers (via hybridization between the aptamer and complementary DNA on the surface of PLNPs) [101].

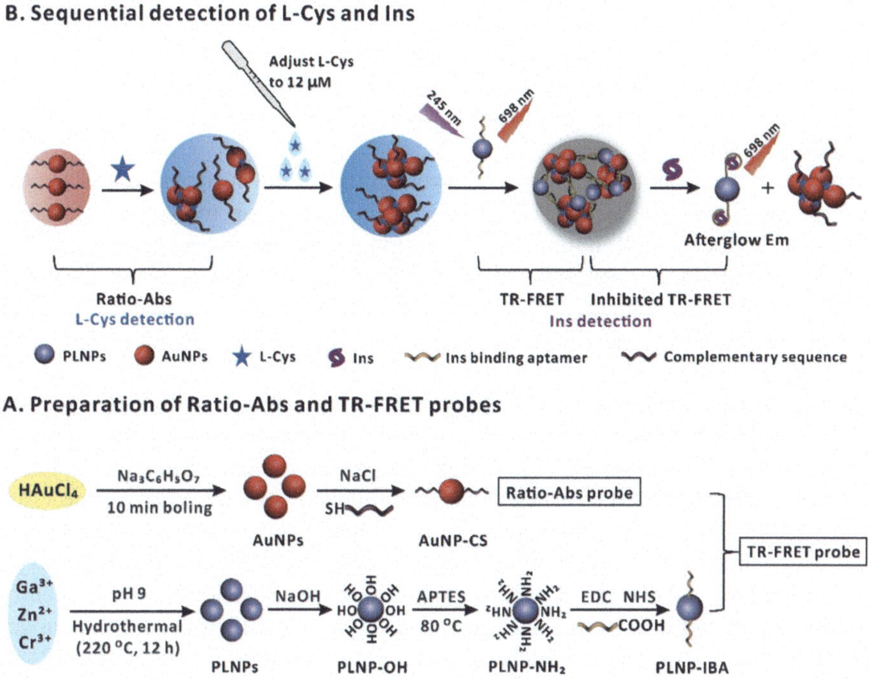

Fig. 13 Ratio-Abs and TR-FRET dual-signaling nanoplatform schematic diagram for sequential detection of L-Cys and Ins. IBA stands for Ins binding aptamer, and complementary sequence stands for IBA's partial complementary sequence. Reprinted with permission from [100]. Copyright (2018) Nanoscale

In another study, Zhang et al. [12] prepared three PLNPs-based nanoprobes via TR-FRET for biosensing. For detection of caspase-3, microRNA, and protein, these biosensors are bio-functionalized by covalently assembling dye-labeled peptides or DNA to carboxyl-functionalized PLNPs.

10 Detection and Biosensing Based on Lateral Flow Assays

Lateral flow assays (LFAs) are a popular and versatile tool for detection and biosensing due to their simplicity, speed, cost-effectiveness, and versatility [102]. They are commonly used for a wide range of applications, including medical diagnostics, environmental monitoring, drug testing, food safety testing, and more [103]. LFAs have demonstrated exceptional capabilities in the analysis, detection, and biosensing of diverse compounds such as proteins, nucleic acids, and tiny molecules [104].

LFAs, also known as lateral flow tests or lateral flow immunoassays, operate on the principle of capillary action. They are typically composed of a porous membrane, such as nitrocellulose or glass fiber, that is impregnated with various materials, such as a sample pad, conjugate pad, test line, and control line [105, 106].

LFAs using photoluminescent nanomaterials including quantum dots [107], upconversion nanoparticles [108], and carbon nanotube [109] have been widely reported.

However, these nanomaterials may have significant downsides, such as photobleaching and the expensive and complex cost of the device that emerges from fluorescence measurements with continuous stimulation. Furthermore, background autofluorescence can reduce sensitivity [110]. In light of all of these issues, researchers discovered that incorporating PLNPs into lateral flow assays had the potential to improve sensitivity, multiplexing capabilities, and stability in biosensing and detection applications.

Paterson et al. [110] used rare-earth doped strontium aluminate PLNPs as reports in lateral flow assays. Wet milling and differential sedimentation were used to separate the PLNPs from commercial materials with bigger particle sizes, and the particles were water-stabilized by silica encapsulation using a modified Stöber process. To insert PLNPs into LAFs, their surfaces were treated with aldehyde silane before reductive amination with heterobifunctional amine-poly (ethylene glycol)-carboxyl, hence using typical carbodiimide chemistry; proteins (NeutrAvidin) were covalently linked to the particles. Biotinylated hen egg lysozyme (bHEL) was used as a model analyte to demonstrate the potential application of strontium aluminate PLNPs as LFA reporters. The limit of detection for LFAs performed with PLNPs was thus determined to be somewhat below 0.1 ng/mL (7 pM), roughly an order of magnitude more sensitive than Streptavidin gold nanoparticles. Despite, the researchers proposed future work to improve the sensitivity of PLNPs LFAs, such as surface modification strategies such as PEG grafting protocol optimization, light detectors (photomultiplier tubes and avalanche photodiodes) rather than a digital camera for reading, and the synthesis of smaller, more monodisperse PLNPs.

Xu et al. [111] developed superbright red afterglow nanoparticles based on luminescent lanthanide complexes and organic molecular materials suitable for procalcitonin detection, with the whole-blood lateral flow assay demonstrating high sensitivity with a naked-eye readout limit of 0.5 ng mL^{-1}, allowing for earlier diagnosis (Fig. 14).

A simultaneous detection of two different analytes (prostate-specific antigen (PSA) and human chorionic gonadotropin (hCG)) in a single sample was reported [42]. The system is based on the integration of two PLNPs for green emission ($SrAl_2O_4$:Eu^{2+}, Dy^{3+}) and bleu emission (($(Sr_{0.625}Ba_{0.375})_2MgSi_2O7$:$Eu^{2+}$, Dy^{3+})) in a single LFA. This device provides various advantages, including high sensitivity, photostability, and great detection limits (0.1 ng $mL^{-}1$ of PSA and 1 ng mL^{-1} of hCG), which are comparable to commercially available LFAs.

Incorporating PLNPs into LFAs can improve the sensitivity, versatility, and usability of the diagnostic tests, making them effective instruments for a variety of applications such as medical diagnostics, environmental monitoring, and food safety testing.

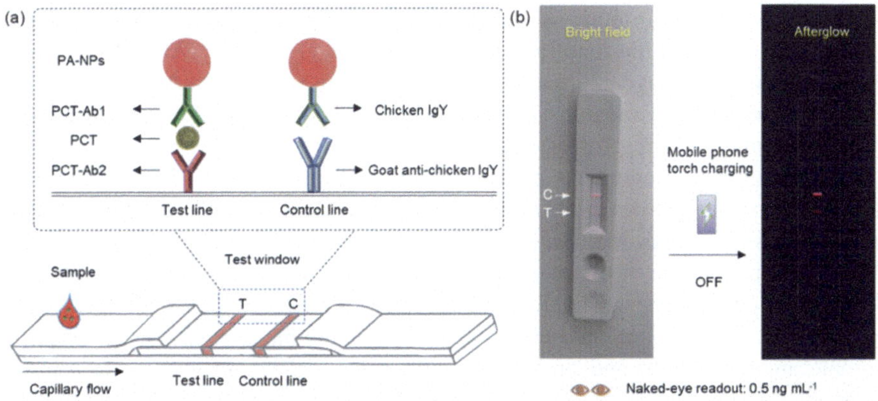

Fig. 14 Procalcitonin (PCT) whole-blood lateral flow assay using photochemical afterglow nanoparticles as dual-model indication. **a** Schematic depiction of the PCT detection lateral flow assay. **b** Photos of the PCT detection result as seen with the naked eye. Reprinted with permission from [111]. Copyright (2021) Science China Chemistry

11 Protein Detection

Nowadays, protein detection is crucial in various fields, including medicine, research, and diagnostics. They typically consist of a biological element (like antibodies or enzymes) and a transducer (converts a biological response into an electrical signal). This pair is capable of detecting certain proteins in a sample [112]. One common type is the enzyme-linked immunosorbent assay (ELISA) [113], where antibodies are used to capture and detect proteins. Another type is surface plasmon resonance (SPR) [114], which measures changes in refractive index as proteins bind to a sensor surface. Each protein detection method has its advantages and limitations. The choice of method depends on factors such as sensitivity, specificity, and the amount of sample available. Researchers often use a combination of techniques for a comprehensive analysis of proteins.

Because of their unique ability of generating light for an extended period of time after being exposed to a light source, PLNPs are an intriguing choice for protein detection.

Interleukin-6 (IL-6) is a pleiotropic cytokine, meaning it has multiple functions, and it plays a complex role in the immune response. While it has some anti-inflammatory effects in certain contexts, it is often associated with pro-inflammatory responses, particularly in the context of various diseases [115]. Thus, the detection of interleukin-6 (IL-6) is crucial for understanding its role in various physiological and pathological processes and for diagnosing and monitoring certain medical conditions. Recently, Olvera et al. [116] developed $Zn_2GeO_4:Mn^{2+}$ PLNPs using one-pot, microwave-assisted hydrothermal method for autofluorescence-free detection of interleukin-6 (IL-6), PAA was used as a synthesis additive. To detect IL-6, Zn_2GeO_4:0.50%Mn^{2+} PLNPs were used as a signal transducer element in autofluorescence-free immunoassay biosensing. To develop the IL-6 detecting kit, an antibody sandwich immunofluorescent assay (Ab_c)—(IL-6)—(Ab_d-PLNPs) was employed based on two capture and detection antibodies (Ab_c and Ab_d). The developed immunosensor is capable of detecting as low as 1 pg·mL–1 of IL-6 spiked in undiluted human serum samples (Fig. 15).

Carcinoembryonic antigen (CEA) and prostate-specific antigen (PSA) are important tumor markers associated with different types of cancers. CEA is related with several types of cancers, including lung, breast, gastrointestinal, and ovarian cancer. PSA is considered one of the most specific and efficient biomarkers for the clinical detection of prostate cancer.

Certainly, the simultaneous detection of multiple biomarkers, such as carcinoembryonic antigen (CEA) and prostate-specific antigen (PSA), can provide a more comprehensive and accurate assessment for the prediction of prostate cancer. Establishing a simple and ultrasensitive sensor for this purpose is a significant challenge but holds great potential for improving diagnostic accuracy [117].

Shi et al. [3] reported two luminescence PLNPs probes based on green emitted nanorodes (ZGO:Mn) and NIR-emitted NPs (ZGC) for simultaneous detection of prostate specific antigens (PSA) and carcinoembryonic antigens (CEA) with superior selectivity. The suggested afterglow biosensor has a low LOD of 8.9 fg mL^{-1} PSA and 72 fg mL^{-1} CEA. In another work [31], Zn_2GeO_4 co-doped with Mn^{2+}, Pr^{3+} (ZGOMP) nanorods PLNPs were synthesized by hydrothermal method for PSA detection where the autofluorescence-free biosensor ZGOMP was used as a probe and the gluconic acid as a quencher. To accomplish efficient signal amplification, the design included a hybridization chain reaction (HCR). With varying concentrations of PSA standard samples (0.001–10 ng/mL), the analytical performance of the autofluorescence-free biosensor based on ZGOMP and HCR method was investigated. As a result, the LOD of the produced biosensor was roughly 0.64 pg/mL,

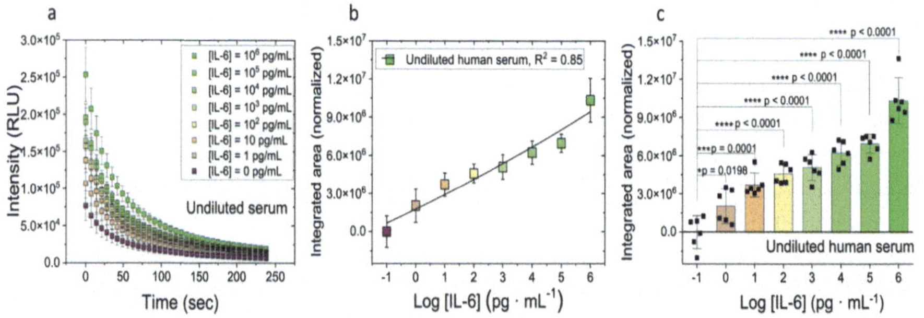

Fig. 15 Immunoassay (Ab_c)—(IL-6)—(Ab_d-PLNPs) sandwich in undiluted human serum ([Ab_c] = 10 μg·mL–1, [Ab_d] = 2.36 μg·mL–1, $n = 6$). **a** The decay curves of persistent luminescence. RLU = Relative luminescence units. **b** The normalized integrated area of obtained decay curves in Fig. 9a in relation to IL-6 concentration. **c** Unpaired and two-tailed t-test retrieved from Fig. 9b data analysis. Reprinted with permission from [116]. Copyright (2023) ACS Applied Materials and Interfaces

which is lower than the clinical diagnostic standard for prostate cancer. The findings of these investigations indicated a promising future application in PSA detection.

The use of PLNPs in protein detection is an exciting area of research, and ongoing developments aim to enhance their performance and broaden their applications.

12 Small Molecule Detection

Small molecule is a term commonly used in the fields of chemistry, biochemistry, and pharmacology to describe a compound that has a low molecular weight. Small-molecule detection is vital across multiple applications; in clinical, environmental, and agricultural contexts [118], as a result, developing sensitive and selective assays for the quick detection of small molecule is an intriguing task.

L-cysteine (L-Cys) and insulin (Ins) are small molecules that are related to diabetes and a series of diseases. L-cysteine is a crucial component in the synthesis of the antioxidant glutathione, which plays a protective role against oxidative stress. In fact, oxidative stress is often elevated in diabetes and related metabolic disorders. The balance between oxidative stress and antioxidant defenses is disrupted in diabetes, and increased levels of L-cysteine can contribute to this imbalance [119]. Insulin is a key hormone that regulates blood glucose levels by promoting the uptake of glucose into cells. In diabetes, either the pancreas does not produce enough insulin (as in type 1 diabetes) or the body's cells become resistant to the effects of insulin (as in type 2 diabetes). This leads to elevated blood glucose levels [120]. In this regard, the detection of L-Cys and Ins is of the interest. Li et al. [100] reported PLNPs based on ratiometric absorption (Ratio-Abs) and time resolved fluorescence resonance energy transfer (TRFRET) for high-throughput sequential detection of L-Cys and Ins. The Ratio-Abs and TR-FRET techniques are designed to mitigate interference from matrix effects and autofluorescence. The as-prepared nanoplatform offers high-throughput detection of L-Cys and Ins (with the LOD of 2.2 nM for L-Cys and 2.06 pM for Ins) due to the remarkable optical properties of AuNPs and PLNPs, such as long afterglow nature of PLNPs, high molar extinction coefficient, and tunable absorption of AuNPs. In addition to biological L-Cys and Ins small molecules, glutathione (GSH) plays crucial roles in various cellular processes and is often referred to as the body's "master antioxidant," it is a tripeptide composed of three amino acids: glutamate, cysteine, and glycine. Glutathione serves as a potent antioxidant by scavenging free radicals and reactive oxygen species (ROS) within cells. This helps protect cells from oxidative damage, which is implicated in aging and various diseases [121]. Feng et al. [122] reported NIR PLNPs based on ZGO@mSiO_2 core–shell structure for the detection of GSH. The authors claim that the addition of Cu^{2+} significantly suppressed the NIR afterglow signal, confirming the use of Cu^{2+} as a quencher for GSH detection. The recovery of NIR afterglow happened due to the prolonged distance quencher (Cu^{2+})-donor (PLNPs). The NIR bioprobes offered an enhanced autofluorescence-free GSH sensing with a low LOD of 0.35 μM.

The quantitative analysis of small molecule active substances is a crucial method for ensuring food safety. Small molecules, such as pesticides, food additives, preservatives, and contaminants, can have significant implications for human health when present in food beyond permissible levels [123].

Kanamycin (KAN) is an aminoglycoside antibiotic with notable antibacterial properties which is used in various fields including livestock husbandry, aquaculture and agriculture [124]. Wang and his co-workers [124] synthesized autofluorescence-free PLNPs aptasensor for the detection of kanamycin in food samples (milk, honey, and milk Powder). First, the aptamer-modified magnetic nanoparticles Fe_3O_4 (MNPs) are created for kanamycin to act as both the recognition and separation unit. Second, as the NIR long afterglow luminescence, cDNA-functionalized $ZnGa_2O_4$:Cr PLNPs were produced. The analytical performance of the synthezized aptasensor was carried out by varying kanamycin concentration of in the range of 1 pg mL^{-1} to 5 ng mL^{-1} where the LOD value is of 0.32 pg mL^{-1} (Fig. 16b). Furthermore, by simply changing the aptamer, the created PLNPs-nanoplatform provides a promising autofluorescence-free sensor for selective and sensitive detection of hazardous analytes in food samples.

Ochratoxin A (OTA) is a concerning mycotoxin that poses a significant health threat to both animals and humans, it is a naturally occurring mycotoxin produced by certain fungi, especially molds belonging to the genera Aspergillus and Penicillium. OTA can emit fluorescence under appropriate conditions. This property is often utilized in analytical methods for its detection [125]. Recently, Wang et al. [126] reported autofluorescence-free biosensors based on $Zn_2GeO_4:Mn^{2+}$, Eu^{3+} (ZGME) PLNPS using ELISA method for OTA determination.

The combination of the complex glucose oxidase (GOx)–OTA was prepared based on the excellent pH-responsive persistent luminescence. The researchers proposed some issues for the development of pH-responsive ZGME biosensors, including (i) which phosphor exhibits the best luminescence among ZGME, ZG, or ZGM,

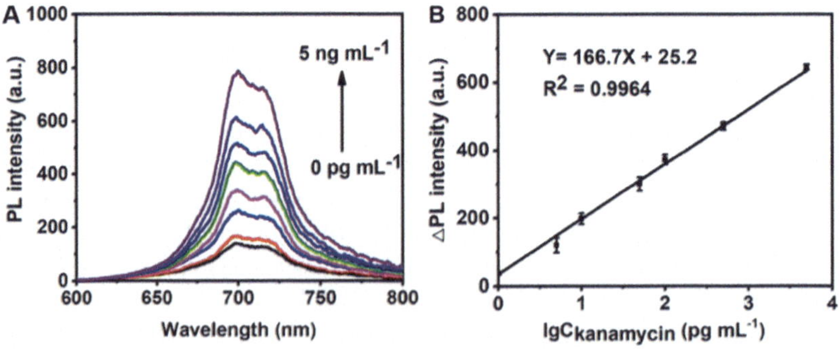

Fig. 16 **a** Variation of PL emission in solution after magnetic separation as a function of kanamycin concentration. **b** Representive ΔPL plot intensity in solution after magnetic separation versus the logarithm of kanamycin concentration in the range of 1 pg mL^{-1}–5 ng mL^{-1}. Reprinted with permission from [124]. Copyright (2021) ACS, Journal of Agricultural and Food Chemistry

(ii) whether the persistent luminescence signal of ZGME promotes pH-responses as a function of different pH solutions, and (iii) whether the competitive ELISA based on pH-responsive PLNPs provides efficient OTA detection. In response to these concerns, the OTA in different grain samples was determined by graphing the increase or reduction in persistent luminescence intensity of ZGME at 535 nm with optimization of experimental conditions. The acquired results demonstrated a successful ELISA approach for detecting OTA based on pH-responsive PLNPs with a reduced LOD of 0.045 g L^{-1}, high recovery (range from 81.3 to 94.4%), and a relative standard deviation (RSD) of less than 4.2%. This technique provides a sensitive, accurate, dependable, and autofluorescence-free biosensor for detecting OTA in various grain samples.

The ability to monitor small molecules in real time opens up new avenues for research, diagnosis, and personalized medicine. Interestingly, the use of persistent luminescence nanoparticles for in vivo small molecule detection holds great promise, offering several perspectives that could significantly impact biomedical research and healthcare.

13 Conclusion and Future Prospects

Persistent luminescence nanoparticles (PLNPs) have received increased attention due to their unique benefits; thus, we discussed recent breakthroughs in PLNPs for potential biosensing applications. The majority of the review focuses on the applications of PLNPs in the detection and biosensing of various biological components such as tumor biomarkers, bioactive molecules, pathogenic microorganisms, pollutants, and so on. We also discuss the persistent luminescence mechanism, which includes organic and inorganic materials and is based on multiple ways for detecting various analytes, such as resonance energy transfer RET process, Time-resolved fluorescence resonance energy transfer (TR-FRET), and Lateral flow assays (LFAs). The unique property of PLNPs is their ability to emit light for an extended period after being exposed to an external light source. This persistent luminescence makes them particularly useful in biosensing because they can be used as long-lasting probes for imaging and detection. In biosensing, PLNPs can be functionalized with specific biomolecules or ligands that selectively bind to target molecules, such as proteins, nucleic acids, or even whole cells. When the PLNPs bind to the target, their luminescence properties can change, allowing for the detection and quantification of the target analyte. The long-lasting luminescence of PLNPs is advantageous because it enables time-delayed detection, reducing background noise and improving the signal-to-noise ratio. This is especially beneficial in in vivo imaging and other applications where real-time monitoring is challenging. Additionally, PLNPs can be excited with non-UV light, which is less harmful to biological samples, making them suitable for sensitive biosensing in living organisms. Overall, the use of persistent luminescence nanoparticles in biosensing opens up exciting possibilities for improved detection sensitivity and imaging in various biological applications. While PLNPs hold great promise for detection and biosensing, there are still some challenges and areas for improvement in their application. Here are a few: (1) Signal intensity and sensitivity: One challenge is optimizing the signal intensity and sensitivity of PLNPs for detection purposes. Enhancing the brightness of the persistent luminescence and improving the detection limits will make them more effective for detecting low concentrations of analytes. (2) Biocompatibility: Ensuring the biocompatibility of PLNPs is crucial, especially for in vivo applications. The nanoparticles should not cause toxicity or adverse reactions in biological systems. Developing surface coatings or modifications that enhance biocompatibility is an ongoing area of research. (3) Targeting specificity: Achieving high specificity in targeting particular biomolecules or cells is essential. Improvements in the design and functionalization of PLNPs to enhance their binding affinity to specific targets will contribute to more accurate and reliable biosensing. (4) Stability: Persistent luminescence can be affected by environmental factors, such as temperature and pH. Ensuring the stability of PLNPs under various conditions, especially in complex biological environments, is a challenge that needs to be addressed for reliable and reproducible biosensing. (5) Synthesis and scalability: The synthesis of PLNPs with consistent properties and at a large scale is a challenge. Ensuring reproducibility in the production process is essential for the widespread adoption of PLNPs in biosensing applications. (6) Integration with Existing Technologies: Integration of PLNPs with existing imaging and detection technologies is crucial for practical applications. This involves developing compatible instrumentation and protocols for incorporating PLNPs into existing biosensing platforms. (7) Cost: The cost of production and synthesis of PLNPs can be a limiting factor for their extensive use. Finding cost-effective methods for large-scale production will make them more accessible for research and commercial applications. While challenges in the synthesis of PLNPs for biosensing, ongoing research and collaboration among scientists and material engineers are expected to continue improving the synthesis techniques, properties, and applications of PLNPs. These advancements hold great potential in the future for a wide range of fields, including medicine, environmental science, and beyond.

References

1. X. Sun, L. Song, N. Liu, J. Shi, Y. Zhang, Chromium-doped zinc gallate near-infrared persistent luminescence nanoparticles in autofluorescence-free biosensing and bioimaging: a review. ACS Appl. Nano Mater. **4**(7), 6497–6514 (2021). https://doi.org/10.1021/acsanm.1c01115
2. L. Liang, J. Chen, K. Shao, X. Qin, Z. Pan, X. Liu, Controlling persistent luminescence in nanocrystalline phosphors. Nat. Mater. **22**(3), 289–304 (2023). https://doi.org/10.1038/s41563-022-01468-y
3. L. Shi, J. Shao, X. Jing, W. Zheng, H. Liu, Y. Zhao, Autoluminescence-free dual tumor marker biosensing by persistent luminescence nanostructures. ACS Sustain. Chem. Eng. **8**(1), 686–694 (2020). https://doi.org/10.1021/acssuschemeng.9b06621
4. K. Rajfur, P. Głuchowski, Design of the persistent luminescence colour of a novel Gd3−xTbxGa3Al2O12phosphor: synthesis methods, spectroscopic properties and mechanism. Dalton Trans. **50**(14), 4830–4839 (2021). https://doi.org/10.1039/d0dt04343e

5. J. Nie, Y. Li, S. Liu, Q. Chen, Q. Xu, J. Qiu, Tunable long persistent luminescence in the second near-infrared window via crystal field control. Sci. Rep. **7**(1), 3–9 (2017). https://doi.org/10.1038/s41598-017-12591-1
6. P. Głuchowski, K. Rajfur, Impact of the synthesis method on the conventional and persistent luminescence in Gd3-xCexGa3Al2O12. Inorg. Chem. **60**(24), 18777–18788 (2021). https://doi.org/10.1021/acs.inorgchem.1c02239
7. X. Zhao, L.J. Chen, K.C. Zhao, Y.S. Liu, J. Liu, lin, & Yan, X. P., Autofluorescence-free chemo/biosensing in complex matrixes based on persistent luminescence nanoparticles. TrAC—Trends Anal. Chem. **118**, 65–72 (2019). https://doi.org/10.1016/j.trac.2019.05.025
8. J. Li, X. Huang, X. Zhao, L. Chen, X. Yan, PH-responsive torpedo-like persistent luminescence nanoparticles for autofluorescence-free biosensing and high-level information encryption. Angew. Chem. **133**(5), 2428–2435 (2021). https://doi.org/10.1002/ange.202011553
9. J. Liu, B. Viana, N. Mignet, D. Scherman, Y. Liu, C. Richard, H2O2-induced persistent luminescence signal enhancement applied to biosensing. Small **2303509**, 1–9 (2023). https://doi.org/10.1002/smll.202303509
10. N. Liu, X. Chen, X. Sun, X. Sun, J. Shi, Persistent luminescence nanoparticles for cancer theranostics application. J. Nanobiotechnol. **19**(1), 1–24 (2021). https://doi.org/10.1186/s12951-021-00862-z
11. K.Y. Zhang, Q. Yu, H. Wei, S. Liu, Q. Zhao, W. Huang, Long-lived emissive probes for time-resolved photoluminescence bioimaging and biosensing. Chem. Rev. **118**(4), 1770–1839 (2018). https://doi.org/10.1021/acs.chemrev.7b00425
12. L. Zhang, J. Lei, J. Liu, F. Ma, H. Ju, Persistent luminescence nanoprobe for biosensing and lifetime imaging of cell apoptosis via time-resolved fluorescence resonance energy transfer. Biomaterials **67**, 323–334 (2015). https://doi.org/10.1016/j.biomaterials.2015.07.037
13. K.H. Lee, D.M. Kim, In vitro use of cellular synthetic machinery for biosensing applications. Front. Pharmacol. **10**(October), 1–9 (2019). https://doi.org/10.3389/fphar.2019.01166
14. X. Huang, Y. Liu, B. Yung, Y. Xiong, X. Chen, Nanotechnology-enhanced no-wash biosensors for in vitro diagnostics of cancer. ACS Nano **11**(6), 5238–5292 (2017). https://doi.org/10.1021/acsnano.7b02618
15. S. Li, Z. Li, G.Y. Tan, Z. Xin, W. Wang, In vitro allosteric transcription factor-based biosensing. Trends Biotechnol. **41**(8), 1080–1095 (2023). https://doi.org/10.1016/j.tibtech.2023.03.001
16. Y. Orooji, H. Sohrabi, N. Hemmat, F. Oroojalian, B. Baradaran, A. Mokhtarzadeh et al., An overview on SARS-CoV-2 (COVID-19) and other human coronaviruses and their detection capability via amplification assay, chemical sensing, biosensing, immunosensing, and clinical assays. Nano-Micro Lett. **13**(1), 1–30 (2021). https://doi.org/10.1007/s40820-020-00533-y
17. A. Molazemhosseini, L. Magagnin, P. Vena, C.C. Liu, Single-use disposable electrochemical label-free immunosensor for detection of glycated hemoglobin (HbA1c) using differential pulse voltammetry (DPV). Sensors (Switzerland) **16**(7), 1–11 (2016). https://doi.org/10.3390/s16071024
18. Q. Wang, J. Wang, Y. Huang, Y. Du, Y. Zhang, Y. Cui, D.M. Kong, Development of the DNA-based biosensors for high performance in detection of molecular biomarkers: more rapid, sensitive, and universal. Biosensors Bioelectron. **197**(October 2021), 113739 (2022). https://doi.org/10.1016/j.bios.2021.113739
19. Y. Hua, J. Ma, D. Li, R. Wang, DNA-based biosensors for the biochemical analysis: a review. Biosensors **12**(3) (2022). https://doi.org/10.3390/bios12030183
20. S. Wu, Y. Li, W. Ding, L. Xu, Y. Ma, L. Zhang, Recent advances of persistent luminescence nanoparticles in bioapplications. Nano-Micro Lett. **12**(1), 1–26 (2020). https://doi.org/10.1007/s40820-020-0404-8
21. J. Shi, X. Sun, J. Zhu, J. Li, H. Zhang, One-step synthesis of amino-functionalized ultra-small near infrared-emitting persistent luminescent nanoparticles for: in vitro and in vivo bioimaging. Nanoscale **8**(18), 9798–9804 (2016). https://doi.org/10.1039/c6nr00590j
22. L.J. Chen, S.K. Sun, Y. Wang, C.X. Yang, S.Q. Wu, X.P. Yan, Activatable multifunctional persistent luminescence nanoparticle/copper sulfide nanoprobe for in vivo luminescence imaging-guided photothermal therapy. ACS Appl. Mater. Interfaces **8**(48), 32667–32674 (2016). https://doi.org/10.1021/acsami.6b10702

23. S. Bhuckory, J.C. Kays, A.M. Dennis, In vivo biosensing using resonance energy transfer. Biosensors **9**(2) (2019). https://doi.org/10.3390/bios9020076
24. Y. Zhu, K. Mandal, A.L. Hernandez, S. Kawakita, W. Huang, P. Bandaru et al., State of the art in integrated biosensors for organ-on-a-chip applications. Curr. Opin. Biomed. Eng. **19**, 100309 (2021). https://doi.org/10.1016/j.cobme.2021.100309
25. G. Rong, S.R. Corrie, H.A. Clark, In vivo biosensing: progress and perspectives. ACS Sensors **2**(3), 327–338 (2017). https://doi.org/10.1021/acssensors.6b00834
26. X. Xuan, C. Pérez-Ràfols, C. Chen, M. Cuartero, G.A. Crespo, Lactate biosensing for reliable on-body sweat analysis. ACS Sensors **6**(7), 2763–2771 (2021). https://doi.org/10.1021/acssensors.1c01009
27. Y. Wei, S. Jiang, J. Li, X. Li, K. Li, J. Li, Z. Fang, A biosensor material with robust mechanical properties, fatigue-resistance, biocompatibility, biodegradability, and anti-freezing capabilities. J. Mater. Chem. A **10**, 8491–8500 (2022). https://doi.org/10.1039/D1TA10998G
28. Y. Li, X. Liu, X. Xu, H. Xin, Y. Zhang, B. Li, Red-blood-cell waveguide as a living biosensor and micromotor. Adv. Func. Mater. **29**(50), 1–10 (2019). https://doi.org/10.1002/adfm.201905568
29. R. Zou, S. Shan, L. Huang, Z. Chen, T. Lawson, M. Lin et al., High-performance intraocular biosensors from chitosan-functionalized nitrogen-containing graphene for the detection of glucose. ACS Biomater. Sci. Eng. **6**(1), 673–679 (2020). https://doi.org/10.1021/acsbiomaterials.9b01149
30. I. Cicha, R. Priefer, P. Severino, E.B. Souto, S. Jain, Biosensor-integrated drug delivery systems as new materials for biomedical applications. Biomolecules **12**(9) (2022). https://doi.org/10.3390/biom12091198
31. Z. Yin, L. Zhu, Z. Lv, M. Li, D. Tang, Persistent luminescence nanorods-based autofluorescence-free biosensor for prostate-specific antigen detection. Talanta **233**(May), 122563 (2021). https://doi.org/10.1016/j.talanta.2021.122563
32. N. Li, W. Diao, Y. Han, W. Pan, T. Zhang, B. Tang, MnO_2-modified persistent luminescence nanoparticles for detection and imaging of glutathione in living cells and in vivo. Chem. Eur. J. **20**(50), 16488–16491 (2014). https://doi.org/10.1002/chem.201404625
33. H. Liu, F. Ren, X. Zhou, C. Ma, T. Wang, H. Zhang et al., Ultra-sensitive detection and inhibition of the metastasis of breast cancer cells to adjacent lymph nodes and distant organs by using long-persistent luminescence nanoparticles. Anal. Chem. **91**(23), 15064–15072 (2019). https://doi.org/10.1021/acs.analchem.9b03739
34. F. Weihs, A. Anderson, S. Trowell, K. Caron, Resonance energy transfer-based biosensors for point-of-need diagnosis—progress and perspectives. Sensors (Switzerland) **21**(2), 1–18 (2021). https://doi.org/10.3390/s21020660
35. J. Zhou, J. Chen, Y. Ge, Y. Shao, Two-dimensional nanomaterials for Förster resonance energy transfer-based sensing applications. Nanophotonics **9**(7), 1855–1875 (2020). https://doi.org/10.1515/nanoph-2020-0065
36. F. Feng, X. Chen, G. Li, S. Liang, Z. Hong, H.F. Wang, Afterglow resonance energy transfer inhibition for fibroblast activation protein-α assay. ACS Sensors **3**(9), 1846–1854 (2018). https://doi.org/10.1021/acssensors.8b00680
37. X. Wang, Y. Wang, S. Chen, P. Fu, Y. Lin, S. Ye, Y. Long, G. Gao, J. Zheng, A persistent luminescence resonance energy transfer-based molecular beacon probe for the highly sensitive detection of microRNA in biological samples. Biosens. Bioelectron. **198**, 113849 (2022). https://doi.org/10.1016/j.bios.2021.113849
38. Y.Y. Jiang, X. Zhao, L.J. Chen, C. Yang, X.B. Yin, X.P. Yan, Persistent luminescence nanorod based luminescence resonance energy transfer aptasensor for autofluorescence-free detection of mycotoxin. Talanta **218**(May), 121101 (2020). https://doi.org/10.1016/j.talanta.2020.121101
39. M.R. Hasan, M.S. Ahommed, M. Daizy, M.S. Bacchu, M.R. Ali, M.R. Al-Mamun et al., Recent development in electrochemical biosensors for cancer biomarkers detection. Biosens. Bioelectron.: X **8**(June), 100075 (2021). https://doi.org/10.1016/j.biosx.2021.100075

40. V.K. Sarhadi, G. Armengol, Molecular biomarkers in cancer. Biomolecules **12**(8), 1–39 (2022). https://doi.org/10.3390/biom12081021
41. A. Kaushal, N. Kaur, S. Sharma, A.K. Sharma, D. Kala, H. Prakash, S. Gupta, Current update on biomarkers for detection of cancer: comprehensive analysis. Vaccines **10**(12), 1–16 (2022). https://doi.org/10.3390/vaccines10122138
42. A.N. Danthanarayana, E. Finley, B. Vu, K. Kourentzi, R.C. Willson, J. Brgoch, A multicolor multiplex lateral flow assay for high-sensitivity analyte detection using persistent luminescent nanophosphors. Anal. Methods **12**(3), 272–280 (2020). https://doi.org/10.1039/c9ay02247c
43. L. Shi, W. Zheng, H. Miao, H. Liu, X. Jing, Y. Zhao, Ratiometric persistent luminescence aptasensors for carcinoembryonic antigen detection. Microchimica Acta **187**(11) (2020). https://doi.org/10.1007/s00604-020-04593-0
44. L. Liu, H. Wang, B. Xie, B. Zhang, Y. Lin, L. Gao, Detection of alpha-fetoprotein using aptamer-based sensors. Biosensors **12**(10) (2022). https://doi.org/10.3390/bios12100780
45. B.Y. Wu, H.F. Wang, J.T. Chen, X.P. Yan, Fluorescence resonance energy transfer inhibition assay for α-fetoprotein excreted during cancer cell growth using functionalized persistent luminescence nanoparticles. J. Am. Chem. Soc. **133**(4), 686–688 (2011). https://doi.org/10.1021/ja108788p
46. D. Su, X. Wang, W. Zhang, P. Li, B. Tang, Fluorescence imaging for visualizing the bioactive molecules of lipid peroxidation within biological systems. TrAC Trends Anal. Chem. **146**, 116484. https://doi.org/10.1016/j.trac.2021.116484
47. D. Giordano, Bioactive molecules from extreme environments ii. Mar. Drugs **19**(11), 1–7 (2021). https://doi.org/10.3390/md19110642
48. S.S. Chu, H.A. Nguyen, J. Zhang, S. Tabassum, H. Cao, Towards multiplexed and multimodal biosensor platforms in real-time monitoring of metabolic disorders. Sensors **22**(14) (2022). https://doi.org/10.3390/s22145200
49. Q. Liu, C. Wu, H. Cai, N. Hu, J. Zhou, P. Wang, Cell-based biosensors and their application in biomedicine. Chem. Rev. **114**(12), 6423–6461 (2014). https://doi.org/10.1021/cr2003129
50. S. Yang, W. Dai, M. Tang et al., Nonstoichiometric Nanocubes with a controllable morphology and persistent luminescence for autofluorescence-free biosensing. Appl. Mater. **5**(32), 38644–38652 (2023). https://doi.org/10.1021/acsami.3c05895
51. G. Rattu, N. Khansili, V.K. Maurya, P.M. Krishna, Lactate detection sensors for food, clinical and biological applications: a review. Environ. Chem. Lett. **19**(2), 1135–1152 (2021). https://doi.org/10.1007/s10311-020-01106-6
52. K. Rathee, V. Dhull, R. Dhull, S. Singh, Biosensors based on electrochemical lactate detection: a comprehensive review. Biochem. Biophys. Rep. **5**, 35–54 (2016). https://doi.org/10.1016/j.bbrep.2015.11.010
53. J. Madden, E. Vaughan, M. Thompson, A. O' Riordan, P. Galvin, D. Iacopino, S.R. Teixeira, Electrochemical sensor for enzymatic lactate detection based on laser-scribed graphitic carbon modified with platinum, chitosan and lactate oxidase. Talanta **246**, 123492 (2022). https://doi.org/10.1016/j.talanta.2022.123492
54. F. Poletti, B. Zanfrognini, L. Favaretto, V. Quintano, J. Sun, E. Treossi et al., Continuous capillary-flow sensing of glucose and lactate in sweat with an electrochemical sensor based on functionalized graphene oxide. Sens. Actuators, B Chem. **344**(June), 130253 (2021). https://doi.org/10.1016/j.snb.2021.130253
55. B.K. Ashley, M.S. Brown, Y. Park, S. Kuan, A. Koh, Skin-inspired, open mesh electrochemical sensors for lactate and oxygen monitoring. Biosens. Bioelectron. **132**(November 2018), 343–351(2019). https://doi.org/10.1016/j.bios.2019.02.041
56. C. Zhu, H. Xue, H. Zhao, T. Fei, S. Liu, Q. Chen, B. Gao, T. Zhang, A dual-functional polyaniline film-based flexible electrochemical sensor for the detection of pH and lactate in sweat of the human body. Talanta **242**, 123289 (2022). https://doi.org/10.1016/j.talanta.2022.123289
57. Y. Wei, Y. Jiao, D. An, D. Li, W. Li, Q. Wei, Review of dissolved oxygen detection technology: from laboratory analysis to online intelligent detection. Sensors (Switzerland), **19**(18) (2019). https://doi.org/10.3390/s19183995

58. Y. Amao, Probes and polymers for optical sensing of oxygen. Microchim. Acta **143**(1), 1–12 (2003). https://doi.org/10.1007/s00604-003-0037-x
59. R. Keruckiene, D. Volyniuk, K. Leitonas, J.V. Grazulevicius, Dual emission fluorescence/room-temperature phosphorescence of phenothiazine and benzotrifluoride derivatives and its application for optical sensing of oxygen. Sens. Actuators, B Chem. **321**(February), 128533 (2020). https://doi.org/10.1016/j.snb.2020.128533
60. C.Y. Zhu, Z. Wang, J.T. Mo, Y.N. Fan, M. Pan, A long persistent phosphorescent metal-organic framework for multi-level sensing of oxygen. J. Mater. Chem. C **8**(29), 9916–9922 (2020). https://doi.org/10.1039/d0tc02391d
61. S.M. Tanner, T.F. Berryhill, J.L. Ellenburg, T. Jilling, D.S. Cleveland, R.G. Lorenz, C.A. Martin, Pathogenesis of necrotizing enterocolitis: modeling the innate immune response. Am. J. Pathol. **185**(1), 4–16 (2015). https://doi.org/10.1016/j.ajpath.2014.08.028
62. R. Mendes, P. Garbeva, J.M. Raaijmakers, The rhizosphere microbiome: significance of plant beneficial, plant pathogenic, and human pathogenic microorganisms. FEMS Microbiol. Rev. **37**(5), 634–663 (2013). https://doi.org/10.1111/1574-6976.12028
63. J. Riu, B. Giussani, Electrochemical biosensors for the detection of pathogenic bacteria in food. TrAC—Trends Anal. Chem. **126**, 115863 (2020). https://doi.org/10.1016/j.trac.2020.115863
64. Y. Jo, J. Park, J.K. Park, Colorimetric detection of Escherichia coli O157:H7 with signal enhancement using size-based filtration on a finger-powered microfluidic device. Sensors (Switzerland) **20**(8) (2020). https://doi.org/10.3390/s20082267
65. M.S. Chang, J.H. Yoo, D.H. Woo, M.S. Chun, Efficient detection of Escherichia coli O157:H7 using a reusable microfluidic chip embedded with antimicrobial peptide-labeled beads. Analyst **140**(23), 7997–8006 (2015). https://doi.org/10.1039/c5an01307k
66. U. Dharmasiri, M.A. Witek, A.A. Adams, J.K. Osiri, M.L. Hupert, T.S. Bianchi et al., Enrichment and detection of escherichia coli O157:H7 from water samples using an antibody modified microfluidic chip. Anal. Chem. **82**(7), 2844–2849 (2010). https://doi.org/10.1021/ac100323k
67. M. Kim, T. Jung, Y. Kim, C. Lee, K. Woo, J.H. Seol, S. Yang, A microfluidic device for label-free detection of Escherichia coli in drinking water using positive dielectrophoretic focusing, capturing, and impedance measurement. Biosens. Bioelectron. **74**, 1011–1015 (2015). https://doi.org/10.1016/j.bios.2015.07.059
68. L.F. Alonzo, T.C. Hinkley, A. Miller, R. Calderon, S. Garing, J. Williford et al., A microfluidic device and instrument prototypes for the detection of Escherichia coli in water samples using a phage-based bioluminescence assay. Lab Chip **22**(11), 2155–2164 (2022). https://doi.org/10.1039/d1lc00888a
69. P.D.K.P. Ananda, A. Tillekaratne, C. Hettiarachchi, N. Lalichchandran, Sensitive detection of E. coli using bioconjugated fluorescent silica nanoparticles. Appl. Surf. Sci. Adv. **6**, 100159 (2021). https://doi.org/10.1016/j.apsadv.2021.100159
70. S. Tok, K. De Haan, D. Tseng, C.F. Usanmaz, H. Ceylan Koydemir, A. Ozcan, Early detection of: E. coli and total coliform using an automated, colorimetric and fluorometric fiber optics-based device. Lab on a Chip **19**(17), 2925–2935 (2019). https://doi.org/10.1039/c9lc00652d
71. M.B. Maas, G.H.C. Maybery, W.J. Perold, D.P. Neveling, L.M.T. Dicks, Borosilicate glass fiber-optic biosensor for the detection of Escherichia coli. Curr. Microbiol. **75**(2), 150–155 (2018). https://doi.org/10.1007/s00284-017-1359-y
72. Y. Shen, L. Xu, Y. Li, Biosensors for rapid detection of Salmonella in food: a review. Comprehensive Rev. Food Sci. Food Safe. **20**(1), 149–197 (2021). https://doi.org/10.1111/1541-4337.12662
73. G.J. Yang, J.L. Huang, W.J. Meng, M. Shen, X.A. Jiao, A reusable capacitive immunosensor for detection of Salmonella spp. based on grafted ethylene diamine and self-assembled gold nanoparticle monolayers. Analytica Chimica Acta **647** (2), 159–166 (2009). https://doi.org/10.1016/j.aca.2009.06.008
74. T. Bu, Q. Huang, L. Yan, L. Huang, M. Zhang, Q. Yang et al., Ultra technically-simple and sensitive detection for Salmonella Enteritidis by immunochromatographic assay based on gold growth. Food Control **84**, 536–543 (2018). https://doi.org/10.1016/j.foodcont.2017.08.036

75. A. Garrido-Maestu, S. Azinheiro, J. Carvalho, S. Abalde-Cela, E. Carbó-Argibay, L. Diéguez, et al., Combination of microfluidic loop-mediated isothermal amplification with gold nanoparticles for rapid detection of Salmonella spp. in food samples. Front. Microbiol. **8**(NOV), 1–8 (2017). https://doi.org/10.3389/fmicb.2017.02159
76. D.K. Soni, R. Ahmad, S.K. Dubey, Biosensor for the detection of Listeria monocytogenes: emerging trends. Crit. Rev. Microbiol. **44**(5), 590–608 (2018). https://doi.org/10.1080/1040841X.2018.1473331
77. O. Zolti, B. Suganthan, R. Maynard, H. Asadi, J. Locklin, R.P. Ramasamy, Electrochemical biosensor for rapid detection of Listeria monocytogenes. J. Electrochem. Soc. **196**, 1945–7111 (2022). https://doi.org/10.1149/1945-7111/ac7a63
78. F. Li, Q. Ye, M. Chen, B. Zhou, J. Zhang, R. Pang et al., An ultrasensitive CRISPR/Cas12a based electrochemical biosensor for Listeria monocytogenes detection. Biosens. Bioelectron. **179**(January), 113073 (2021). https://doi.org/10.1016/j.bios.2021.113073
79. Q. Chen, F. Huang, G. Cai, M. Wang, J. Lin, An optical biosensor using immunomagnetic separation, urease catalysis and pH indication for rapid and sensitive detection of Listeria monocytogenes. Sens. Actuators, B Chem. **258**, 447–453 (2018). https://doi.org/10.1016/j.snb.2017.11.087
80. H. Hossein-Nejad-Ariani, T. Kim, K. Kaur, Peptide-based biosensor utilizing fluorescent gold nanoclusters for detection of listeria monocytogenes. ACS Appl. Nano Mater. **1** (2018). https://doi.org/10.1021/acsanm.8b00600
81. B.B. Wang, X. Zhao, L.J. Chen, C. Yang, X.P. Yan, Functionalized persistent luminescence nanoparticle-based aptasensor for autofluorescence-free determination of kanamycin in food samples. Anal. Chem. **93**(4), 2589–2595 (2021). https://doi.org/10.1021/acs.analchem.0c04648
82. Z. Ma, C. Meliana, H.S.H. Munawaroh, C. Karaman, H. Karimi-Maleh, S.S. Low, P.L. Show, Recent advances in the analytical strategies of microbial biosensor for detection of pollutants. Chemosphere **306**, 135515 (2022). https://doi.org/10.1016/j.chemosphere.2022.135515
83. P. Gupta, C.E. Rahm, D. Jiang, V.K. Gupta, W.R. Heineman, G. Justin, N.T. Alvarez, Parts per trillion detection of heavy metals in as-is tap water using carbon nanotube microelectrodes. Anal. Chim. Acta **1155**, 338353 (2021). https://doi.org/10.1016/j.aca.2021.338353
84. Q. Luo, L. Qin, P. Zhang, B. Feng, X. Ye, T. Qing, Z. Qing, A persistent luminescent nanobeacon for practical detection of lead ions via avoiding background interference. Anal. Chim. Acta **1198**, 339555 (2022). https://doi.org/10.1016/j.aca.2022.339555
85. K. Ge, J. Liu, P. Wang, G. Fang, D. Zhang, S. Wang, Near-infrared-emitting persistent luminescent nanoparticles modified with gold nanorods as multifunctional probes for detection of arsenic(III). Microchim. Acta **186**(3), 1–9 (2019). https://doi.org/10.1007/s00604-019-3294-z
86. A. Gupta, N.C. Verma, S. Khan, C.K. Nandi, Carbon dots for naked eye colorimetric ultrasensitive arsenic and glutathione detection. Biosens. Bioelectron. **81**, 465–472 (2016). https://doi.org/10.1016/j.bios.2016.03.018
87. N. Butwong, T. Noipa, R. Burakham, S. Srijaranai, W. Ngeontae, Determination of arsenic based on quenching of CdS quantum dots fluorescence using the gas-diffusion flow injection method. Talanta **85**(2), 1063–1069 (2011). https://doi.org/10.1016/j.talanta.2011.05.023
88. T.O. Hara, B. Singh, Electrochemical biosensors for detection of pesticides and heavy metal toxicants in water: recent trends and progress. ACS Environ. Sci. Technol. Water **1**(3), 462–478 (2021). https://doi.org/10.1021/acsestwater.0c00125
89. D. Tsounidi, D. Soulis, F. Manoli, A. Klinakis, G. Tsekenis, AChE-based electrochemical biosensor for pesticide detection in vegetable oils: matrix effects and synergistic inhibition of the immobilized enzyme. Anal. Bioanal. Chem. **415**(4), 615–625 (2023). https://doi.org/10.1007/s00216-022-04448-y
90. V. Caratelli, G. Fegatelli, D. Moscone, F. Arduini, A paper-based electrochemical device for the detection of pesticides in aerosol phase inspired by nature: a flower-like origami biosensor for precision agriculture. Biosens. Bioelectron. **205**, 114119 (2022). https://doi.org/10.1016/j.bios.2022.114119

91. K. Aliqab, J. Wekalao, M. Alsharari, A. Armghan, D. Agravat, S.K. Patel, Designing a graphene metasurface organic material sensor for detection of organic compounds in wastewater. Biosensors **13**(8), 1–16 (2023). https://doi.org/10.3390/bios13080759
92. N. Singh, J. Prakash, M. Misra, A. Sharma, R.K. Gupta, Dual functional Ta-doped electrospun TiO_2 nanofibers with enhanced photocatalysis and SERS detection for organic compounds. ACS Appl. Mater. Interfaces **9**(34), 28495–28507 (2017). https://doi.org/10.1021/acsami.7b07571
93. C. Hurot, S. Brenet, A. Buhot, E. Barou, C. Belloir, L. Briand, Y. Hou, Highly sensitive olfactory biosensors for the detection of volatile organic compounds by surface plasmon resonance imaging. Biosens. Bioelectron. **123**(June 2018), 230–236 (2019). https://doi.org/10.1016/j.bios.2018.08.072
94. D. Zhu, W. Li, L. Ma, Y. Lei, Glutathione-functionalized Mn:ZnS/ZnO core/shell quantum dots as potential time-resolved FRET bioprobes. RSC Adv. **4**(18), 9372–9378 (2014). https://doi.org/10.1039/c3ra45491f
95. L. Mattera, S. Bhuckory, K.D. Wegner, X. Qiu, F. Agnese, C. Lincheneau et al., Compact quantum dot-antibody conjugates for FRET immunoassays with subnanomolar detection limits. Nanoscale **8**(21), 11275–11283 (2016). https://doi.org/10.1039/c6nr03261c
96. K.D. Wegner, P.T. Lanh, T. Jennings, E. Oh, V. Jain, S.M. Fairclough et al., Influence of luminescence quantum yield, surface coating, and functionalization of quantum dots on the sensitivity of time-resolved FRET bioassays. ACS Appl. Mater. Interfaces **5**(8), 2881–2892 (2013). https://doi.org/10.1021/am3030728
97. D. Tu, L. Liu, Q. Ju, Y. Liu, H. Zhu, R. Li, X. Chen, Time-resolved FRET biosensor based on amine-functionalized lanthanide-doped NaYF4 nanocrystals. Angewandte Chemie—International Edition **50**(28), 6306–6310 (2011). https://doi.org/10.1002/anie.201100303
98. Q. Ju, D. Tu, Y. Liu, R. Li, H. Zhu, J. Chen et al., Amine-functionalized lanthanide-doped KGdF 4 nanocrystals as potential optical/magnetic multimodal bioprobes. J. Am. Chem. Soc. **134**(2), 1323–1330 (2012). https://doi.org/10.1021/ja2102604
99. M. Wang, H. Ye, L. You, X. Chen, A supramolecular sensor array using lanthanide-doped nanoparticles for sensitive detection of glyphosate and proteins. ACS Appl. Mater. Interfaces **8**(1), 574–581 (2016). https://doi.org/10.1021/acsami.5b09607
100. J. Li, C. Yang, W.L. Wang, X.P. Yan, Functionalized gold and persistent luminescence nanoparticle-based ratiometric absorption and TR-FRET nanoplatform for high-throughput sequential detection of l-cysteine and insulin. Nanoscale **10**(31), 14931–14937 (2018). https://doi.org/10.1039/c8nr04414g
101. Z. Pan, D. Yang, J. Lin, K. Shao, S. Shi, Y.J. Teng, H. Liu, Y. She, Autofluorescence free detection of carcinoembryonic antigen in pleural effusion by persistent luminescence nanoparticle-based aptasensors. Anal. Chim. Acta **1194**, 339408 (2022). https://doi.org/10.1016/j.aca.2021.339408
102. X. Wang, N. Choi, Z. Cheng, J. Ko, L. Chen, J. Choo, Simultaneous detection of dual nucleic acids using a SERS-based lateral flow assay biosensor. Anal. Chem. **89**(2), 1163–1169 (2017). https://doi.org/10.1021/acs.analchem.6b03536
103. J. Hu, J.R. Choi, S. Wang, Y. Gong, S. Feng, B. Pingguan-Murphy et al., Multiple test zones for improved detection performance in lateral flow assays. Sens. Actuators, B Chem. **243**, 484–488 (2017). https://doi.org/10.1016/j.snb.2016.12.008
104. Z. Yang, C. Yi, S. Lv, Y. Sheng, W. Wen, X. Zhang, S. Wang, Development of a lateral flow strip biosensor based on copper oxide nanoparticles for rapid and sensitive detection of HPV16 DNA. Sens. Actuators, B: Chem. **285**(December 2018), 326–332 (2019). https://doi.org/10.1016/j.snb.2019.01.056
105. N. Ariffin, N.A. Yusof, J. Abdullah, S.F. Abd Rahman, N.H. Ahmad Raston, N. Kusnin, S. Suraiya, Lateral flow immunoassay for naked eye detection of mycobacterium tuberculosis. J. Sens. **2020** (2020).https://doi.org/10.1155/2020/1365983
106. Y. Xu, Y. Liu, Y. Wu, X. Xia, Y. Liao, Q. Li, Fluorescent probe-based lateral flow assay for multiplex nucleic acid detection. Anal. Chem. **86**(12), 5611–5614 (2014). https://doi.org/10.1021/ac5010458

107. J. Chen, Z. Huang, H. Meng, L. Zhang, D. Ji, J. Liu et al., A facile fluorescence lateral flow biosensor for glutathione detection based on quantum dots-MnO_2 nanocomposites. Sens. Actuators, B Chem. **260**, 770–777 (2018). https://doi.org/10.1016/j.snb.2018.01.101
108. Y. Gong, Y. Zheng, B. Jin, M. You, J. Wang, X.J. Li et al., A portable and universal upconversion nanoparticle-based lateral flow assay platform for point-of-care testing. Talanta **201**(March), 126–133 (2019). https://doi.org/10.1016/j.talanta.2019.03.105
109. W. Qiu, K. Baryeh, S. Takalkar, W. Chen, G. Liu, Carbon nanotube-based lateral flow immunoassay for ultrasensitive detection of proteins: application to the determination of IgG. Microchimica Acta **186**(7) (2019). https://doi.org/10.1007/s00604-019-3508-4
110. A.S. Paterson, B. Raja, G. Garvey, A. Kolhatkar, A.E.V. Hagstrom, K. Kourentzi et al., Persistent luminescence strontium aluminate nanoparticles as reporters in lateral flow assays. Anal. Chem. **86**(19), 9481–9488 (2014). https://doi.org/10.1021/ac5012624
111. M. Xu, J. Liu, X. Su, Q. Zhou, H. Yuan, Y. Wen et al., Lanthanide-containing persistent luminescence materials with superbright red afterglow and excellent solution processability. Sci. China Chem. **64**(12), 2125–2133 (2021). https://doi.org/10.1007/s11426-021-1099-x
112. J. Li, J.L. Wang, W.L. Zhang, Z. Tu, X.F. Cai, Y.W. Wang, C.-Y. Gan, H.J. Deng et al., Protein sensors combining both on-and-off model for antibody homogeneous assay. Biosens. Bioelectron. **209**, 114226 (2022). https://doi.org/10.1016/j.bios.2022.114226
113. S.K. Arya, P. Estrela, Electrochemical ELISA protein biosensing in. Sensors **20** (2020). https://doi.org/10.3390/s20102857
114. Q. Wu, W. Wu, F. Chen, P. Ren, Highly sensitive and selective surface plasmon resonance biosensor for the detection of SARS-CoV-2 spike S1 protein. Analyst **147**, 2809–2818 (2022). https://doi.org/10.1039/D2AN00426G
115. M. Majdinasab, M.L. Chapelle, J.L. Marty, Recent progresses in optical biosensors for interleukin 6 detection. Biosensors **13**(9), 898 (2023). https://doi.org/10.3390/bios13090898
116. R.M. Calderón-Olvera, E. Arroyo, A.M. Jankelow, R. Bashir, E. Valera, M. Ocaña, A.I. Becerro, Persistent luminescence Zn_2GeO_4:Mn^{2+} nanoparticles functionalized with polyacrylic acid: one-pot synthesis and biosensing applications. ACS Appl. Mater. Interfaces **15**(17), 20613–20624 (2023). https://doi.org/10.1021/acsami.2c21735
117. Y. Sun, J. Fan, L. Cui, W. Ke, F. Zheng, Y. Zhao, Fluorometric nanoprobes for simultaneous aptamer-based detection of carcinoembryonic antigen and prostate specific antigen. Microchim. Acta **186**(3), 4–13 (2019). https://doi.org/10.1007/s00604-019-3281-4
118. X. Wang, L. Cohen, J. Wang, D.R. Walt, Competitive immunoassays for the detection of small molecules using single molecule arrays. J. Am. Chem. Soc. **140**(51), 18132–18139 (2018). research-article. https://doi.org/10.1021/jacs.8b11185
119. D.H. Kwon, H.J. Cha, H. Lee, S.H. Hong, C. Park, S.H. Park, et al., Protective effect of glutathione against oxidative stress-induced cytotoxicity in RAW 264.7 macrophages through activating the nuclear factor erythroid 2-related factor-2/heme oxygenase-1 pathway. Antioxidants **8**(4) (2019). https://doi.org/10.3390/antiox8040082
120. M.M. Qaid, M.M. Abdelrahman, Role of insulin and other related hormones in energy metabolism—a review. Cogent Food Agric. **2**(1), 1–18 (2016). https://doi.org/10.1080/23311932.2016.1267691
121. Q. Ma, M. Wang, H. Cai, F. Li, S. Fu, Y. Liu, Y. Zhao, A sensitive and rapid detection of glutathione based on a fluorescence-enhanced "turn-on" strategy. J. Mater. Chem. B **9**, 3563–3572 (2021). https://doi.org/10.1039/d1tb00232e
122. Y. Feng, L. Zhang, R. Liu, Y. Lv, Modulating near-infrared persistent luminescence of core-shell nanoplatform for imaging of glutathione in tumor mouse model. Biosens. Bioelectron. **144**(August) (2019). https://doi.org/10.1016/j.bios.2019.111671
123. R. Alexander Reese, B. Xu, Single-molecule detection of proteins and toxins in food using atomic force microscopy. Trends Food Sci. Technol. **87**, 26–34 (2019). https://doi.org/10.1016/j.tifs.2019.03.031
124. Y. Tang, Y. Hu, P. Zhou, C. Wang, H. Tao, Y. Wu, Colorimetric detection of kanamycin residue in foods based on the aptamer-enhanced peroxidase-mimicking activity of layered WS_2 nanosheets. J. Agric. Food Chem. **2021**(69), 2884–2893 (2021). https://doi.org/10.1021/acs.jafc.1c00925

125. D.E. Armstrong-Price, P.S. Deore, R.A. Manderville, Intrinsic "turn-On" aptasensor detection of ochratoxin A using energy-transfer fluorescence. J. Agric. Food Chem. **68**(7), 2249–2255 (2020). https://doi.org/10.1021/acs.jafc.9b07391
126. Y. Wang, L. Yu, H. Zhang, R. Zhu, Z. Meng, Competitive ELISA based on pH-responsive persistent luminescence nanoparticles for autofluorescence-free biosensor determination of ochratoxin A in cereals. Anal. Bioanal. Chem. **415**(10), 1877–1887 (2023). https://doi.org/10.1007/s00216-023-04591-0

Engineering of Persistent Luminescence Nanoparticles for Bioimaging

Khouloud Hamraoui, Mounir Ferhi, and Karima Horchani-Naifer

Abstract The quest for biocompatible and efficient photoluminescent nanoparticles (PLNPs) for bioimaging and disease treatment has intensified, driven by their ultra-slow persistent emission and ability to suppress autofluorescence interferences. This chapter provides an updated assessment of the current achievements of various PLNPs in imaging and therapeutic technologies, focusing on their use in delivering compounds to targets and influencing cellular absorption in living systems. Various excitation sources for in vivo optical imaging are proposed, requiring appropriate excitation wavelengths for super-resolution bioimaging. Hybrid PLNPs, combining optical and physical features, have been developed to enhance bioimaging modalities, leading to intriguing multimodal imaging as a nanoplatform for disease detection. Additionally, PLNPs coupled with functional units have found significant usage as theranostic materials, enabling simultaneous treatment and diagnosis of multiple diseases, including chemotherapy, photothermal therapy, and photodynamic therapy. Despite progress in developing smart nanoprobes based on PLNPs, further efforts are needed to engineer efficient PLNPs. Selecting the right activator and host matrix and employing co-doping techniques are crucial for improving luminescence properties. Understanding the physics underlying persistent luminescence mechanisms requires future investigation. Additionally, ensuring biocompatibility is essential for in vivo applications, considering size, structure, surface modification, and formulation. Innovative development of PLNPs holds substantial potential for a wide range of bio applications, including diagnosing and treating antibacterial and viral infections, tissue engineering, and cardiovascular disorders.

Keywords Persistent photoluminescence · NIR-emitting PLNPs · UV light · Bioimaging · Sensitivity · Long-lasting emission · Background free · Drug delivery · Photodynamic therapy

---

K. Hamraoui · M. Ferhi · K. Horchani-Naifer (✉)
Laboratory of Physical Chemistry of Mineral Materials and Their Applications, National Research Center in Materials Science, Technopole Borj Cedria, Soliman, Tunisia
e-mail: karima.horchani@cnrsm.rnrt.tn

V. Kumar et al. (eds.), *Persistent Luminescence*, Progress in Optical Science and Photonics 30, https://doi.org/10.1007/978-981-97-4943-0_10

1 Introduction

Optics for bioimaging has sparked a lot of interest in nanomedicine and nanotechnology research because of its high sensitivity, selectivity, remote controllability, and reliable non-damage method compared to traditional techniques [1]. Diverse optical approaches have proven useful for imaging cells and tissues; yet, these techniques confront several hurdles, such as real-time imaging and noninvasive monitoring, both of which are key qualities for clinical treatments [2]. Previous research has shown that assessing autofluorescence and photodamage caused by light radiation are important topics in tissue histology and cell biology [3]. Because of the presence of many organic molecules in ultraviolet (UV), such as proteins, DNA, and collagen, scattering increases for shorter wavelengths. To overcome this constraint, the infrared IR region is actively sought after to fundamentally increase tissue penetration depth [4]. Intensive research has widely used fluorescent probes and nanoparticle NPs of many sorts in the field of active bioimaging. These fluorescent NPs have the potential to modify the optical properties by themselves. For example, upconversion nanoparticles (UCNPs), downshifting nanoparticles (DSNPs), polymer and quantum dots, or fluorescent NPs based on a labeling strategy, using a fluorophore as a label to be visualized and monitored by fluorescence, such as dye-doped nanoporous polymer matrix, liposomes and proteins NPs, and fluorescence mesoporous silica NPs (MSNs) [5]. Fluorescent NPs for bioimaging are categorized into three types based on their chemical composition: organic, inorganic, and organic/inorganic hybrid nanoparticles. Some fluorescent NPs are still limited by undesirable problems. Organic dyes and gold NPs have the same photothermal heating capacity, even though the majority of organic NIR dyes are generally overlooked due to photobleaching and poor stability [6]. Moreover, lifetime is a critical parameter in bioimaging. Fluorescent probes with short lifetimes typically have low fluorescence intensity, making it difficult to attenuate the probe signal owing to background noise interference, resulting in significantly lower detection sensitivity and imaging resolution [7]. Several major factors could considerably influence the effectiveness of cellular imaging utilizing fluorescent NPs from the standpoint of bioimaging proficiency. (i) excitation light; for example, it is widely known that at 980 nm excitation, nearly all biological samples can absorb the light and convert it into significant heating, causing tissue and cell damage [8]. Aside from this wavelength, the absorption spectra of water in live cells coincide with the 980 nm wavelength, significantly reducing diffusion into biological material and resulting in limited penetration depth [9]. (ii) Fluorescence imaging is heavily reliant on nanoparticle size and shape, particularly during biodistribution into biological materials [10], gene delivery [11], receptor-mediated cellular uptake [12], and blood circulation times [13]. Indeed, nanoparticles of smaller size (10 nm) have the benefit of being quickly eliminated by the renal excretion system, which is not the case for NPs larger than 100 nm. Above 100 nm, the likelihood of undesired buildup in organs such as lymph nodes, liver, spleen, and lungs is significant [10]. However, it is stated in the literature that small NPs are more harmful than their larger counterparts. The investigation of the size-dependent activity of nanosilver on mammalian cells and microorganisms reveals that increasing NP size increases colloidal stability, resulting in a strong barrier beside the articles impacting aggregation [14]. Furthermore, various researchers discovered that NPs with sizes ranging from 10 to 60 nm have been proposed to boost cellular uptake [10]. Gold NPs with 50 nm diameters are taken up in greater quantities by cancer cells (HeLa) than smaller and larger gold NPs; however, these latter require precise functionalization such as surface modification to provide similarly efficient delivery into cells [15]. (iii) Modifying the surface chemistry of NPs might directly alter their photoluminescence PL properties by preventing inexplicit interactions and introducing precise binding to their targets for usage in bioimaging applications [16]. Amjad et al. [17] investigated the effect of surface chemical composition on the PL properties of CG (bovine gelatin) and CA (PHM3 algae) nanodots. They proposed that PL emission could be caused by surface defect states of functional groups, and they detected a red shift in the PL spectrum of CA (PHM3 algae) nanodots, which could be associated to amino groups and a high degree of surface oxidation. Gutierrez et al. [18] used a surface chemical approach to transfer nanoparticles into aqueous solutions, yielding water-dispersible Ag_2S NPs with good PL properties for optical bioimaging. Due to their unique optical features, particularly the long-lasting luminescence after halting illumination, persistent luminescence nanoparticles (PLNPs) have recently dominated the field of fluorescent NPs [19]. Without continuous illumination, the emission background is reduced and depth penetration issues are reduced, allowing for highly sensitive fluorescent probes for in vivo imaging applications [20]. Furthermore, PLNPs can be induced by any wavelength; however, NIR activation is preferred for deeper penetration in biological tissues and high image quality [21]. In the bioimaging field, a variety of distinct PLNPs chemical compositions have been developed, including organic-polymer NPs [22–25], NPs doped with rare-earth or transition metal cations [26–32]. In comparison with other organic materials, inorganic NPs have important features such as high PL quantum yield [33], robust stability [34], large Stokes shift and tunable emissions [35].

Nonetheless, fluorescent labels should have certain advantages in optical imaging, such as strong water solubility [36], and clinical applications requiring low toxicity and regulated mechanical qualities [37]. To improve the efficacy of inorganic PLNP probes, various researchers have developed inorganic/organic hybrid structures [38]. The surface of NPs was frequently modified from hydrophobic to hydrophilic utilizing functional organic capping agents or polymer surfactants [39]. Jiang et al. [27], for example, treated $Zn_{1.3}Ga_{1.4}Sn_{0.3}O_4:Cr^{3+}$, Y^{3+} PLNPs with PEG (Polyethylene glycol) to minimize toxicity. To increase the stability, targeted imaging capabilities, and biocompatibility of $ZnGa_2O_4$: Cr PLNPs, Liu et al. [40] used N-hydroxysuccinimide (NHS) and 2,3-dimethylmaleic anhydride (DMMA) functionalized polyethylene glycol (PEG) (NHS-PEG-DMMA). Lv et al. [41] developed the first core–shell multifunctional nanoplatform using chromium-doped zinc gallogermanate PLNPs and zeolitic imidazolate framework-8 (Metal–organic frameworks (MOFs)) to improve biocompatibility and serve as a smart drug delivery vehicle. According to Roxana et al. [42], polyacrylic acid (PAA) connected to the surface of $Zn_2GeO_4:Mn^{2+}$ PLNPs via electrostatic and steric interactions has great colloidal stability, providing carboxylate groups that can act as anchor sites for the eventual biomolecules binding to the surface. Overall, developing great prospective PLNPs probes by chemical and physical stimulation remains a significant issue for future optical imaging. Imaging-guided phototherapies such as chemotherapy, photodynamic therapy (PDT), and photothermal therapy (PTT) have been developed in response to the requirement for clinical diagnosis and guiding treatment. The use of fluorescence imaging for precision tumor diagnosis has grown in popularity. Zou et al. [43] designed multifunctional hybrid mesoporous NIR–PLNPs (HMNPs) made of (Ga_2O_3:Cr^{3+}, Nd^{3+}), magnetic nanoparticles (Gd_2O_3), and radionuclides (^{68}Ga) loaded with Doxorubicin (DOX chemotherapy agent) and silicon phthalocyanine (Si-Pc photosensitizer) for chemotherapy and NIR-PL-sensitized PDT. They demonstrated that DOX/Si Pc loaded HMNPs provide excellent cancer cell killing ability and tumor inhibition effects. Chen et al. [44] designed intelligent, biocompatible nanoscale liposome-coated NIR-PLNPs (Lipo-PLNPs)-based theranostic nanoplatforms for imaging-guided precision chemotherapy. The investigation of multifunctional nanoplatforms $mSiO_2@Gd_3Ga_5O_{12}$:Cr^{3+}, Nd^{3+} revealed that they had a strong fluorescence and a high r_1 value (10.70 mM^{-1} s^{-1}), making them appropriate for T_1 MR Imaging and cancer therapy [45]. The combination of imaging-guided phototherapies and diagnostic tools, such as magnetic resonance imaging (MRI), ultrasound imaging, and positron emission tomography (PET), in a single nanoplatform has recently opened the door to an intelligent therapeutic strategy in the biosciences. The combined PLNP-HSA-Fe^{3+}-IR780 shows excellent features such as long-last-luminescence, high longitudinal relaxivity, and obvious photoacoustic contrast signals, making it suitable for persistent luminescence multimodal imaging, cancer magnetic resonance imaging, and photoacoustic imaging [46]. Lanthanide-doped nanoparticles have been successfully produced as a theranostic agent with multimodal imaging capacity, including high-contrast multiplexed imaging of viscera and multimodal NIR-II PL-magnetic resonance-positron emission tomography imaging of murine tumors [47]. Despite this, some future prospects and attempts toward the development of well-chemical-architecture agents for improved pH and reactive oxygen species sensitive to particularly target tumor locations are envisaged. This review focuses on the most recent advances in persistent luminescent nanoparticles (PLNPs) and the relationship between them and bioimaging. We also discuss the benefits and drawbacks of PLNPs charged by different light sources (UV, Visible, and NIR). We describe the classification of PLNPs based on their chemical structure, as well as their growing uses in optical imaging. This review also analyzes the limitations of current PLNPs as well as the inherent limitations of imaging techniques. We focus on multimodal bioimaging with hybrid PLNPs, which has received a lot of interest in the search for high-quality bioimaging solutions. Finally, the multifunctional theranostic nanoplatforms are thoroughly discussed to increase tumor-targeted imaging, which could aid clinical research and medical science precision.

2 Correlation Between PLNPs and Bioimaging

The long-lasting luminescence capability of persistent luminescence was generally referred to as a phosphorescence phenomenon, however, physically; the principle is not the same. In phosphorescence, the longer lifetime of the excited state results from the change in electron spin; thus, a triplet state can occur, expanding the emission of light in time [48]. The extended emission time concept in persistent luminescence is based on two factors: emission centers and trap centers [49]. The mechanism of the persistent luminescence process is as follows: Following an excitation source, liberated charge carriers such as electrons and holes are trapped in trap centers; this phase is responsible for the persistence of luminescence's extended decay period. Following that, the electrons may be subject to re-trapping via an electron tunneling mechanism, heating or stimulating electron emission, and recombination at the emission center [50, 51]. Having long-lasting luminescence because of their unique optical properties, nanoparticles show considerable potential for the bioimaging sector, particularly long-lasting luminescence (from minutes to hours) with no requirement for continuous stimulation, resulting in a longer decay time [52]. This property enables for autofluorescence-free measurements and reduces background noise in biological samples, boosting sensitivity for optical imaging measurements [53]. Furthermore, eliminating continual excitation light prevents photobleaching while maintaining high sensitivity and preventing injury to living tissues [54]. As a result, strong signal-to-noise ratios and deep tissue imaging have revealed the ability of PLNPs to make a significant effect in in-vivo bioimaging, implying a promising future for in vivo bioimaging [55]. Researchers have been attracted by the production of PLNPs with various chemical substances in recent years. Some of the most recent PLNP for bioimaging are discussed in Table 1.

3 Bioimaging with UV Light-Charged PLNPs

Excitation light is a vital component of the bioimaging discipline. PLNPs can emit light with a wide range of excitation wavelengths, from UV to visible to IR and NIR. Ex vivo UV light-charged PLNPs represent the first generation of in vivo optical imaging as a high-energy excitation source for efficient charge carrier production [60]. However, UV excitation light exposure to cells might be detrimental to biological tissues, restricting the use of injected PLNPs in vivo. For the first time, researchers used PLNPs in bioimaging by charging them before injecting them into biological samples [60].

Trivalent chromium (Cr^{3+}) doped zinc gallogermanate (ZGGO) and zinc gallate (ZGO) compounds have received a lot of attention as persistent luminescence nanophosphors in PLNPs charged with UV light. ZGO and ZGGO serve as hosts in these structures; Cr^{3+} ions act as luminescent centers with broadband NIR emission

Table 1 Recent PLNPs for bioimaging application

PLNPs	Synthesis method	Functionalization of PLNPs	Refs.	Bioimaging applications
$Y_3Al_2Ga_3O_{12}:Ce^{3+}, Cr^{3+}, Nd^{3+}$ Nanospheres (184 nm)	Solvothermal method	$Y_3Al_2Ga_3O_{12}:Ce^{3+}, Cr^{3+}, Nd^{3+}$@polyacrylic acid Nanospheres (163 nm)	Arroyo et al. [54]	Bioimaging probes
$LaGaO_3:Sb^{3+}, Cr^{3+}$ (180 nm)	Hydrothermal method	–	Liu et al. [56]	In vivo bioimaging
$ZnGa_2O_4$:Cr Square particles (17.nm)	Surfactant-assisted hydrothermal method	$ZnGa_2O_4$:Cr@polydopamine@ 2,3-dimethylmaleic anhydride Square particles (18 nm)	Liu et al. [40]	In vivo bioimaging-chemo-photothermal synergistic therapy
Triple-doped zinc gallogermanate(ZGGO) nanostructures (53 nm)	Solvent thermal method in combination with calcination in air	ZGGO@mSiO_2@Red blood cell (RBC) Nanospheres (136 nm)	Liu et al. [57]	In vivo bioimaging- in situ real-time monitoring-drug delivery
$Zn_{1.3}Ga_{1.4}Sn_{0.3}O_4:Cr^{3+}, Y^{3+}$ (ZGSCY) Nanostructures (18 nm)	Combustion	ZGSCY@PEG	Jiang et al. [27]	In vivo Bioimaging -tumor-targeted imaging
$ZnGaGeO_4$:Cr (ZGGC) (24 nm)	Hydrothermal method	ZGGC@APTES@CTAB Nanospheres (10 nm)	Fu et al. [58]	In vivo bioimaging
ZGGC Cube (5.7 nm)	Hydrothermal method	ZGGC@ZIF-8	Zhao et al. [59]	In vitro and in vivo tumor imaging- drug delivery

(650–950 nm) via its ^2E (^2G) à $^4A^2$ (^4F) d-d transition, which overlaps with the biologically transparent window; and Ge^{4+} ions act as traps (defect centers) responsible for long afterglow time in the PLNPs [61].

A powerful and broad NIR photoluminescence ranging from 650 to 900 nm under UV irradiation (260 nm) is one of the well-studied benefits of (Cr^{3+}) doped ZGGO/ZGO for bioimaging applications. Furthermore, they produce an ultra-long afterglow persistent light with a considerable lifetime of roughly 90 h for ZGGO and 150 h for ZGO PLNPs, which is owing to the presence of Ge^{4+} ions in the lattices [61].

The presence of traps with modest trap depth fluctuation was discovered to be dependent on the excitation energy. Under lower wavelength excitation (680 nm), for example, the persistent luminescence process is probably occurring in certain areas around Cr^{3+} ions for higher energy due to the $^4A_2 \rightarrow {}^4T_2$ Cr^{3+} transition. The localization of both excited ($^4T_2$) and emission (^2E) states inside the band gap explains this phenomenon. During the higher energy excitation, there is no charge propagation via conduction and (or) valence bands, which may separate the charge over longer distances and/or at longer distances from Cr^{3+} centers [30]. It should be noted that in the case of nanoparticles, high-energy stimulation might induce local defects on the surface of the NPs; this issue should be addressed in bioimaging applications. Wang et al. [62] created useful ZGGO: Cr^{3+}, Eu^{3+} powder PLNPs for bioimaging. The experiment utilizing a Kodak In-Vivo Imaging System FX Pro (Fig. 1) demonstrates that NIR afterglow decays at different intervals even after deleting UV irradiation with an extended afterglow that can persist for more than 120 min displaying continuous high NIR emission intensity.

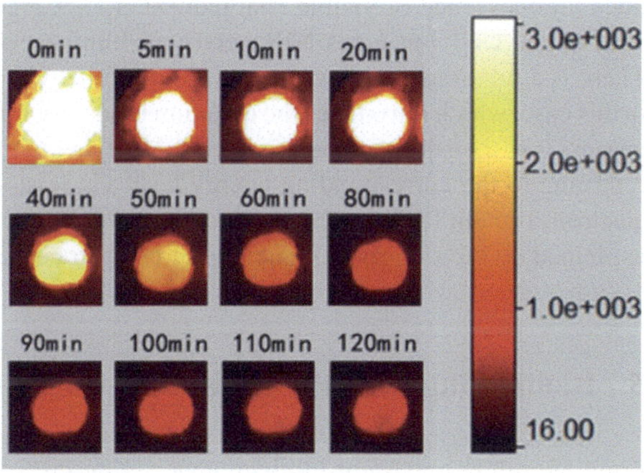

Fig. 1 NIR afterglow decay images of ZGGO: Cr^{3+}, Eu^{3+} powders captured by a Kodak in-vivo imaging system FX pro at various intervals after UV irradiation was stopped

4 Bioimaging with Visible Light-Charged PLNPs

The advantage of persistent emission is that it may be controlled to excite a potential acceptor; similar tunability is also demonstrated with visible light-charged PLNPs in this paper. This approach is simple to utilize, and it is unique in that it allows for in-vivo re-excitation imaging of PLNPs without causing extra tissue damage, which is not the case with UV-charged PLNPs [28]. With these features, visible light-charged PLNPs have a high potential for in vivo tracing and bioimaging applications in the long run. This charging process, however, produces substantially lower persistent luminescence efficiency (luminescence intensity and lifetime) than UV. As a result, developing PLNPs charged with low-energy light that achieves the magnitude produced by high-energy UV light remains a barrier to practical applications in the field of bioimaging [63].

By charging PLNPS with visible light, the low energy confines the delocalization state of luminescent ions to fill the electron ions, traps operate as important parameters in the persistent luminescence mechanism [63]. To address this limitation, researchers doped PLNPs based on visible light absorption by metal ions to control the traps. Pan et al. [64] developed effective NIR persistent phosphors (Cr^{3+} doped zinc gallogermanates) that are visible light active and have a two-week afterglow luminescence.

Due to its intense NIR persistent luminescence upon UV and visible excitation, $ZnGa_2O_4:Cr^{3+}$ charged by visible light has recently gained a lot of interest. However, the mechanism of visible active PNPLs differs slightly from that of UV-charged PNPLs. The presence of antisite defects Cr^{3+} around the ion causes continuous luminescence of visible charged $ZnGa_2O_4:Cr^{3+}$ [28]. The transition $^4A_2$ $^4T2$ (t^2e) of the Cr^{3+} ion gives NIR persistent luminescence under visible irradiation. There is a relationship between persistent luminescence intensity (I^{-1}) and time (t) in this case, which introduces the quantum tunneling effect during the operation. This process is responsible for electrons in $^4T_2$ (t^2e) being filled in traps and subsequently returning to the corresponding energy level via reverse tunneling recombination if electrons remain far from the conduction band. The extended persistence of light is explained by the sluggish kinetic rate of both processes [52].

5 Bioimaging with Photostimulated PLNPs

Long afterglow persistent luminescence, as described above, is greatly dependent on the number and type of traps. There were two kinds of traps: shallow traps and deep traps. Researchers discovered that shallow traps with bandgaps ranging from 0.5 to 1 eV are more effective than deeper traps ($E > 1$ eV) [65]. Deep traps must be activated by visible or NIR light stimulations to cause persistent luminescence; otherwise, the photostimulation induced by the PLNPs can repeat the luminescence [60].

Chuang et al. [66] employed the light stimulation technique to produce an extremely long-lasting NIR persistent luminescence. They experimented with LGO: Cr nanoparticles for multiplexed imaging since these NPs glow in the near-infrared range and might function in conjunction with visible bioluminescence probes. The injection of PEI-LGO: Cr into the back (Fig. 2) and into the brain (Fig. 3) of a nude mouse was used to test cell tracking. Figure 3's Row A shows that the intensity of NIR signals declined throughout time, from 10 s to 4 h. To regenerate the NIR signals, the mouse was exposed to a white LED spotlight for 15 s each time (row B, Fig. 2). The LED lamp was then presented to the mouse an extra 8 times per day from 72 to 240 h. The irradiation of PEI-LGO is depicted in Fig. 2: Cr NPs were exposed to a 254 nm UV lamp for 15 min (before photo-stimulated persistent luminescence), then the mouse head was exposed to a white LED bulb for 15 s at 6 h (**A**), 72 h (**B**), and 120 h (**C**), and the image was captured with a 1-min delay on an IVIS imaging system. Figures 2 and 3 show that NIR signals from UV-irradiated LGO:Cr nanoparticles may be regenerated and modulated in vivo by a white LED flashlight with long-afterglow luminescence (from several hours to more than 10 days).

Using the combination of visible light irradiation and an orange/red LED source, researchers increased the persistent luminescence images of tumor-bearing mice (Fig. 4). The images clearly show the tumors 4 h after PEGylated $ZnGa_2O_4$:Cr^{3+} injection [67].

The red/NIR light could potentially be used to release traps and regenerate persistent luminescence for bioimaging [68–70].

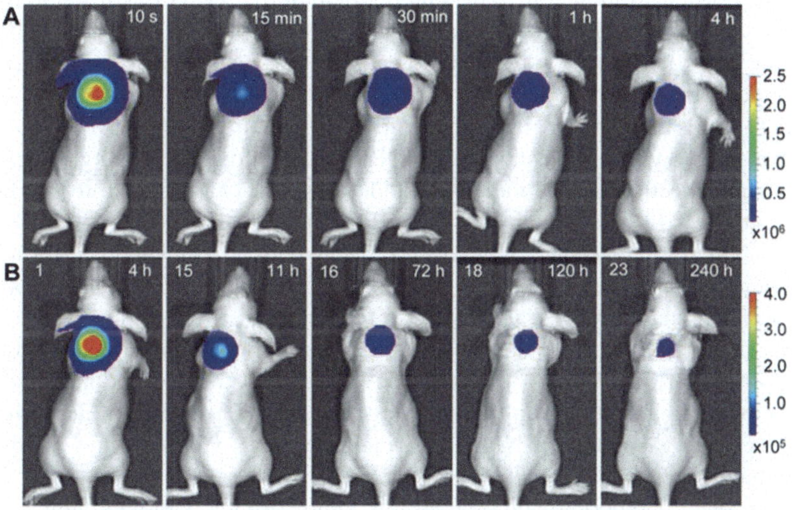

Fig. 2 Acquisition of NIR persistent luminescence and NIR PSPL signals from PEI-LGO:Cr NPs-labeled 4T1 cells subcutaneously implanted onto the back of a nude mouse before and after white LED flashlight irradiation. Reprinted with permission from [66]. Copyright (2014), ©Ivyspring International Publisher

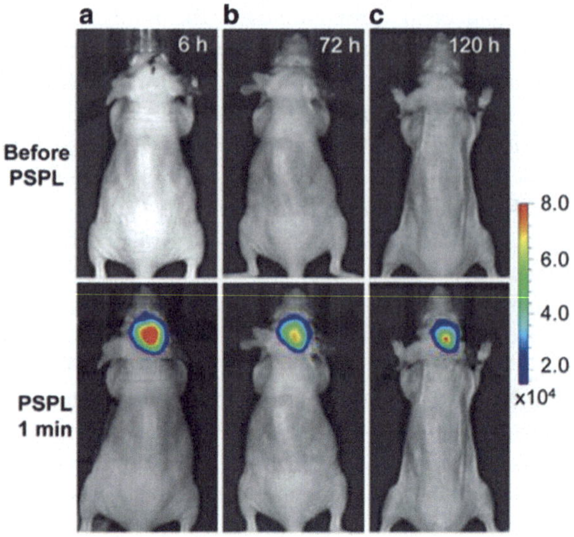

Fig. 3 NIR PSPL signals were obtained from PEI-LGO:Cr nanoparticles injected into the brain of a naked mouse before and after photo-stimulated persistent luminescence (PSPL). Reprinted with permission from [66]. Copyright (2014), ©Ivyspring International Publisher

6 Bioimaging with NIR-Emitting PLNPs

According to the literature, in vivo tracking is still hampered by its poor penetration in biological tissue, which is connected to biological tissue absorption and scattering qualities. Currently, researchers used NIR-emitting PLNPs in different tissue transparency windows for imaging techniques: NIR-I, 650–950 nm; NIR-II, 1000–1350 nm; and NIR-III, 1500–1800 nm. Because of its excellent signal-to-noise ratio (SNR), deep tissue penetration, and lack of the need for in situ activation, NIR-emitting PLNPs have shown considerable promise in bioimaging [60]. These advantages pave the way for the development of NIR-PLPs for optical imaging, which has promising applications in the biomedical field. Kang et al. [71] successfully produced $SrAl_{12}O_{19}$: Fe^{3+} PLNPs that emit in the NIR-I for more than 90 min. The surface of the PLNPs was then coated with APTES, which demonstrated good performance in ambient aqueous stability, making them viable candidates for bioanalysis and monitoring cellular networks.

Jiang et al. [27] recently increased the persistent luminescence in NIR-I by co-doping $Zn_{1.3}Ga_{1.4}Sn_{0.3}O_4$:$Cr^{3+}$ ions with Y^{3+} ions (ZGSCY) (Fig. 5); this improvement is owing to an increase in trap concentration in the host matrix. Furthermore, for in vivo imaging, ZGSCY NPs@PEG increased SNR approximately 1.8 times higher than ZGSC.

Lin et al. [72] reported $MgGe_{0.8}Ga_{0.2}O_3$: Yb^{3+} (MGGO) PLNPs with NIR-II persistent luminescence for decreased scattering and deep tissue penetration depth in biological tissues. They investigated the optical penetration depth of PLNPs prior to in vivo applications (Fig. 6). Figure 6a depicts images of Yb^{3+} persistent luminescence obtained by covering the sample with various thicknesses of tissue (0–10 mm).

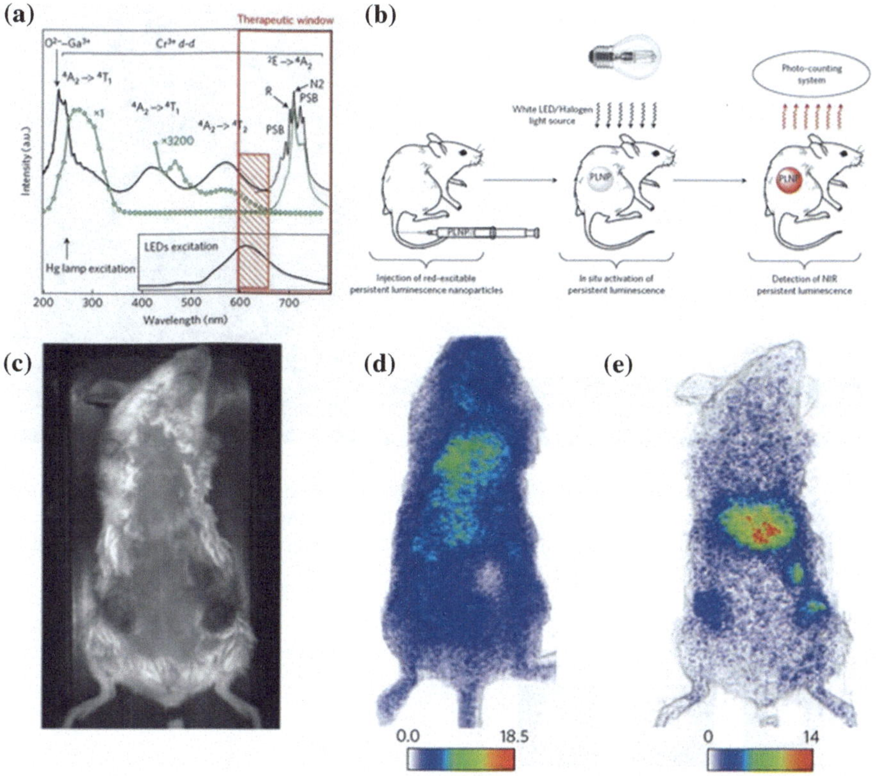

Fig. 4 **a** UV–Visible excitation (black solid line) and NIR emission (black dotted line) spectra of $ZnGa_2O_4$:Cr^{3+}. **b** Schematic illustration of in vivo imaging following PLNP activation in situ. **c** Optical representation of a tumor-bearing mouse. **d** Image of a tumor-bearing mouse with persistent luminescence 2 h after injection of ZGO-PEG NPs. **e** Image of a tumor-bearing mouse with persistent luminescence immediately after LED illumination, 4 h after injection of ZGO-PEG NPs. Reprinted with permission from [67]. Copyright (2014), Nature Materials

From Fig. 6b, they demonstrated that the optical penetration length of persistent luminescence of Yb^{3+} ions is about 3.9 mm through tissue.

Xu et al. [73] presented a novel garnet persistent phosphor of $Y_3Al_2Ga_3O_{12}$ co-doped with Er^{3+} and Cr^{3+} (YAGG:Er-Cr) exhibiting high Cr^{3+} persistent luminescence in the first biological window (690 nm) and Er^{3+} persistent luminescence in the third biological window (1532 nm). The optical imaging capability of both Cr^{3+} and Er^{3+} persistent luminescence in PNPLs was validated by capping the sample with raw-pork tissues 1 cm thick. Figure 7 shows those 30 min after stopping excitation, both YAGG: Cr and YAGG: Er-Cr samples create intense optical signals of Cr^{3+} persistent luminescence through pork tissues, which the Si camera clearly captures. Both samples have a high SNR at 30 min, however the substantial light scattering in the NIR-I window confuses signal localization in pork tissues from each sample. To

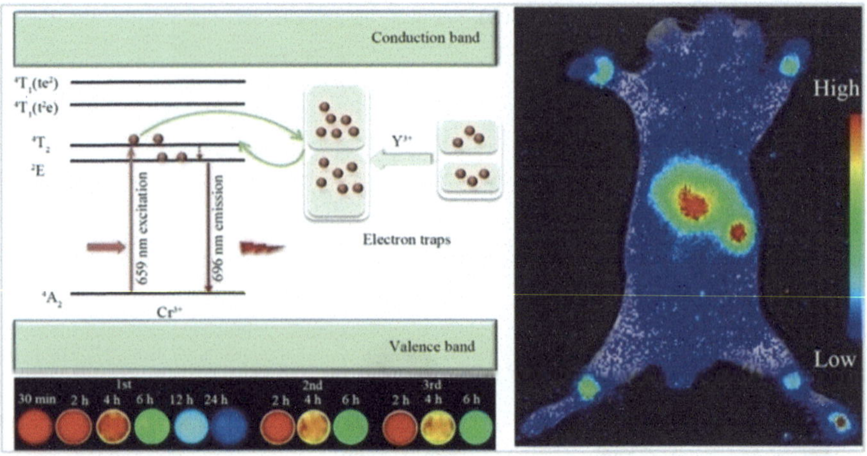

Fig. 5 $Zn_{1.3}Ga_{1.4}Sn_{0.3}O_4:Cr^{3+}$, Y^{3+} NIR PLNPs exhibit excellent and repeatable persistent luminescence performance after being activated by light with a wavelength in the biological window for in vivo imaging. Reprinted with permission from [27]. Copyright (2022), Journal of Rare Earths

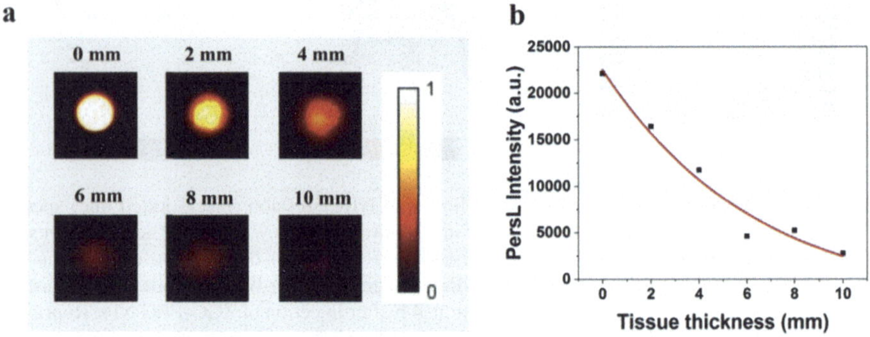

Fig. 6 **a** Persistent luminescence images of MGGO: Yb^{3+} PLNPs capping with varied tissue thicknesses (0–10 mm) after 10 min of X-ray irradiation. **b** The intensity of Yb^{3+} doped MGGO's persistent luminescence as a function of tissue thickness. Reprinted with permission from [72]. Copyright (2023), Journal of Luminescence

address these challenges, YAGG:Er-Cr was employed to reduce scattering by emitting in the NIR-III window (1500 nm), which enables greater imaging contrast and higher spatial resolution deep tissue bioimaging while suppressing "halo" pictures surrounding samples.

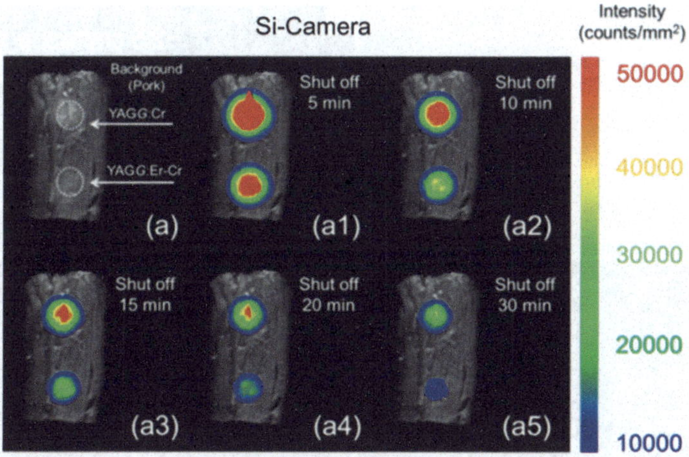

Fig. 7 Optical imaging of Cr^{3+}, Er^{3+} persistent luminescence in YAGG:Er–Cr and YAGG:Cr samples through raw-pork tissues after 254 nm (6 W output) illumination for 10 min. Reprinted with permission from [73]. Copyright (2018), ACS Inorganic Chemistry

7 Bioimaging for Disease Diagnosis

Bioimaging has acquired a lot of attention in recent years as a potent tool in the field of diagnostic techniques for accurate disease diagnosis and treatment. Researchers worldwide have concentrated on the development of various fluorescent microscopic devices that could aid in the early detection of serious diseases such as cancer. The unique optical properties, tailorable surface modifications, and ability for carrying various diseases of fluorescent nanoparticles enable them to improve the diagnosis and therapy of several diseases [5].

Various luminescence nanomaterials have been employed in this context, with persistent luminescence nanoparticles emerging as possible probes for illness diagnostics due to their long-lasting luminescence without continual stimulation. The degree of tissue penetration is crucial for in vivo biomedical applications in clinical diagnostics.

Song et al. [74] investigated the depth tissue penetration of a piece of pork by injecting it with PEG-$SrAl_2O_4:Eu^{2+}$ PLNPs. Figure 8 depicts the deep tissue penetration simulation experiment; first, a piece of pork was inserted between the PEG-$SrAl_2O_4:Eu^{2+}$ PLNPs and the soft X-ray source (Fig. 8a). Figure 8b illustrates that there is no persistent luminescence before the lighting of a soft X-ray source. However, when a mild X-ray source was activated (Fig. 8c), a notable persistent luminescence was seen despite the thickness of the meat being up to 2.5 cm. Furthermore, exposing a piece of pork to a blue laser blocks the PLNPs' luminescence; hence, the results demonstrate that the persistent luminescence of PLNPs can be induced in deep tissue imaging by soft X-ray excitation. Furthermore, the PEG-$SrAl_2O_4:Eu^{2+}$ PLNPs had a greater signal-to-background ratio (19.1) than CdSe QDs (1.4). This

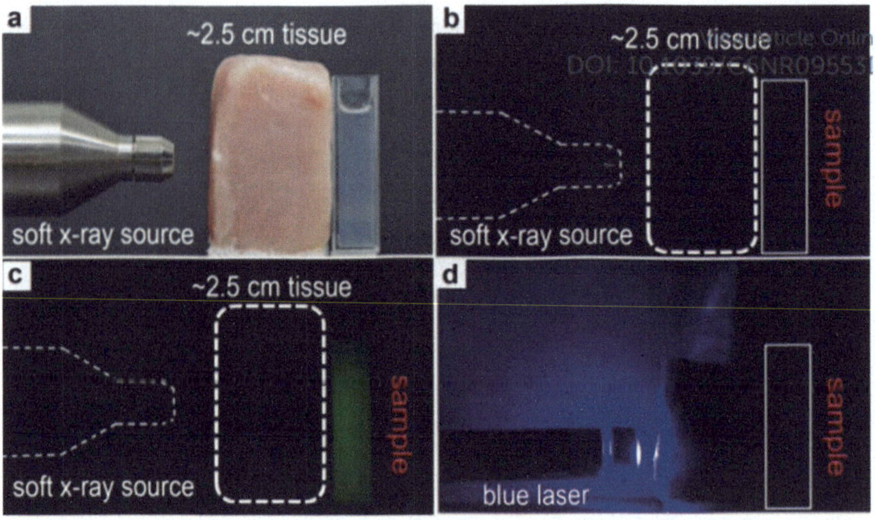

Fig. 8 Photographs presenting the experimental process for deep tissue penetration of piece of pork. **a** Photograph 2.5 cm tissue between soft X-ray source and PEG-$SrAl_2O_4:Eu^{2+}$ PLNPs. **b** Dark field photograph, $SrAl_2O_4:Eu^{2+}$ PLNPs are not activated with soft X-ray source. **c** Photograph showing the persistent luminescence under soft X-ray source. **d** Dark field photograph of $SrAl_2O_4:Eu^{2+}$ PLNPs under blue laser irradiation

study indicates the feasibility of using soft X-ray-activated PLNPs for deep tissue therapy.

Rapid and recent advancement has demonstrated that NIR-emitting PLNPs activated by X-ray are promising probes for in vivo deep-tissue bioimaging due to their great sensitivity and ability to accurately diagnose disease [75].

Xue et al. [75] developed biocompatible PEG-ZnGa2O4: Cr PLNPs with efficient NIR persistent emission under soft X-ray excitation and rechargeable activation properties for in vivo deep-tissue bioimaging. The strong persistent brightness of PLNPs demonstrated deep penetration depths of up to 20 mm with great sensitivity, potentially improving disease diagnosis efficiency.

Calatayud et al. [76] recently produced SrAlO: Eu, Dy@SiO_2 probes with long persistence luminescence response that could enable time-gated imaging. In vitro experiments show that core–shell particles do not harm cells after 48 h of incubation. However, the process of synthesis of SrAlO: Eu, Dy@SiO_2 particles should be considered later to avoid agglomeration in order to harness the potential of these probes in bioimaging for disease detection and treatment.

In recent years, researchers have concentrated their efforts on developing high-quality bioimaging technologies. They propose multimodal probes with combined functionality, such as one probe that offers both NIR persistent luminescence and magnetic resonance imaging (MRI) for effective medical diagnostics.

Shi et al. [45] synthesized multimodal nanoplatforms $mSiO_2$@$Gd_3Ga_5O_{12}:Cr^{3+}$, Nd^{3+} ($mSiO_2$@GGO) for multimodal imaging and cancer therapy with success.

These PLNPs are highly fascinating since they exhibit NIR-I (745 nm) persistent luminescence after UV irradiation and good NIR-II (1067 nm) under 808 nm excitation wavelength. Furthermore, because of the high concentration of Gd^{3+} ions in the host matrix ($Gd_3Ga_5O_{12}$), these probes demonstrated a high r_1 value (10.70 $mM^{-1}\ s^{-1}$), making them ideal for T1 MR imaging. Figure 9a depicts a significant persistent luminescence of $mSiO_2$@GGO in the first biological window that began after UV excitation (four broad excitation peak bands). Even 3 h after 5 min of 254 nm UV lamp irradiation, the persistent luminescence intensity of $mSiO_2$@GGO PLNPs was observed to be higher than the background (Fig. 9b). The in vivo persistent luminescence investigation demonstrates that the signal of $mSiO_2$@GGO NPs appears in the entire body in 5 min. After 5 min, the strong persistent luminescence is exclusively found in the liver location, which was explained by the reticuloendothelial system capping $mSiO_2$@GGO NPs. The persistent luminescence signal decreased over time, but it was still detectable in the liver after 60 min, suggesting the potential application of $mSiO_2$@GGO NPs to generate NIR-I persistent luminescence without continuous irradiation for optical imaging. Deep tissue penetration was used to study the SNR (Fig. 9d and e), and researchers discovered that the NIR-I persistent luminescence of the $mSiO_2$@GGO NPs can achieve a high signal-to-noise ratio at up to 2 cm tissue thickness.

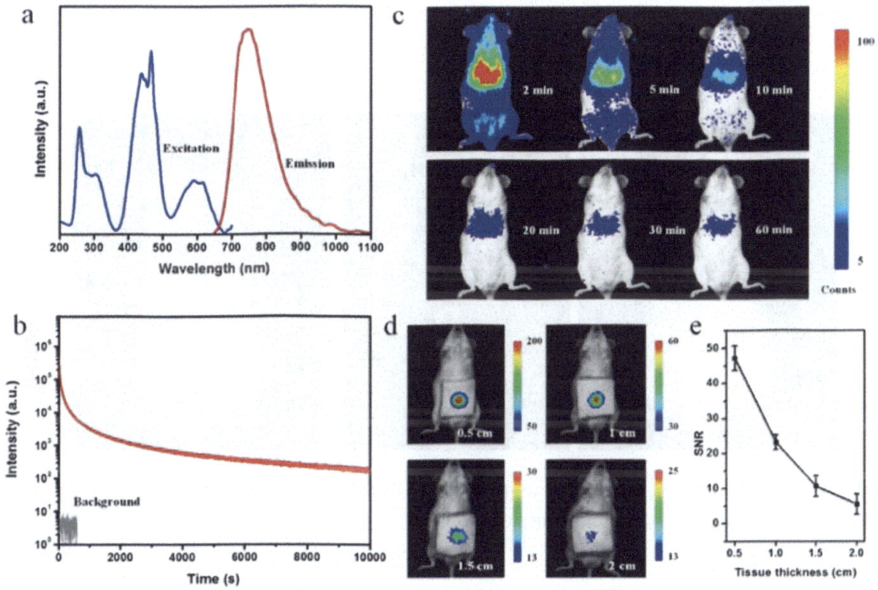

Fig. 9 **a** Excitation and emission spectra of $mSiO_2$@GGO persistent luminescence NPs. **b** Decay curves of persistent luminescence (745 nm) after 5 min irradiation with a 254 nm UV lamp. **c** In vivo imaging showing the persistent luminescence NPs over time. **d** In vivo deep tissue penetration covered with piece of pork. **e** SNR of $mSiO_2$@GGO persistent luminescence NPs as a function of tissue thickness. Reprinted with permission from [45]. Copyright (2018), Biomaterials

Figure 10 depicts the evaluation of NIR-II luminescence properties and in vivo imaging of $mSiO_2$@GGO NPs. The $mSiO_2$@GGO NPs generates intense, persistent luminescence in NIR-I and NIR-II when excited in NIR-I (808 nm) (Fig. 10b). The in vivo experiment on a normal mouse following subcutaneous injection of PLNPs exhibits a significant NIR-II luminescence signal only at the injection site, confirming the intriguing applicability of the researched PLNPs in vivo imaging in the NIR-II region.

By injecting the persistent luminescence $mSiO_2$@GGO NPs into mice, magnetic properties and in vivo MR imaging of the $mSiO_2$@GGO NPs were examined (Fig. 11). The results show that the presence of Gd ions in the host causes contrast enhancement in the liver, which is considered a promising MRI contrast agent.

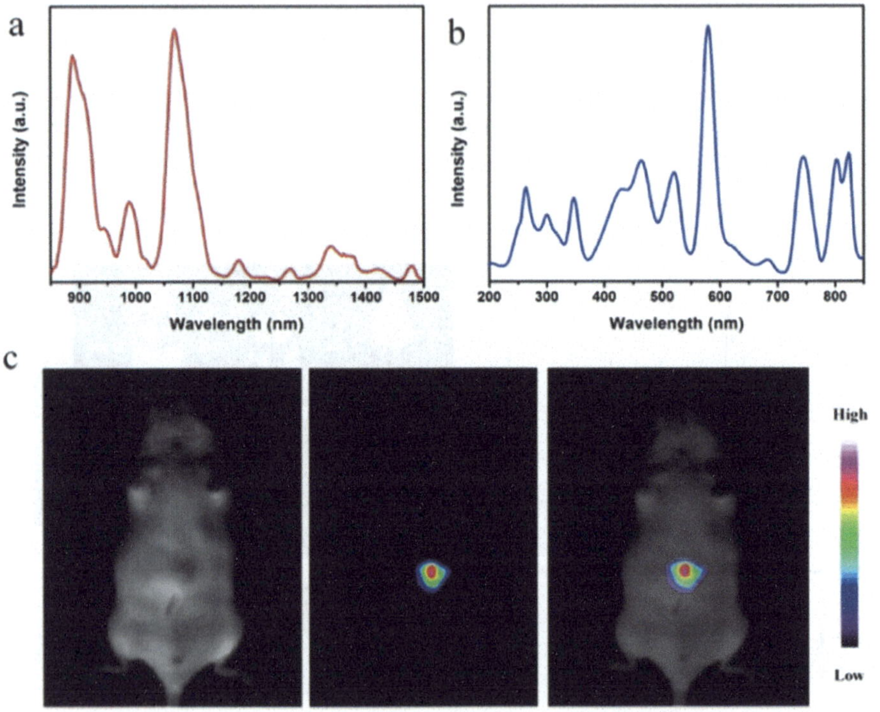

Fig. 10 **a** Emission spectra of PLNPs under 808 nm excitation in NIR-I. **b** Excitation spectra with 1067 nm wavelength. **c** In vivo imaging of mouse showing NIR-II persistent luminescence after subcutaneous injection of $mSiO_2$@GGO NPs. Reprinted with permission from [45]. Copyright (2018), Biomaterials

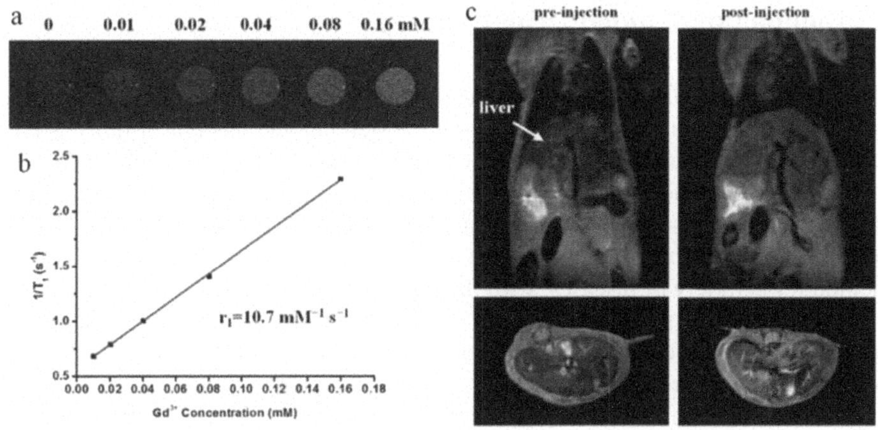

Fig. 11 **a** T1-weighted MR images analyzed with different concentration of PLNPs. **b** The dependence of Gd^{3+} concentration on T1-weighted MRI. **c** In vivo T1-weighted MR imaging before and after the injection of mSiO2@GGO NPs (higher image: coronal plane, down image: transverse plane). Reprinted with permission from [45]. Copyright (2018), Biomaterials

8 Bioimaging for Guiding Treatment

Because of its highly sensitive detection property, which permits analysis of specific diseased locations in real-time and encourages accurate monitoring of precise treatment processes, bioimaging for guiding treatment is a promising technology for improving cancer detection at the earliest possible stage [77]. A single nanoplatform that enables optical imaging and therapy has recently received a lot of attention in biomedicine. This nanoplatform possessed numerous advantages over therapeutic efficiency, including an intelligent tailored drug delivery system in the living body [78]. A system combining NIR-emitting nanoparticles and photothermal agents that absorb NIR light and convert the irradiation photon into heat with low-energy light, on the other hand, could minimize extra damage to normal cells and favor deep tissue penetration [79].

Because of their noninvasive manner for tumor diagnosis and treatment, as well as the ease with which their physical and optical properties may be tuned by structural modification, fluorescent nanoparticles have been widely used in diagnostic imaging and targeted therapy (cancer theranostics) [80]. Nanotheranostic and therapy based on PLNPs with a high signal-to-noise ratio for visualization cancer treatments [78] with long-lasting luminescence without continuous stimulation and low autofluorescence can be very useful in bioimaging for treatment guidance [81]. The adaptable technique for designing PLNP-based nanoprobes should be based on some surface engineering principles and adjusting energy transfer for appropriate emission wavelength in the biological transparent window.

9 Bioimaging-Guided Drug Delivery

Nanoparticle-based drug delivery has been extensively developed in recent decades because to significant advantages such as good stability and biocompatibility, accurate targeting, regulated method, increased permeability, and surface chemical modification [82, 83]. The mechanism of drug delivery may suffer from poor bioavailability and aqueous solubility which disadvantage the in vivo pharmacokinetic properties [82]. Bioimaging-guided drug delivery has emerged as an effective therapeutic method for a variety of diseases, with the potential to reduce side effects in cancer treatment [84]. This technology has clinical promise because it can act as a targeted vehicle for controlled drug release via in vivo imaging, as well as evaluate disease areas in real-time through specific diagnosis and treatment [78–85].

Liu et al. [57] fabricated a functional nanostructure based on triple-doped zinc gallogermanate (ZGGO) PLNPs for long-luminescence, mesoporous silica ($mSiO_2$) for drug delivery, and Red blood cell (RBC) membrane vesicle for biomimetic functionalization.

The researchers carried out an in vivo biodistribution experiment of ZGGO, ZGGO@$mSiO_2$, and RBC-ZGGO@$mSiO_2$ nanoprobes by tail vein injection to estimate the use of cell-mimicking nanocarriers as the qualified drug-delivery vehicle. They demonstrated that, when compared to ZGGO and ZGGO@$mSiO_2$, RBC-ZGGO@$mSiO_2$ NPs revealed better retention efficiency; additionally, the efficiency of nanoparticle circulation as measured by 700-nm luminescence showed that both ZGGO and Doxorubicin-ZGGO@$mSiO_2$ (DZGGO@$mSiO_2$) present faster clearance. Then, RBC-DZGGO@mSiO2 was shown to have a longer circulation period with a half-life duration approaching 9 h. Tumor-growth suppression effects observed in the 4T1 orthotopic mammary tumor model verified RBC-DZGGO@$mSiO_2$'s greatest performance as a tumor-growth inhibitor. Zhang et al. [86] developed a polyamide-amine (PAMAM) and Doxorubicin (DOX)-based persistent luminescent for aptamer-guided tumor imaging and acid-responsive drug administration (PLNPs-PAMAM-AS1411/DOX). After ceasing excitation with a 254 nm UV lamp for 5 min, the PLNPs demonstrated good persistent luminescence performance at 700 nm, which was then reproduced under irradiation with a 650 nm red LED lamp. Figure 12a reveals that all PLNPs, PLNPs-PAMAM, PLNPs-PAMAM-AS1411, and PLNPs-PAMAM-AS1411/DOX exhibit an emission band at 700, which is attributed to the Cr^{3+} ion $^2E_4 \rightarrow A_2$ transition in gallogermanate.

On mice, PLNPs were tested for in vivo target-specific cancer imaging (first with 254 nm UV light irradiation 10 min before injection, followed by 650 nm LED light irradiation for 2 min). The PLNPs-PAMAM-AS1411/DOX generate excellent persistent luminescence in the tumor area of HeLa tumor-bearing mice for at least 8 h, as illustrated in Fig. 12c. However, the PLNPs-PAMAM-DOX did not exhibit an optical signal, demonstrating the targeting impact of the AS1411 aptamer functionalized nanoprobe in promoting nanoparticle accumulation at the tumor site. Gao et al. [87] recently developed a high-performance $ZnGa_2O_4:Cr^{3+}/Sn^{4+}$@MSNs multifunctional nanoplatform with excellent rechargeable NIR persistent luminescence as well

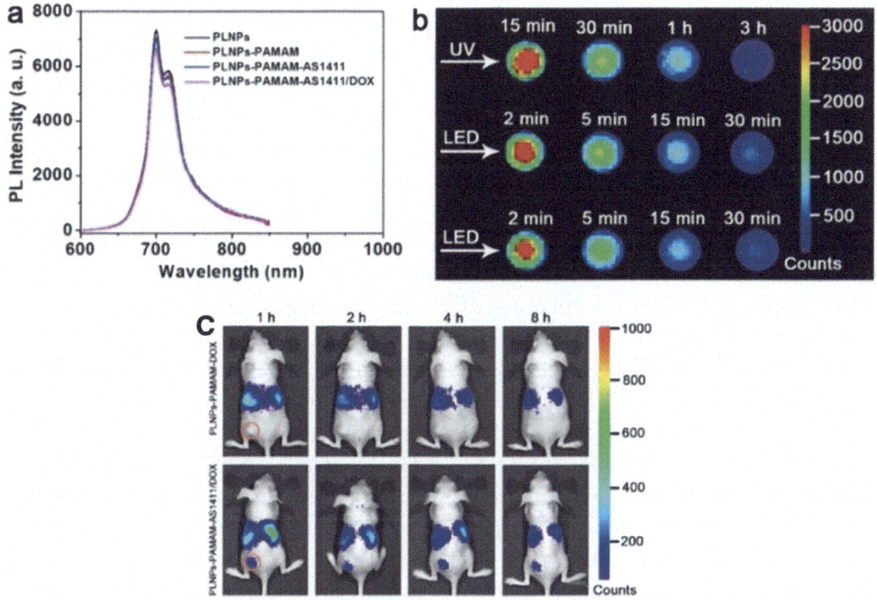

Fig. 12 **a** Emission spectra of all PLNPs. **b** Images showing Persistent luminescence of the PLNPs-PAMAM-AS1411/DOX aqueous dispersion after 5 min excitation with a 254 nm UV lamp and 2 min re-activation with red LED **c** In vivo persistence luminescence images of HeLa tumor-bearing mice post intravenous injection of PLNPs-PAMAM-DOX and PLNPs-PAMAM-AS1411/DOX (254 nm UV light irradiation 10 min before injection followed by 650 nm LED light irradiation for 2 min). Reprinted with permission from [86]. Copyright (2020), Talanta

as very efficient drug (Dox) loading and releasing for effective cancer cell killing. According to the authors, this study addressed the problem of low drug loading efficiency, which could pave the way for the creation of nanoprobes based on NIR persistent luminescence as theranostic nanoplatforms.

10 Multimodal Bioimaging with Hybrid PLNPs

Various high-resolution imaging technologies are used to collect the maximum number of signals from the target to increase the final imaging quality for a better therapeutic response to diseases, primarily cancer. We mention fluorescence imaging, computed tomography CT, magnetic resonance imaging MRI, ultrasonography, X-rays, and so on [88]. When compared to single-modal imaging, multimodal bioimaging provides more precise, complete, and reliable information for diagnosis [44].

Traditional probes may have disadvantages such as high toxicity, low sensitivity, and resolution, which may restrict their use in in vivo imaging. To address these shortcomings, researchers investigated effective multifunctional hybrid nanoparticles with physical and chemical manipulation as challenging possibilities for multimodal bioimaging [89].

Due to their intriguing optical properties such as persistent emission for minutes to hours without an excitation source, optical rechargeable features, high SNR, and high storage capability, hybrid PLNPs are considered efficient multimodal imaging nanoprobes among currently utilized nanoprobes. Furthermore, the eradication of PLNPs' in vivo tissue autofluorescence and long-term luminescence benefits their application for efficient long-term in vitro and in vivo bioimaging [90].

Furthermore, in vivo CT imaging of HepG2 tumor-bearing mice using cyclic-NGR peptides; NGR ZGGO: Cr, Pr@TaO@SiO_2 by tail vein injection verified the nanoprobe's tumor-targeting capacity. All results suggested that the synthesized platform might be used for multifunctional bioimaging due to its high SNR, NIR emission that reduces tissue damage, and good spatial resolution.

Multifunctional NPLNP $mSiO_2@Gd_3Ga_5O_{12}:Cr^{3+}$, Nd^{3+} ($mSiO_2$@GGO) nanoplatforms have demonstrated good multimodality imaging and cancer therapy [45]. First, in vivo NIR-I imaging revealed that the $mSiO_2$@GGO NPs exhibited persistent luminescence with a high SNR at up to 2 cm tissue thickness. Second, experimental in vivo imaging revealed persistent fluorescence (at 1067 nm) upon excitation at 808 nm, demonstrating the nanoplatform's utility for in vivo imaging in the NIR-II area. Furthermore, the authors tested the potential of $mSiO_2$@GGO NPs as positive MRI contrast agents for in vivo T1-weighted imaging by injecting them intravenously into healthy Kunming mice. Previous research has used PLNPs surface modification to fulfill the diverse needs of PLNPs-based multi-imaging. Liu et al. [91] used Mn^{4+} and Ge^{4+} co-doped gadolinium aluminate ($GdAlO_3:Mn^{4+}$, Ge^{4+}) PLNPs as the NIR persistent luminescence emission center, and their surface was coated with gold nanoshell to improve plasmon resonance efficiency. Figure 13 depicts the various synthesis processes of $GdAlO_3:Mn^{4+}$, Ge^{4+}@Au, beginning with the hydrothermal/calcination process and progressing to multimodal imaging. Magnetic resonance imaging is provided by Gd ions in the host, optical imaging is provided by PLNPs, and computed tomography is provided by gold nanoshell.

11 Tumor-Targeted Imaging via Intravenous Administration

In vivo imaging is a non-invasive technology that delivers detailed information on tumor localization by targeting hard-to-reach cancer locations with luminescent nanoparticle agents, which improves cancer detection by delivering therapy at the earliest possible stage [92].

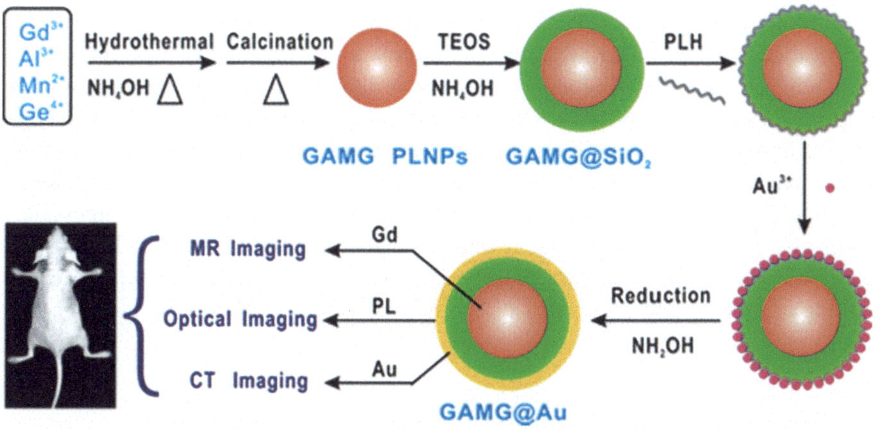

Fig. 13 Schematic showing different synthesis steps of GAMG@Au core–shell nanoplatform for in vivo trimodality bioimaging. Reprinted with permission from [91]. Copyright (2016), ACS Applied Materials

For tumor-targeted imaging via intravenous administration, luminescent nanoparticles should have prolonged circulation times in the blood because, following intravenous administration, the first type of cells they will mostly encounter are the cells circulating in the bloodstream of all body organs and tissues [93].

As a result, developing effective and safe delivery systems capable of targeting tumors via imaging techniques via intravenous administration has become an attractive topic for prospective biomedical research improvements. Among the luminescent NPs utilized for tumor-targeted imaging by intravenous administration, the so-called "optical battery" PLNPs have demonstrated significant advantages in this field.

Liu et al. [94] demonstrated in 2020 that the radiopharmaceutical ^{18}F-FDG can operate as an internal light source to stimulate intravenously injected $ZnGa_2O_4:Cr^{3+}$ nanoparticles (ZGCs) for tumor-targeted imaging (over 3 h) via intravenous injection. In this case, energy is transferred by Cerenkov resonance as well as ionizing radiation. The team also investigated the in vivo radiopharmaceutical luminescence of $ZnGa_2O_4:Cr^{3+}$ nanoparticles for tumor imaging. In vivo, experiments on 4T1 tumor-bearing mice reveal that treatment of ^{18}F-FDG for varying lengths of time (24, 48, and 72 h) enhances the luminescence of a small amount of ZGCs arriving at the tumor.

The tumor fluorescence can remain for more than 240 min following 24 h of ^{18}F-FDG activation of the implanted ZGCs nanoparticles. Two further injections of ^{18}F-FDG were performed at 48 and 72 h time points, demonstrating long-lasting luminescence in the tumor. These findings support the in vivo repeatedly chargeable luminescence of ZGCs by radiopharmaceuticals, which could be useful in long-term optical tumor imaging.

Jiang et al. [27] developed NIR PLNPs based on Cr^{3+}/Y^{3+} co-doped ZGSC in 2022, which had good optical properties after excitation by light with a wavelength in the biological window. The authors demonstrated that following PEG surface modification, NIR PLNPs have low cytotoxicity, allowing them to examine in vivo bioimaging. In vivo, imaging and passive tumor-targeted imaging experiments were performed by injecting NIR PLNPs-PEG into mice via the tail vein and activating with the light of a wavelength in the biological window. The liver, spleen, and lung of a normal mouse present a significant persistent luminescence signal, according to in vivo and ex vivo data. Furthermore, in vivo and ex vivo persistent luminescence results of 4T1 tumor-bearing mice revealed the potential utility of NIR PLNPs-PEG nanoprobe for reproducible and highly sensitive in vivo persistent luminescence imaging of tumors.

12 Tumor-Targeted Imaging via Oral Administration

Because of its ease, self-administration, flexibility in the dosing regimen, and physiological reasons, among other things, the oral route has become the most important technique for drug delivery [95]. Oral administration of bioimaging contrast agents is the most preferred route, but it poses a problem in the creation of effective contrast agents due to the harsh environment in the gastrointestinal tract [96].

One advantage of persistent luminescence nanoparticles is their ability to give targeted and/or localized medication administration via blood circulation and accumulation in tumors, in addition to their high and persistent luminescence intensity, which ensures sensitivity imaging at low agent dosage [80–97].

In 2023, Giordano et al. [98] suggested new rechargeable PLNPs with nanocomposite structure (–$NaGdYbErF_4/ZnGaSnCrO_4$) powered by an infrared laser (980 nm) and an upconversion mechanism in the visible region (700 nm). The in vivo imaging in deeper tissue investigation in mice was examined using the oral route via capsule administration. The results revealed that for ex-situ assays, the capsule can be pre-excited, and then a signal generated by the capsule was clearly recognized following delivery of the capsule orally via a cannula. This study demonstrates the viability of an oral delivery method based on the emission of a persistent luminescence signal from an energy transfer nanocomposite that can be recharged using a 980 nm excitation (1 W cm^{-2}, 2 min). The kinetic degradation of the capsule in the stomach, on the other hand, should be researched in the future.

13 Imaging-Guided Therapy Based on PLNPs

Nanotheranostics have recently become popular among sophisticated nanomedicine strategies for disease diagnosis and treatment. In reality, nanotheranostics is a subfield of theranostics that combines therapeutic and current diagnostic tools or imaging into a single nanoplatform (hybrids) [99, 100].

To be used in vivo, these nanotheranostics must have certain characteristics, including (i) very low toxicity, (ii) accumulation of the nanotheranostics in a specific organ to ensure effective targeting and imaging, and (iii) the ability to deliver the therapeutic agent at the preferred site for effective disease treatment [65]. This chapter highlights the development of nanotheranostics based on persistent luminescence for the diagnosis and treatment of a variety of diseases.

13.1 Imaging-Guided Chemotherapy

In recent years, the storing of a chemotherapeutic drug inside nanoparticles is a promising technique that has drawn considerable attention for suppressing tumor growth by shrinking tumors using smaller doses of the drug; as a result, the harmful side effects are reduced, which improves patient survival and clinical symptoms. [83].

Notably, PLNPs can be effective as a chemo drug loading and delivery mechanism. As luminescence drug carriers for chemotherapy, many hybrid structures based on persistent luminescence have been investigated. Due to benefits such as the mesoporous structure and specific surface area, silica integrated into the PLNPs platform has mainly been used for chemo-drug delivery in prior works [44].

However, this approach has significant drawbacks that limit its implementation for drug administration, such as (i) the time-consuming and expensive manufacturing method, (ii) limited loading efficiency, and (iii) challenging surface modification in the case of hydrophobic drugs [44].

In 2017, Chen et al. [44] developed biocompatible nanovehicle liposomes coated NIR-PLNPs (Lipo-PLNPs) as a potential persistent luminescence nanoplatform for imaging-guided chemotherapy.

The intravenous administration of paclitaxel (PTX) loaded Lipo-PLNP into normal mice and MCF-7 tumor-bearing animals resulted in vivo imaging research. The luminescence signal is shown in both circumstances in Fig. 14. The signal increased in the tumor area of tumor-bearing animals 24 h after injection, confirming the accumulation of the composite into the tumor site due to long-term circulation.

To investigate the in vivo therapeutic efficacy of Lipo-PLNP-PTX, experiments were performed on MCF-7 tumor-bearing athymic mice, first with Lipo-PLNP-PTX, then with two control groups (Lipo-PLNPs or PBS only). The suppression of tumor growth accomplished by Lipo-PLNP-PTX is demonstrated in Fig. 15a, b. Furthermore, the histology analysis shown in Fig. 15c demonstrates an important necrotic reaction of Lipo-PLNP-PTX administered intravenously. These findings indicate that Lipo-PLNPs are a viable drug delivery carrier in biological applications.

Liu et al. [40] employed a surfactant-assisted hydrothermal technique to synthesize PLNPs (ZnGa2O4: Cr) for bioimaging using a post-calcination procedure. To provide long-term precision for in vivo imaging, the persistent luminescence signal was recharged utilizing an excitation by an LED lamp. Theranostic nanoprobe

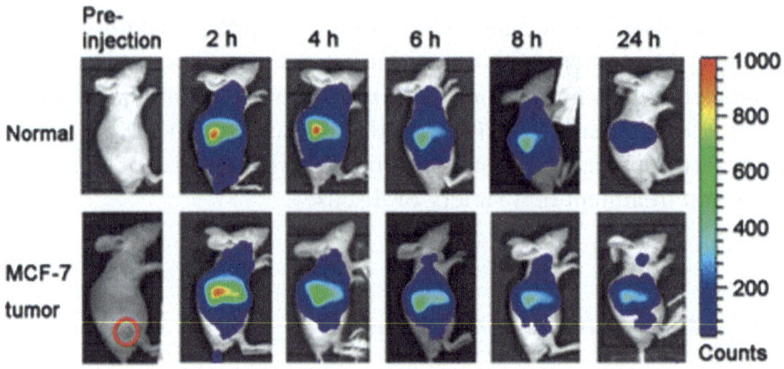

Fig. 14 In vivo images showing persistent luminescence after intravenously injected Lipo-PLNP-PTX in normal mice and MCF-7 tumor-bearing mice (10 min excitation with a 254 nm UV light before intravenous injection, 2 min irradiation with LED lamp before each acquisition). Reprinted with permission from [44]. Copyright (2017), ACS Analytical Chemistry

imaging and diagnosis are confirmed by in vivo persistent luminescence imaging, relative tumor volume changes, and histological labeling.

13.2 Imaging-Guided Photothermal Therapy

Photothermal therapy (PTT) has sparked growing in pre-clinical investigations because of its advantages such as non-invasiveness, deep tissue penetration, high specificity, and efficacy [101]. PTT based on light-absorbing materials, where hyperthermia is produced upon excitation in the near-infrared (NIR) range, has opened up new avenues for cancer therapy in recent years. When the temperature is precisely maintained during PTT, this procedure becomes more successful and ensures less injury to normal organs [102].

Several research studies have demonstrated temperature control for PTT utilizing optical agents such as gold (Au) NPs [103] CuS NPs [104] boron quantum dots (BQDs) [105] rare earth based NPs [106] etc. However, some of these NPs exhibit significant autofluorescence background signals during in situ excitation, limited quantum yield, and tissue photothermal toxicity due to continuous NIR light irradiation, all of which limit their utility in vivo imaging and therapy [107].

Thanks to the long persistent luminescence property with ceasing the excitation light that eliminates the autofluorescene and allows a high SNR favoring the application of PLNPs in bio-applications particularly in PTT [108].

Meng et al. [109] used hexadecyl trimethyl ammonium bromide (CTAB) and phosphotungstic acid (PW12) molecules to modify the surface of a biocompatible PLNP and gold nanorods (GNRs) as illustrated in Fig. 16. Under 635 nm excitation, the biocompatible composite nanoplatforms show a high photothermal conversion efficiency (0.8 W·cm^{-2}) derived from the photothermal agent (GNRs). Furthermore,

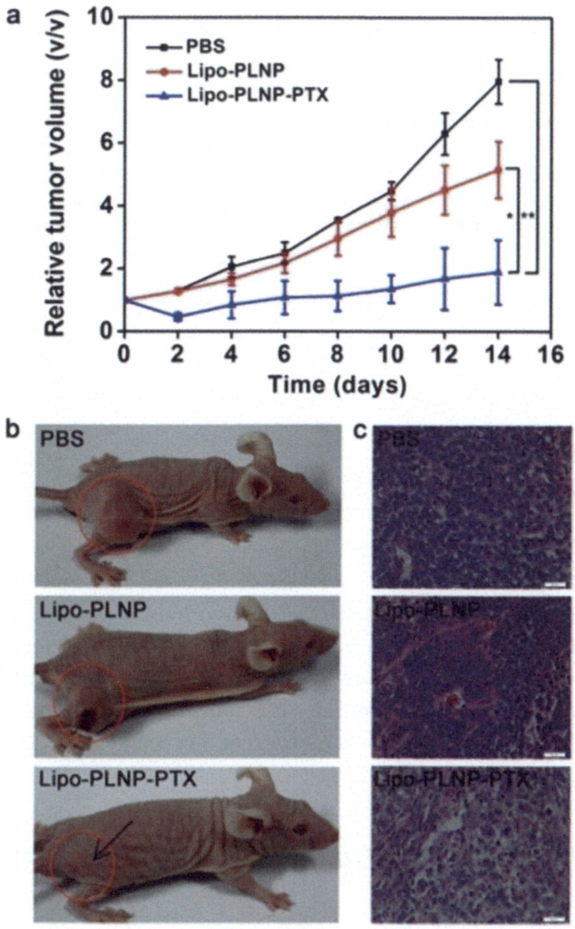

Fig. 15 **a** MCF-7 tumor growth curves performed in three mice groups with different treatments. *P < 0.05, **P < 0.01. **b** Photographs showing mice with three injections of groups after a 14-day treatment. **c** Tumor tissues identified by histological staining from different treated mice groups. Reprinted with permission from [44]. Copyright (2017), ACS Analytical Chemistry

the composite produces efficient NIR afterglow emissions coming from PLNPs, allowing for accurate control of excitation light strength as well as temperature change, resulting in less photodamage to normal tissue during the deep tissue PTT process.

Wang et al. [110] reported mesoporous silica and photothermal fluorescent dye IR825 functionalized persistent luminescence nanoparticles $ZnGaGeO_4:Cr^{3+}$, Yb^{3+}, Er^{3+} (ZGGO) PLNPs for precise cancer treatment of a biomimetic nanodelivery system.

Figure 17 demonstrates the entire process of imaging-guided photothermal therapy for colorectal cancer. To begin, PLNPs were synthesized as an imaging-guided agent utilizing a bi-phase solvent thermal technique covered with mesoporous silica to ensure drug loading. The researchers then combined the photothermal fluorescent dye IR825 and the chemotherapeutic drug irinotecan (Ir) into the ZGGO@SiO_2 composite to achieve both photothermal and chemotherapeutic effects.

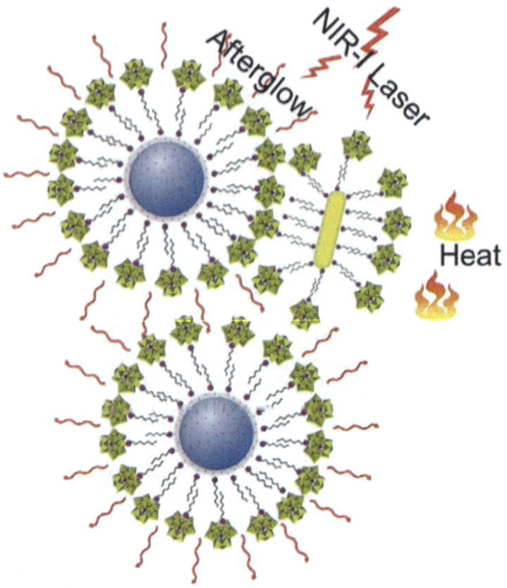

Fig. 16 Schematic representation of Biocompatible PLNP-GNR composite nanoplatforms for afterglow luminescence and PTT. Reprinted with permission from [109]. Copyright (2021), Applied Surface Science

The connection of cell-macrophage hybrid membranes is significant in increasing drug half-life in vivo as well as facilitating nanomaterial distribution into tumor locations. In vivo, imaging investigations revealed that PLNPs produce efficient persistent luminescence, and in vivo, anti-tumor trials proved that chemotherapeutic and photothermal therapies have great therapeutic effects.

13.3 Imaging-Guided Photodynamic Therapy

Image-guided photodynamic therapy (PDT) is a potential strategy for cancer treatment, with the basic idea being that light activates photosensitizers (PS), resulting in reactive oxygen species (ROS) that destroy cancer cells [111]. PDT is a treatment that relies on three main factors: (i) the type, concentration, and localization of photosensitizing agents at the tumor site, which are determined by the chemical nature of the PS, (ii) light sources and wavelengths, and (iii) molecular oxygen, which produces cytotoxic species and induces tumor death [112].

A PDT strategy based on persistent luminescence is being explored as a smart approach for tumor afterglow imaging and chemical/photodynamic therapy [113].

Unlike PDT caused by low-intensity illumination, which may suffer from temporary oxygen deficiencies due to oxygen surrounding consumption by PDT itself, PDT induced by high-intensity illumination allows the system to recover its oxygen often [114]. The researchers offered a preliminary technique to address the PDT-induced oxygen shortage, which is the building of a suitable PLNPs/PS platform for imaging-guided photodynamic treatment.

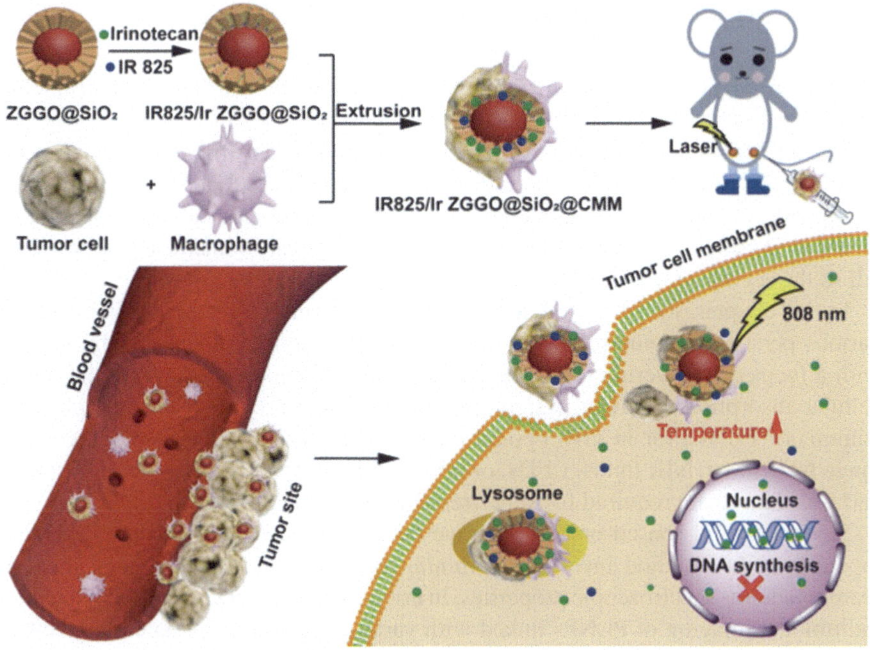

Fig. 17 Schematic diagram showing the association of cancer-macrophage hybrid membrane with persistent luminescence agents IR825/Ir ZGGO@SiO_2 for imaging guided photothermal therapy of Colorectal Cancer. Reprinted with permission from [110]. Copyright (2020), ACS Applied Nano Materials

Chang et al. [115] developed a new blue persistent luminescence material (PLM) based on CaA12O4: Eu, Nd for afterglow in vivo imaging after in vitro pre-excitation. Blue-emitting PLM can activate cyanobacterial cells and verteporfin, which produce singlet oxygen, without the requirement for long-term external excitation, resulting in controlled tumor cell growth and improved cancer photodynamic therapy efficiency. Recently, Zhao et al. [116] developed a DNA nanocomplex integrated into PLNPs to increase the tissue-penetration depths for PDT treatment of breast cancer.

They explained how PLNP-integrating DNA nano-complexes based on glutathione (GSH) trigger PDT in mice with MCF-7 xenografts works without the use of an external laser. The AS1411 aptamer serves as the scaffold of a nano-complex, containing a guanine-rich sequence used to load the photosensitizer silicon phthalocyanine dichloride ($SiPcCl_2$) as well as target nucleolin, which can inhibit tumor cell proliferation. MnO_2-coated PLNPs (S-PLNP@MnO_2) were used as a catalyst for the intracellular decomposition of H_2O_2 into O_2 as well as the extension of energy storage. When the S-PLNP@MnO_2 nanoplatform enters cancer cells, MnO_2 decomposes, allowing PLNPs to generate persistent luminescence, which triggers photosensitizer $SiPcCl_2$ to catalytically convert O_2 into $^1O_2$ and destroy cancer cells.

14 Future Perspective and Conclusion

The search for biocompatible and efficient luminescent nanoparticles for bioimaging and disease treatment is intensifying.

The ultra-slow persistent emission induced by higher and efficient trap states and lower charge carrier mobilities promotes long luminescence that can last for tens of hours without constant in situ excitation, which suppresses autofluorescence interferences in persistent luminescence materials for optical biomedical applications. All of these characteristics hold a lot of promise for bioapplications.

In this chapter, we presented an updated assessment of current achievements of various persistent luminescent materials in imaging and therapeutic technologies, with a focus on delivering to targets and activating compounds that may influence cellular absorption in a living system. Initially, different excitation sources (physical stimuli) of PLNPs for in vivo optical imaging were proposed. These sources can range from red to NIR lights, LEDs, or lasers; thus, an appropriate excitation source and wavelength are required for super-resolution bioimaging.

Several hybrids based on PLNPs have been developed to improve bioimaging modalities; this method involves combining and modifying optical and physical features such as spectroscopic properties. In this paper, we offer current and intriguing multimodal imaging of PLNPs linked with various materials as a nanoplatform for disease detection. PLNPs coupled with several functional units have found considerable usage as theranostics materials capable of simultaneously treating and diagnosing multiple diseases. Thus, the purpose of this chapter is to describe and discuss imaging-guided non-invasive therapies such as chemotherapy (the use of drugs to treat cancer), photothermal therapy (based on heating impact), and photodynamic therapy (based on the creation of reactive oxygen species).

Despite advances in the development of smart nanoprobes based on PLNPs for bio applications, more efforts are required, with an emphasis on engineering methodologies for producing efficient PLNPs. To increase the luminescence property of PLNPs for a good imaging depth and resolution prior to in vivo imaging experiments, the right activator and host matrix must be chosen, moreover, the co-doping technique can assure effective energy transfer to the emitters. This allows for the manipulation of the trapping process within PLNPs by tuning the luminescence spectrum. However, the physics underlying persistent luminescence mechanisms is a highly intricate field that will require further investigation in the future. Paying great attention to the selection of materials forms PLNPs for in vivo use, which must be as biocompatible as feasible taking into account their size (nanometric), structure, surface modification, and formulation. Researchers need to address issues related to potential toxicity, immunogenicity, and long-term effects on living tissues. Penetrating deep tissues with persistent luminescence can be challenging due to photon scattering and absorption. Improving the ability of PLNPs to emit and retain light in deep tissues is a hurdle that needs to be overcome for broader biomedical applications. Another advantage of PLNP optical properties that should be thoroughly researched is the development of biological thermometers for hyperthermia applications. Innovative development of PLNPs will continue to emerge as a substantial potential field in a wide range of bioapplications for diagnosing and treating various serious health problems such as antibacterial and viral infections, tissue engineering, and cardiovascular disorders. However, additional research and effort are required to fully understand the mechanisms underlying persistent luminescence nanoparticles in order to develop efficient afterglow nanomaterials for a variety of bio-applications.

References

1. S.S. Syamchand, G. Sony, Europium enabled luminescent nanoparticles for biomedical applications. J. Lumin. **165**, 190–215 (2015). https://doi.org/10.1016/j.jlumin.2015.04.042
2. V. Parodi, E. Jacchetti, R. Osellame, G. Cerullo, D. Polli, M.T. Raimondi, Nonlinear optical microscopy: from fundamentals to applications in live bioimaging. Front. Bioeng. Biotechnol. **8**(October), 1–18 (2020). https://doi.org/10.3389/fbioe.2020.585363
3. F. Jamme, S. Kascakova, S. Villette, F. Allouche, S. Pallu, V. Rouam, M. Réfrégiers, Deep UV autofluorescence microscopy for cell biology and tissue histology. Biol. Cell **105**(7), 277–288 (2013). https://doi.org/10.1111/boc.201200075
4. X. Dang, N.M. Bardhan, J. Qi, L. Gu, N.A. Eze, C.W. Lin et al., Deep-tissue optical imaging of near cellular-sized features. Sci. Rep. **9**(1), 1–12 (2019). https://doi.org/10.1038/s41598-019-39502-w
5. F.W. Pratiwi, C.W. Kuo, B.C. Chen, P. Chen, Recent advances in the use of fluorescent nanoparticles for bioimaging. Nanomedicine **14**(13), 1759–1769 (2019). https://doi.org/10.2217/nnm-2019-0105
6. S. Guha, S.K. Shaw, G.T. Spence, F.M. Roland, B.D. Smith, Clean photothermal heating and controlled release from near-infrared dye doped nanoparticles without oxygen photosensitization. Langmuir **31**(28), 7826–7834 (2015). https://doi.org/10.1021/acs.langmuir.5b01878
7. Q. Ma, J. Wang, Z. Li, X. Lv, L. Liang, Q. Yuan, Recent progress in time-resolved biosensing and bioimaging based on lanthanide-doped nanoparticles. Small **15**(32), 1–22 (2019). https://doi.org/10.1002/smll.201804969
8. X. Li, R. Wang, F. Zhang, L. Zhou, D. Shen, C. Yao, D. Zhao, Nd3+ sensitized up/down converting dual-mode nanomaterials for efficient in-vitro and in-vivo bioimaging excited at 800 nm. Sci. Rep. **3**, 1–7 (2013). https://doi.org/10.1038/srep03536
9. L. Francés-Soriano, N. Peruffo, M.M. Natile, N. Hildebrandt, Er3+-to-dye energy transfer in DNA-coated core and core/shell/shell upconverting nanoparticles with 980 nm and 808 nm excitation of Yb3+ and Nd3+. Analyst **145**(7), 2543–2553 (2020). https://doi.org/10.1039/c9an02532d
10. X. Han, K. Xu, O. Taratula, K. Farsad, Applications of nanoparticles in biomedical imaging. Nanoscale **11**(3), 799–819 (2019). https://doi.org/10.1039/c8nr07769j
11. S. Prabha, G. Arya, R. Chandra, B. Ahmed, S. Nimesh, Effect of size on biological properties of nanoparticles employed in gene delivery. Artif. Cells, Nanomed. Biotechnol. **44**(1), 83–91 (2016). https://doi.org/10.3109/21691401.2014.913054
12. C. Pastore, Size-dependent nano–bio interactions. Nat. Nanotechnol. **16**(10), 1052 (2021). https://doi.org/10.1038/s41565-021-00996-z
13. H. Peng, Z. Xu, Y. Wang, N. Feng, W. Yang, J. Tang, Biomimetic mesoporous silica nanoparticles for enhanced blood circulation and cancer therapy. ACS Appl. Bio Mater. **3**(11), 7849–7857 (2020). https://doi.org/10.1021/acsabm.0c01014
14. P. Bélteky, A. Rónavári, D. Zakupszky, E. Boka, N. Igaz, B. Szerencsés et al., Are smaller nanoparticles always better? Understanding the biological effect of size-dependent silver

nanoparticle aggregation under biorelevant conditions. Int. J. Nanomed. **16**, 3021–3040 (2021). https://doi.org/10.2147/IJN.S304138
15. J. Mosquera, I. García, L.M. Liz-Marzán, Cellular uptake of nanoparticles versus small molecules: a matter of size. Acc. Chem. Res. **51**(9), 2305–2313 (2018). https://doi.org/10.1021/acs.accounts.8b00292
16. A. Reisch, A.S. Klymchenko, Fluorescent polymer nanoparticles based on dyes: seeking brighter tools for bioimaging. Small **12**(15), 1968–1992 (2016). https://doi.org/10.1002/smll.201503396
17. M. Amjad, M. Iqbal, A. Faisal, A.M. Junjua, I. Hussain, S.Z. Hussain et al., Hydrothermal synthesis of carbon nanodots from bovine gelatin and PHM3 microalgae strain for anticancer and bioimaging applications. Nanoscale Adv. **1**(8), 2924–2936 (2019). https://doi.org/10.1039/c9na00164f
18. I.Z. Gutierrez, C. Gerke, Y. Shen, E. Ximendes, M.M. Silvan, R. Marin et al., Boosting the near-infrared emission of Ag2S nanoparticles by a controllable surface treatment for bioimaging applications. ACS Appl. Mater. Interfaces **14**(4), 4871–4881 (2022). https://doi.org/10.1021/acsami.1c19344
19. N. Liu, X. Chen, X. Sun, X. Sun, J. Shi, Persistent luminescence nanoparticles for cancer theranostics application. J. Nanobiotechnol. **19**(1), 1–24 (2021). https://doi.org/10.1186/s12951-021-00862-z
20. D.L. Fritzen, L. Giordano, L.C.V. Rodrigues, J.H.S.K. Monteiro, Opportunities for persistent luminescent nanoparticles in luminescence imaging of biological systems and photodynamic therapy. Nanomaterials **10**(10), 1–38 (2020). https://doi.org/10.3390/nano10102015
21. M. Shi, S. Lan, C. Zhang, W. Xiong, Q. Li, T. Maimaiti et al., Biocompatible zinc gallogermanate persistent luminescent nanoparticles for fast tumor drainage lymph node imaging in vivo. Colloids Surf., B **205**(May), 111887 (2021). https://doi.org/10.1016/j.colsurfb.2021.111887
22. C. Xie, X. Zhen, Q. Miao, Y. Lyu, K. Pu, Self-assembled semiconducting polymer nanoparticles for ultrasensitive near-infrared afterglow imaging of metastatic tumors. Adv. Mater. **30**(21), 1–9 (2018). https://doi.org/10.1002/adma.201801331
23. F. Lu, L. Li, T. Zhao, B. Ding, J. Liu, Q. Wang, C. Xie, Q. Fan, W. Huang, Fabrication of semiconducting polymer-blend dots with strong near-infrared fluorescence and afterglow luminescence for bioimaging. Dyes Pigm. **200**, 110124 (2022). https://doi.org/10.1016/j.dyepig.2022.110124
24. X. Zhao, L.J. Chen, K.C. Zhao, Y.S. Liu, J.L. Liu, X.P. Yan, Autofluorescence-free chemo/biosensing in complex matrixes based on persistent luminescence nanoparticles. TrAC—Trends Anal. Chem. **118**, 65–72 (2019). https://doi.org/10.1016/j.trac.2019.05.025
25. T.H. Qazi, D.J. Mooney, M. Pumberger, S. Geißler, G.N. Duda, Biomaterials based strategies for skeletal muscle tissue engineering: existing technologies and future trends. Biomaterials **53**, 502–521 (2015). https://doi.org/10.1016/j.biomaterials.2015.02.110
26. Y. Wu, M. Wang, S. Sun, S. Shao, Z. Pan, J. Li, H. Wei, B. Cao, Enhanced X-ray-induced luminescence and persistent luminescence of NaLuF4:Tb3+ nanoparticles with Li+ and Gd3+ co-doped. J. Lumin. **252**, 119393 (2022)
27. H. Jiang, L. Liu, K. Yu, X. Yin, S. Zheng, L. Song et al., Cr3+/Y3+ co-doped persistent luminescence nanoparticles with biological window activation for in vivo repeatable imaging. J. Rare Earths **40**(9), 1389–1398 (2022). https://doi.org/10.1016/j.jre.2021.05.007
28. X. Sun, L. Song, N. Liu, J. Shi, Y. Zhang, Chromium-doped zinc gallate near-infrared persistent luminescence nanoparticles in autofluorescence-free biosensing and bioimaging: a review. ACS Appl. Nano Mater. **4**(7), 6497–6514 (2021). https://doi.org/10.1021/acsanm.1c01115
29. T. Maldiney, G. Sraiki, B. Viana, D. Gourier, C. Richard, D. Scherman et al., In vivo optical imaging with rare earth doped Ca_2Si_5N_8 persistent luminescence nanoparticles. Optical Mate. Express **2**(3), 261 (2012). https://doi.org/10.1364/ome.2.000261
30. B. Viana, S.K. Sharma, D. Gourier, T. Maldiney, E. Teston, D. Scherman, C. Richard, Long term in vivo imaging with Cr3+ doped spinel nanoparticles exhibiting persistent luminescence. J. Lumin. **170**, 879–887 (2016). https://doi.org/10.1016/j.jlumin.2015.09.014

31. L. Li, F. Pan, P.A. Tanner, K.L. Wong, Tunable dual visible and near-infrared persistent luminescence in doped zinc gallogermanate nanoparticles for simultaneous photosensitization and bioimaging. ACS Appl. Nano Mater. **3**(2), 1961–1971 (2020). https://doi.org/10.1021/acsanm.9b02613
32. X. Wei, X. Huang, Y. Zeng, L. Jing, W. Tang, X. Li et al., Longer and stronger: improving persistent luminescence in size-tuned zinc gallate nanoparticles by alcohol-mediated chromium doping. ACS Nano **14**(9), 12113–12124 (2020). https://doi.org/10.1021/acsnano.0c05655
33. Y. Braeken, S. Cheruku, A. Ethirajan, W. Maes, Conjugated polymer nanoparticles for bioimaging. Materials **10**(12), 1–23 (2017). https://doi.org/10.3390/ma10121420
34. S. Orbay, O. Kocaturk, R. Sanyal, A. Sanyal, Molecularly imprinted polymer-coated inorganic nanoparticles: fabrication and biomedical applications. Micromachines **13**(9) (2022). https://doi.org/10.3390/mi13091464
35. R. Bilan, I. Nabiev, A. Sukhanova, Quantum dot-based nanotools for bioimaging, diagnostics, and drug delivery. Chembiochem **357** (2016). https://doi.org/10.1002/cbic.201600357
36. D. Xiao, H. Qi, Y. Teng, D. Pierre, P.T. Kutoka, D. Liu, Advances and challenges of fluorescent nanomaterials for synthesis and biomedical applications. Nanoscale Res. Lett. (2021). https://doi.org/10.1186/s11671-021-03613-z
37. M. Luo, K. Shaitan, X. Qu, A.P. Bonartsev, B. Lei, Bioactive rare earth-based inorganic-organic hybrid biomaterials for wound healing and repair. Appl. Mater. Today **26**, 101304 (2022). https://doi.org/10.1016/j.apmt.2021.101304
38. Z. Wei, L. Sun, J. Liu, J.Z. Zhang, H. Yang, Y. Yang, L. Shi, Cysteine modified rare-earth up-converting nanoparticles for invitro and invivo bioimaging. Biomaterials **35**(1), 387–392 (2014). https://doi.org/10.1016/j.biomaterials.2013.09.110
39. E. Andresen, U. Resch-Genger, M. Schäferling, Surface modifications for photon-upconversion-based energy-transfer nanoprobes. Langmuir **35**(15), 5093–5113 (2019). https://doi.org/10.1021/acs.langmuir.9b00238
40. J.L. Liu, X. Zhao, L.J. Chen, L.M. Pan, X.P. Yan, Dual-emissive persistent luminescence nanoparticle-based charge-reversible intelligent nanoprobe for persistent luminescence-ratio bioimaging along with chemo-photothermal synergic therapy. Anal. Chem. **93**(19), 7348–7354 (2021). https://doi.org/10.1021/acs.analchem.1c01220
41. Y. Lv, D. Ding, Y. Zhuang, Y. Feng, J. Shi, H. Zhang et al., Chromium-doped zinc Gallogermanate@Zeolitic Imidazolate framework-8: a multifunctional nanoplatform for rechargeable in vivo persistent luminescence imaging and pH-responsive drug release. ACS Appl. Mater. Interfaces **11**(2), 1907–1916 (2019). https://doi.org/10.1021/acsami.8b19172
42. R.M. Calderón-Olvera, E. Arroyo, A.M. Jankelow, R. Bashir, E. Valera, M. Ocaña, A.I. Becerro, Persistent luminescence Zn2GeO4:Mn2+ nanoparticles functionalized with poly-acrylic acid: one-pot synthesis and biosensing applications. ACS Appl. Mater. Interfaces **15**(17), 20613–20624 (2023). https://doi.org/10.1021/acsami.2c21735
43. R. Zou, Y. Gao, Y. Zhang, J. Jiao, K.L. Wong, J. Wang, 68Ga-labeled magnetic-NIR persistent luminescent hybrid mesoporous nanoparticles for multimodal imaging-guided chemotherapy and photodynamic therapy. ACS Appl. Mater. Interfaces **13**(8), 9667–9680 (2021). https://doi.org/10.1021/acsami.0c21623
44. L.J. Chen, C.X. Yang, X.P. Yan, Liposome-coated persistent luminescence nanoparticles as luminescence trackable drug carrier for chemotherapy. Anal. Chem. **89**(13), 6936–6939 (2017). https://doi.org/10.1021/acs.analchem.7b01397
45. J. Shi, X. Sun, S. Zheng, J. Li, X. Fu, H. Zhang, A new near-infrared persistent luminescence nanoparticle as a multifunctional nanoplatform for multimodal imaging and cancer therapy. Biomaterials **152**, 15–23 (2018). https://doi.org/10.1016/j.biomaterials.2017.10.032
46. S. Wu, Z. Qiao, Y. Li, S. Hu, Y. Ma, S. Wei, L. Zhang, Persistent luminescence nanoplatform with fenton-like catalytic activity for tumor multimodal imaging and photoenhanced combination therapy. ACS Appl. Mater. Interfaces **12**(23), 25572–25580 (2020). https://doi.org/10.1021/acsami.0c04438

47. P. Pei, Y. Chen, C. Sun, Y. Fan, Y. Yang, X. Liu et al., X-ray-activated persistent luminescence nanomaterials for NIR-II imaging. Nat. Nanotechnol. **16**(9), 1011–1018 (2021). https://doi.org/10.1038/s41565-021-00922-3
48. Q. Mu, K. Zhang, H. Zou, H. Liu, Y. Song, C.K. Wang, L. Lin, J. Fan, Theoretical insights into room temperature phosphorescence emission with anti-Kasha behavior in aggregate. Dyes Pigm. **205**, 110560 (2022). https://doi.org/10.1016/j.dyepig.2022.110560
49. J. Xu, S. Tanabe, Persistent luminescence instead of phosphorescence: history, mechanism, and perspective. J. Lumin. **205**, 581–620 (2019). https://doi.org/10.1016/j.jlumin.2018.09.047
50. A. Jain, A. Kumar, S.J. Dhoble, D.R. Peshwe, Persistent luminescence: an insight. Renew. Sustain. Energy Rev. **65**, 135–153 (2016). https://doi.org/10.1016/j.rser.2016.06.081
51. A. Antuzevics, G. Krieke, G. Doke, B. Berzina, The origin of bright cyan persistent luminescence in Ca2SnO4:La3+. Materialia **21**, 101374 (2022). https://doi.org/10.1016/j.mtla.2022.101374
52. Q. Lin, Z. Li, C. Ji, Q. Yuan, Electronic structure engineering and biomedical applications of low energy-excited persistent luminescence nanoparticles. Nanoscale Adv. **2**(4), 1380–1394 (2020). https://doi.org/10.1039/c9na00817a
53. B. Viana, C. Richard, V. Castaing, E. Glais, M. Pellerin, J. Liu, C. Chanéac, NIR-persistent luminescence nanoparticles for bioimaging, principle and perspectives. Springer Nature 163–197 (2020). https://doi.org/10.1007/978-3-030-32036-2_8
54. E. Arroyo, B. Torres Herrero, J.M. de la Fuente, M. Ocaña, A.I. Becerro, Highly uniform Y3Al2Ga3O12-based nanophosphors for persistent luminescence bioimaging in the visible and NIR regions. Inorganic Chem. Front. (2022). https://doi.org/10.1039/d2qi00480a
55. Z. Wu, A.C. Midgley, D. Kong, D. Ding, Organic persistent luminescence imaging for biomedical applications. Mater. Today Bio **17**(September), 100481 (2022). https://doi.org/10.1016/j.mtbio.2022.100481
56. B.M. Liu, R. Zou, S.Q. Lou, Y.F. Gao, L. Ma, K.L. Wong, J. Wang, Low-dose X-ray-stimulated LaGaO3:Sb,Cr near-infrared persistent luminescence nanoparticles for deep-tissue and renewable in vivo bioimaging. Chem. Eng. J. **404** (2021). https://doi.org/10.1016/j.cej.2020.127133
57. J.M. Liu, D.D. Zhang, G.Z. Fang, S. Wang, Erythrocyte membrane bioinspired near-infrared persistent luminescence nanocarriers for in vivo long-circulating bioimaging and drug delivery. Biomaterials **165**, 39–47 (2018). https://doi.org/10.1016/j.biomaterials.2018.02.042
58. J. Fu, Q.Y. Lv, Y.S. Li, X. Song, Q. Zhu, X. Ren, H.F. Cui, Bright, small sizes and hydro-dispersive NIR persistent luminescence nanoparticles modified with Si and amino groups for enhanced bioimaging. Nanotechnology **34**, 175601 (2023). https://doi.org/10.1088/1361-6528/acb69c
59. H. Zhao, G. Shu, J. Zhu, Y. Fu, Z. Gu, D. Yang, Persistent luminescent metal-organic frameworks with long-lasting near infrared emission for tumor site activated imaging and drug delivery. Biomaterials **217**(March) (2019). https://doi.org/10.1016/j.biomaterials.2019.119332
60. J. Wang, Q. Ma, Y. Wang, H. Shen, Q. Yuan, Recent progress in biomedical applications of persistent luminescence nanoparticles. Nanoscale **9**(19), 6204–6218 (2017). https://doi.org/10.1039/c7nr01488k
61. Y. Liu, J.M. Liu, D. Zhang, K. Ge, P. Wang, H. Liu et al., Persistent luminescence nanophosphor involved near-infrared optical bioimaging for investigation of foodborne probiotics biodistribution in vivo: a proof-of-concept study. J. Agric. Food Chem. **65**(37), 8229–8240 (2017). https://doi.org/10.1021/acs.jafc.7b02870
62. Q. Wang, S. Zhang, Z. Li, Q. Zhu, Near infrared-emitting Cr3+/Eu3+ Co-doped zinc gallogermanate persistence luminescent nanoparticles for cell imaging. Nanoscale Res. Lett. **13**, 15–21 (2018). https://doi.org/10.1186/s11671-018-2477-6
63. Y. Chen, F. Liu, Y. Liang, X. Wang, J. Bi, X.J. Wang, Z. Pan, A new up-conversion charging concept for effectively charging persistent phosphors using low-energy visible-light laser diodes. J. Mater. Chem. C **6**(30), 8002–8010 (2018). https://doi.org/10.1039/c8tc02419g

64. Z. Pan, Y.Y. Lu, F. Liu, Sunlight-activated long-persistent luminescence in the near-infrared from Cr3+-doped zinc gallogermanates. Nat. Mater. **11**(1), 58–63 (2012). https://doi.org/10.1038/nmat3173
65. J. Liu, T. Lécuyer, J. Seguin, N. Mignet, D. Scherman, B. Viana, C. Richard, Imaging and therapeutic applications of persistent luminescence nanomaterials. Adv. Drug Deliv. Rev. **138**, 193–210 (2019). https://doi.org/10.1016/j.addr.2018.10.015
66. Y.J. Chuang, Z. Zhen, F. Zhang, F. Liu, J.P. Mishra, W. Tang et al., Photostimulable near-infrared persistent luminescent nanoprobes for ultrasensitive and longitudinal deep-tissue bio-imaging. Theranostics **4**(11), 1112–1122 (2014). https://doi.org/10.7150/thno.9710
67. T. Maldiney, A. Bessiere, J. Seguin, E. Teston, S.K. Sharma, B. Viana, A.J.J. Bos, P. Dorenbos, M. Bessodes, D. Gourier, D. Scherman, C. Richard, The in vivo activation of persistent nanophosphors for optical imaging of vascularization, tumours and grafted cells. Nat. Mater. **13**(4), 418–426 (2014). https://doi.org/10.1038/nmat3908
68. Y. Zhuang, J. Ueda, S. Tanabe, Tunable trap depth in Zn(Ga1-xAlx)2O 4:Cr, Bi red persistent phosphors: considerations of high-temperature persistent luminescence and photostimulated persistent luminescence. J. Mater. Chem. C **1**(47), 7849–7855 (2013). https://doi.org/10.1039/c3tc31462f
69. D. Chen, W. Xu, X. Li, Y. Zhou, S. Liu, Y. Chen, Near-infrared photo-stimulated red luminescence in Eu2+/Ln3+ (Ln = La-Lu):SrS persistent phosphors. J. Alloy. Compd. **720**, 239–244 (2017). https://doi.org/10.1016/j.jallcom.2017.05.276
70. D.C. Rodríguez Burbano, S.K. Sharma, P. Dorenbos, B. Viana, J.A. Capobianco, Persistent and photostimulated red emission in CaS: Eu2+, Dy3+ nanophosphors. Adv. Opt. Mater. **3**(4), 551–557 (2015). https://doi.org/10.1002/adom.201400562
71. R. Kang, X. Dou, H. Lian, Y. Li, SrAl12O19: Fe3+ @3-aminopropyl triethoxysilane: ambient aqueous stable near-infrared persistent luminescent nanocomposites. J. Am. Ceram. Soc. **103**(1), 258–265 (2020). https://doi.org/10.1111/jace.16707
72. Y. Lin, J. Hu, Y. Guo, Q. Zou, D. Chen, K. Xu, S. Liang, X. Yi, H. Lu, S.B. Wang, H. Zhu, Enhanced persistent luminescence of MgGeO3: Yb3+ nanoparticles via substitution of Ge4+ by Ga3+ ions for biological applications. J. Lumi. **257**, 119741 (2023). https://doi.org/10.1016/j.jlumin.2023.119741
73. J. Xu, D. Murata, J. Ueda, B. Viana, S. Tanabe, Toward rechargeable persistent luminescence for the first and third biological windows via persistent energy transfer and electron trap redistribution. Inorg. Chem. **57**(9), 5194–5203 (2018). https://doi.org/10.1021/acs.inorgchem.8b00218
74. L. Song, X.H. Lin, X.R. Song, S. Chen, X.F. Chen, J. Li, H.H. Yang, Repeatable deep-tissue activation of persistent luminescent nanoparticles by soft X-ray for high sensitivity long-term in vivo bioimaging. Nanoscale **9**(8), 2718–2722 (2017). https://doi.org/10.1039/c6nr09553d
75. Z. Xue, X. Li, Y. Li, M. Jiang, H. Liu, S. Zeng, J. Hao, X-ray-activated near-infrared persistent luminescent probe for deep-tissue and renewable in vivo bioimaging. ACS Appl. Mater. Interfaces **9**(27), 22132–22142 (2017). https://doi.org/10.1021/acsami.7b03802
76. D.G. Calatayud, T. Jardiel, E. Cordero-Oyonarte, A.C. Caballero, M. Villegas, A. Valle-Noguera, et al., Biocompatible probes based on rare-earth doped strontium aluminates with long-lasting phosphorescent properties for in vitro optical imaging. Int. J. Mol. Sci. **23**(6) (2022). https://doi.org/10.3390/ijms23063410
77. D. Chen, C. Zhao, J. Ye, Q. Li, X. Liu, M. Su et al., In situ biosynthesis of fluorescent platinum nanoclusters: toward self-bioimaging-guided cancer theranostics. ACS Appl. Mater. Interfaces **7**(32), 18163–18169 (2015). https://doi.org/10.1021/acsami.5b05805
78. Q. Lin, Z. Li, Q. Yuan, Recent advances in autofluorescence-free biosensing and bioimaging based on persistent luminescence nanoparticles. Chin. Chem. Lett. **30**(9), 1547–1556 (2019). https://doi.org/10.1016/j.cclet.2019.06.016
79. Y. Zhang, N. Song, Y. Li, Z. Yang, L. Chen, T. Sun, Z. Xie, Comparative study of two near-infrared coumarin-BODIPY dyes for bioimaging and photothermal therapy of cancer. J. Mater. Chem. B **7**(30), 4717–4724 (2019). https://doi.org/10.1039/c9tb01165j

80. Y. Yang, L. Wang, B. Wan, Y. Gu, X. Li, Optically active nanomaterials for bioimaging and targeted therapy. Front. Bioeng. Biotechnol. **7**(November), 1–24 (2019). https://doi.org/10.3389/fbioe.2019.00320
81. S.K. Sun, H.F. Wang, X.P. Yan, Engineering persistent luminescence nanoparticles for biological applications: from biosensing/bioimaging to theranostics. Acc. Chem. Res. **51**(5), 1131–1143 (2018). research-article. https://doi.org/10.1021/acs.accounts.7b00619
82. A. Yusuf, A.R.Z. Almotairy, H. Henidi, O.Y. Alshehri, M.S. Aldughaim, Nanoparticles as Drug delivery systems: a review of the implication of nanoparticles' physicochemical properties on responses in biological systems. Polymers **15**(7) (2023). https://doi.org/10.3390/polym15071596
83. Y. Yao, Y. Zhou, L. Liu, Y. Xu, Q. Chen, Y. Wang et al., Nanoparticle-based drug delivery in cancer therapy and its role in overcoming drug resistance. Front. Mol. Biosci. **7**(August), 1–14 (2020). https://doi.org/10.3389/fmolb.2020.00193
84. Y. Yang, J. Mu, B. Xing, Photoactivated drug delivery and bioimaging. Wiley Interdisc. Rev.: Nanomed. Nanobiotechnol. **9**(2) (2017). https://doi.org/10.1002/wnan.1408
85. G. Zhang, R. Du, L. Zhang, D. Cai, X. Sun, Y. Zhou et al., Gadolinium-doped iron oxide nanoprobe as multifunctional bioimaging agent and drug delivery system. Adv. Func. Mater. **25**(38), 6101–6111 (2015). https://doi.org/10.1002/adfm.201502868
86. H.J. Zhang, X. Zhao, L.J. Chen, C.X. Yang, X.P. Yan, Dendrimer grafted persistent luminescent nanoplatform for aptamer guided tumor imaging and acid-responsive drug delivery. Talanta **219**(May), 121209 (2020). https://doi.org/10.1016/j.talanta.2020.121209
87. Y.F. Gao, R. Zou, G.F. Chen, B.M. Liu, Y. Zhang, J. Jiao et al., Large-pore mesoporous-silica-assisted synthesis of high-performance ZnGa2O4:Cr3+/Sn4+@MSNs multifunctional nanoplatform with optimized optical probe mass ratio and superior residual pore volume for improved bioimaging and drug delivery. Chem. Eng. J. **420**(P1), 130021 (2021). https://doi.org/10.1016/j.cej.2021.130021
88. J. Lee, J. Kim, J. Cheon, Magnetic nanoparticles for multi-imaging and drug delivery, 274–284 (2013). https://doi.org/10.1007/s10059-013-0103-0
89. M.K. Tsang, Y.T. Wong, J. Hao, Cutting-edge nanomaterials for advanced multimodal bioimaging applications. Small Methods **2**(1), 1–14 (2018). https://doi.org/10.1002/smtd.201700265
90. H. Tan, T. Wang, Y. Shao, C. Yu, L. Hu, Crucial breakthrough of functional persistent luminescence materials for biomedical and information technological applications. Front. Chem. **7**(MAY) (2019). https://doi.org/10.3389/fchem.2019.00387
91. J.M. Liu, Y.Y. Liu, D.D. Zhang, G.Z. Fang, S. Wang, Synthesis of GdAlO3:Mn4+, Ge4+@Au core-shell nanoprobes with plasmon-enhanced near-infrared persistent luminescence for in vivo trimodality bioimaging. ACS Appl. Mater. Interfaces **8**(44), 29939–29949 (2016). https://doi.org/10.1021/acsami.6b09580
92. S.J. Shin, J.R. Beech, K.A. Kelly, Targeted nanoparticles in imaging: paving the way for personalized medicine in the battle against cancer. Integr. Biol. (United Kingdom) **5**(1), 29–42 (2013). https://doi.org/10.1039/c2ib20047c
93. Q. Lin, P. Fathi, X. Chen, Nanoparticle delivery in vivo: a fresh look from intravital imaging. EBioMedicine **59**(2020). https://doi.org/10.1016/j.ebiom.2020.102958
94. N. Liu, J. Shi, Q. Wang, J. Guo, Z. Hou, X. Su et al., In vivo repeatedly activated persistent luminescence nanoparticles by radiopharmaceuticals for long-lasting tumor optical imaging. Small **16**(26), 1–8 (2020). https://doi.org/10.1002/smll.202001494
95. A.A. Date, J. Hanes, L.M. Ensign, Nanoparticles for oral delivery: design, evaluation and state-of-the-art. J. Control. Release **240**, 504–526 (2016). https://doi.org/10.1016/j.jconrel.2016.06.016
96. Y.J. Li, X.P. Yan, Synthesis of functionalized triple-doped zinc gallogermanate nanoparticles with superlong near-infrared persistent luminescence for long-term orally administrated bioimaging. Nanoscale **8**(32), 14965–14970 (2016). https://doi.org/10.1039/c6nr04950h
97. H. Liu, F. Ren, H. Zhang, Y. Han, H. Qin, J. Zeng, Y. Wang, Q. Sun, Z. Li, M. Gao, Oral administration of highly bright Cr3+ doped ZnGa2O4 nanocrystals for in vivo targeted imaging of

orthotopic breast cancer. J. Mater. Chem. B. **6**(10) (2018). https://doi.org/10.1039/C7TB03148C
98. L. Giordano, G. Cai, J. Seguin, J. Liu, C. Richard, L.C.V. Rodrigues, B. Viana, Persistent luminescence induced by upconversion: an alternative approach for rechargeable bio-emitters. Opt. Bioimaging Therapy **11**, 2201468 (2023). https://doi.org/10.1002/adom.202201468
99. C. Enrico, *Nanotheranostics and Theranostic Nanomedicine for Diseases and Cancer Treatment* (Elsevier Inc., Design of nanostructures for theranostics applications, 2018). https://doi.org/10.1016/B978-0-12-813669-0.00002-6
100. R.B. Ladju, Z.S. Ulhaq, G.V. Soraya, Nanotheranostics: a powerful next-generation solution to tackle hepatocellular carcinoma. World J. Gastroenterol. **28**(2), 176–187 (2022). https://doi.org/10.3748/wjg.v28.i2.176
101. H. Zhu, B. Li, C.Y. Chan, B.L.Q. Ling, J. Tor, X.Y. Oh, W. Jiang, E. Ye, Z. Li, Loh., X. J., Advances in single-component inorganic nanostructures for photoacoustic imaging guided photothermal therapy. Adv. Drug Deliv. Rev. **129**, 114644 (2023). https://doi.org/10.1016/j.addr.2022.114644
102. X. Zhen, C. Xie, K. Pu, Temperature-correlated afterglow of a semiconducting polymer nanococktail for imaging-guided photothermal therapy. Angew. Chem. **130**(15), 4002–4006 (2018). https://doi.org/10.1002/ange.201712550
103. M.R.K. Ali, Y. Wu, M.A. El-Sayed, Gold-nanoparticle-assisted plasmonic photothermal therapy advances toward clinical application. J. Phys. Chem. C **123**(25), 15375–15393 (2019). review-article. https://doi.org/10.1021/acs.jpcc.9b01961
104. X. Wu, K. Liu, Q. Huang, Q. Zhang, X. Yang, X. Liu, R. Wang, Photothermal therapy based on CuS nanoparticles for alleviating arterial restenosis induced by mechanical injury of endovascular treatment. Front. Mater. **7**(January), 1–10 (2021). https://doi.org/10.3389/fmats.2020.591281
105. T. Guo, Q. Tang, Y. Guo, H. Qiu, J. Dai, C. Xing et al., Boron quantum dots for photoacoustic imaging-guided photothermal therapy. ACS Appl. Mater. Interfaces **13**(1), 306–311 (2021). https://doi.org/10.1021/acsami.0c21198
106. T.K.L. Rezende, H.P. Barbosa, L.F. dos Santos, O.K. de Lima, P. Alves de Matos, T.M. Tsubone, et al., Upconversion rare earths nanomaterials applied to photodynamic therapy and bioimaging. Front. Chem. **10**(November), 1–15 (2022). https://doi.org/10.3389/fchem.2022.1035449
107. S. Wu, Y. Li, R. Zhang, K. Fan, W. Ding, L. Xu, L. Zhang, Persistent luminescence-polypyrrole nanocomposite for dual-modal imaging and photothermal therapy of mammary cancer. Talanta **221**(July 2020), 121435 (2021). https://doi.org/10.1016/j.talanta.2020.121435
108. Y. Wang, H.M. Meng, G. Song, Z. Li, X.B. Zhang, Conjugated-polymer-based nanomaterials for photothermal therapy. ACS Appl. Polymer Mater. **2**(10), 4258–4272 (2020). https://doi.org/10.1021/acsapm.0c00680
109. Y. Meng, J. Yang, R. Jiang, S. Wang, L. Zheng, G. Wang et al., Biocompatible PLNP-GNR composite nanoplatforms for monitoring deep-tissue photothermal therapy process. Appl. Surf. Sci. **562**(February), 150189 (2021). https://doi.org/10.1016/j.apsusc.2021.150189
110. Z.H. Wang, J.M. Liu, N. Zhao, C.Y. Li, S.W. Lv, Y. Hu et al., Cancer cell macrophage membrane camouflaged persistent luminescent nanoparticles for imaging-guided photothermal therapy of colorectal cancer. ACS Appl. Nano Mater. **3**(7), 7105–7118 (2020). https://doi.org/10.1021/acsanm.0c01433
111. P. Sarbadhikary, B.P. George, H. Abrahamse, Recent advances in photosensitizers as multifunctional theranostic agents for imaging-guided photodynamic therapy of cancer. Theranostics **11**(18), 9054–9088 (2021). https://doi.org/10.7150/thno.62479
112. G. Gunaydin, M.E. Gedik, S. Ayan, Photodynamic therapy—current limitations and novel approaches. Front. Chem. **9**(June), 1–25 (2021). https://doi.org/10.3389/fchem.2021.691697
113. J. Wang, J. Li, J. Yu, H. Zhang, B. Zhang, Large hollow cavity luminous nanoparticles with near-infrared persistent luminescence and tunable sizes for tumor afterglow imaging and chemo-/photodynamic therapies. ACS Nano **12**(5), 4246–4258 (2018). https://doi.org/10.1021/acsnano.7b07606

114. A. Bessière, J.O. Durand, C. Noûs, Persistent luminescence materials for deep photodynamic therapy. Nanophotonics **10**(12), 2999–3029 (2021). https://doi.org/10.1515/nanoph-2021-0254
115. M. Chang, W. Feng, L. Ding, H. Zhang, C. Dong, Y. Chen, J. Shi, Persistent luminescence phosphor as in-vivo light source for tumoral cyanobacterial photosynthetic oxygenation and photodynamic therapy. Bioactive Mater. **10**(July 2021), 131–144 (2022). https://doi.org/10.1016/j.bioactmat.2021.08.030
116. H. Zhao, L. Li, F. Li, C. Liu, M. Huang, J. Li, F. Gao, X. Ruan, D. Yang, An energy-storing DNA-based nanocomplex for laser-free photodynamic therapy. Adv. Mat. **34**, 2109920 (2022). https://doi.org/10.1002/adma.202109920

Multicolor Persistent Luminescence

Amba Mondal, Jairam Manam, and Sourav Das

Abstract Persistent phosphors are very demanding materials nowadays owing to their various applications. The emission wavelength determines the color of the emission. An activator can exhibit different emission colors in different hosts. Several approaches have been carried out to produce multicolor long-lasting phosphors, but their short persistent time and color fading issues restricted them from being used for potential applications. However, the results of the color conversion method are appreciable. This chapter discusses the multicolor persistent luminescence based on the color conversion method. The color conversion method is based on two parameters: phosphorescent light source and color conversion agent. The excitation spectrum of the color conversion agent should be matched with the emission spectrum of the phosphorescent light source. In the color conversion method, long-lasting phosphor (LLP) can be used as a phosphorescent light source, while the organic/ inorganic fluorescent dye can be used as a color conversion agent. The light emitted by LLP is absorbed by the dyes and converted into long-lasting phosphorescence with different colors. The mixture of LLPs and dye-encapsulated mesoporous silica nanoparticles showed multicolor persistent luminescence. Traditional LLPs have large grains with a small surface area, limiting the absorption of dye molecules on the particle surface. Furthermore, high concentrations of dyes can lead to self-quenching, reducing conversion efficiency. Also, some dyes are unstable and prone to bleaching under light exposure. To address these issues, fluorescent dyes are encapsulated into mesoporous silica. Mesoporous silica provides a significant surface area, allowing a larger number of dye molecules to be absorbed by one particle, thus increasing adsorption efficiency. Encapsulation enhances the stability of fluorescent dyes, preventing bleaching in light

A. Mondal · J. Manam · S. Das
Department of Physics, Indian Institute of Technology (Indian School of Mines), Dhanbad, India

A. Mondal
Department of Physics, Raniganj Girls' College, Raniganj, West Bengal, India

S. Das (✉)
Central Research Facility, Indian Institute of Technology (Indian School of Mines), Dhanbad, India
e-mail: sourav@iitism.ac.in

V. Kumar et al. (eds.), *Persistent Luminescence*, Progress in Optical Science and Photonics 30, https://doi.org/10.1007/978-981-97-4943-0_11

environments. The encapsulated fluorescent dyes produce characteristic emissions, contributing to improved color conversion efficiency.

Keywords Persistent luminescence · Long persistent phosphor · Color conversion process · Photoluminescence · Phosphorescence

1 Introduction

Long persistent luminescence is an interesting phenomenon of getting emission for a long time after removal of excitation. Various persistent luminescence phosphors have been discovered to date and continue. They are of rare earth or transition metal ion doped oxide, fluorides, and sulfide host materials which can be crystals or glasses for various practical applications. It was a challenging task to increase the persistent luminescence time from minutes to tens of hours. Proper traps should be created to get such long persistent luminescence. However, significant progress has been made in this field since the early twentieth century. We are now familiar with various colors emitting long persistent luminescent phosphors. Eu^{2+} doped alkaline-earth aluminate phosphors showed appreciable performance. They emit various colors like, $SrAl_2O_4$: Eu^{2+}, Dy^{3+} emits green, $BaAl_2O_4$: Eu^{2+}, Dy^{3+} emits cyan, $CaAl_2O_4$: Eu^{2+}, and Nd^{3+} emits blue. Eu^{2+} doped silicate phosphor $Sr_2MgSi_2O_7$: Eu^{2+}, Dy^{3+} is blue emitting long persistent phosphor, Eu^{3+} doped yttrium oxysulfide Y_2O_2S:Eu^{3+}, Ti^{4+}, $Mg^{2+}$is red emitting persistent phosphor. However, among the three primary colors, i.e., red, green, and blue, the persistent time of red-emitting phosphor is much less than that of blue and green. Also, there is a lack of other colors emitting long persistent phosphors, i.e., red, pink, violet, and yellow. Apart from the primary color emitting phosphors, the white light emitting long persistent phosphors are also needed for potential applications. Hence, multicolor long persistent phosphors are in great demand. They can be prepared by multiple doping in a single host and persistent energy transfer among different luminescent centers. Any primary color emitting long persistent phosphor's emission color can be converted with the addition of a color conversion agent. This color conversion agent can be an organic or inorganic fluorescent dye or any phosphor. A composite phosphor is prepared by mixing the long persistent phosphor and color conversion agent. In the composite phosphor, energy transfers from a long persistent phosphor to this color conversion agent satisfying some conditions and the emission color can be tuned. In this chapter, the multicolor long persistent phosphor based on the color conversion process is discussed in detail.

2 Various Long Persistent Phosphors

2.1 *Lanthanides Ion Doped Long Persistent Phosphors*

2.1.1 Eu^{2+} Doped Long Persistent Phosphors

Eu^{2+} ions serving as emitting centers of persistent phosphors have received interest in the past decades. Eu^{2+} ions can be doped into different host materials, including aluminates, silicates, and other oxides. The divalent Eu^{2+} lanthanide ion has $4f^{n-1}5d$ electronic configurations in its excited state, which is strongly influenced by lattice structures, as the 5d orbital is sensitive to the surrounding environment of the host lattice. The transitions occur during optical absorption and emissions are $4f^n \leftrightarrow 4f^{n-1}5d$, which leads to broad absorption and emission bands. The absorption and emission bands of Eu^{2+} ions show strong changes from one host to another.

Since the discovery of $SrAl_2O_4$: Eu^{2+}, Dy^{3+}, the aluminate hosts have been given attention in Eu^{2+} doped long persistent phosphor materials. The prolonged persistent luminescence of MAl_2O_4: Eu^{2+}, RE^{3+} (M = Sr, Ca, Ba; RE = Dy, Nd) are due to codopant RE^{3+} which induced defect level. In MAl_2O_4: Eu^{2+}, RE^{3+} long persistent phosphors, Eu^{2+} acts as the emitting center while Dy^{3+}/Nd^{3+} acts as a trapping center. Silicate-based phosphors have attracted much due to their high chemical and physical stability and weather resistance. The persistence color of silicate phosphor varies from blue to green region depending strongly upon the surrounding crystal field. Among them, $Sr_2MgSi_2O_7$: Eu^{2+}, Dy^{3+} long persistent phosphor emits bright blue emission.

2.1.2 Eu^{3+} Activated Long Persistent Phosphor

For Eu^{3+} doped phosphors, the main emission peaks are due to $^5D_J(J = 0,1) \rightarrow {}^7F_J$ (J = 0, 1, 2,3,4) transitions of Eu^{3+}. Sometimes several narrow emission bands can also be observed due to f-f transition. When Eu^{3+} is doped in the Y_2O_2S/Gd_2O_2S host, Eu^{3+} replaces the site of Y^{3+}/Gd^{3+}, and the codoped Mg^{2+}, Ti^{3+} also replaces the site of Y^{3+}/Gd^{3+} to form traps to store the energy and gradually transfer it to Eu^{3+}.

2.1.3 Sm^{3+} Activated Long Persistent Phosphor

Sm^{3+} is a suitable host to emit red light when incorporated into a given host. Various oxides (CaO, $La_2Zr_2O_7$, $Sr_2ZnSi_2O_7$) oxysulfides (La_2O_2S), fluoride (KY_3F_{10}), Ca_2SnO_4, Sr_2SnO_4, $MSnO_4$ (M = Ca, Sr, Ba) can exhibit reddish orange afterglow of Sm^{3+} ions with an afterglow for 7 h.

2.1.4 Pr^{3+} Activated Long Persistent Phosphor

Perovskite type oxides ($CaTiO_3$, $NaNbO_3$, $SrZrO_3$) host are used for Pr^{3+} activator to achieve red afterglow. It shows an emission owing to Pr^{3+} $^1D_2 \rightarrow {}^3H_4$, $^3P_0 \rightarrow {}^3F_2$ transitions. $CdGeO_3$, $ZnTa_2O_6$, $Y_3Al_5O_{12}$, $La_2Ti_2O_7$, $Ca_2Zn_4Ti_{16}O_{38}$, $CaSnSiO_5$, Ca_2SnO_4, YPO_4 are some reported host for Pr^{3+} activator with red afterglow.

2.1.5 Ce^{3+} Activated Long Persistent Phosphor

Ce^{3+} is the second important afterglow activator of the lanthanide series that has $4f^1$ configuration. The 5d-4f transition energy experiences a strong crystal field and gives rise to a broad emission band. Lu_2SiO_5, $Ca_2Al_2SiO_7$, $SrAl_2O_4$, $CaAl_2O_4$, $CaAl_4O_7$, $Sr_4Al_{14}O_{25}$, $Sr_3Al_2O_5Cl_2$ are some of the host for Ce^{3+} dopant. They show blue-green LPP with more than 3 h persistence time. The green-yellow LPP are shown in Ce^{3+} doped $Y_3Sc_2Ga_3O_{12}$, $LuAl_2Ga_3O_{12}$, $Y_3Al_2Ga_3O_{12}$, and $Gd_3Al_2Ga_3O_{12}$ under excitation by 460 nm blue light along with UV light.

2.1.6 Dy^{3+} Activated Long Persistent Phosphor

Dy^{3+} ions are usually served as codopant into Eu^{2+} doped aluminates and silicates to increase the afterglow lifetime of Eu^{2+}. However, only Dy^{3+} can also act as an activator to get white light. It is well known that white light-emitting phosphor can be made by mixing three primary colors, red, green, and blue phosphor. But this is a challenging task as the persistent white color cannot be retained for all the time in this case. Different phosphors have different lifetimes. Hence, Dy^{3+} affords for white light afterglow emission. $^4F_{9/2} \rightarrow {}^6H_{15/2}$ and $^4F_{9/2} \rightarrow {}^6H_{13/2}$ are two main transitions responsible for emission lines ranging from 470–500 nm (blue) and 570–600 (yellow) of Dy^{3+}. Dy^{3+} ions are capable of emitting blue, green, and red light simultaneously. Dy^{3+}-activated phosphors are highly promising for the realization of white afterglow for the illumination application. $CdSiO_3:Dy^{3+}$ emits white persistent luminescence for 5 h. Similarly, $SrSiO_3:Dy^{3+}$ also produces white afterglow for more than 1 h. Other hosts include $Ca_3SnSi_2O_9$, $Sr_2Al_2SiO_7$, $CaAl_2O_4$, $CaMgSi_2O_6$, $CaMgSi_2O_6$, $CaSnSiO_5$, Sr_2SiO_4 for white long persistent luminescence with Dy^{3+} activator.

2.1.7 Tb^{3+}, Tm^{3+}, Yb^{3+}, Yb^{2+} Activated Long Persistent Phosphors

Long green afterglow can be obtained from Tb^{3+} activated phosphors owing to $^5D_4 \rightarrow {}^7F_J$ (J = 6, 5, 4,3) transitions. Blue LPP was obtained from Tm^{3+} doped $Zn_2P_2O_7$ due to $^1D_2 \rightarrow {}^3H_6$, $^1D_2 \rightarrow {}^3H_4$, and $^1G_4 \rightarrow {}^3H_6$ transitions of Tm^{3+}. Blue LPP was observed for 1 h after the removal of UV excitation. $Sr_4Al_{14}O_{25}$: Yb^{2+}, $Dy^{3+}$and $SrAl_xO_{(1+1.5x)}$: Yb^{2+} (x = 3, 4, 5) phosphors exhibit green LPP. On the other hand, red LPP was observed in SrS: Yb^{2+}. The afterglow emission is due to the $4f^{13}5d^1 \rightarrow$

$4f^{14}$ transition of Yb^{2+}. $Gd_{2x}Yb_xO_2CO_3{:}Yb^{3+}$ exhibited NIR long persistent emission with a persistence time of 144 h.

2.2 *Transition Metal Ion Doped Long Persistent Phosphors:*

Transition metal ions have an incomplete d shell. The d electrons are largely affected by the outer crystal field. Among the transition metal ions, Mn^{2+}, Mn^{4+}, and Cr^{3+} have been investigated thoroughly.

2.2.1 Mn^{2+}, Mn^{4+} Activated Long Persistent Phosphor

Green as well as red persistent luminescence can be obtained from Mn^{2+} activated phosphor. Parity-forbidden 3d-3d intra-atomic transition from the level $^4T_1$ (^4G), which is the lowest excited level to the ground state $^6A_1$ (^6S) is responsible for the emission. Tetrahedral coordinated Mn^{2+} ions give green emission and octahedral coordinated Mn^{2+} ions exhibit an orange to red emission. Hence, the emission color is dependent upon the coordination number of Mn^{2+} in the host. For example, Zn_2GeO_4: Mn^{2+} shows green afterglow while $CdSiO_3$: Mn^{2+}, $CaZnGe_2O_6{:}Mn^{2+}$, $MgGeO_3$: Mn^{2+}, $Ca_3(PO_4)_2{:}Mn^{2+}$, $CaMgSi_2O_6{:}Mn^{2+}$, etc. presents an orange-red LPP.

Mn^{4+} is another potential activator. It has a $3d^3$ electronic configuration. This results in an NIR emission from 600 to 800 nm in different hosts. This is strongly dependent on the crystal field environment of the host lattices. $LaAlO_3{:}Mn^{4+}$, Ge^{4+}, $GdAlO_3{:}Mn^{4+}$, Ge^{4+} shows the NIR long persistent emission band with a peak at 730 nm and persistent time for more than 20 h.

2.2.2 Cr^{3+} Activated Long Persistent Phosphor

Cr^{3+} ion has electronic configuration $1s^22s^22p^63s^23p^64s^23d^3$. Thus Cr^{3+} is a d^3 ion. These three electrons are distributed among the five 3d orbitals depending upon the surrounding environment. ^4F, ^4P, ^2G, ^2H, etc. are ground state terms. These levels split into two or more levels whose degeneracy is represented by $2L + 1$. These crystal field-modified degenerate energy levels are represented by ^{2S+1}X, X corresponds to A for no degeneracy, E for two-fold degeneracy, and T for three-fold degeneracy. The NIR emission in Cr^{3+} is observed because of spin forbidden $^2E \rightarrow {}^4A_2$ transition or a broad-band emission (650–1600 nm) corresponding to spin-allowed $^4T_2 \rightarrow {}^4A_2$ transition depending upon the host crystal-field environment. For a strong crystal field environment, $^2E \rightarrow {}^4A_2$ transition dominates, and for a weak crystal field environment, $^4T_2 \rightarrow {}^4A_2$ transition dominates, and broadband is achieved. Hence, the NIR emission wavelength from Cr^{3+} can be controlled by modifying the ligand field strength. Further, defect levels/traps play an important role in making the emission long persistent. Keeping this in view, Cr^{3+} has been studied for NIR persistent luminescence in different crystal systems.

For developing Cr^{3+} activated NIR persistent phosphor, gallates are frequently used as hosts due to the excellent ability of Cr^{3+} to replace the Ga^{3+} because of the nearly equal ionic radii of the two ions. Replacement of Ga^{3+} by Cr^{3+} in a distorted octahedral site creates suitable host crystal field strength around Cr^{3+} ions to achieve intense NIR emission. Because of this, Cr^{3+} is widely used as a dopant in gallate-based hosts for NIR long-lasting phosphorescence. Table 1 shows the different colors emitting reported long persistent phosphors.

Hence, from Table 1, it is seen that most of the blue emitting phosphors are rare earth-doped silicates and aluminates. The best performance is exhibited by $Sr_2MgSi_2O_7$:Eu^{2+}, Dy^{3+}, and $Sr_3MgSi_2O_8$:Eu^{2+}, Dy^{3+} phosphor with more than 10-h persistent time. $CaAl_2Si_2O_8$: Eu^{2+}, Dy^{3+}, $CaMgSi_2O_6$:Eu^{2+}, Dy^{3+}, $Ca_3MgSi_2O_8$: Eu^{2+}, Dy^{3+} exhibit persistent blue phosphorescence for around 4–5 h. Among aluminates, $CaAl_2O_4$:Eu^{2+}, and Nd^{3+} show blue LLP with a persistent time of 16 h. Also, $CaAl_2O_4$:Ce^{3+}, $BaAl_2O_4$:Ce^{3+}, $Sr_4Al_{14}O_{25}$:Eu^{2+}, Dy^{3+}, $SrAl_2O_4$:Ce^{3+} show appreciable results. Among the phosphate hosts, $Sr_2P_2O_7$:Eu^{2+}, Y^{3+}, $Ca_2P_2O_7$:Eu^{2+}, Y^{3+} $Ca_9Bi(PO_4)_7$:Eu^{2+}, Dy^{3+}emit LLP for 8 h, 6 h, and 5 h, respectively. Ca_2SnO_4:Tm^{3+} shows blue blue-long persistent signal for 5 h. $Sr_3Al_2O_5Cl_2$: Ce^{3+} emits for 3 h. From the table, it can be seen that there are abundant hosts that emit blue LLP with appreciable persistent time.

Table 2 also shows that most of the green-emitting phosphors are silicates and aluminates. Ca_2MgSiO_7:Eu^{2+}, Dy^{3+}, Nd^{3+}, $Ba_2MgSi_2O_7$:Eu^{2+} show persistent green signal for 5 h. $Ca_2MgSi_2O_7$:Eu^{2+} shows the best result of 12 h green persistent signal. $Ca_2Al_2SiO_7$:Ce^{3+}, Mn^{2+} exhibit LLP for 10 h. $Ba_4(Si_3O_8)_2$: Eu^{2+}, Ho^{3+}, $CdSiO_3$:Tb^{3+} emits for 4 h and 5 h, respectively. Among aluminates, the best result is shown by $SrAl_2O_4$:Eu^{2+}, Dy^{3+} with a significant persistent time of 30 h. $CaAl_2O_4$:Ce^{3+}, Mn^{2+}, $CaAl_2O_4$:Ce^{3+}, Tb^{3+} give green LLP for 10 h. Among the gallates the $ZnGa_2Ge_2O_{10}$:Mn^{2+} has appreciable green LPL for 8 h. $CaZnGe_2O_6$:Tb^{3+}, Li_2ZnGeO_4:Mn^{2+}, Mg_2SnO_4:Mn^{2+}, Ca_2SnO_4:Tb^{3+} are also good long persistent green phosphor with persistent time ~3–5 h. The binary oxide Lu_2O_3:Tb^{3+}, Ca^{2+}, and Sr^{3+} exhibit green LPL for 20–30 h, which is a significant result. Hence, these materials are of great importance from an application point of view.

Table 3 shows that for yellow emitting long persistent phosphors, Sr_3SiO_5:Eu^{2+}, Lu^{3+}, $Ca_2ZnSi_2O_7$: Eu^{2+}, Dy^{3+} show luminescence for 20 h and 12 h, respectively, which are very significant results for yellow LPP. $Cd_2Ge_7O_{16}$: Mn^{2+} gives persistent signal for 3 h. Among oxysulfides, Y_2O_2S: Ti^{3+}, Mg^{2+}, Gd^{3+}, Lu^{3+}, Y_2O_2S: Tm^{3+}, Mg^{2+}, Ti^{4+}, Gd_2O_2S: Ti^{3+} emit for 5 h, 1 h and 2 h, respectively. Ca_2BO_3Cl: Eu^{2+}, Dy^{3+} is proved to be the best yellow long persistent phosphor with a persistent time of 48 h.

From Table 4, it is seen that silicates, titanates, and stannates are substantial hosts for getting orange long-lasting emissions. Pr^{3+}, Eu^{3+}, and Sm^{3+} doped $CdSiO_3$ show orange LLP for 5 h. Tb^{3+}, Mn^{2+}, and Eu^{3+}, Mn^{2+} co-doped $CdSiO_3$ emit for 1 h. Sm^{3+} doped CaSnO4, $Sr_3Sn_2O_7$ exhibit for 7 h and 1 h. $CaTiO_3$:Pr^{3+} emit orange

Table 1 Blue emitting long-lasting phosphors

Host materials		Dopants	Afterglow emission wavelength (nm)	Afterglow duration	References
Silicates	$Sr_2MgSi_2O_7$	Eu^{2+}, Dy^{3}	465	20 h	Shi et al. [1]
	$Sr_3MgSi_2O_8$	Eu^{2+}, Dy^{3+}	460	10 h	Lin et al. [2]
	$Sr_2Al_2SiO_7$	Eu^{2+}, Dy^{3+}	484	2 h	Ding et al. [3]
	$Sr_2ZnSi_2O_7$	Eu^{2+}, Dy^{3+}	460	5 min	Jiang et al. [4]
	$Sr_3Al_{10}SiO_{20}$	Eu^{2+},Ho^{3+}	465	6 h	Kuang et al. [5]
	$CaAl_2Si_2O_8$	Eu^{2+}, Dy^{3+}	440	3 h	Wang et al. [6]
	$Ca_3MgSi_2O_8$	Eu^{2+}, Dy^{3+}	470	5 h	Lin et al. [2]
	$CaMgSi_2O_6$	Eu^{2+}, Dy^{3+}	445	4 h	Jiang et al. [7]
	$Ba_3MgSi_2O_8$	Eu^{2+}, Dy^{3+}	440	1 h	Lin et al. [2]
	$Ba_{13}Al_{22}Si_{10}O_{66}$	Eu^{2+}	375–700	43 min	Shen et al. [8]
	Lu_2SiO_5	Ce^{3+}	410	3 h	Dorenbos et al. [9]
	$Ca_2Al_2SiO_7$	Ce^{3+}	400–417	1 h	Kodama et al. [10]
	$SrMgSi_2O_6$	Dy^{3+}	455 + 576	5 min	He et al. [11]
	$Ca_3SnSi_2O_9$	Defect	426	700 s	Xu et al. [12]
Aluminates	$CaAl_2O_4$	Eu^{2+}, Nd^{3+}	440	16 h	Zhang et al. [13]
	$Ca_{12}Al_{14}O_{33}$	Eu^{2+}, Nd^{3+}	440	10 min	Zhang et al. [14]
	$Sr_4Al_{14}O_{25}$	Eu^{2+}, Dy^{3+}	490	20 h	Suriyamurthy and Panigrahi [15]
	$SrA_{14}O_7$	Eu^{2+}, Dy^{3+}	480	3 h	Jianhui et al. [16]
	$SrAl_{12}O_{19}$	Eu^{2+}, Dy^{3+}	400	3 h	Nag and Kutty [17]
	$BaMgAl_{10}O_{17}$	Eu^{2+}, Co^{3+}	450	5 min	Zhang et al. [18]
	$BaAl_2O_4$	Ce^{3+}	402 + 450	10 h	Jia et al. [19]
	$CaAl_2O_4$	Ce^{3+}	413	10 h	Jia and Yen [20]
	$CaAl_4O_7$	Ce^{3+}	325, 420	10 h	Jia et al. [21], Van den Eeckhout et al. [22]
	$SrAl_2O_4$	Ce^{3+}	375–385, 427	10 h	Jia et al. [23]
	$Sr_4Al_{14}O_{25}$	Ce^{3+}	472 + 511	10 min	Sharma et al. [24]
	$Sr_4Al_{14}O_{25}$	Tb^{3+}	380	Disputed	Su et al. [25]

(continued)

Table 1 (continued)

Host materials		Dopants	Afterglow emission wavelength (nm)	Afterglow duration	References
Phosphate	$Sr_2P_2O_7$	Eu^{2+}, Y^{3+}	420	8 h	Pang et al. [26]
	$Ca_2P_2O_7$	Eu^{2+}, Y^{3+}	415	6 h	Pang et al. [27]
	$SrMg_2(PO_4)_2$	Eu^{2+}, Tb^{3+}	411	2 h	Liu et al. [28]
	$BaMg_2(PO_4)_2$	Eu^{2+}, Gd^{3+}	444	2 h	Liu et al. [28]
	$Ca_3Mg_3(PO_4)_4$	Eu^{2+}	433	1000 s	Ju et al. [29]
	$Ca_9Bi(PO_4)_7$	Eu^{2+}, Dy^{3+}	475	5 h	Li et al. [30]
	$Zn_2P_2O_7$	Tm^{3+}	365, 452, 478	1 h	Pang et al. [31, 32]
Stannate	Ca_2SnO_4	Tm^{3+}	468	5 h	Li et al. [33]
	Mg_2SnO_4	Ti^{4+}	458	2 h	Zhang et al. [34, 35]
	Ca_2SnO_4	Defect, Gd^{3+}	410, 466	3 h	Shi et al. [36], Li et al. [37]
Tungstate	$CaWO_4$	Bi^{3+}	440	1000 s	Kubus et al. [38], Kang, and Peng [39]
Zirconate	$BaZrO_3$	Defect, Mg^{2+}	350–500	2000s	Ma et al. [40]
Germanate	Zn_2GeO_4	Bi^{3+}	455	500 s	Zhang et al. [41]
Others	$Sr_5(PO_4)_3Cl$	Eu^{2+}	467	1040 s	Wu et al. [42]
	$Ba_5(PO_4)_3Cl$	Eu^{2+}, Gd^{3+}	435	1000 s	Ju et al. [43]
	$Sr_3Al_2O_5Cl_2$	Ce^{3+}	435	3 h	Ning et al. [44]

LLP for 2 h. $La_2Ti_2O_7$:Pr^{3+} has a persistent time of 1 h. Y_2O_2S: Eu^{3+}, Mg^{2+}, Ti^{4+} emits for 3 h. $Ca_2Si_5N_8$: Eu^{2+}, Tm^{3+} shows orange LLP for more than 3 h. $Ga_2Si_5N_8$: Eu^{2+}, Tm^{3+} produce LPP of 1 h. Eu^{2+}, Pr^{3+} codoped SrS shows the best result of 16 h persistent time. Among others $Sr_3Al_2O_5Cl_2$:Eu^{2+}, Tm^{3+}, SrO: Eu^{3+}, and CaO: Eu^{3+} are good phosphors with persistent time around 1–3 h.

From Table 5, we can see that though various hosts with rare earth doping exhibit red LLP but the number is less compared to blue and green emitting phosphors. Tm^{3+}, Sm^{3+} doped $CdSiO_3$ exhibit red LLP for 5 h. Eu^{2+}, Mn^{2+} co-doped MgSiO3, $Sr_3MgSi_2O_8$ emit for 4 h and 2 h, respectively. Ce^{3+}, Mn^{2+} co-doped $BaMg_2Si_2O_7$ also shows red LLP for 2 h. $Ca_3(PO_4)_2$: Mn^{2+}, Dy^{3+}, $Ca_{9.5}Mn(PO_4)_7$: Mn^{2+}, Ca_2SnO_4: Sm^{3+}, Sr_2SnO_4: Sm^{3+}, Dy^{3+}, $Sr_3Sn_2O_7$: Sm^{3+} phosphors have persistent time of 1 h. Ca_2SnO_4: Eu^{3+} has persistent time of almost 2 h. $Na_2CaSn_2Ge_3O_{12}$: Sm^{3+}, $MgGeO_3$:Mn^{2+}, Bi^{3+}, $CaZnGe_2O_6$: Mn^{2+} exhibit LLP for 4.8 h, 1.8 h, and 3 h, respectively. CaS: Eu^{2+}, Tm^{3+}, Y_2O_2S: Eu^{3+}, Mg^{2+}, Ti^{4+} have persistent time of 1 h

Table 2 Green emitting long-lasting phosphors

Host materials		Dopants	Afterglow emission wavelength (nm)	Afterglow duration	References
Silicates	$Sr_2MgSi_2O_7$	Eu^{2+}, Tb^{3+}	515, 530	Disputed	Li et al. [45]
	Ca_2MgSiO_7	Eu^{2+}, Dy^{3+} Nd^{3+}	518	5 h	Jiang et al. [46]
	$Ba_2MgSi_2O_7$	Eu^{2+}	500, 515	5 h	Tam et al. [47]
	Sr_2SiO_4	Eu^{2+}, Dy^{3+}	480	5 min	Lakshminarasimhan and Varadaraju[48]
	$Sr_2Al_2SiO_7$	Eu^{2+}, Dy^{3+}	485	2 h	Gong et al. [49]
	$Ba_4(Si_3O_8)_2$	Eu^{2+},Ho^{3+}	496	4 h	Gong et al. [50]
	$BaMgSiO_4$	Eu^{2+}	537	1.5 h	Li et al. [51]
	$Ca_2MgSi_2O_7$	Eu^{2+}	535	12 h	He et al. [52]
	$CaAl_2Si_2O_8$	Eu^{2+}, Pr^{3+}	375–600	2 h	Ju et al. [53], Dai et al. [54]
	$CdSiO_3$	Tb^{3+}	540	5 h	Rodrigueset al. [55]
	$CdSiO_3$	Pb^{2+}	498	2 h	Kuang and Liu [56]
	$Ca_2Al_2SiO_7$	Ce^{3+}, Mn^{2+}	550	10 h	Wang [57]
	$Ca_3SnSi_2O_9$	Tb^{3+}	426, 495, 542, 590	10 min	Matsuzawa et al. [58]
Aluminates	$SrAl_2O_4$	Eu^{2+}, Dy^{3+}	520	30 h	Xu et al. [12]
	$BaAl_2O_4$	Eu^{2+}, Dy^{3+}	500	2 h	Ji et al. [59]
	$Sr_3Al_2O_6$	Eu^{2+}, Dy^{3+}	512	Few hours	Chang et al. [60]
	$SrMgAl_{10}O_{17}$	Eu^{2+}, Dy^{3+}	515	3 min	Tang et al. [61]
	$CaAl_2O_4$	$Ce^{3+} \rightarrow Mn^{2+}$	525	10 h	Xu et al. [62]
	$SrAl_2O_4$	$Ce^{3+} \rightarrow Mn^{2+}$	375, 515	5 h	Matsuzawa et al. [58]
	$CaAl_2O_4$	$Ce^{3+} \rightarrow Tb^{3+}$	543	10 h	Jia et al. [63]
	$Y_3Al_5O_{12}$	Ce^{3+}	525	2 min	Zhong-Fei et al. [64]

(continued)

Table 2 (continued)

Host materials		Dopants	Afterglow emission wavelength (nm)	Afterglow duration	References
	$Y_3Al_{2.5}Ga_{2.5}O_{12}$	Ce^{3+}	505	200–1000 min	Ueda et al. [65]
	$CaAl_2O_4$	Tb^{3+}	443, 543, 590, 621	1 h	Satapathy et al. [66]
	$Y_3Al_2Ga_3O_{12}$	$Ce^{3+} \rightarrow Cr^{3+}$	505, 700	5 min	Ueda et al. [67], Xu et al. [68]
	$SrAl_2O_4$	$Eu^{2+} \rightarrow Nd^{3+}, Dy^{3+}$	525, 882	15 min	Teng et al. [69]
	$SrAl_2O_4$	$Eu^{2+} \rightarrow Er^{3+}, Dy^{3+}$	515, 1530	100 min	Yu et al. [70]
	$Sr_4Al_{14}O_{25}$	$Eu^{2+} \rightarrow Cr^{3+} Dy^{3+}$	495, 695	10 h, 20 min	Zhong et al. [71]
	$Sr_4Al_{14}O_{25}$	Ce^{3+}	472, 511	600 s	Jia et al. [23], Van den Eeckhout et al. [72]
Gallate	$Y_3Sc_2Ga_3O_{12}$	Ce^{3+}	500	100 min	Luo and Xia [73], Zhuang et al. [74]
	$ZnGa_2O_4$	Mn^{2+}	504	15 min	Satapathy et al. [66]
	$Zn_3Ga_2Ge_2O_{10}$	Mn^{2+}	520	8 h	Xu et al. [75]
Phosphate	$Zn_3(PO_4)_2$	Mn^{2+}, Ga^{3+}	507, 620	2 h	Liu and Zhong [76]
	$Ca_9Tb(PO_4)_7$	Tb^{3+}	490 + 545	5 h	Bessière et al. [77]
Zirconate	$SrZrO_3$	Pr^{3+}	350, 499, 612	600 s	Jin et al. [78]
Stannate	$CaSnO_3$	Tb^{3+}	491, 545, 588, 622	4 h	Liu and Liu [79]
	Mg_2SnO_4	Mn^{2+}	458	5 h	Lei et al. [80]
	Ca_2SnO_4	Tb^{3+}	483, 545	3 h	Jin et al. [81]
	Mg_2SnO_4	Defect	470–500	5 h	Li et al. [82]
Tungstate	$CaWO_4$	Tb^{3+}	550	400 s	Jin et al. [83], Su et al. [84], Wu et al. [85]
Germanate	$CaZnGe_2O_6$	Tb^{3+}	488, 552, 583	4 h	Woo et al. [86]
	Zn_2GeO_4	Mn^{2+}	528	2 h	Takahashi et al. [87]

(continued)

Table 2 (continued)

Host materials		Dopants	Afterglow emission wavelength (nm)	Afterglow duration	References
	Li_2ZnGeO_4	Mn^{2+}	530	8 h	Jin et al. [88]
Sulfide	ZnS	Cu^+, Co^{2+}	525	3 h	Yang et al. [89]
Nitride	$SrSi_2O_2N_2$	Eu^{2+}	539	2.5 h	Zhang et al. [90]
	$BaSi_2O_2N_2$	Eu^{2+}	498	500 min	Qin et al. [91]
	$BaAlSi_5N_7O_2$	Eu^{2+}	400–650	40 min	Zhang et al. [92]
	C_3N_4	Defect	553	1.5 h	Tang et al. [93]
	BCNO	Defect	520	2 h	Liu et al. [94]
Binary oxide	Lu_2O_3	Tb^{3+}, Ca^{2+}, Sr^{2+}	543	20–30 h	Trojan-Piegza et al. [95]
	ZrO_2	Ti^{3+}	496	800 s	Chandra [96]
Others	$CaGa_2S_4$	Eu^{2+}, Ce^{3+} or Ho^{3+}	556	0.5 h	Guo et al. [97]
	$YTaO_4$	Tb^{3+}	492 + 543 + 590 + 624	2 h	Takayama et al. [98]
	$Ca_8Mg(SiO_4)_4Cl_2$	Eu^{2+}	505	3 h	Lei et al. [99]
	KY_3F_{10}	Tb^{3+}	542	108 s	Cao et al. [100]
	$Sr_3Al_2O_5Cl_2$	Tb^{3+}	544	1 h	Zhang et al. [101]

Table 3 Yellow emitting long persistent phosphors

Host materials		Dopants	Afterglow emission wavelength (nm)	Afterglow duration	References
Silicates	Sr_2SiO_4	Eu^{2+}, Dy^{3+}	570	60 min, 10 min	Szczodrowski et al. [102], Madej and Zych [103]
	Sr_3SiO_5	Eu^{2+}, Lu^{3+}	574	20 h	Li et al. [104], Liu et al. [105]
	$Ca_2ZnSi_2O_7$	Eu^{2+}, Dy^{3+}	580	12 h	Jiang et al. [106]
	$CdSiO_3$	Mn^{2+}, Nd^{3+}	587	300 s	Qu et al. [107]
Aluminate	$Gd_3Al_2Ga_3O_{12}$	Ce^{3+}, Cr^{3+}	540	120 min	Xu et al. [108], Ueda et al. [109]
	$Y_3Al_5O_{12}$	Mn^{2+}	580	400 s	Mu et al. [110]
	$Y_3Al_5O_{12}$	Pr^{3+}	490, 610	600 s	Zhang et al. [111]
Gallate	$ZnGa_2O_4$	Bi^{3+}	410, 550	40 min	Zhuang et al. [112]
Germanate	$Cd_2Ge_7O_{16}$	Mn^{2+}	585	3 h	Che et al. [113]
Sulfides	Y_2O_2S	Tm^{3+}, Mg^{2+}, Ti^{4+}	495, 545, 588	1 h	Lei et al. [114]
	Y_2O_2S	Ti^{3+}, Mg^{2+}, Gd^{3+}, Lu^{3+}	540–594	5 h	Wang et al. [115]
	Gd_2O_2S	Ti^{3+}	590	2 h	Zhang et al. [116]
Others	Ca_2BO_3Cl	Eu^{2+}, Dy^{3+}	580	48 h	Zeng et al. [117]

and 3 h, respectively. AlN: Mn^{2+}, $Ca_2Si_5N_8$: Eu^{2+} emit for 1 h. Most of the phosphors have persistent time less than 1 h. As compared to the primary color emitting persistent phosphors, the red emitting phosphors are less in number and with less persistent time.

From Table 6, we see that most of the white emitting long persistent phosphors are Dy^{3+} doped. Dy^{3+} doped $Ca_2MgSi_2O_7$ exhibits persistent luminescence for 3 h. $CdSiO_3$ is a better host with a persistent time of 5 h. Sr_2SiO_4, $CaSnSiO_5$, $SrSiO_3$ have persistent time of 1 h. By co-doping Dy^{3+} with Eu^{2+}, Mn^{2+} in the $CaAl_2Si_2O_8$ host and with Ce^{3+} in the $Sr_2Al_2SiO_7$ host shows the white LLP for 1 h. $CaZnGe_2O_6$:Dy^{3+} and $CaAl_2O_4$:Dy^{3+} have persistent times of 3 h and 0.5 h, respectively. Dy^{3+}, Mg^{2+}, Si^{4+} co-doped Y_2O_2S have persistent time of 36 min. Among the other dopants, Pr^{3+} doped $CaSnO_3$ is shown to have persistent time of 3 h, and Eu^{3+}, Zr^{4+} co-doped $SrMg(PO_4)_2$ gives emission for 1.5 h.

From Table 7, it can be seen that most of the NIR emitting persistent phosphors are Cr^{3+} doped gallate phosphors. They have excellent persistent luminescence behavior also. Pan et al. [205] have demonstrated sunlight-activated Cr^{3+} doped $Zn_3Ga_2Ge_2O_{10}$ with NIR LLP for 360 h. Liu et al. [207] reported $LiGa_5O_8$:Cr^{3+} to have persistent NIR emission for 1000 h, which is the best result so far. Cr^{3+} doped

Table 4 Orange emitting long persistent phosphors

Host materials		Dopants	Afterglow emission wavelength (nm)	Afterglow duration	References
Silicates	$CaSnSiO_5$	Defect → Sm^{3+} (ET)	600	700 s	Xu et al. [118]
	$Ca_3SnSi_2O_9$	Defect → Sm^{3+} (ET)	600	1200 s	Feng et al. [119]
	$Ca_2ZrSi_4O_{12}$	Defect → Sm^{3+} (ET)	616	800 s	Jia et al. [120]
	$CdSiO_3$	Pr^{3+}	402, 600	5 h	Xu et al. [12]
	$CdSiO_3$	Eu^{3+}	420, 615	5 h	Xu et al. [12]
	$CdSiO_3$	Sm^{3+}	400, 565 + 600	5 h	Xu et al. [12]
	$CdSiO_3$	Tb^{3+}, Mn^{2+}	486 + 548, 587	1 h	Qu et al. [121]
	$CdSiO_3$	Eu^{3+}, Mn^{2+}	597, 610	1 h	Kuang et al. [122]
Aluminate	$Sr_3Al_2O_6$	Eu^{2+}	612	Disputed	Zhang et al. [123]
Phosphate	YPO_4	Pr^{3+}, Nd^{3+}, Er^{3+}, Ho^{3+}, Dy^{3+}	612	1800s	Lecointre et al. [124]
Stannate	Ba_2SnO_4	Sm^{3+}	580, 611, 623	20 min	Xu et al. [125]
	Sr_2SnO_4	Sm^{3+}	582, 624, 672	1 h	Xu et al. [125]
	Ca_2SnO_4	Eu^{3+}	585, 618, 633	50 min	Gao et al. [126]
	Ca_2SnO_4	Sm^{3+}	566, 609, 653	1–7 h	Xu et al. [125]
	$Sr_3Sn_2O_7$	Sm^{3+}	580, 621, 665,735	1 h	Lei et al. [127]
Titanate	$CaTiO_3$	Pr^{3+}	612	2 h	Haranath et al. [128], Noto et al. [129]
	$Ca_2Zn_4Ti_{16}O_{38}$	Pr^{3+}	614	30 min	Lian et al. [130]
	$ZnTiO_3$	Pr^{3+}	612	20 min	Haranath et al. [131]
	$SrTiO_3$	Pr^{3+},Al^{3+}	614		Kutty and Nag [132]
	$La_2Ti_2O_7$	Pr^{3+}	611	1 h	Minghui et al. [133]
Tungstate	$CaWO_4$	Eu^{2+}	618	40 min	Buckmaster and Chatterjee [134]

(continued)

Table 4 (continued)

Host materials		Dopants	Afterglow emission wavelength (nm)	Afterglow duration	References
Zirconate	$BaZrO_3$	Eu^{3+}, Ti^{4+}	574, 596, 614	10 min	Sun et al. [135]
Molybdate	$CaMoO_4$	Eu^{3+}	616	5 min	Feng-Wen et al. [136]
Binary oxide	Y_2O_3	Eu^{3+}	612	90 min	Lin et al. [137]
	Lu_2O_3	Eu^{3+}	611	3 min	Zych and Trojan-Piegza [138]
	ZrO_2	Sm^{3+}	570, 614	15 min	Zhao and Wang [139]
Sulfides	SrS	Eu^{2+}, Pr^{3+}	611	16 h	Luo et al. [140]
	Y_2O_2S	Eu^{3+}, Mg^{2+}, Ti^{4+}	590, 614, 627, 710	3 h	Li et al. [141], Cui et al. [142]
	La_2O_2S	Eu^{2+}	590, 614, 627, 710		Song et al. [143]
	Y_2O_2S	Sm^{3+}, Mg^{2+}, Ti^{4+}	570, 606, 659	2.5 h	Li et al. [141]
Nitrides	$Ca_2Si_5N_8$	Eu^{2+}, Tm^{3+}	620	200 min	Zhang et al. [144], Wang et al. [145], Smet et al. [146]
	$Ga_2Si_5N_8$	Eu^{2+}, Tm^{3+}	610	1 h	Smet et al. [147], Miyamoto et al. [148]
Others	$Sr_3Al_2O_5Cl_2$	Eu^{2+}, Tm^{3+}	620	220 min	Li et al. [149]
	$Sr_5(BO_3)_3Cl$	Eu^{2+}	610	15 min	Jin et al. [150]
	$Sr_3Al_2Cl_2O_5$	$Ce^{3+} \rightarrow Eu^{2+}$ (ET)	435, 620	3 h	Fu [151]
	SrO	Eu^{3+}	594, 616	1 h	Fu [152]
	CaO	Eu^{3+}	594, 616	2 h	Ning et al. [44]
	$Ca_2Sb_2O_7$	Sm^{3+}	600	770 s	Shi et al. [153]

Table 5 Red emitting long persistent phosphors

Host materials		Dopants	Afterglow emission wavelength (nm)	Afterglow duration	References
Silicates	$BaMg_2Si_2O_7$	Ce^{3+}, Mn^{2+}	408, 680	2 h	Gong et al. [154]
	$MgSiO_3$	$Eu^{2+} \rightarrow Mn^{2+}$ (ET)	665	4 h	Lin et al. [155]
	$Sr_3MgSi_2O_8$	$Eu^{2+} \rightarrow Mn^{2+}$ (ET)	457, 670	2 h	Gong et al. [156]
	$Sr_2ZnSi_2O_7$	Sm^{3+}	561, 598, 644	30 min	Zhang et al. [157]
	$CdSiO_3$	Defect $\rightarrow Sm^{3+}$	530–680	5 h	Rodrigues et al. [158]
	$CdSiO_3$	Defect $\rightarrow Tm^{3+}$	580–680	5 h	Rodrigues et al. [158]
Phosphate	$Ca_3(PO_4)_2$	Mn^{2+}, Dy^{3+}	660	1 h	Bessiere et al. [159]
	$Ca_{9.5}Mn(PO_4)_7$	Mn^{2+}	645	1 h	Lecointre et al. [160]
Stannate	Ca_2SnO_4	Sm^{3+}	566, 609, 633	1 h	Ju et al. [161]
	Sr_2SnO_4	Sm^{3+}, Dy^{3+}	582, 624, 672	1 h	Xu et al. [162]
	$Sr_3Sn_2O_7$	Sm^{3+}	580, 621, 665, 735	1 h	Lei et al. [163]
	Ca_2SnO_4	Pr^{3+}	496, 605, 625, 658	20 min	Jin et al. [164]
	Ca_2SnO_4	Eu^{3+}	617, 633	100 min	Gao et al. [165]
Niobate	$NaNbO_3$	Pr^{3+}	620	40 s	Boutinaud et al. [166]
Borate	$LiSi_4(BO_3)_3$	Eu^{2+}	630	20 min	Zhang et al. [167], Wu et al. [168]
Titanate	$La_2Ti_2O_7$	Sm^{3+}, Ti^{4+}	460,570, 605, 660 (Blue, red, white)	540 s	Shao et al. [169]
Germanate	$Na_2CaSn_2Ge_3O_{12}$	Sm^{3+}	566, 605, 664	4.8 h	Xu et al. [170]
	$CdGeO_3$	Pr^{3+}	470–510, 580–620	1000 s	Jin et al. [171]

(continued)

Table 5 (continued)

Host materials		Dopants	Afterglow emission wavelength (nm)	Afterglow duration	References
	$MgGeO_3$	Mn^{2+},Bi^{3+}	680	100 min	Zhuang et al. [172]
	$CaZnGe_2O_6$	Mn^{2+}	648	3 h	Che et al. [173]
Sulfide	CaS	Eu^{2+}, Tm^{3+}	650	1 h	Rodriguez Burbano et al. [174], Suda et al. [175]
	Ca_2SiS_4	Eu^{2+}, Nd^{3+}	660	30 min	Suda et al. [175]
	Y_2O_2S	Eu^{3+}, Mg^{2+}, Ti^{4+}	590, 614, 627, 710	3 h	Li et al. [176], Hang et al. [177]
	Gd_2O_2S	Eu^{3+}	627	600 s	Liu et al. [178]
	La_2O_2S	Sm^{3+}	605, 665	120 s	Zhang et al. [144]
	La_2O_2S	Eu^{2+}	590, 614, 627, 710		Song et al. [143]
Nitrides	$Sr_2Si_5N_8$	Eu^{2+}, Tm^{3+}	612	10 min	Teng et al. [179]
	$SrCaSi_5N_8$	Eu^{2+}, Tm^{3+}	647	30 min	Wang et al. [180]
	AlN	Mn^{2+}	570–700	1 h	Zhang et al. [181]
	$Ca_2Si_5N_8$	Eu^{2+}	570–680	1 h	Li et al. [182]
Others	$ZnTa_2O_6$	Pr^{3+}	590–670	1200 s	Noto et al. [183], Wu et al. [184]

$ZnGa_2O_4$, $MgGa_2O_4$, $SrGa_{12}O_{19}$, $La_3Ga_5GeO_{14}$, $Zn_3Ga_2SnO_8$, LaGaO3 have NIR emission with persistent time of 5 h, 1 h, 2 h, 8 h, 300 h, 1 h, respectively. Ce^{3+} co-doped with Cr^{3+} in the $Y_3Al_{5-x}Ga_xO_{12}$ host shows a persistent signal for 16 h. Ga_2O_3: Cr^{3+} emits for 4 h. Apart from the gallate host, Zn_2SnO_4: Cr^{3+} also exhibits appreciable NIR emission for 35 h. Among other dopants, some rare earth dopants are also used for getting NIR LPP. $Y_3Al_2Ga_3O_{12}$:Ce^{3+}, Nd^{3+} has persistent time of 10 h. Mn^{4+}, Ge^{4+} co-doped $LaAlO_3$, and $GdAlO_3$ show emission for 20 h. $SrAl_2O_4$:Eu^{2+},

Table 6 White emitting long persistent phosphors

Host materials		Dopants	Afterglow emission wavelength (nm)	Afterglow duration	References
Silicates	$CaAl_2Si_2O_8$	Eu^{2+},Mn^{2+} Dy^{3+}	418, 580	1 h	Dai [185], Zhang et al. [186]
	$Sr_2Al_2SiO_7$	Ce^{3+}, Dy^{3+}	408, 491, 573	1 h	Chen et al. [187]
	$CaMgSi_2O_6$	Dy^{3+}	480, 575, 667	2 h	Gong et al. [50], Sahu et al. [188]
	$Ca_2MgSi_2O_7$	Dy^{3+}	458, 580, 660, 755	3 h	Gong et al. [50], Revannasiddappa et al. [189], Rosticher et al. [190]
	Sr_2SiO_4	Dy^{3+}	480, 575, 665	1 h	Tshabalala et al. [191]
	$Sr_2MgSi_2O_7$	Dy^{3+}	470, 575, 678	0.7 h	Sahu et al. [192]
	$CaSnSiO_5$	Defect → Dy^{3+} (ET)	482, 572	1 h	Xu et al. [193]
	$Ca_3SnSi_2O_9$	Defect → Dy^{3+} (ET)	572	1200 s	Kuang et al. [194]
	$CdSiO_3$	Defect → Dy^{3+}	480, 580	5 h	Rodrigues et al. [158]
	$SrSiO_3$	Dy^{3+}	480, 572, 664	1 h	Ito et al. [195]
	$Ca_{0.5}Sr_{1.5}Al_2SiO_7$	Ce^{3+}, Tb^{3+}	386, 483, 542, 591	1 min	Xu et al. [12]
	$Ca_3MgSi_2O_8$	Dy^{3+}	480, 575, 667	5 min	Sahu et al. [188]
Aluminate	$CaAl_2O_4$	Dy^{3+}	477, 491, 577, 668	30 min	Liu et al. [196]
Germanate	$CaZnGe_2O_6$	Dy^{3+}	488, 552, 583	3 h	Che et al. [197]
Phosphate	$SrMg(PO_4)_2$	Eu^{3+}, Zr^{4+}	500, 588	1.5 h	Wang et al. [198]
Stannate	$CaSnO_3$	Pr^{3+}	488, 541, 620, 653	3 h	Nakamura et al. [199]
Sulfides	Y_2O_2S	Dy^{3+}, Mg^{2+}, Si^{4+}	486, 557	36 min	Huang et al. [200]

Er^{3+}, and Dy^{3+} have persistent time greater than 1 h. Sr_2SnO_4: Nd^{3+} also shows NIR LLP for 1 h.

Table 7 NIR emitting long persistent phosphors

Host materials		Dopants	Afterglow emission wavelength (nm)	Afterglow duration	References
Gallates	$La_3Ga_5GeO_{14}$	Cr^{3+}	700–1300	8 h	Wu et al. [201], Yi et al. [202], Yan et al. [203], Jia et al. [204]
	$Zn_3Ga_2Ge_2O_{10}$	Cr^{3+}	650–1000	360 h	Pan et al. [205], Wu et al. [206]
	$LiGa_5O_8$	Cr^{3+}	650–850	1000 h	Liu et al. [207], Chuang et al. [208], Chen [209]
	$SrGa_{12}O_{19}$	Cr^{3+}	650–950	2 h	Xu et al. [210]
	$MgGa_2O_4$	Cr^{3+}	650–770	1 h	Mondal et al. [211]
	$ZnGa_2O_4$	Cr^{3+}	600–800	5 h	Zhuang et al. [212], Allix et al. [213], Bessiere et al. [214], Gourier et al. [215], Li et al. [216], Zhuang et al. [217], Bessiere et al. [218], Mondal et al. [219]
	$Zn_3Ga_2SnO_8$	Cr^{3+}	650–1000	300 h	Li et al. [220]
	$Ca_3Ga_2Ge_3O_{12}$	Cr^{3+},Yb^{3+}, Tm^{3+}	650–850	7000 s	Liu et al. [221], Chen et al. [222]
	$LaGaO_3$	Cr^{3+}	650–850	1 h	Mondal and Manam [223]
	$SrGa_2O_4$	Cr^{3+}	650–850	200 s	Mondal and Manam [224]
	$Y_3Al_{5-x}Ga_xO_{12}$	$Ce^{3+} \rightarrow Cr^{3+}$	600–800	1000 min	Xu et al. [225], Brenier et al. [226]
	$Y_3Al_{5-x}Ga_xO_{12}$	$Ce^{3+} \rightarrow Nd^{3+}$	600–1400	1000 min	Brenier et al. [226]
	$Gd_3Ga_5O_{12}$	Cr^{3+}, Tm^{3+}	694		le Masne et al. [227]

(continued)

Table 7 (continued)

Host materials		Dopants	Afterglow emission wavelength (nm)	Afterglow duration	References
Silicates	$Ca_{0.2}Zn_{0.9}Mg_{0.9}Si_2O_6$	$Eu^{2+} \rightarrow Mn^{2+}$	600–800	1 h	Li et al. [228]
Aluminate	$LaAlO_3$	Mn^{4+}, Ge^{4+}	600–800	20 h	Kamimura et al. [229]
	$GdAlO_3$	Mn^{4+}, Ge^{4+}	600–800	20 h	Kamimura et al. [229]
	$Sr_4A_{14}O_{25}$	Eu^{2+}, Cr^{3+}, Dy^{3+}	495, 695	20 min	Zhong et al. [71]
	$SrAl_2O_4$	$Eu^{2+} \rightarrow Nd^{3+}$, Dy^{3+}	515, 1530	15 min	Teng et al. [69]
	$SrAl_2O_4$	$Eu^{2+} \rightarrow Er^{3+}$, Dy^{3+}	525, 882	100 min	Yu et al. [70]
	$Y_3Al_2Ga_3O_{12}$	$Ce^{3+} \rightarrow Cr^{3+}$	505, 700	5 min	Ueda et al. [67], Xu et al. [68]
	$Y_3Al_2Ga_3O_{12}$	$Ce^{3+} \rightarrow Nd^{3+}$	808, 1064, 1335	10 h	Xu et al. [68]
Stannate	Sr_2SnO_4	Nd^{3+}	850–1400	1 h	Li et al. [230]
	Zn_2SnO_4	Cr^{3+}	650–1200	35 h	Qin et al. [231]
	$SrSnO_3$	Bi^{2+}	700–850	30 min	Teng et al. [69]
Binary oxide	Ga_2O_3	Cr^{3+}	650–850 (713)	4 h	Wang et al. [232], Lu et al. [233]

3 Multicolor Long Persistent Phosphor Based on Color Conversion Process

Any multicolor long-lasting phosphor can be prepared by mixing the three primary colors emitting long-lasting phosphors. Comparing the existing long-lasting phosphors which emit three primary colors, green and blue ones are much better in terms of afterglow performance than red ones. For example, the afterglow time of $SrAl_2O_4$: Eu^{2+}, Dy^{3+} (green light emitting), and $CaAl_2O_4$: Eu^{2+}, Dy^{3+} (blue light emitting) is longer than 10 h, while that of Y_2O_2S: Eu^{3+}, Ti^{4+}, Mg^{2+} (red light emitting) is only 5 h. Therefore, the search for red long-lasting phosphors (LLP) with longer persistence time is still in progress. Although many red LLPs, such as $CaTiO_3$: Pr, CaS: Eu, and $CaWO_4$: Eu have been reported their applications are limited due to the short persisting time compared to other color emitting long lasing phosphors. Most of the researchers focused their works on exploring various luminescent centers, changing doping and co-doping ions, optimization of synthesis conditions, and so on. The color conversion principle has been successfully used in the field of light-emitting diodes. However, fewer woks have been reported on the application of the color conversion process for developing long-lasting phosphors. In this section, the color conversion process of long-lasting phosphor to generate multicolor persistent luminescence is discussed.

3.1 Color Conversion Principle

Organic, inorganic fluorescent dye with high luminescence efficiency and absorption band overlapped with the phosphorescent band of LLP can be used as a color conversion agent. These dyes can absorb the light emitted by LLP and convert it into its characteristic emission color to LLP phosphorescence. Generally, traditional LLPs encompass big grains and small surface areas. Therefore, only a few dye molecules are absorbed on the particle surface. Furthermore, dyes with high concentration will suffer from self-quenching. These cause lower color conversion efficiency. In addition, some dyes are very unstable and are easy to bleach under a light environment. To produce highly efficient color conversion, the fluorescent dyes should be encapsulated into mesoporous silica nanoparticles. Due to the large surface area and small particulate diameter of mesoporous silica, a large amount of dye molecules can be absorbed by one particle increasing adsorption efficiency. At the same time, fluorescent dye encapsulated by mesoporous silica becomes more stable and produces highly efficient fluorescent emission. When the LLP material and organic dye encapsulated mesoporous silica will be mixed in a weight ratio to make a composite form, the phosphorescent light of LLP will excite the organic dye and as an output, the continuous emission of mixed color is achieved. Thus, based on the concept of the color conversion process, the long-lasting phosphor can be used as a phosphorescent light source, while fluorescent dye, which is encapsulated inside the mesoporous silica nanoparticles, is used as the color conversion agent to generate multicolor long-lasting phosphors.

3.2 Color Conversion Process

3.2.1 Synthesis Strategies

As discussed earlier, the color conversion method involves two parameters—phosphorescent light source and color converting agent. Hence, at first LLP is prepared by high-temperature solid state reaction or combustion reactions which are familiar and discussed elsewhere. Mesoporous silica nanoparticles (MSNp) are synthesized according to the modified Strober method. Dried mesoporous silica nanoparticles are dispersed in a solution of organic dye in a conical flask and stirred at room temperature. After adsorption, the as-obtained precipitate is filtered and washed thoroughly with deionized water to ensure that unabsorbed dyes are completely removed. Finally, dye-encapsulated mesoporous silica nanoparticles are mixed with long-lasting phosphors in a defined weight ratio mechanically with a mortar and pestle to make the composite phosphors.

$Sr_2MgSi_2O_7$: Eu^{2+}, Dy^{3+} is synthesized by solid-state reaction at 1300°C for 5 h in a reducing atmospheric environment (5% H_2 + 95% N_2). The $Y_3Al_5O_{12}$: Ce^{3+} is synthesized by combustion synthesis method followed by high temperature sintering at 1400°C for 5 h in a reducing atmosphere. The $Sr_2MgSi_2O_7$: Eu^{2+}, Dy^{3+}/ $Y_3Al_5O_{12}$: Ce^{3+} composite phosphor is prepared by mixing them mechanically in an agate mortar in different weight ratios.

3.2.2 3.2.2. Color Conversion of $SrAl_2O_4$:Eu^{2+}, Dy^{3+} Long-Lasting Phosphor by Using Rhodamine 6G Dye Encapsulated Mesoporous Silica (MSNp) as Color Conversion Agent

The excitation spectrum of MSNp-Rh6G is overlapped with the emission spectrum of SAO. So here, SAO is used as an excitation light source to excite the MSNp-Rh6G nanoparticles and emit the characteristic emission of rhodamine 6G, which helps to get color converted MSNp-Rh6G/SAO composite yellow long-lasting phosphor.

The phosphorescence emission spectra of SAO excited at 360 nm show a peak at 517 nm, which is the emission spectra of SAO green long-lasting phosphor. Now when MSNp-Rh6G dye is mixed in SAO to form a composite phosphor that serves as color color-converted phosphor, two emission bands are observed, one between 440 and 520 nm (peak I) and another between 520 and 650 nm (peak II). The first one is due to the phosphorescence emission of SAO and the second one is due to the fluorescence emission of rhodamine 6G, which occurs because of the excitation by the phosphorescence emission of SAO. When composite phosphor is prepared, we take certain mass ratios of MSNp-Rh6G and SAO for their mixing. A certain trend is observed in the emission spectra with increasing mass ratio. From the experimental data, it has been seen that when the mass ratio is increased from 0.5, the peak intensities decrease and peak I is blue-shifted while peak II is red-shifted. The decrease in peak intensity is explained by two factors. The first one is that as the stacking density of mesoporous silica is lower than SAO so when the mass ratio is increased more volumetric mesoporous silica is mixed with SAO which enhances the scattering by MSNp with an increasing mass ratio of MSNp. The second one is that as the emission spectra of SAO overlap with the excitation spectra of MSNp-Rh6G MSNp-Rh6G will absorb the part of the emission light of SAO, resulting in the decrease of phosphorescence intensity of the composite long-lasting phosphor. The blue shift of peak-I is due to the nonuniform absorption of MSNp-Rh6G as the MSNp-Rh6G nanoparticles covered the surface of SAO in the composite long-lasting phosphor. The excitation spectrum intensity of MSNp-Rh6G is in between 425 and 500 nm, which leads to the blue shift of peak-I. With decreasing mass ratio, the blue shifting of peak-II indicates that a part of phosphorescence spectra of SAO and fluorescence spectra of MSNp-Rh6G overlap. So when the mass ratio gradually decreased from 1 to 0, the position of peak II goes closure to emission of peak SAO.

For getting the color-converted long-lasting phosphor, the intensity ratio of both peak-II and peak-I is an important factor that can be obtained by optimizing the mixing of MSNp-Rh6G and SAO in the proper ratio. It is seen from the experimental result, that with increasing the mass ratio MSNp-Rh6G/SAO, the intensity ratio of peak II/peak I is increasing indicating the color conversion of composite phosphor from green to yellow. The proper mass ratio of SAO and MSNp-Rh6G should be chosen to get the color-converted yellow long-lasting phosphor with optimized phosphorescence brightness. The CIE color coordinates indicate the color conversion of the well-known green emitting long-lasting phosphor SAO to yellow long-lasting phosphor by mixing with rhodamine 6G dye encapsulated mesoporous silica nanoparticles.

The afterglow decay curves recorded by exciting at 360 nm and monitoring the emission wavelength at 498 nm and 565 nm shows that the decay rate of peak-I (phosphorescence emission of SAO) and peak-II (fluorescence emission of MSNp-Rh6G) are equal which indicate the conversion of fluorescence characteristic to phosphorescence behavior of rhodamine 6G fluorescence dye due to continuous energy transfer from SAO to MSNp-Rh6G in the MSNp-Rh6G/SAO composite long-lasting phosphor. So the energy of SAO successfully transfers to the MSNp-Rh6G nanoparticles and color converted yellow long-lasting phosphor is obtained.

Just like rhodamine 6G, rhodamine B dye encapsulated mesoporous silica (MSNp-RhB) can also be used as color conversion agent for SAO. The excitation spectrum of MSNp-RhB (450–550 nm) also overlapped with the emission spectrum of SAO. Hence, SAO can be used as phosphorescent light source to excite the MSNp-RhB nanoparticles and emit the characteristic emission of rhodamine B, which helps to get color converted composite long-lasting phosphor. So, by using MSNp-RhB, near white light can be achieved from SAO green LLP for commercial applications.

3.2.3 Color Conversion of $CaAl_2O_4$: Eu^{2+}, Nd^{3+} (CAO) Long-Lasting Phosphor

The emission spectrum of CAO overlapped with the excitation spectra of MSNp-Rh6G. CAO serves as phosphorescent light source to excite the MSNp-Rh6G color conversion agent in the composite of CAO/MSNp-Rh6G phosphor.

The LPL emission spectra are recorded by exciting the samples at 365 nm UV lamp for 5 min and then switching off the light source. The mass of CAO is fixed and MSNp-Rh6G is varied. Two emission bands are observed in between 400–520 nm (peak-I) and 520–700 nm (peak-II). CAO sows long persistent phosphorescence due to the $4f^65d^1 \rightarrow 4f^7$ transition of Eu^{2+} in the CAO. The rhodamine 6G dye is generally a fluorescent dye but here in this composite phosphor, it shows phosphorescence behavior which is due to radiative energy transfer from CAO to MSnp-Rh6G. The afterglow decay curves of the two emission wavelengths show same decay rates which confirm the continuous excitation of MSNp-Rh6G by CAO. The afterglow emission with different time interval shows same behavior. The CIE diagram of the CAO/MSNp-Rh6G composite for different mass ratios indicates that with the increase of the content (in weight) of MSNp-Rh6G, the color can be tuned from blue to purple. Therefore, this is an effective way of converting the color of existing commercial long-lasting phosphor to get the multicolor long-lasting phosphor.

3.2.4 Color Conversion of $BaAl_2O_4$:Eu^{2+}, Dy^{3+} (BAO) Long-Lasting Phosphor

Under UV excitation at 342 nm, BAO exhibits a broad emission from 425 to 600 nm with a maximum of 497 nm. The excitation spectra of MSNp-Rh6G overlapped with the emission spectra of BAO. So radiative energy transfer occurs from BAO to MSNp-Rh6G.

The LPL spectra of the BAO/MSNp-Rh6G composite is recorded after irradiating by 365 nm UV lamp for 5 min. Two emission ranges are observed. Peak-I (cyan) is due to $4f^65d^1 \rightarrow 4f^7$ transition of Eu^{2+} and peak-II (yellow) corresponds to fluorescence emission of MSNp-Rh6G nanoparticles. As usual, here also peak intensity decreased with increasing the amount of MSNp-Rh6G and peak shift is observed for the same reason discussed earlier. The afterglow decay is recorded by exciting the samples at 365 nm UV light for 5 min and then ceasing off the source, monitoring the emission wavelength at 485 nm and 555 nm. For both the emission, decay rates are equal indicating that the color of the samples would not be changed during the afterglow.

The CIE diagram of BAO/MSNp-Rh6G shows that color can be tuned from cyan region to yellow region with increasing the mass of MSNp-Rh6G in BAO/MSNp-Rh6G composites. Therefore, yellow color can be obtained from cyan emitting BAO by mixing MSNp-Rh6G dye, which is the color conversion agent (Fig. 6).

3.2.5 Color Conversion of $Sr_2MgSi_2O_7$: Eu^{2+}, Dy^{3+} Long-Lasting Phosphor Using $Y_3Al_5O_{12}$:Ce^{3+} Phosphor as Color Conversion Agent

Apart from aluminates, silicate-based phosphors are also very much investigated due to their high chemical and physical stability. The persistent phosphorescence color of the silicate phosphors can be tuned from blue to green region depending on the surrounding crystal field. Most of the silicate phosphors are Eu^{2+} doped. Among silicate phosphors, $Sr_2MgSi_2O_7$: Eu^{2+}, Dy^{3+} is a well-known blue emitting long persistent phosphor with persistent time more than 10 h.

On the other hand, $Y_3Al_5O_{12}$: Ce^{3+} has strong absorption in the blue region, broad emission in yellow region, high quantum efficiency and thermal stability which make it suitable for converting color of blue LED into white LED.

$Sr_2MgSi_2O_7$: Eu^{2+}, Dy^{3+} has excitation peaks at 285 nm and 340 nm and emission peak at 461 nm due to $4f^7 \leftrightarrow 4f^65d^1$ transition of Eu^{2+}. Here, the codoped Dy^{3+} has the role to create deep trap for getting long persistent luminescence.

$Y_3Al_5O_{12}$: Ce^{3+} has excitation bands peaking at 342 nm and 458 nm and emission at 549 nm. Hence, the excitation spectra of $Y_3Al_5O_{12}$: Ce^{3+} match with the emission spectra of $Sr_2MgSi_2O_7$: Eu^{2+}, Dy^{3+}. So energy transfer can happen from $Sr_2MgSi_2O_7$: Eu^{2+}, Dy^{3+} to $Y_3Al_5O_{12}$: Ce^{3+}. So $Sr_2MgSi_2O_7$: Eu^{2+}, Dy^{3+} can be used as phosphorescent light source to excite the $Y_3Al_5O_{12}$: Ce^{3+} phosphor to get color tuned LLP.

The LPL spectra of SMSO/YAG composite phosphor is recorded after exciting at 365 nm UV lamp for 5 min and then ceasing off the excitation source. The LPL spectra of composite phosphor with different S:Y (mixing ratio according to weight) show that when Y component is zero, the LPL of only SMSO is observed which is due to the $4f^65d^1 \rightarrow 4f^7$ transition of Eu^{2+}. With increasing the Y component to the composite phosphor, the yellow emission component increases which is due to the transition from Ce^{3+} from the lowest crystal field components of $5d^1$ state to the ground state levels $^2F_{7/2}$, $^2F_{5/2}$. The deconvoluted emission spectrum shows three peaks at 462, 496 and 545 nm for S:Y ratio 10:15. The first and second peaks are due to the transition of Eu^{2+} ions and the third one is due to the transition of Ce^{3+} ions. It can be ascribed to the radiative energy transfer from Eu^{2+} in SMSO to Ce^{3+} in YAG. The luminescence intensity decreases with increasing Y component in S:Y ratio of composite phosphor. Hence, the color tuned long-lasting phosphor can be obtained by radiative energy transfer from long persistent phosphor materials to non-LPL remote phosphor. The long afterglow decay curves show the same decay rate for three emission wavelengths. The CIE diagram shows the conversion from blue to near white LPL with increasing the Y component in SMSO/YAG composite phosphor.

Therefore, the problem of color fading with time for composite long persistent phosphor is overcome.

4 Conclusions

In summary, persistent luminescence happens because of the temperature-induced release of the trapped charges. Afterglow emission time and emission color are two crucial parameters to characterize any long-lasting phosphor. Afterglow emission time can be increased by creating traps in the host. Again, the fading of emission color throughout the decay can be prevented. Usually, long-lasting phosphors are designed either by using a single activator or by multiple activators and sensitizer. Persistent energy transfer is a key factor in the case of multiple doping. For multicolor phosphor, multiple centers are doped in a single host. Again, mixing of two or more long persistent phosphors at different emission colors can give multicolor long persistent phosphor. In this case, the emission color fades away as the emission of the independent components in the composite are time dependent. In persistent energy transfer process energy of a long-lasting phosphor which acts as donor is transferred to another luminescent center which acts as acceptor. The acceptor is chosen based on the required afterglow color and its absorption wavelength must

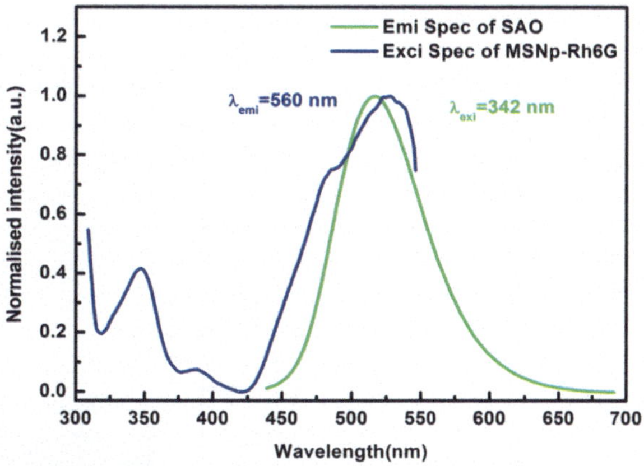

Fig. 1 Photoluminescence emission spectra of SAO and excitation spectra of MSNp-Rh6G (Reproduced from ECS Journal of Solid State Science and Technology, 2016, 5, R98-R103 with permission from the IOP Publishing)

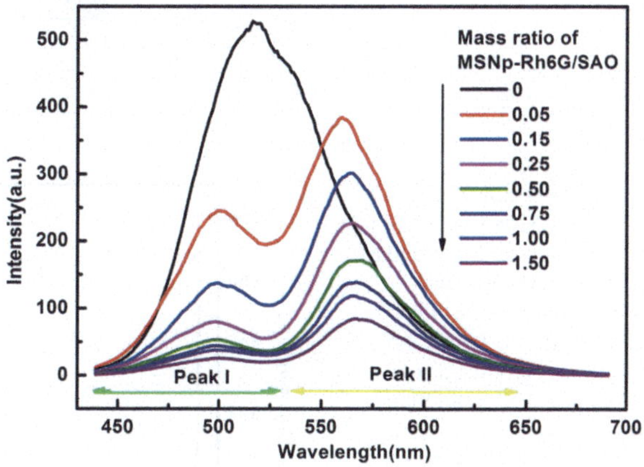

Fig. 2 Phosphorescence emission spectra of MSNp-Rh6G/SAO composite long-lasting phosphors (λexc 360 nm) with different mass ratios. (Reproduced from ECS Journal of Solid State Science and Technology, 2016, 5, R98-R103 with permission from the IOP Publishing)

overlap the donor emission wavelength. Hence, the long persistent luminescence of donor is converted into the long persistent luminescence of activator at required wavelength and accordingly color is converted.

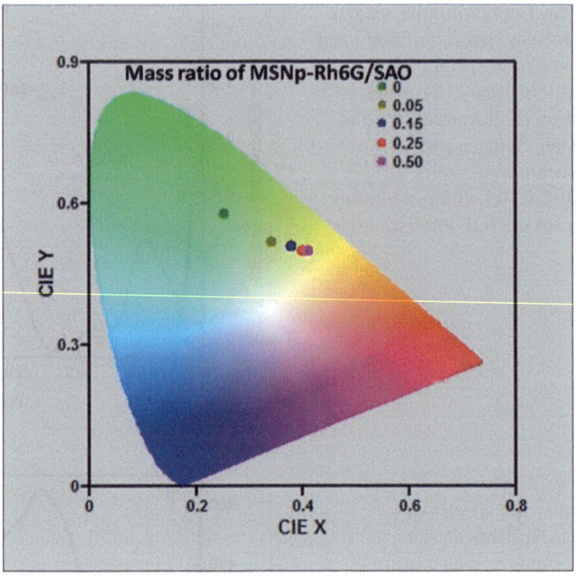

Fig. 3 CIE diagram of the MSNp-Rh6G/SAO composite long-lasting phosphor with different mass ratio (Reproduced from ECS Journal of Solid State Science and Technology, 2016, 5, R98-R103 with permission from the IOP Publishing)

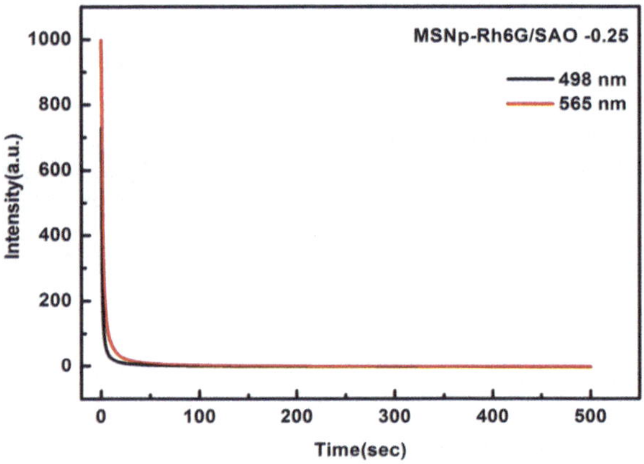

Fig. 4 Afterglow decay curve of MSNp-Rh6G/SAO (0.25 mass ratio) recorded by exciting at 360 nm and monitoring the emission wavelength at 498 nm and 565 nm (Reproduced from ECS Journal of Solid State Science and Technology, 2016, 5, R98-R103 with permission from the IOP Publishing)

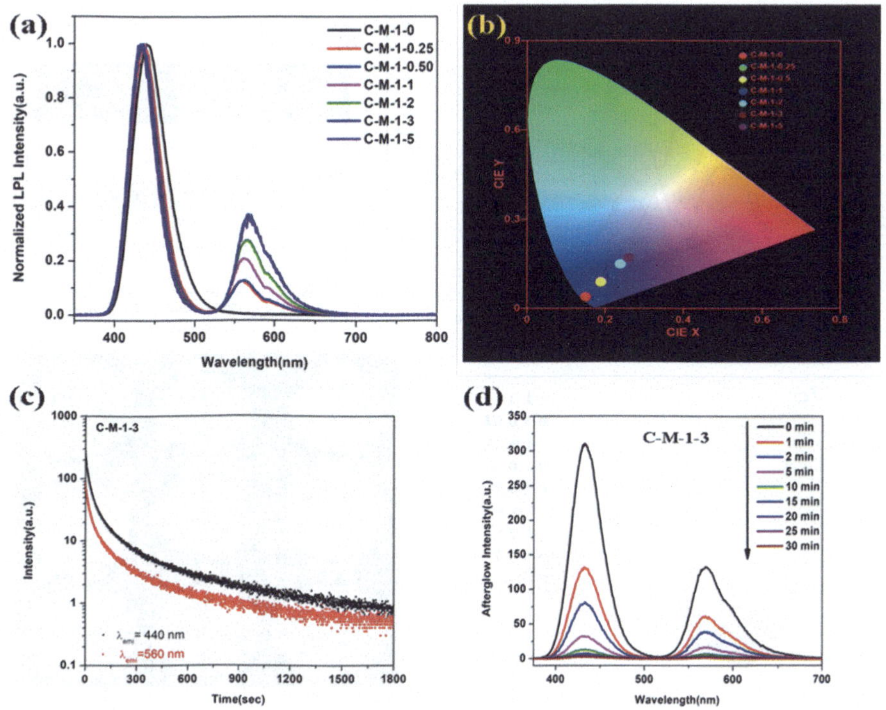

Fig. 5 **a** Long persistent luminescence (LPL) spectra, **b** CIE diagram, **c** afterglow decay curve, and **d** afterglow emission spectra of CAO/MSNp-Rh6G composite long-lasting phosphor.(Author's unpublished work)

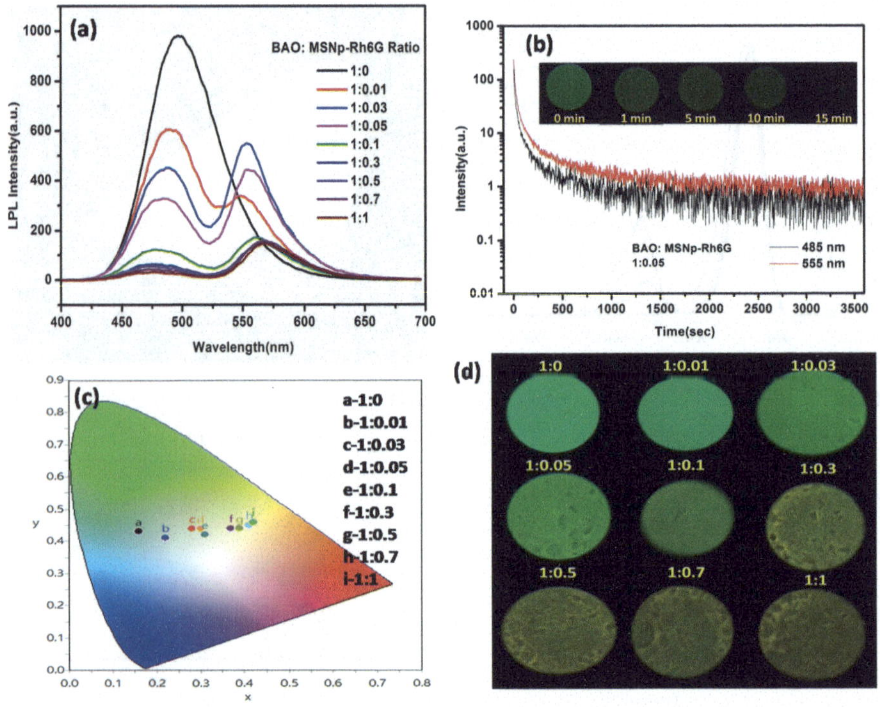

Fig. 6 **a** Long persistent luminescence (LPL) spectra, **b** afterglow decay curves, **c** CIE diagram, and **d** photographs of BAO/MSNP-Rh6G composites with different ratios. (Reproduced from New Journal of Chemistry, 2017, 41, 5934 with permission from the RSC)

Fig. 7 Emission spectrum of $Sr_2MgSi_2O_7$: Eu^{2+}, Dy^{3+} and excitation spectrum of YAG: Ce^{3+} phosphors (Reproduced from ECS Journal of Solid State Science and Technology, 2017, 6, R105-R109 with permission from the IOP Publishing)

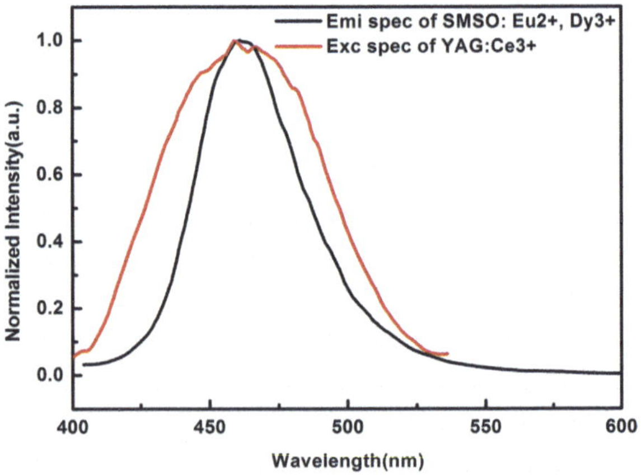

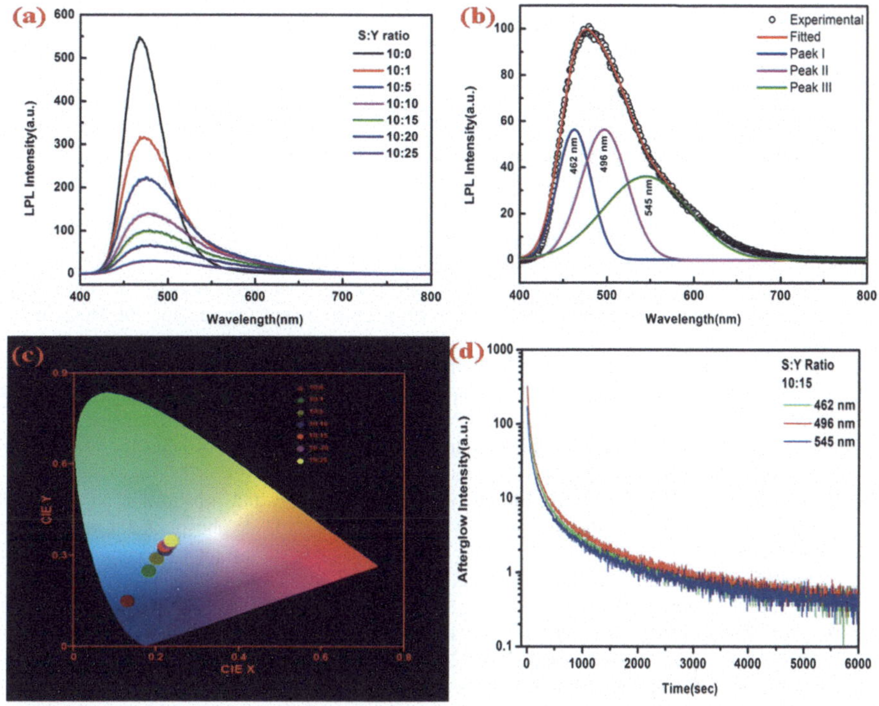

Fig. 8 **a** Long persistent luminescence (LPL) emission spectra, **b** deconvoluted LPL emission spectra, **c** CIE diagram and **d** Afterglow decay curves of $Sr_2MgSi_2O_7$: Eu^{2+}, Dy^{3+}/$Y_3Al_5O_{12}$: Ce^{3+} composite phosphor with different weight mixing ratio (S: Y). (Reproduced from ECS Journal of Solid State Science and Technology, 2017, 6, R105-R109 with permission from the IOP Publishing)

References

1. C. Shi, Y. Fu, B. Liu, G. Zhang, Y. Chen, Z. Qi, X. Luo, The roles of Eu^{2+} and Dy^{3+} in the blue long-lasting phosphor $Sr_2MgSi_2O_7$:Eu^{2+}, D^{y3}+. J. Lumin. **122–123**, 11–13 (2007)
2. Y. Lin, Z. Tang, Z. Zhang, C.W. Nan, Luminescence of Eu^{2+} and Dy^{3+} activated $R_3MgSi_2O_8$-based (R= Ca, Sr, Ba) phosphors. J. Alloys Compd. **348**, 76–79 (2003)
3. Y. Ding, Y. Zhang, Z. Wang, W. Li, D. Mao, H. Han, C. Chang, Photoluminescence of Eu single doped and Eu/Dy codoped $Sr_2Al_2SiO_7$ phosphors with long persistence. J. Lumin. **129**, 294–299 (2009)
4. L. Jiang, C. Chang, D. Mao, B. Zhang, A new long persistent blue-emitting $Sr_2ZnSi_2O_7$:Eu^{2+}, D^{y3}+ prepared by sol–gel method. Mater. Lett. **58**, 1825–1829 (2004)
5. J. Kuang, Y. Liu, J. Zhang, L. Huang, J. Rong, D. Yuan, Blue-emitting long-lasting phosphor, $Sr_3Al_{10}SiO_{20}$:Eu^{2+}, H^{o3}+. Solid State Commun. **136**, 6–10 (2005)
6. Y. Wang, Z. Wang, P. Zhang, Z. Hong, X. Fan, G. Qian, Preparation of Eu^{2+} and Dy^{3+} coactivated $CaAl_2Si_2O_8$-based phosphor and its optical properties. Mater. Lett. **58**, 3308–3311 (2004)
7. L. Jiang, C. Chang, D. Mao, C. Feng, Luminescent properties of $CaMgSi_2O_6$-based phosphors co-doped with different rare earth ions. J. Alloys Compd. **377**, 211–215 (2004)

8. W.H. Shen, X.H. Xu, B.H. Zhang, T. Wang, J.B. Qiu, X. Yu, Distribution of luminescence centers and trapping centers in $Ba_{13}Al_{22}Si_{10}O_{66}$:$Eu^{2+}$, Tm^{3+}. ECS Solid State Lett. **4**, R31–R33 (2015)
9. P. Dorenbos, C.W.E.V. Eijk, A.J.J. Bos, C.L. Melcher, Afterglow and thermoluminescence properties of Lu_2SiO_5:Ce scintillation crystals. J. Phys. Condens. Matter **6**, 4167–4180 (1994)
10. N. Kodama, T. Takahashi, M. Yamaga, Y. Tanii, J. Qiu, K. Hirao, Long-lasting phosphorescence in Ce^{3+}-doped $Ca_2Al_2SiO_7$ and $CaYAl_3O_7$ crystals. Appl. Phys. Lett. **75**, 1715–1717 (1999)
11. Z. He, X.J. Wang, W.M. Yen, Behavior of Mn^{2+} Ions in the trapping process of $SrMg(SiO_3)_2$:Mn, Dy. J. Lumin. **122**, 381–384 (2007)
12. X. Xu, Y.H. Wang, W. Zeng, Y. Gong, B.T. Liu, Luminescent properties of the multicolor afterglow phosphors $Ca_3SnSi_2O_9$:Re^{3+} (Re = Pr, Tb, Sm). J. Am. Ceram. Soc. **94**, 3632–3635 (2011)
13. X. Zhang, W. Jiang, X. Liu, L. Lu, X. Wang, Z. Bai, Q. Liu, Z. Cao, Preparation and characterization of $CaAl_2O_4$:Eu^{2+}, Nd^{3+} nanoparticles. Spectrosc. Spectr. Anal. **25**, 1560–1563 (2006)
14. J. Zhang, Z. Zhang, T. Wang, W. Hao, Preparation and characterization of a new long afterglow indigo phosphor $Ca_{12}Al_{14}O_{33}$:Nd, Eu. Mater. Lett. **57**, 4315–4318 (2003)
15. N. Suriyamurthy, B.S. Panigrahi, Effects of non-stoichiometry and substitution on photoluminescence and afterglow luminescence of $Sr_4Al_{14}O_{25}$:Eu^{2+}, Dy^{3+} phosphor. J. Lumin. **128**, 1809–1814 (2008)
16. Y. Jianhui, Y. Honghui, Z. Jihua, Synthesis of blue long afterglow phosphors $SrAl_4O_7$:Eu^{2+}, $D^{y3}+$ and its luminescent properties. J. Optoelectron. Laser. **17**, 444–449 (2006)
17. A. Nag, T.R.N. Kutty, Role of B_2O_3 on the phase stability and long phosphorescence of $SrAl_2O_4$:Eu, Dy. J. Alloys Compd. **354**, 221–231 (2003)
18. P. Zhang, M. Xu, Z. Zheng, B. Sun, Y. Zhang, Rapid formation of red long afterglow phosphor $Sr_3Al_2O_6$:Eu^{2+}, $D^{y3}+$ by microwave irradiation. Mater. Sci. Eng. B **136**, 159–164 (2007)
19. D. Jia, X.-J. Wang, E. Van der Kolk, W. Yen, Site dependent thermoluminescence of long persistent phosphorescence of $BaAl_2O_4$:Ce^{3+}. Opt. Commun. **204**, 247–251 (2002)
20. D. Jia, W. Yen, Trapping mechanism associated with electron delocalization and tunneling of $CaAl_2O_4$: Ce^{3+}, a persistent phosphor. J. Electrochem. Soc. **150**, H61–H65 (2003)
21. D.D. Jia, X.J. Wang, W.M. Yen, Delocalization, thermal ionization, and energy transfer in singly doped and codoped $CaAl_4O_7$ and Y_2O_3. Phys. Rev. B: Condens. Matter Mater. Phys. **69**, 235113 (2004)
22. K. Van den Eeckhout, P.F. Smet, D. Poelman, Persistent luminescence in Eu^{2+}-doped compounds: a review. Materials **3**, 2536–2566 (2010)
23. D. Jia, X.-J. Wang, W. Jia, W. Yen, Trapping processes of 5d electrons in Ce^{3+} doped $SrAl_2O_4$. J. Lumin. **122**, 311–314 (2007)
24. S. Sharma, S. Pitale, M. Manzarmalik, R. Dubey, M. Qureshi, Luminescence studies on the blue–green emitting $Sr_4Al_{14}O_{25}$:Ce^{3+} phosphor synthesized through solution combustion route. J. Lumin. **129**, 140–147 (2009)
25. Z. Su, P. Ran, C. Li, S. Qiang, Green photoluminescence, but blue afterglow of Tb^{3+} activated $Sr_4Al_{14}O_{25}$. J. Lumin. **130**, 2223–2225 (2010)
26. R. Pang, C. Li, L. Shi, Q. Su, A novel blue-emitting long-lasting proyphosphate phosphor $Sr_2P_2O_7$:Eu^{2+}, $^{Y3}+$. J. Phys. Chem. Solids **70**, 303–306 (2009)
27. R. Pang, C. Li, S. Zhang, Q. Su, Luminescent properties of a new blue long-lasting phosphor $Ca_2P_2O_7$:Eu^{2+}, $^{Y3}+$. Mater. Chem. Phys. **113**, 215–218 (2009)
28. L. Liu, C. Li, S. Wang, Q. Su, Redshift phenomenon of the excitation light of long life emission phosphor. Appl. Phys. Lett. **88**, 1107 (2006)
29. G.F. Ju, Y.H. Hu, L. Chen, X.J. Wang, Z.F. Mu, Opt. Mater. **36**, 1183–1188 (2014)
30. K. Li, M.M. Shang, Y. Zhang, J. Fan, H.Z. Lian, J. Lin, Photoluminescence properties of single-component white-emitting Ca 9 Bi (PO 4) 7: Ce 3+, Tb 3+, Mn 2+ phosphors for UV LEDs. J. Mater. Chem. C **3**, 7096–7104 (2015)

31. R. Pang, W.Z. Sun, J.P. Fu, H.F. Li, Y.L. Jia, D. Li, L.H. Jiang, S. Zhang, C.Y. Li, Luminescence properties of a novel reddish orange long-lasting phosphorescence phosphor $Zn_2P_2O_7$:Sm^{3+}, L^i+. RSC Adv. **5**, 82704–82710 (2015)
32. R. Pang, Y.L. Jia, R. Zhao, H.F. Li, J.P. Fu, W.Z. Sun, L.H. Jiang, S. Zhang, C.Y. Li, Q. Su, Tunable long lasting phosphorescence due to the selective energy transfer from defects to luminescent centres *via* tunnelling in Mn^{2+} and Tm^{3+} co-doped zinc pyrophosphate. Dalton Trans. **43**, 9661–9668 (2014)
33. H.-F. Li, W.-Z. Sun, Y.-L. Jia, T.-F. Ma, J.-P. Fu, D. Li, S. Zhang, L.-H. Jiang, R. Pang, C.-Y. Li, Investigation on luminescence properties of a long afterglow phosphor Ca_2SnO_4:Tm^{3+}. Chem. Asian J. **10**, 2361–2367 (2015)
34. J.C. Zhang, Q.S. Qin, M.H. Yu, M.J. Zhou, Y.H. Wang, The photoluminescence, afterglow and up conversion photostimulated luminescence of Eu^{3+} doped Mg_2SnO_4 phosphors. J. Lumin. **132**, 23–26 (2012)
35. J.C. Zhang, M.H. Yu, Q.S. Qin, H.L. Zhou, M.J. Zhou, X.H. Xu, Y.H. Wang, The persistent luminescence and up conversion photostimulated luminescence properties of nondoped Mg_2SnO_4 material. J. Appl. Phys. **108**, 123518 (2010)
36. M.M. Shi, D.Y. Zhang, C.K. Chang, Dy^{3+}:Ca_2SnO_4, a new yellow phosphor with afterglow behavior. J. Alloys Compd. **639**, 168–172 (2015)
37. Y. Li, M. Geceviciusa, J. Qiu, Long persistent phosphors—from fundamentals to applications. Chem. Soc. Rev. **45**, 2090–2136 (2016)
38. M. Kubus, A.M. Klonkowski, A. Lotnyk, L. Kienle, Luminescence enhancement in composite material: $CaWO_4$: Tb^{3+} nanocrystals incorporated into silica xerogel. Mater. Chem. Phys. **149**, 424–429 (2015)
39. F.W. Kang, M.Y. Peng, A new study on the energy transfer in the color-tunable phosphor $CaWO_4$: Bi. Dalton Trans. **43**, 277–284 (2014)
40. X.L. Ma, J.C. Zhang, H.H. Li, B.C. Duan, L.N. Guo, M.D. Que, Y.H. Wang, Violet blue long-lasting phosphorescence properties of Mg-doped BaZrO3 and its ability to assist photocatalysis. J. Alloys Compd. **580**, 564–569 (2013)
41. L. Zhang, X.F. Cao, X.T. Chen, Z.L. Xue, BiOBr hierarchical microspheres: microwave-assisted solvothermal synthesis, strong adsorption and excellent photocatalytic properties. J. Colloid Interface Sci. **354**, 630–636 (2011)
42. C.Q. Wu, J.C. Zhang, P.F. Feng, Y.M. Duan, Z.Y. Zhang, Y.H. Wang, Blue photoluminescence and long lasting phosphorescence properties of a novel chloride phosphate phosphor: $Sr_5(PO_4)_3Cl$: Eu^{2+}. J. Lumin. **147**, 229–234 (2014)
43. G.F. Ju, Y.H. Hu, L. Chen, X.J. Wang, Persistent luminescence and its mechanism of $Ba_5(PO_4)_3Cl$: Ce^{3+}, Eu^{2+}. J. Appl. Phys. **111**, 113508 (2012)
44. L.X. Ning, C.C. Zhou, W.P. Chen, Y.C. Huang, C.K. Duan, P. Dorenbos, Y. Tao, H.B. Liang, Electronic properties of Ce^{3+}-doped $Sr_3Al_2O_5Cl_2$: a combined spectroscopic and theoretical study. J. Phys. Chem. C **119**, 6785–6792 (2015)
45. B. Li, J. Yang, J. Wang, M. Wu, Two-color emitting of Eu^{2+} and Tb^{3+} co-doped $Sr_2MgSi_2O_7$ for UV LEDs. Opt. Mater. **36**(2014), 1649–1654.S
46. L. Jiang, C. Chang, D. Mao, C. Feng, Luminescent properties of $Ca_2MgSi_2O_7$ phosphor activated by Eu^{2+}, Dy^{3+} and Nd^{3+}. Opt. Mater. **27**, 51–55 (2004)
47. T.T.H. Tam, N.V. Du, N.D.T. Kien, C.X. Thang, N.D. Cuong, P.T. Huy, N.D. Chien, D.H. Nguyen, Co-precipitation synthesis and optical properties of green-emitting $Ba_2MgSi_2O_7$:Eu^{2+} phosphor. J. Lumin. **147**, 358–362 (2014)
48. N. Lakshminarasimhan, U.V. Varadaraju, Luminescence and afterglow in Sr_2SiO_4:Eu^{2+}, R^{E3}+ [RE = Ce, Nd, Sm and Dy] phosphors—role of co-dopants in search for afterglow. Mater. Res. Bull. **43**, 2946–2953 (2008)
49. Y. Gong, Y.H. Wang, Y.Q. Li, X.H. Xu, Ce^{3+}, Dy^{3+} Co-doped white-light long-lasting phosphor: $Sr_2Al_2SiO_7$ through energy transfer. J. Electrochem. Soc. **157**, J208–J211 (2010)
50. Y. Gong, Y. Wang, Y. Li, X. Xu, W. Zeng, Fluorescence and phosphorescence properties of new long-lasting phosphor $Ba_4(Si_3O_8)_2$:Eu^{2+},Dy^{3+}. Opt. Express **19**(5), 4310–5

51. Y. Li, Y. Wang, Y. Gong, X. Xu, F. Zhang, Photoionization behavior of Eu^{2+}-doped $BaMgSiO_4$ long-persisting phosphor upon UV irradiation. Acta Mater. **59**, 3174–3183 (2011)
52. H. He, R.L. Fu, X.F. Song, R. Li, Z.W. Pan, X.R. Zhao, Z.H. Deng, Y.G. Cao, Observation of fluorescence and phosphorescence in $Ca_2MgSi_2O_7$: Eu^{2+}, Dy^{3+} phosphors. J. Electrochem. Soc. **157**, J69–J73 (2010)
53. G.F. Ju, Y.H. Hu, L. Chen, X.J. Wang, Z.F. Mu, Persistent luminescence in $CaAl_2Si_2O_8$: Eu^{2+}, R^{3+} (R= Pr, Nd, Dy, Ho and Er). J. Lumin. **146**, 102–108 (2014)
54. W.B. Dai, M. Zhou, Z.Y. Xian, L.K. Zeng, Structure and photoluminescence characteristics of europium (III) doped in $CaAl_2Si_2O_8$ phosphors. RSC Adv. **4**, 25470–25478 (2014)
55. L.C. Rodrigues, H.F. Brito, J. Hölsä, M. Lastusaari, Persistent luminescence behavior of materials doped with Eu^{2+} and Tb^{3+}. Opt. Mater. Express **2**, 382–390 (2012)
56. J. Kuang, Y. Liu, Luminescence properties of a Pb^{2+} activated long-afterglow phosphor. J. Electrochem. Soc. **153**, G245–G247 (2005)
57. X. Wang, Mn^{2+} activated green, yellow, and red long persistent phosphors. J. Lumin. **102–103**, 34–37 (2003)
58. T. Matsuzawa, Y. Aoki, N. Takeuchi, Y. Murayama, A new long phosphorescent phosphor with high brightness, $SrAl_2O_4$:Eu^{2+}, D^{y3}+. J. Electrochem. Soc. **143**, 2670–2673 (1996)
59. P.T. Ji, X.Y. Chen, Y.Q. Wu, Encapsulating MAl_2O_4:Eu^{2+}, D^{y3}+ (M=Sr, Ca, Ba) phosphors with triethanolamine to enhance water resistance. Appl. Surf. Sci. **258**, 1888–1893 (2011)
60. C. Chang, W. Li, X. Huang, Z. Wang, X. Chen, X. Qian, R. Guo, Y. Ding, D. Mao, Photoluminescence and afterglow behavior of Eu^{2+}, D^{y3}+ and E^{u3}+, ^{Dy}3+ in $_{S}r3_{A}l_2O6$ matrix. J. Lumin. **130**, 347–350 (2010)
61. W. Tang, D. Chen, M. Wu, W. Tang, D. Chen, M. Wu, Luminescence studies on $SrMgAl_{10}O_{17}$:Eu, Dy phosphor crystals. Opt. Laser Technol. **41**, 81–84 (2009)
62. X. Xu, Y. Wang, X. Yu, Y. Li, Y. Gong, Investigation of Ce-Mn energy transfer in $SrAl_2O_4$:Ce^{3+}, Mn^{2+}. J. Am. Ceram. Soc. **94**, 160–163 (2011)
63. D. Jia, R.S. Meltzer, W.M. Yen, W. Jia, X. Wang, Green phosphorescence of $CaAl_2O_4$:Tb^{3+},Ce^{3+} through persistence energy transfer. Appl. Phys. Lett. **80**, 1535–1537 (2002)
64. M. Zhong-Fei, W. Yin-Hai, H. Yi-Hua, The afterglow and thermoluminescence properties of $Y_3Al_5O_{12}$:Ce^{3+}. Acta Phys. Sin. **60**, 013201–12110 (2011)
65. J. Ueda, K. Aishima, S. Nishiura, S. Tanabe, Afterglow luminescence in Ce^{3+}-doped $Y_3Sc_2Ga_3O_{12}$ ceramics. Appl. Phys. Express **4**, 042602 (2011)
66. K.K. Satapathy, G.C. Mishra, R.S. Kher, S.J. Dhoble, Mechanoluminescence and thermoluminescence characterization of Tb^{3+} doped $CaAl_2O_4$: a theoretical and experimental study. RSC Adv. **5**, 79391–79396 (2015)
67. J. Ueda, M. Katayama, K. Asami, J. Xu, Y. Inada, S. Tanabe, Evidence of valence state change of Ce^{3+} and Cr^{3+} during UV charging process in $Y_3Al_2Ga_3O_{12}$ persistent phosphor. Opt. Mater. Express **7**, 2471–2476 (2017)
68. J. Xu, S. Tanabe, A.D. Sontakke, J. Ueda, Near-infrared multi-wavelengths long persistent luminescence of Nd^{3+} ion through persistent energy transfer in Ce^{3+}, Cr^{3+} co-doped $Y_3Al_2Ga_3O_{12}$ for the first and second bio-imaging windows. Appl. Phys. Lett. **107**, 081903 (2015)
69. Y. Teng, J.J. Zhou, Z.J. Ma, M.M. Smedskjaer, J.R. Qiu, Persistent near infrared phosphorescence from rare earth ions co-doped strontium aluminate phosphors. J. Electrochem. Soc. **158**, K17–K19 (2011)
70. N.Y. Yu, F. Liu, X.F. Li, Z.W. Pan, Near infrared longpersistent phosphorescence in $SrAl_2O_4$:Eu^{2+}, Dy^{3+}, Er^{3+} phosphors based on persistent energy transfer. Appl. Phys. Lett. **95**, 231110 (2009)
71. R.X. Zhong, J.H. Zhang, X. Zhang, S.Z. Lu, X.J. Wang, Red phosphorescence in $Sr_4Al_{14}O_{25}$:Cr^{3+}, E^{u2}+, ^{Dy}3+ through persistent energy transfer. Appl. Phys. Lett. **88**, 201916 (2006)
72. K. Van den Eeckhout, D. Poelman, P.F. Smet, Persistent luminescence in non-Eu^{2+}-doped compounds: a review. Materials **6**, 2789–2818 (2013)

73. Y. Luo, Z.G. Xia, Effect of Al/Ga substitution on photoluminescence and phosphorescence properties of garnet-type $Y_3Sc_2Ga_{3-x}Al_xO_{12}$:$Ce^{3+}$ phosphor. J. Phys. Chem. C **118**, 23297–23305 (2014)
74. Y.X. Zhuang, J. Ueda, S. Tanabe, Multi-color persistent luminescence in transparent glass ceramics containing spinel nano-crystals with Mn^{2+} ions. Appl. Phys. Lett. **105**, 191904 (2014)
75. X.Q. Xu, J. Ren, G.R. Chen, D.S. Kong, C.J. Gu, C.M. Chen, L.R. Kong, Bright green emission from the Mn^{2+}-doped zinc gallogermanate phosphors. Opt. Mater. Express **3**, 1727–1732 (2013)
76. Z.R. Liu, R.X. Zhong, Green and red long lasting phosphorescence (LLP) in γ-$Zn_3(PO_4)_2$: Mn^{2+}/Ga^{3+}. J. Alloys Compd. **556**, 6–11 (2013)
77. A. Bessière, R.A.T. Benhamou, G. Wallez, A. Lecointre, B. Viana, Site occupancy and mechanisms of thermally stimulated luminescence in $Ca_9Ln(PO_4)_7$ (Ln = lanthanide). Acta Mater. **60**, 6641–6649 (2012)
78. Y.H. Jin, Y.H. Hu, L. Chen, X.J. Wang, G.F. Ju, Z.F. Mou, Luminescence properties of dual-emission (UV/Visible) long afterglow phosphor $SrZrO_3$: Pr^{3+}. J. Am. Ceram. Soc. **96**, 3821–3827 (2013)
79. Z.W. Liu, Y.L. Liu, Synthesis and luminescent properties of a new green afterglow phosphor $CaSnO_3$: Tb. Mater. Chem. Phys. **93**, 129–132 (2005)
80. B.F. Lei, B. Li, X.J. Wang, W. Li, Green emitting long lasting phosphorescence (LLP) properties of Mg_2SnO_4:Mn^{2+} phosphor. J. Lumin. **118**, 173–178 (2006)
81. Y. Jin, Y. Hu, L. Chen, X. Wang, G. Ju, Z. Mu, Luminescent properties of Tb^{3+}-doped Ca_2SnO_4 phosphor. J. Lumin. **138**, 83–88 (2013)
82. G. Li, Y. H. Wang, S.C. Han, W. Zeng, W.B. Chen, Y. Gong, A red long lasting phosphorescence material $Gd_{9.33}(SiO_4)_6O_2$:Sm^{3+} and effect of oxygen vacancies on its performance. ECS J. Solid State Sci. Technol. **2**, R161–R164 (2013)
83. Y.H. Jin, Y.H. Hu, L. Chen, X.J. Wang, G.F. Ju, Z.F. Mu, Persistent luminescence in Bi^{3+} doped $CaWO_4$ matrix. Radiat. Meas. **51–52**, 18–24 (2013)
84. Y.G. Su, L.P. Li, G.S. Li, Synthesis and optimum luminescence of $CaWO_4$-based red phosphors with codoping of Eu^{3+} and Na^+. Chem. Mater. **20**, 6060–6067 (2008)
85. H. Wu, Y. Hu, F. Kang, L. Chen, X. Wang, G. Ju, Z. Mu, Observation on long afterglow of Tb^{3+} in $CaWO_4$. Mater. Res. Bull. **46**, 2489–2493 (2011)
86. B.K. Woo, Z. Luo, Y. Li, S.P. Singh, A.G. Joly, M. Hossu, Z. Liu, Luminescence enhancement of $CaZnGe_2O_6$:Tb^{3+} afterglow phosphors synthesized using ZnO nanopowders. Opt. Mater. **33**(8), 1283–1290 (2011)
87. Y. Takahashi, M. Ando, R. Ihara, T. Fujiwara, Green-emissive Mn-activated nanocrystallized glass with willemite-type Zn_2GeO_4. Opt. Mater. Express **1**, 372–378 (2011)
88. Y.H. Jin, Y.H. Hu, H. Duan, L. Chen, X.J. Wang, The long persistent luminescence properties of phosphors: Li_2ZnGeO_4 and Li_2ZnGeO_4: Mn^{2+}. RSC Adv. **4**, 11360–11366 (2014)
89. P. Yang, M. Lü, G. Zhou, D. Yuan, D. Xu, Photoluminescence characteristics of ZnS nanocrystallites co-doped with Co^{2+} and Cu^{2+} **4**, 734–737 (2001)
90. H.R. Zhang, B.F. Lei, J.L. Qin, J.F. Li, Y.L. Liu, X.T. Liu, M.T. Zheng, Y. Xiao, H.W. Dong, Preparation of $SrSi_2O_2N_2$:Eu^{2+} phosphor by SrSi alloy precursor and its long-lasting phosphorescence properties. J. Am. Ceram. Soc. **96**, 1810–1814 (2013)
91. J.L. Qin, H.R. Zhang, B.F. Lei, C.F. Hu, J.F. Li, Y.L. Liu, J.X. Meng, J. Wang, M.T. Zheng, Y. Xiao, Thermoluminescence and temperature-dependent afterglow properties in $BaSi_2O_2N_2$:Eu^{2+}. J. Am. Ceram. Soc. **96**, 3149–3154 (2013)
92. H.R. Zhang, M.T. Zheng, B.F. Lei, Y. Xiao, H.W. Dong, Y.L. Liu, X.T. Liu, J.J. Deng, J.W. Deng, Z.L. Huang, Preparation and long-lasting phosphorescence properties of $BaAlSi_5N_7O_2$:Eu^{2+}. ECS Solid State Lett. **2**, R16–R18 (2013)
93. Y.R. Tang, H.J. Song, Y.Y. Su, Y. Lv, Turn-on persistent luminescence probe based on graphitic carbon nitride for imaging detection of biothiols in biological fluids. Anal. Chem. **85**, 11876–11884 (2013)

94. X.F. Liu, Y.B. Qiao, G.P. Dong, S. Ye, B. Zhu, Y.X. Zhuang, J.R. Qiu, BCNO-based long-persistent phosphor. J. Electrochem. Soc. **156**, P81–P84 (2009)
95. J. Trojan-Piegza, J. Niittykoski, J. Holsa, E. Zych, Thermoluminescence and kinetics of persistent luminescence of vacuum-sintered Tb^{3+}-Doped and Tb^{3+}, C^{a2}+-codoped L_u2_O3 materials. Chem. Mater. **20**, 2252–2261 (2008)
96. B.P. Chandra, Persistent mechanoluminescence induced by elastic deformation of ZrO2: Ti phosphors. J. Lumin. **130**, 2218–2222 (2010)
97. C. Guo, Q. Tang, D. Huang, C. Zhang, Q. Su, Influence of co-doping different rare earth ions on $CaGa_2S_4$: Eu^{2+}, RE^{3+}(RE=Ln) phosphors. J. Phys. Chem. Sold. **68**, 217–223 (2007)
98. T. Takayama, T. Katsumata, S. Komuro, T. Morikawa, Growth and characteristics of a new long afterglow phosphorescent yttrium tantalate crystal. J. Cryst. Growth **275**, e2013–e2017 (2005)
99. B.F. Lei, L. Sha, H.R. Zhang, Y.L. Liu, S.Q. Man, S. Yue, Preparation and luminescence properties of green-light-emitting afterglow phosphor $Ca_8Mg(SiO_4)_4Cl_2$:Eu^{2+}. Solid State Sci. **12**, 2177–2181 (2010)
100. C. Cao, S.L. Guo, B.K. Moon, B.C. Choi, J.H. Jeong, Synthesis, grouping, and optical properties of REF3–KF nanocrystals. Mater. Chem. Phys. **139**, 609–615 (2013)
101. X.M. Zhang, N. Choi, K. Park, J. Kim, Orange emissive $Sr_3Al_2O_5Cl_2$: Eu^{2+} phosphor for warm-white light-emitting diodes. Solid State Commun. **149**, 1017–1020 (2009)
102. K. Szczodrowski, A. Chruscinska, J. Barzowska, K. Przegietka, K. Anders, R. Piramidowicz, M. Grinberg, Influence of Ti^{4+} on the long lasting luminescence of Sr_2SiO_4: Eu^{2+}. RSC Adv. **5**, 65236–65244 (2015)
103. A. Madej, E. Zych, Controlled synthesis of the monoclinic and orthorhombic polymorphs of Sr_2SiO_4 activated with Ce^{3+} or Eu^{2+}. RSC Adv. **5**, 104441 (2015)
104. Y. Li, B.H. Li, C.C. Ni, S.X. Yuan, J. Wang, Q. Tang, Q. Su, Synthesis, persistent luminescence and thermoluminescence properties of yellow Sr_3SiO_5:Eu^{3+}, RE^{3+} (RE = Ce, Nd, Dy, Ho, Er, Tm, Yb) and orange-Red $Sr_{3-x}Ba_xSiO_5$:Eu^{2+}, Dy^{3+} phosphor. Chem.—Asian J. **9**, 494–499 (2014)
105. X. Liu, J.H. Zhang, X. Zhang, Z.D. Hao, J. Qiao, X.L. Dong, Strongly enhancing photostimulated luminescence by doping Tm^{3+} in Sr_3SiO_5: Eu^{2+}. Opt. Lett. **38**, 148–150 (2013)
106. L.L. Jiang, S.G. Xiao, X.L. Yang, X.A. Zhang, X.H. Liu, B.Y. Zhou, X.L. Jin, Preparation and luminescence properties of yellow long-lasting phosphor $Ca_2ZnSi_2O_7$: Eu^{2+}, Dy^{3+}. Mater. Sci. Eng. B **178**, 123–126 (2013)
107. X.F. Qu, L.X. Cao, W. Liu, G. Su, P.P. Wang, I. Schultz, Sol–gel synthesis of long-lasting phosphors $CdSiO_3$: Mn^{2+}, RE^{3+} (RE= Tb, Eu, Nd) and luminescence mechanism research. Mater. Res. Bull. **47**, 1598–1603 (2012)
108. J. Xu, J. Ueda, Y.X. Zhuang, B. Viana, S. Tanabe, $Y_3Al_{5-x}Ga_xO_{12}$:Cr^{3+}: a novel red persistent phosphor with high brightness. Appl. Phys. Express **8**, 042602 (2015)
109. J. Ueda, K. Kuroishi, S. Tanabe, Yellow persistent luminescence in Ce3+–Cr3+-codoped gadolinium aluminum gallium garnet transparent ceramics after blue-light excitation. Appl. Phys. Express **7**, 062201 (2014)
110. Z.F. Mu, Y.H. Hu, Y.H. Wang, H.Y. Wu, C.J. Fu, F.W. Kang, The structure and luminescence properties of long afterglow phosphor $Y_{3-x}Mn_xAl_{5-x}Si_xO_{12}$. J. Lumin. **131**, 676–681 (2011)
111. S. Zhang, C.Y. Li, R. Pang, L.H. Jiang, L.L. Shi, Q. Su, Energy transfer and excitation wavelength dependent long-lasting phosphorescence in Pr^{3+} activated $Y_3Al_5O_{12}$. J. Lumin. **131**, 2730–2734 (2011)
112. Y.X. Zhuang, J. Ueda, S. Tanabe, Photochromism and white long-lasting persistent luminescence in Bi^{3+}-doped $ZnGa_2O_4$ ceramics. Opt. Mater. Express **2**, 1378–1383 (2012)
113. G.B. Che, X.Y. Li, C.B. Liu, H. Wang, Y. Liu, Z.L. Xu, Long-lasting phosphorescence properties of Mn^{2+}-doped $Cd_2Ge_7O_{16}$ orange light-emitting phosphor. Phys. Status Solidi A **205**, 194–198 (2008)
114. B.F. Lei, Y.L. Liu, G.B. Tang, Z.R. Ye, C.S. Shi, Upconversion of Y_2O_2S: Tm, Mg, Ti phosphor and its long-lasting phosphorescence. Electrochem. Solid-State Lett. **7**, G225–G227 (2004)

115. L.X. Wang, L. Zhang, Y.D. Huang, D.Z. Jia, J.J. Lu, Effects of Gd^{3+} and Lu^{3+} co-doping on the long afterglow properties of yellowish-orange phosphor Y_2O_2S: Ti^{4+}, Mg^{2+}. J. Lumin. **129**, 1032–1035 (2009)
116. J.W. Zhang, Y.L. Liu, J.X. Zhang, D.S. Yuan, J.H. Rong, L.H. Huang, Long afterglow property and mechanism on Gd_2O_2S:Ti,. Rare Met. Mater. Eng. **35**, 766–769 (2006)
117. W. Zeng, Y.H. Wang, S.C. Han, W.B. Chen, G. Li, Y.Z. Wang, Y. Wen, Design, synthesis and characterization of a novel yellow long-persistent phosphor: Ca_2BO_3Cl: Eu^{2+}, Dy^{3+}. J. Mater. Chem. C **1**, 3004–3011 (2013)
118. X.H. Xu, J.B. Qiu, Y. Xue, Multicolor long persistent luminescence phosphors $CaSnSiO_5$:$R3^+$ (R = Pr, Tb, Sm, Dy). ECS Solid State Lett. **2**, R9–R11 (2013)
119. P.F. Feng, J.C. Zhang, C.Q. Wu, X. Liu, Y.H. Wang, Self-activated afterglow luminescence of un-doped $Ca_2ZrSi_4O_{12}$ material and explorations of new afterglow phosphors in a rare earth element-doped $Ca_2ZrSi_4O_{12}$ system. Mater. Chem. Phys. **141**, 495–501 (2013)
120. W. Jia, H. Yuan, L. Lu, H. Liu, W.M. Yen, Crystal growth and characterization of Eu^{2+}, D^{y3}+:$SrA_{l}2_{O}4$ and E^{u2}+, $^{Nd}3$+:$Ca_{A}l_2O4$ by the LHPG method. J. Cryst. Growth **200**, 179–184 (1999)
121. X. Qu, L. Cao, W. Liu, G. Su, Preparation and properties of $CdSiO_3$:Mn^{2+}, T^{b3}+ phosphor. Ceram. Int. **38**, 1765–1769 (2012)
122. J. Kuang, Y. Liu, B. Lei, Effect of RE^{3+} as a co-dopant in long-lasting phosphorescence CdSiO3:Mn^{2+} (RE=Y, La, Gd, Lu). J. Lumin. **118**, 33–38 (1006)
123. P. Zhang, L. Li, M. Xu, L. Liu, The new red luminescent $Sr_3Al_2O_6$:Eu^{2+} phosphor powders synthesized via sol–gel route by microwave-assisted. J. Alloys Compd. **456**, 216–219 (2008)
124. A. Lecointre, A. Bessiere, A.J.J. Bos, P. Dorenbos, B. Viana, S. Jacquart, Designing a red persistent luminescence phosphor: the example of YPO_4:Pr^{3+}, L^{n3}+ (Ln = Nd, Er, Ho, Dy). J. Phys. Chem. C **115**, 4217–4227 (2011)
125. X. Xu, Y. Wang, Z. Wei, G. Yu, Luminescence and storage properties of Sm-doped alkalineearth atannates. J. Electrochem. Soc. **158**, J305–J309 (2011)
126. X. Gao, Z. Zhang, C. Wang et al., The persistent energy transfer and effect of oxygen vacancies on red long-persistent phosphorescence phosphors Ca_2SnO_4:Gd^{3+}, E^{u3}+. J. Electrochem. Soc. **158**, J405–J408 (2011)
127. B. Lei, S.Q. Man, Y. Liu, S. Yue, Luminescence properties of Sm^{3+}-doped $Sr_3Sn_2O_7$ phosphor. Mater. Chem. Phys. **124**, 912–915 (2010)
128. D. Haranath, A.F. Khan, H. Chander, Luminescence enhancement of (Ca, Zn)$Ti_{O}3$:P^{r3}+ phosphor using nanosized silica powder. Appl. Phys. Lett. **89**, 091903 (2006)
129. L.L. Noto, S.S. Pitale, M.A. Gusowki, J.J. Terblans, O.M. Ntwaeaborwa, H.C. Swart, Afterglow enhancement with In^{3+} codoping in $CaTiO_3$: Pr^{3+} red phosphor. Powder Technol. **237**, 141–146 (2013)
130. S.X. Lian, Y. Qi, C.Y. Rong, L.P. Yu, A.L. Zhu, D.L. Yin, S.B. Liu, Effectively leveraging solar energy through persistent dual red phosphorescence: preparation, characterization, and density functional theory study of $Ca_2Zn_4Ti_{16}O_{38}$:Pr^{3+}. J. Phys. Chem. C **114**, 7196–7204 (2010)
131. D. Haranath, A.F. Khan, H. Chander, Bright red luminescence and energy transfer of Pr^{3+}—doped (Ca,Zn)TiO_3 phosphor for long decay applications. J. Phys. D: Appl. Phys. **39**, 4956–4960(5) (2006)
132. T.R.N. Kutty, A. Nag, Role of interface states associated with transitional nanophase precipitates in the photoluminescence enhancement of $SrTiO_3$:Pr^{3+}, A^{l3}+. J. Mater. Chem. **13**, 2271–2278 (2003)
133. C. Minghui, J. Dapeng, Z. Chengjiu, L. Bin, Long-lasting phosphorescence properties of Pyrochlore $La_2Ti_2O_7$:Pr^{3+} phosphor. Chin. Phys. Lett. **27**, 208–211 (2004)
134. H.A. Buckmaster, R. Chatterjee, Generalized spin-hamiltonian for Eu^{2+} in the tetragonal crystalline electric field of $CaWO_4$. J. Mag. Res. **5**, 1–10 (1971)
135. D. Sun, D. Li, Z. Zhu, J. Xiao, Z. Tao, W. Liu, Photoluminescence properties of europium and titanium Co-doped $BaZrO_3$ phosphors powders synthesized by the solid-state reaction method. Opt. Mater. **34**, 1890–1896 (2012)

136. K. Feng-Wen, H. Yi-Hua, W. Hao-Yi, J. Gui-Fang, Red afterglow properties of Eu^{3+} in $CaMoO_4$ phosphor. Chin. Phys. Lett. **28**, 107201–107204 (2011)
137. Y. Lin, C.W. Nan, N. Cai, X. Zhou, H. Wang, D. Chen, Anomalous afterglow from Y_2O_3-based phosphor. J. Alloys Compd. **361**, 92–95 (2003)
138. E. Zych, J. Trojan-Piegza, Anomalous activity of Eu^{3+} in S_6 site of Lu_2O_3 in persistent luminescence. J. Lumin. **122**, 335–338 (2007)
139. Z. Zhao, Y. Wang, The synthesis and afterglow luminescence properties of a novel red afterglow phosphor: ZrO_2:Sm^{3+}, Sn^{4+}. J. Lumin. **132**, 2842–2846 (2012)
140. H.D. Luo, A.J.J. Bos, A. Dobrowolska, P. Dorenbos, Low-temperature VUV photoluminescence and thermoluminescence of UV excited afterglow phosphor $Sr_3Al_xSi_{1-x}O_5$: Ce^{3+}, Ln^{3+}(Ln= Er, Nd, Sm, Dy and Tm). Phys. Chem. Chem. Phys. **17**, 15419–15427 (2015)
141. Y.J. Li, M.W. Wang, L.D. Zhang, D. Gao, S.X. Liu, Soft chemical synthesis and luminescence properties of red long-lasting phosphors Y_2O_2S:Sm^{3+}. Int. J. Miner. Metall. Mater. **20**, 972–977 (2013)
142. C.E. Cui, G.W. Jiang, P. Huang, L. Wang, D. Liu, Synthesis and characterization of Y_2O_2S: Eu^{3+}, Mg^{2+}, Ti^{4+} nanotubes via hydrothermal method. J. Lumin. **145**, 665–668 (2014)
143. C.-Y. Song, B.-F. Lei, Y.-L. Liu, C.-S. Shi, J.-X. Zhang, L.-H. Huang, D.-S. Yuan, The long afterglow of Eu^{2+}-activated La_2O_2S. Chin. J. Inorg. Chem. **20**, 89–93 (2004)
144. H. Zhang, H. Dong, B. Lei, P. Wang, J. Li, Y. Liu, J. Wang, Y. Xiao, M. Zheng, J. Meng, Enhanced performance of $Ca_2Si_5N_8$:Eu^{2+}, Tm^{3+} reddish-orange afterglow phosphor by co-doping with Dy^{3+}. Opt. Mater. **36**, 1846–1849 (2014)
145. J. Wang, H.R. Zhang, B.F. Lei, Z.G. Xia, H.W. Dong, Y.L. Liu, M.T. Zheng, Y. Xiao, Enhanced photoluminescence and phosphorescence properties of red $CaAlSiN_3$: Eu^{2+} phosphor via simultaneous UV-NIR stimulation. J. Mater. Chem. C **3**, 4445–4451 (2015)
146. P.F. Smet, K. Van den Eeckhout, A.J.J. Bos, E. van der Kolk, P. Dorenbos, Temperature and wavelength dependent trap filling in $M_2Si_5N_8$: Eu (M= Ca, Sr, Ba) persistent phosphors. J. Lumin. **132**, 682–689 (2012)
147. P.F. Smet, J. Botterman, K. Van den Eeckhout, K. Korthout, D. Poelman, Persistent luminescence in nitride and oxynitride phosphors: a review. Opt. Mater. **36**, 1913–1919 (2014)
148. Y. Miyamoto, H. Kato, Y. Honna, H. Yamamoto, K. Ohmi, An orange-emitting, long persistent phosphor, $Ca_2Si_5N_8$:Eu^{2+},Tm^{3+}. J. Electrochem. Soc. **156** (2009)
149. Y.Q. Li, Y.H. Wang, Y. Gong, X.H. Xu, M.J. Zhou, Design, synthesis and characterization of an orange-yellow long persistent phosphor: $Sr_3Al_2O_5Cl_2$:Eu^{2+}, T^{m3}+. Opt. Express **18**, 24853–24858 (2010)
150. Y.H. Jin, Y.H. Hu, L. Chen, X.J. Wang, Synthesis and persistent luminescence mechanism of a novel orange emitting persistent phosphor $Sr_5(BO_3)_3Cl$:Eu^{2+}. J. Am. Ceram. Soc. **97**, 2573–2579 (2014)
151. J. Fu, ChemInform abstract: orange and violet-emitting long-lasting phosphors. ChemInform **33**, 255–257 (2002)
152. J. Fu, Orange and red emitting long-lasting phosphors MO:Eu^{3+} (M = Ca, Sr, Ba). Electrochem. Solid-State Lett. **3**, 350–351 (2000)
153. L. Shi, J. Zhang, H. Li, P. Feng, X. Liu, Z. Fu, Y. Wang, Afterglow luminescence properties and mechanism of novel orange afterglow phosphor: $Ca_2Sb_2O_7$:Sm^{3+}. J. Alloys Compd. **579**, 82–85 (2013)
154. Y. Gong, X. Xu, W. Zeng, C. Wu, Y. Wang, Ce^{3+}, Mn^{2+} co-doped red–light long–lasting phosphor: $BaMg_2Si_2O_7$ through energy transfer. Phys. Procedia **29**, 86–90 (2012)
155. L. Lin, C.S. Shi, Z.F. Wang, W.P. Zhang, M. Yin, A kinetics model of red long-lasting phosphorescence in $MgSiO_3$: Eu^{2+}, Dy^{3+}, Mn^{2+}. J. Alloys Compd. **466**, 546–550 (2008)
156. Y. Gong, Y.H. Wang, X.H. Xu, Y.Q. Li, S.Y. Xin, L.R. Shi, The persistent energy transfer of Eu^{2+} and Mn^{2+} and the thermoluminescence properties of long-lasting phosphor $Sr_3MgSi_2O_8$: Eu^{2+}, Mn^{2+}, Dy^{3+}. Opt. Mater. **33**, 1781–1785 (2011)
157. Y.Y. Zhang, R. Pang, C.Y. Li, C.Y. Zang, Q.A. Su, Reddish orange long lasting phosphorescence of Sm^{3+} in $Sr_2ZnSi_2O_7$: Sm^{3+} phosphors. J. Rare Earths **28**, 705–708 (2010)

158. L.C.V. Rodrigues, J. Holsa, M. Lastusaari, M.C.F.C. Felinto, H.F. Brito, Defect to R^{3+} energy transfer: colour tuning of persistent luminescence in $CdSiO_3$. J. Mater. Chem. C **2**, 1612–1618 (2014)
159. A. Bessiere, A. Lecointre, R.A. Benhamou, E. Suard, G. Wallez, B. Viana, How to induce red persistent luminescence in biocompatible $Ca_3(PO_4)_2$. J. Mater. Chem. C **1**, 1252–1259 (2013)
160. A. Lecointre, R. Ait Benhamou, A. Bessierre, G. Wallez, M. Elaatmani, B. Viana, Red long-lasting phosphorescence (LLP) in β-TCP type $Ca_{9.5}Mn(PO_4)_7$ compounds. Opt. Mater. **34**, 376–380 (2011)
161. Z.H. Ju, R.P. Wei, J.R. Zheng, X.P. Gao, S.H. Zhang, W.S. Liu, Synthesis and phosphorescence mechanism of a reddish orange emissive long afterglow phosphor Sm^{3+}-doped Ca_2SnO_4. Appl. Phys. Lett. **98**, 1906 (2011)
162. X.H. Xu, Y.H. Wang, Y. Gong, W. Zeng, Y.Q. Li, Effect of oxygen vacancies on the red phosphorescence of Sr_2SnO_4:Sm^{3+} phosphor. Opt. Express **18**, 16989–16994 (2010)
163. B.F. Lei, S.Q. Man, Y.L. Du, S. Yue, Luminescence properties of Sm^{3+}-doped $Sr_3Sn_2O_7$ phosphor. Mater. Chem. Phys. **124**, 912–915 (2010)
164. Y.H. Jin, Y.H. Hu, L. Chen, X.J. Wang, G.F. Ju, Luminescent properties of a red afterglow phosphor Ca_2SnO_4:Pr^{3+}. Opt. Mater. **35**, 1378–1384 (2013)
165. X.P. Gao, Z.Y. Zhang, C. Wang, J. Xu, Z.H. Ju, Y.Q. An, W.S. Liu, The persistent energy transfer and effect of oxygen vacancies on red long-persistent phosphorescence phosphors Ca_2SnO_4: Gd^{3+}, Eu^{3+}. J. Electrochem. Soc. **158**, J405–J408 (2011)
166. P. Boutinaud, L. Sarakha, R. Mahiou, $NaNbO_3$: Pr^{3+}: a new red phosphor showing persistent luminescence. J. Phys.: Condens. Matter **21**, 025901 (2009)
167. X.G. Zhang, L.Y. Zhou, M.L. Gong, Thermal quenching of luminescence of $LiSr_4(BO_3)_3$:Eu^{2+} orange-emitting phosphor. Luminescence **29**, 104–108 (2014)
168. L. Wu, X.L. Chen, H. Li, M. He, Y.P. Xu, X.Z. Li, Structure determination and relative properties of novel cubic borates $MM'_4(BO_3)_3$ (M = Li, M' = Sr; M = Na, M' = Sr, Ba). Inorg. Chem. **44**, 6409–6414 (2005)
169. Z.M. Shao, S. Saitzek, P. Roussel, R. Desfeux, Stability limit of the layered-perovskite structure in $Ln_2Ti_2O_7$ (Ln= lanthanide) thin films grown on (110)-oriented $SrTiO_3$ substrates by the sol–gel route. J. Mater. Chem. **22**, 24894–24901 (2012)
170. J. Xu, Z.H. Ju, X.P. Gao, Y.Q. An, X.L. Tang, W.S. Liu, $Na_2CaSn_2Ge_3O_{12}$: a novel host lattice for Sm^{3+}-doped long-persistent phosphorescence materials emitting reddish orange light. Inorg. Chem. **52**, 13875–13881 (2013)
171. Y.H. Jin, Y.H. Hu, L. Chen, Y.R. Fu, Z.F. Mu, T. Wang, J. Lin, Photoluminescence, reddish orange long persistent luminescence and photostimulated luminescence properties of praseodymium doped $CdGeO_3$ phosphor. J. Alloys Compd. **616**, 159–165 (2014)
172. Y.X. Zhuang, Y. Katayama, J. Ueda, S. Tanabe, A brief review on red to near-infrared persistent luminescence in transition-metal-activated phosphors. Opt. Mater. **36**, 1907–1912 (2014)
173. G.B. Che, C.B. Liu, X.Y. Li, Z.L. Xu, Y. Liu, H. Wang, Luminescence properties of a new Mn^{2+}-activated red long-afterglow phosphor. J. Phys. Chem. Solids **69**, 2091–2095 (2008)
174. D.C. Rodriguez Burbano, S.K. Sharma, P. Dorenbos, B. Viana, J.A. Capobianco, Persistent and photostimulated red emission in CaS:Eu^{2+},Dy^{3+} nanophosphors. Adv. Opt. Mater. **3**, 551–557 (2015)
175. Y. Suda, Y. Tamura, S. Yamaguchi, Y. Nanai, T. Okuno, Red afterglow and luminescence arising from defects in CaS:Eu^{2+}, Tm^{3+}. J. Phys. D Appl. Phys. **54**, 415103 (2021)
176. Y.J. Li, M.W. Wang, L.D. Zhang, D. Gao, S.X. Liu, Soft chemical synthesis and luminescence properties of red long-lasting phosphors Y_2O_2S:Sm^{3+}. Int. J. Miner., Metall. Mater. **20**, 972–977 (2013)
177. T. Hang, Q. Liu, D.L. Mao, C.K. Chang, Long lasting behavior of Gd_2O_2S: Eu^{3+} phosphor synthesized by hydrothermal routine. Mater. Chem. Phys. **107**, 142–147 (2008)
178. G.D. Liu, Q.H. Zhang, H.Z. Wang, Y.G. Li, A reddish La_2O_2S-based long-afterglow phosphor with effective absorption in the visible light region. Mater. Sci. Eng., B **177**, 316–320 (2012)

179. X. Teng, Y. Liu, Y. Liu, Y. Hu, H. He, W. Zhuang, Luminescence properties of Tm^{3+} co-doped $Sr_2Si_5N_8$:Eu^{2+} red phosphor. J. Lumin. **130**, 851–854 (2010)
180. J. Wang, H.R. Zhang, B.F. Lei, H.W. Dong, H.M. Zhang, Y.L. Liu, N.L. Lai, Y. Fang, Z.J. Chen, The red persistent luminescence of (Sr, Ca) $AlSiN_3$: Eu^{2+} and mechanism different to $SrAl_2O_4$: Eu^{2+}, Dy^{3+}. Opt. Mater. **36**, 1855–1858 (2014)
181. H.R. Zhang, M.T. Zheng, B.F. Lei, Y.L. Liu, Y. Xiao, H.W. Dong, Y.C. Zhang, S.P. Ye, Luminescence properties of red long-lasting phosphorescence phosphor AlN: Mn^{2+}. ECS J. Solid State Sci. Technol. **2**, R117–R120 (2013)
182. Y.Q. Li, J.E.J. van Steen, J.W.H van Krevel, G. Botty, A.C.A. Delsing, F.J. DiSalvo, G. de With, H.T. Hintzen, Luminescence properties of red-emitting $M_2Si_5N_8$:Eu^{2+} (M=Ca, Sr, Ba) LED conversion phosphors. J. Alloys Compd. **417**, 273–279 (2006)
183. L.L. Noto, M.L. Chitambo, O.M. Ntwaeaborwa, H.C. Swart, Photoluminescence and thermoluminescence properties of Pr^{3+} doped $ZnTa_2O_6$ phosphor. Powder Technol. **247**, 147–150 (2013)
184. S. Wu, Z. Pan, R. Chen, X. Liu, *Long Afterglow Phosphorescent Materials*. Springer (2017). https://doi.org/10.1007/978-3-319-60421-3
185. W.B. Dai, Investigation of the luminescent properties of Ce^{3+} doped and Ce^{3+}/Mn^{2+} co-doped $CaAl_2Si_2O_8$ phosphors. RSC Adv. **4**, 11206–11215 (2014)
186. J.S. Zhang, B.J. Chen, J.S. Sun, X.P. Li, L.H. Cheng, H.Y. Zhong, White long-lasting phosphorescence generation in a CaAl2Si2O8: Eu2+, Mn2+, Dy3+ system through persistent energy transfer. J. Phys. D Appl. Phys. **45**, 325105 (2012)
187. Y.H. Chen, X.R. Cheng, M. Liu, Z.M. Qi, C.S. Shi, Comparison study of the luminescent properties of the white-light long afterglow phosphors: $Ca_xMgSi_2O_{5+x}$: Dy^{3+} (x= 1, 2, 3). J. Lumin. **129**, 531–535 (2009)
188. I.P. Sahu, P. Chandrakar, R.N. Baghel, D.P. Bisen, N. Brahme, R.K. Tamrakar, Luminescence properties of dysprosium doped calcium magnesium silicate phosphor by solid state reaction method. J. Alloys Compd. **649**, 1329–1338 (2015)
189. G.R. Revannasiddappa, R.B. Basavaraj, M.S. Rudresha, G. Nagaraju, Santosh Kumar, N. Sasidhar, White-light emitting $Ca_2MgSi_2O_7$:Dy^{3+} nanopowders: Structural, spectroscopic investigations and advanced forensic applications. Vacuum **184**, 109940 (2021)
190. C. Rosticher, B. Viana, G. Laurent, P. Le Griel, C. Chaneac, Insight into $CaMgSi_2O_6$:Eu^{2+},Mn^{2+},Dy^{3+} nanoprobes: influence of chemical composition and crystallinity on persistent red luminescence. Eur. J. Inorg. Chem. 3681–3687(2015)
191. M.A. Tshabalala, F.B. Dejene, S.S. Pitale, H.C. Swart, O.M. Ntwaeaborwa, Generation of white-light from Dy^{3+} doped Sr_2SiO_4 phosphor. Physica B **439**, 126–129 (2014)
192. I.P. Sahu, D.P. Bisen, N. Brahme, Dysprosium doped di-strontium magnesium di-silicate white light emitting phosphor by solid state reaction method. Displays **35**, 279–286 (2014)
193. X.H. Xu, J.B. Qiu, Y. Xue, Multicolor long persistent luminescence phosphors CaSnSiO5:R^{3+} (R ¼ Pr, Tb, Sm, Dy). ECS Solid State Lett. **2**, R9–R11 (2013)
194. J. Kuang, Y. Liu, J. Zhang, White-light-emitting long-lasting phosphorescence in Dy^{3+} doped $SrSiO_3$. J. Solid State Chem. **179**, 266–269 (2006)
195. Y. Ito, A. Komeno, K. Uematsu, K. Toda, M. Sato, Luminescence properties of long persistence silicate phosphors. J. Alloys Compd. **408–412**, 907–910 (2006)
196. B. Liu, C. S. Shi Z. M. Qi, Potential white-light long-lasting phosphor: Dy^{3+}-doped aluminate. Appl. Phys. Lett. **86**, 191111 (2005)
197. G.B. Che, C.B. Liu, Q.W. Wang, Z.L. Xu, White-light-emission afterglow phosphor. $CaZnGe_2O_6$:Dy^{3+}. Chem. Lett. **37**, 136–137 (2008)
198. X.G. Wang, F.P. Du, D.L. Wei, Y.L. Huang, H.J. Seo, A new long-lasting phosphor Zr^{4+} and Eu^{3+} co-doped $SrMg_2(PO_4)_2$. Sens. Actuators B **158**, 171–175 (2011)
199. T. Nakamura, M. Shima, M. Yasukawa, K. Ueda, Synthesis of Pr^{3+} doped or Tb^{3+}–Mg codoped $CaSnO_3$ perovskite phosphor by the polymerized complex method. J. Sol-Gel Sci. Technol. **61**, 362–366 (2012)
200. P. Huang, X.Q. He, C. Cui, L. Wang, Synthesis and luminescence properties of Y_2O_2S:Dy^{3+}, Mg^{2+}, Ti^{4+} phosphors prepared by sol-gel process. Ceram. Int. **40**, 2663–2668 (2014)

201. Y.L. Wu, Y. Li, X.X. Qin, R.C. Chen, D.K. Wu, S.J. Liu, J.R. Qiu, Dual mode NIR long persistent phosphorescence and NIR-to-NIR Stokes luminescence in $La_3Ga_5GeO_{14}$: Cr^{3+}, Nd^{3+} phosphor. J. Alloys Compd. **649**, 62–66 (2015)
202. X. Yi, Z.T. Chen, S. Ye, Y. Li, E.H. Song, Q.Y. Zhang, Multifunctionalities of near-infrared upconversion luminescence, optical temperature sensing and long persistent luminescence in $La_3Ga_5GeO_{14}$:Cr^{3+}, Y^{b3}+, $^{Er}3$+ and their potential coupling. RSC Adv. **5**, 49680–49687 (2015)
203. W.Z. Yan, F. Liu, Y.Y. Lu, X.J. Wang, M. Yin, Z.W. Pan, Self-activated long persistent luminescence from different trapping centers of calcium germanate. Opt. Express **18**, 20215–20221 (2010)
204. D. Jia, L.A. Lewis, X.J. Wang, Cr^{3+}-doped lanthanum gallogermanate phosphors with long persistent IR emission. Electrochem. Solid-State Lett. **13**, J32–J34 (2010)
205. Z.W. Pan, Y.Y. Lu, F. Liu, Sunlight-activated long-persistent luminescence in the near-infrared from Cr^{3+}-doped zinc gallogermanates. Nat. Mater. **11**, 58–63 (2012)
206. Y.L. Wu, Y. Li, X.X. Qin, R.C. Chen, D.K. Wu, S.J. Liu, J.R. Qiu, Near-infrared long-persistent phosphor of $Zn_3Ga_2Ge_2O_{10}$: Cr^{3+} sintered in different atmosphere. Spectrochim. Acta, Part A **151**, 385–389 (2015)
207. F. Liu, W.Z. Yan, Y.J. Chuang, Z.P. Zhen, J. Xie, Z.W. Pan, Photostimulated near-infrared persistent luminescence as a new optical read-out from Cr^{3+}-doped $LiGa_5O_8$. Sci. Rep. **3**, 01554 (2013)
208. Y.J. Chuang, Z.P. Zhen, F. Zhang, F. Liu, J.P. Mishra, W. Tang, H.M. Chen, X.L. Huang, L.C. Wang, X.Y. Chen, J. Xie, Z.W. Pan, Photostimulable Near-infrared persistent luminescent nanoprobes for ultrasensitive and longitudinal deep-tissue bio-imaging. Theranostics **4**, 1112–1122 (2014)
209. D.Q. Chen, Near-infrared long-lasting phosphorescence in transparent glass ceramics embedding Cr^{3+}-doped $LiGa_5O_8$ nanocrystals. J. Eur. Ceram. Soc. **34**, 4069–4075 (2014)
210. J. Xu, D.Q. Chen, Y.L. Yu, W.J. Zhu, J.C. Zhou, Y.S. Wang, Cr^{3+}:$SrGa_{12}O_{19}$: a. Broadband near-infrared long-persistent phosphor. Chem.—Asian J. **9**, 1020–1025 (2014)
211. A. Mondal, S. Das, J. Manam, Hydrothermal synthesis, structural and luminescent properties of a Cr^{3+} doped $MgGa_2O_4$ near-infrared long lasting nanophospor. RSC Adv. **6**, 82484–82495 (2016)
212. Y.X. Zhuang, J. Ueda, S. Tanabe, P. Dorenbos, Band-gap variation and a self-redox effect induced by compositional deviation in $Zn_xGa_2O_{3+x}$:Cr^{3+} persistent phosphors. J. Mater. Chem. C **2**, 5502–5509 (2014)
213. M. Allix, S. Chenu, E. Veron, T. Poumeyrol, E.A. KouadriBoudjelthia, S. Alahrache, F. Porcher, D. Massiot, F. Fayon, Considerable improvement of long-persistent luminescence in germanium and tin substituted $ZnGa_2O_4$. Chem. Mater. **25**, 1600–1606 (2013)
214. A. Bessiere, S.K. Sharma, N. Basavaraju, K.R. Priolkar, L. Binet, B. Viana, A.J.J. Bos, T. Maldiney, C. Richard, D. Scherman, D. Gourier, Storage of visible light for long-lasting phosphorescence in chromium-doped zinc gallate. Chem. Mater. **26**, 1365–1373 (2014)
215. D. Gourier, A. Bessiere, S.K. Sharma, L. Binet, B. Viana, N. Basavaraju, K.R. Priolkar, Origin of the visible light induced persistent luminescence of Cr^{3+}-doped zinc gallate. J. Phys. Chem. Solids **75**, 826–837 (2014)
216. D.R. Li, Y.H. Wang, K. Xu, H. Zhao, Z.F. Hu, Persistent luminescent and photocatalytic properties of $Zn_xGa_2O_{3+x}$ ($0.8 <= x <= 1$) phosphors. RSC Adv. **5**, 20972–20975 (2015)
217. Y.X. Zhuang, J. Ueda, S. Tanabe, Tunable trap depth in $Zn(Ga_{1-x}Al_x)_2O_4$:Cr, Bi red persistent phosphors: considerations of high-temperature persistent luminescence and photostimulated persistent luminescence. J. Mater. Chem. C **1**, 7849–7855 (2013)
218. A. Bessiere, S. Jacquart, K. Priolkar, A. Lecointre, B. Viana, D. Gourier, $ZnGa_2O_4$:Cr^{3+}: a new red long-lasting phosphor with high brightness. Opt. Express **19**, 10131–10137 (2011)
219. A. Mondal, S. Das, J. Manam, Investigation on spectroscopic properties and temperature dependent photoluminescence of NIR emitting Cr^{3+} doped zinc gallate long persistent nanophosphor. Physica B **569**, 20–30 (2019)

220. Y. Li, S.F. Zhou, Y.Y. Li, K. Sharafudeen, Z.J. Ma, G.P. Dong, M.Y. Peng, J.R. Qiu, Tailoring of the trap distribution and crystal field in Cr^{3+}-doped non-gallate phosphors with near-infrared long-persistence phosphorescence. J. Mater. Chem. C **2**, 2657–2663 (2014)
221. C.Y. Liu, Z.G. Xia, M.S. Molokeev, Q.L. Liu, Synthesis, crystal structure, and enhanced luminescence of garnet-type $Ca_3Ga_2Ge_3O_{12}$:Cr^{3+} by codoping Bi^{3+}. J. Am. Ceram. Soc. **98**, 1870–1876 (2015)
222. D.Q. Chen, Y. Chen, H.W. Lu, Z.G. Ji, A. Bifunctional, Cr/Yb/Tm:$Ca_3Ga_2Ge_3O_{12}$ phosphor with near-infrared long-lasting phosphorescence and upconversion luminescence. Inorg. Chem. **53**, 8638–8645 (2014)
223. A. Mondal, J. Manam, Structural, optical and temperature dependent photoluminescence properties of Cr^{3+}-activated $LaGaO_3$ persistent phosphor for optical thermometry. Ceram. Int. **46**, 23972–23984 (2020)
224. A. Mondal, J. Manam, Citrate sol-gel synthesis and luminescent properties of Cr^{3+}-activated strontium gallate phosphor. AIP Conf. Proc. **2115**, 030193 (2019)
225. J. Xu, J. Ueda, K. Kuroishi, S. Tanabe, Fabrication of Ce^{3+}–Cr^{3+} co-doped yttrium aluminium gallium garnet transparent ceramic phosphors with super long persistent luminescence. Scr. Mater. **102**, 47–50 (2015)
226. A. Brenier, G. Boulon, C. Pédrini, C. Madej, Cr^{3+}-Tm^{3+} ion pairs in $Gd_3Ga_5O_{12}$ garnet. Opt. Mater. **1**, 299–306 (1992)
227. Q. le Masne, C. de Chermont, J. Chaneac, F. Seguin, S. Pelle, J.P. Maitrejean, D. Jolivet, M. Gourier, D.S. Bessodes, Nanoprobes with near-infrared persistent luminescence for *in vivo* imaging. Proc. Natl. Acad. Sci. U. S. A. **104**, 9266–9271 (2007)
228. Y. Li, Y.Y. Li, K. Sharafudeen, G.P. Dong, S.F. Zhou, Z.J. Ma, M.Y. Peng, J.R. Qiu, A strategy for developing near infrared long-persistent phosphors: taking $MAlO_3$:Mn^{4+}, $G^{e4}+$ (M = La, Gd) as an example. J. Mater. Chem. C **2**, 2019–2027 (2014)
229. S. Kamimura, C.N. Xu, H. Yamada, N. Terasaki, M. Fujihala, Long persistent luminescence in the near-infrared from Nd^{3+} -doped Sr_2SnO_4 for in vivo optical imaging. Jpn. J. Appl. Phys. **53**, 062403 (2014)
230. Y. Li, Y.Y. Li, R.C. Chen, K. Sharafudeen, S.F. Zhou, M. Gecevicius, H.H. Wang, G.P. Dong, Y.L. Wu, X.X. Qin, J.R. Qiu, Tailoring of the trap distribution and crystal field in Cr^{3+}-doped non-gallate phosphors with near-infrared long-persistence phosphorescence, NPG Asia Mater. **7,** e180 (2015)
231. X.X. Qin, Y. Li, D.K. Wu, Y.L. Wu, R.C. Chen, Z.J. Ma, S.J. Liu, J.R. Qiu, A novel NIR long phosphorescent phosphor:$SrSnO_3$:Bi^{2+}. RSC Adv. **5**, 101347 (2015)
232. X.S. Wang, W.S. Li, J.Q. Situ, X.Y. Ying, H. Chen, Y. Jin, Y.Z. Du, Multi-functional mesoporous β-Ga_2O_3: Cr^{3+} nanorod with long lasting near infrared luminescence for in vivo imaging and drug delivery. RSC Adv. **5**, 12886–12889 (2015)
233. Y.Y. Lu, F. Liu, Z.J. Gu, Z.W. Pan, Long-lasting near-infrared persistent luminescence from β-Ga_2O_3: Cr^{3+} nanowire assemblies. J. Lumin. **131**, 2784–2787 (2011)

Security Applications of Persistent Luminescence Nanoparticles

Manisha Prasad and Vineet Kumar Rai

Abstract Persistence luminescence nanomaterials are innovative materials that have been investigated and explored by several researchers. They have a variety of special characteristics such as long-lasting intense afterglow emission with long persistence time. The most favorable transition in persistence luminescence occurred due to the trap distribution by activating dopant ions and varying host lattices combination, and the developed trap depths were examined by thermoluminescence analysis. Different color dynamics have been attained after numerous excitations in the NIR to UV wavelength range. Dynamic anti-counterfeiting has been attained by adjusting the concentration of doping ions, types of codopants, trap depths, etc. Because these nanoparticles can continue to emit light even after the excitation source has stopped, they are excellent options for a range of security-related applications. The lanthanide and transition metal ions are quite useful in this field. In the area of security and anti-counterfeiting applications, nanomaterials such as downconverting and upconverting materials are most useful. The present chapter provides an overview of the design, synthesis, and structure of novel long-lasting phosphors, as well as their multiple uses, including security, information protection, data encryption, latent fingerprinting, lip printing, and anti-counterfeiting applications. Persistence luminescence nanoparticles have the potential to provide a growing number of complex and efficient solutions for many security issues as long as research and development in this field continue.

Keywords Persistence phosphors · Security applications · Information protection · Data encryption · Latent fingerprinting · Lip printing · Anti-counterfeiting

M. Prasad (✉) · V. K. Rai
Laser and Spectroscopy Laboratory, Department of Physics, Indian Institute of Technology (Indian School of Mines), Dhanbad, Jharkhand 826004, India
e-mail: manishaprasad92@gmail.com

V. Kumar et al. (eds.), *Persistent Luminescence*, Progress in Optical Science and Photonics 30, https://doi.org/10.1007/978-981-97-4943-0_12

1 Introduction to Persistent Luminescence for Security

The phenomenon known as persistence luminescence occurs when holes and electrons are thermally encouraged to recombine at traps or in a stable excited state at ambient temperature. This process allows the material to retain its luminescence for an extended period, even after the excitation source is removed. It refers to a situation where the luminescence process has a longer lifetime or decay (usually lasting between 10^{-8} s and a few hours). Numerous adjectives, including long persistence, persistence luminescence (persL), phosphorescence, long-lasting, long duration, long decay, and the long afterglow, can be found in the literature to describe this occurrence. Generally, the persistent phosphors are classified as inorganic and organic phosphors [1]. The inorganic persistence phosphors are lanthanide or transition metals ions activated aluminates [2], sulfides [3], silicates [4], germanates [5], gallates [6], titanates [7], nitrides [8] and oxynitrides [9], phosphates, or other host materials [10]. Currently, a large investigation has been performed in rare earths and transition metal ions doped phosphors. The $Eu^{2+/3+}$, Dy^{3+}, Ce^{3+}, Pr^{3+}, Nd^{3+}, Tb^{3+}, etc. dopants are very reliable, which have fixed emissions due to f^n to $f^{n-1}5d$ transition [11–18]. Also, some transition metal and pure metal ions like Mn^{2+}, Mn^{4+}, Cr^{3+}, Pb^{2+}, Bi^{3+}, etc. give fascinating results [19–24]. At present, the rising materials in this field are (Sr, Ca, Mg)$AlSiN_3$: Eu^{2+} phosphors [25], $Sr_2MgSi_2O_7$:Eu^{2+}, Dy^{3+} [26] $LiTaO_3$ [27] $CaAl_2O_4$:Eu^{2+}, Nd^{3+} [28] and Y_2O_2S: Eu^{3+}, Mg^{2+}, Ti^{4+} [29] etc. Another class of persL phosphors are organic materials [30]. Organic persistence luminescence occurs due to deactivations of organic molecules in their excited state and is dissimilar from the f/d-electron induced radiative transitions of the rare earth metals in inorganic luminescence. Polymers, carbon dots, and metal–organic frameworks are some examples of organic phosphors [31, 32]. Organic PersL materials are widely used for OLEDs, anticounterfeiting, optical recording, sensing bioimaging, and applications. Recently, PMMA/G8, PVA/ACDs, MOF-5 are the most rising materials in this field [33–35]. The emission from these phosphors can be either red, blue, or green and thus motivates the researchers to achieve white light emitting long luminescent phosphors. $CdSiO_3$: Dy^{3+} and $Ca_3SnSi_2O_9$: Dy^{3+} are some reported white persL phosphors [36, 37].

The wide applications of inorganic and organic persistence phosphors in different fields are display and lighting, fire safety signs, road traffic signs, luminous watch and fibers, molecular sensing, lateral flow assay, optical information storage, security displays, biological applications like in fluorescence labeling, vivo imaging, and optical anti-counterfeiting, etc. [10, 38–40]. The security of priceless documents, certificates, money, banknotes, well-known brands of goods, information processes, passports, and fingerprint verification are the necessary threats in current times that must be resolved. To overcome these problems, various encryption technologies are available in the market such as laser holograms, labels, bar codes, QR (Quick Response) codes, plasma tags, watermarks, biometric devices, luminescence, chemical inks, and photochromic materials, among other things. Particularly, luminescence has drawn a lot of interest due to its advantages of visibility, simplicity of construction, and easy identification. In addition to harming the reputation of products and violating the company's lawful rights and interests, counterfeiting also puts the company's survival in danger. Encoding particular product information into a separate unit helps improve information security. It has been shown that the emission colors of fluorescent materials are very well-suited for optical multiplexing and information encoding. Several reports deal with persistence phosphors for information protection, data encryption, latent fingerprinting, anti-counterfeiting, etc. and will be discussed in upcoming sections.

A long persistence phosphor can be prepared by various synthesis techniques such as combustion technique, hydrothermal, co-precipitation, sol–gel, solid state reaction method, etc. [25]. Synthesis techniques decides the size, structure, phase, homogeneity, morphology and optical properties of nanoparticles. In general, after synthesis the structural information and optical investigations of persistence phosphors with X-ray diffraction (XRD), field emission scanning electron microscopy (FESEM), energy dispersive X-ray microscopy (EDX), Raman and Fourier transform infrared (FTIR) spectroscopy, UV–vis-NIR spectroscopy, photoluminescence study, thermoluminescence techniques etc. reveal the characteristics of phosphors. Considerable thermal and chemical stability, suitable bandgap, high refractive index, low phonon energy, etc. are some of the basic requirements of the persistence host materials. While persistent luminescence nanoparticles present immense potential for security applications, several challenges and limitations need to be addressed. These include synthesis of efficient long persistence phosphorescence material, development of cost-effective production methods, broad excitation spectrum, desirable emission spectrum, the compatibility with existing security technologies, fabrication, industrial trials, and the environmental impact of these nanoparticles.

The properties of long phosphorescence or persistence phosphors can be identified in terms intense emission, long persistent time with some threshold value of excitation power, presence of more trap levels with suitable depths, and exposure time/pre irradiation time. Performance of PerL is mostly affected by trap characteristics, such as trap types, depth, distribution of the traps, and concentration of dopants [41]. The emission from persistence phosphors may lie in UV, VIS, or NIR spectral regions. Usually, 254 nm, 365 nm, 980 nm, and 808 nm excitations are most popular due matching energy levels with the dopants in the phosphors. Material can be utilized for a variety of purposes, such as energy storage, by adjusting the range of trap depths. For example, in cryopreservation, very shallow traps could be employed to monitor temperature. For dosimetry and x-ray imaging, deeper traps are helpful. It is very well known that trap depth is important for long-lasting sustained luminescence. For example, a longer-lasting luminescence in security displays is aided by a shallower trap depth of 0.7 eV, whereas a larger trap depth of 1 eV is detrimental to long-term luminescence but advantageous to optical information storage as it guarantees a higher storage rate at room temperature [25]. Based on their range of depth, the traps are divided into two categories: shallow and deep. If the depth of the trap is excessively shallow (that is, less than 0.6 eV), the charge carriers will be readily released and their persistence period will be much shorter. Thermoluminescence (TL) tests can be performed to assess the trap depth in certain phosphors.

The concentration of charge carriers caught by the traps is reflected in the TL intensity, and the trap depth is assessed by the TL peak position [42]. Figure 1 shows the TL glow curves (a) and associated Gaussian peaks (b: x = 1) of the prepared $Sr_xCa_{1-x}AlSiN_3$:0.2 at% Eu^{2+} phosphors as reported by Zhu et al. [25] (c) The trap depths of $Sr_xCa_{1-x}AlSiN_3$:0.2 at% Eu^{2+} (x = 0–1) phosphors show that the phosphors contain both shallow traps (P1) and deep traps (P2). The proposed persistent luminescence mechanism for $SrAlSiN_3$:0.2 at% Eu^{2+} phosphors is shown in Fig. 1d [25].

For the trapping process, which is the first and most crucial stage, two main strategies are envisaged with material chemistry: firstly, the inherent defects and the optimization of the trapping by stoichiometric variation and thermal annealing, and secondly, trap optimization by codoping. This latter method, which often involves co-doping with one or two lanthanide cations, has undergone extensive testing to improve the persistent luminescence. The physics underlying the persistent luminescence mechanisms, which follow these two independent approaches, is complex, and numerous models have been put out over the past few years. A global model in which the valence or conduction bands, or both, are used to transmit the trapped/

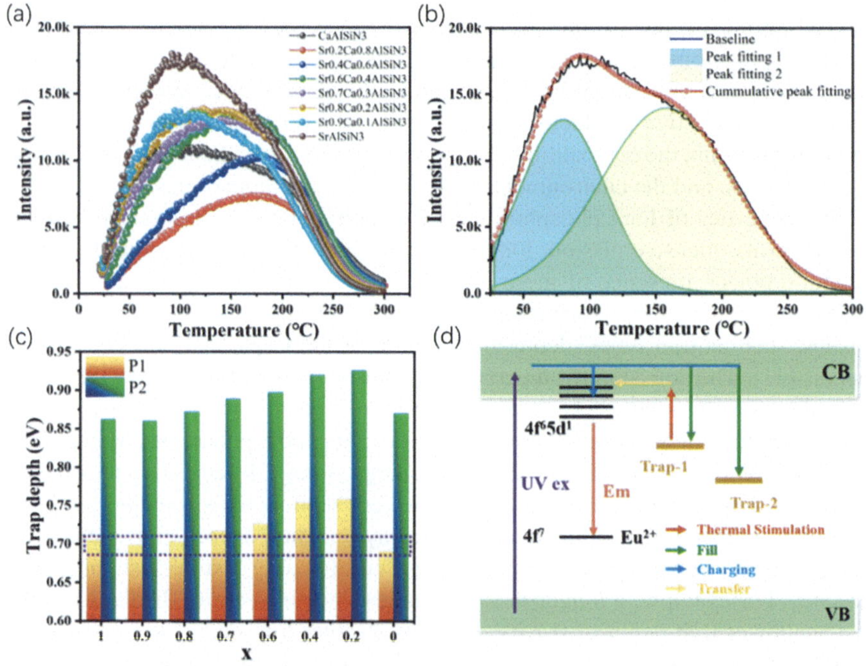

Fig. 1 TL glow curves (**a**) and corresponding Gaussian peaks (**b**: x = 1); **c** Trap depths of $Sr_xCa_{1-x}AlSiN_3$:0.2 at% Eu^{2+} phosphors (x = 0–1) **d** Proposed persistent luminescence mechanism for $SrAlSiN_3$:0.2 at% Eu^{2+}. (*Source* Zhu et al. [25]. Reprinted with permission of Journal of Luminescence)

detrapped charges. It is possible to refer to this model as a global model as the carriers' block functions are delocalized in such circumstances. To effectively charge through the bandgap, materials having intermediate bandgap values between 3 and 7 eV, such as sulfides, oxysulfides, and oxides, such as garnets, gallates, phosphates, and aluminates, required ultraviolet (UV)/blue light. In these circumstances, bandgap engineering could generally be used. However, a direct electron transfer between the activator and the trapping centers is considered in a local model [43].

The ongoing research in the area of long-persistence phosphors, its results, and its applicability make people curious. Zheng et al. [44] reported the cyan emitting persistence phosphorescence using $CaSnO_3$: Lu^{3+} phosphors. The possible mechanism has been explained with TL and charge compensation [44]. The decay curve is another characteristic that describes the afterglow characteristics of phosphor as reported in $SrAl_2O_4$: Eu^{2+}, Dy^{3+}, and $M_2MgSi_2O_7$: Eu^{2+}, Dy^{3+} phosphors [10]. The prolonged lifetime and enhanced brightness, stability against moisture are key features of afterglow phosphors. Jiang et al. [45] reported long phosphorescence in $SrAl_2O_4$: Eu, Dy, $Sr_4Al_{14}O_{25}$: Eu, Dy, $Sr_2MgSi_2O_7$: Eu, Dy, and ZnS: Cu, Co phosphors. The decay time has been recorded for 15–16 h. The afterglow decay consists of slow and fast stages and the mechanism has been explained with rate equations [45]. Wang et al. [46] investigated the $SrCaSi_5N_8$: Eu^{2+}, Tm^{3+} red emitting persistent and photo-stimulated phosphors upon 980 nm laser diode excitation. The thermoluminescence properties of the phosphors with UC excitation show TL peaks at 345 and 360 K [46]. The effect of trap depth has been explained by Zhuang et al. [9] in $SrSi_2O_2N_2$: Ln^{2+}, Ln^{3+} (Ln^{2+} = Yb, Eu; Ln^{3+} = Dy, Ho, Er) phosphors. With the addition of Er^{3+}, Ho^{3+}, and Dy^{3+} codopants, the trap depth has been increased from 0.90 to 1.18 eV and found suitable for information storage applications [9]. Wang J. et al. [47] investigated the optical energy storage properties of $(CaSr)_2Si_5N_8$: Eu^{2+}, Tm^{3+} phosphors. The proper trap depths have been adjusted by cation substitution [47]. The persistence and photo-stimulated properties upon 254 and 980 nm excitation have been investigated by Zhang et al. [8]. The synthesized $SrLiAl_3N_4$: Eu^{2+} phosphor exhibits red emission with trap depths of 0.47 and 0.81 eV, making it a promising candidate for use in optical information storage and anticounterfeiting applications. Zhuang et al. [48] explored deep-trap multicolor emitting persistent luminescence in $(SrBa)Si_2O_2N_2$: Eu/Yb, Dy phosphors for innovative display technology and information security systems [48].

2 Different Persistent Luminescent Materials Used for Security Applications

A lot of studies have been performed in the development of persL phosphors say inorganic and organic materials. In 1996, Matsuzawa et al. [49] and Takasaki et al. [50] parallelly reported persistence luminescence from Eu^{2+} doped aluminate phosphors [49, 50]. Eu^{2+}, Dy^{3+} doped $SrAl_2O_4$ green-emitting phosphors are one of the first commercially available phosphors. Later several studies have been performed to explore the $Eu^{2+/3+}$ ions with different dopants and hosts. Depending on the luminescence time, several works have been described. Codoping of another lanthanide such as Dy^{3+}, Tm^{3+} shows higher brightness and long afterglow than the singly doped Eu^{3+} ions. A typical persistent phosphor is $Sr_4Al_{14}O_{25}$: Eu^{2+}, Dy^{3+}, which has a 490 nm-centered blue emission with a persistence duration of more than 20 h [51, 52]. Chang et al. [44] used triboluminescence and photoluminescence tests to detect a brilliant green emission from the $Sr_3Al_2O_6$: Eu^{2+}, Dy^{3+} phosphor at 510 nm. For example, in Eu^{2+}, Dy^{3+}: $SrAl_2O_4$ and Eu^{2+}, Nd^{3+}: $CaAl_2O_4$ phosphors, Eu^{2+} ions are the emission center and Dy^{3+} and Nd^{3+} ions are the trapping centers in various Eu^{2+} codoped phosphors. Eu^{2+} can either absorb one electron from the valence band and stay in a metastable state when exposed to an ultraviolet source, or it can radiatively revert to its ground state. The resulting hole has two possible outcomes: it can either spread inside the valence band or be trapped by codopant-induced traps. The hole emerges from the traps after the proper temperature activation, goes back to the valence band, and combines with the extra electron. However, synthesis methods also affect the afterglow emission. The $Sr_3A_{12}O_6$:Eu^{2+}, Dy^{3+} phosphors made utilizing the sol–gel and microwave techniques exhibit red emission centered at 612 nm [53, 54]. The primary emission peaks are caused by the transition of Eu^{3+} ions from ^5D_J ($J = 0, 1$) to ^7F_J ($J = 0, 1, 2, 3, 4$). Similar to this, the main optical transitions from the lowest $5d^1$ excited state to the $^2F_{5/2}$ and $^2F_{7/2}$ levels of the $4f^1$ ground state are responsible for luminescence in Ce^{3+}-doped materials. Dy^{3+} ions are employed to produce white light phosphorescence because they can generate blue, green, and red light all at once. In contrast to producing a monochromatic color, Liu et al. [55] found that $CdSiO_3$:Dy^{3+} phosphor was capable of producing white afterglow, which is produced by combining three distinct emissions (420, 486, and 580 nm). When included in a certain host lattice, Sm^{3+} ions are superior emitters that emit red light. Among the phosphors being studied, M_2SnO_4 hosts (M = Ca, Sr, and Ba) enable reddish-orange Sm^{3+} ion afterglows with lifetimes of >7 h [55]. In PersL, praseodymium (Pr^{3+}) ions additionally act as activators. Pr^{3+}-doped ABO_3 type perovskite-type oxides, for example, typically produce light because Pr^{3+} ions exhibit a stronger electron–phonon interaction and many energy level transitions within 4f configuration. Red PersL originates from the dominant $^1D_2 \rightarrow {}^3H_4$ transition (610 nm) [56], such as $CaTiO_3$: Pr^{3+} phosphors [57] and $Ca_3Ti_2O_7$: Pr^{3+} phosphors [58]. Blue-green $^3P_0 \rightarrow {}^3H_4$ emission (495 nm) may also be prominent under certain crystal fields [59], such as white emitting $Ba_2Ga_2GeO_7$: Pr^{3+} phosphors [60], blue emitting $Sr_2Ta_2O_7$: Pr^{3+} phosphors [61] and yellow emitting $CaGa_2O_4$: Pr^{3+} phosphors [62]; additionally, several transitions can be found in the near-infrared (NIR), such as $^1D_2 \rightarrow {}^3F_J$ (J = 2, 3, and 4), $^1D_2 \rightarrow {}^3H_6$, and $^1I_6 \rightarrow {}^1G_4$ [63], that can be used in biological imaging [64].

Among the transition metals, the Mn^{2+} ion is coordinated tetrahedrally or octahedrally, it can emit green and red light, respectively [65–67]. Wang et al. [68] reported the existence of three Mn^{2+}-activated persistent phosphors namely $Ca_2Al_2SiO_7$:Ce^{3+}, Mn^{2+}, $CaAl_2O_4$: Ce^{3+}, Mn^{2+}, and $MgSiO_3$: Mn^{2+}, Eu^{2+}, Dy^{3+}. The Mn^{2+} ions can be combined with a variety of lanthanides, including Ce^{3+}, Eu^{2+}, and Dy^{3+}. It has been discovered that trivalent chromium Cr^{3+} ions can emit NIR extended afterglow when doped into inorganic materials. The spin-forbidden $^2E \rightarrow {}^4A_2$ transition is generally responsible for the observed narrow-band emission centered at 700 nm, whereas the spin-allowed $^4T_2 \rightarrow {}^4A_2$ transition results in broadband emission from 650 to 1600 nm [69, 70].

These reported results can be used in various security applications, some of the works have been discussed in this chapter. Shen et al. [71] developed tunable afterglow phosphors using three different red emitting Y_2O_2S: Eu^{3+}, Mg^{2+}, Ti^{4+} (marked R), green-emitting $SrAl_2O_4$: Eu^{2+}, Dy^{3+} (marked G), and blue-emitting $Sr_2MgSi_2O_7$: Eu^{2+}, Dy^{3+} (marked B) phosphors [71]. The color emission from the finally developed phosphors has been adjusted by varying the ratio of RGB. The phosphors have been excited with 365 nm excitation. The security ink has been prepared by mixing the phosphors with poly-dimethylsiloxane (PDMS). In this work, the need for tunable afterglow phosphors has been realized because the 1D and 2D security techniques are relatively less reliable such as bar codes. A multicolor emitting phosphor enhanced the capability of 3D encryption technologies. Using the developed phosphors, the produced code emits a vibrant luminescence pattern when exposed to a 365 nm UV lamp and may be scanned with a laser to decrypt the data using a mobile phone with color-coding software. Figure 2 shows processes for developing and designing a PDMS 3D code (a) and using the luminous code for security purposes (b).

Gupta [72] reported long-persistent $Sr_3SiAl_4O_{11}$:Eu^{2+}/Dy^{3+} phosphor prepared by solid-state reaction method. The phosphors have been excited by 340 nm excitation and green emission was found at 510 nm due to the $4f^6 5d^1 \rightarrow 4f^7$ transition from Eu^{3+} ions. With the incorporation of Dy^{3+}, the trap depths adjusted and there was an enhancement in the persistence luminescence occurred. The afterglow decays for 4 h. The developed phosphors have been applied on the photo identity card as shown in Fig. 3a under room light (b) after 5 s when the 365 nm wavelength UV excitation source was removed; it exhibits a phosphorescent lotus flower image with emitting green color, which cannot be seen under room light [72].

Zhang et al. [73] developed polymer-based persistence material. Polyvinyl alcohol (PVA) films with vanilla doping display ultralong phosphorescence emission lasting 369.8 ms at 254 nm with a maximum 14.36% phosphorescence quantum yield, whereas methyl vanillate has the longest emission period up to seven seconds. Using PVA materials doped with vanilla to create invisible security inks which have been applied to hand-painted lotus flower patterns. It is impressive since the inks exhibit good luminous emission stability under ambient circumstances and flawless reversibility between room temperature and 65 °C across many cycles [73].

Broad spectrum color-tunable organic DTA-DQD/TPP phosphorescence materials were developed by Lei et al. [74] for printable and written security inks. The ability to dynamically tune color from cyan (502 nm) to orange red (608 nm) is provided by a number of new host–guest organic phosphors. When kept at normal temperature, neither the host nor the guest molecules glow. The phosphorescent color was adjusted using guest materials and the phosphorescence emission was initiated using host materials that work in concert with the guest. These organic phosphors have an extremely long lifespan of 0.7 s and a maximum phosphorescence efficiency

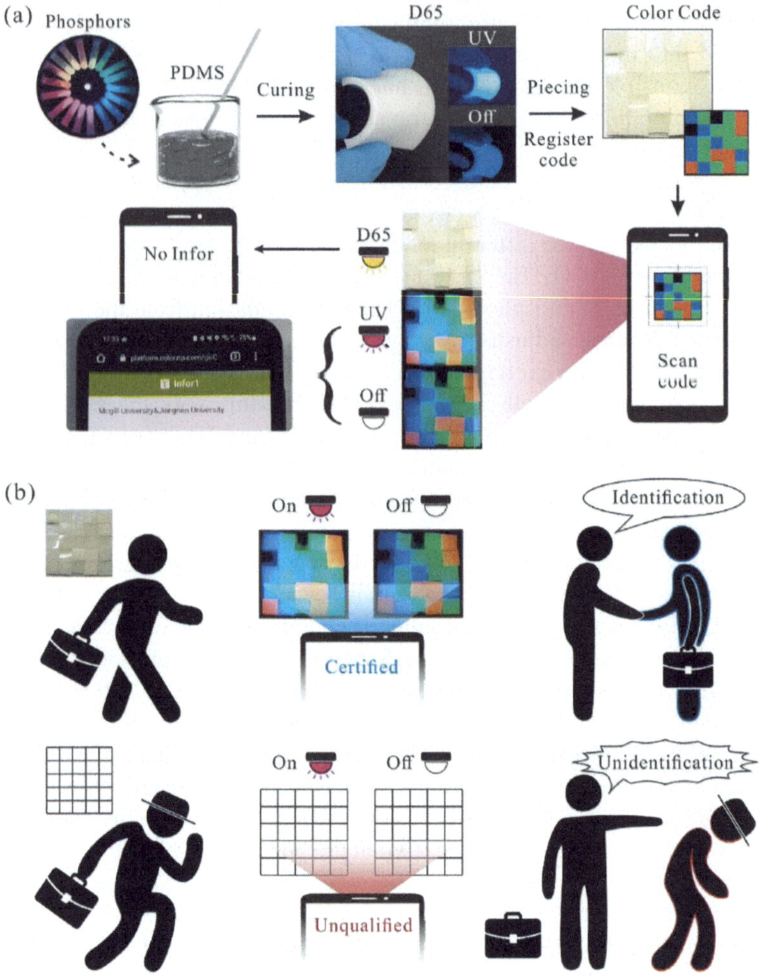

Fig. 2 Processes for developing and designing a PDMS 3D code (**a**) and using the luminous code for security purposes (**b**). (*Source* Shen et al. [71]. Reprinted with permission of Ceramics International)

of 18.2%. The floral pattern shows different information at various temperatures thus, it has been used for thermochromic anti-counterfeiting (Fig. 4) [74].

In another work, organic phosphorescence afterglow material has been developed by Bhatia et al. [75]. Three twisted derivatives of terephthalonitrile (TOC, TOF, and TOM) were created by them. These derivatives have 2,3,5,6-positions covalently linked to 2-chloro-phenoxy, 2-fluorophenoxy, and 2-methoxy-phenoxy groups, respectively. It is observed that when TOC and TOF are implanted in the polyvinyl alcohol matrix, they display persistent RTP (PRTP) with lifetimes of 139 and 736 ms, respectively. A similar solution mixture of dopant and PVA (30%, w/w)

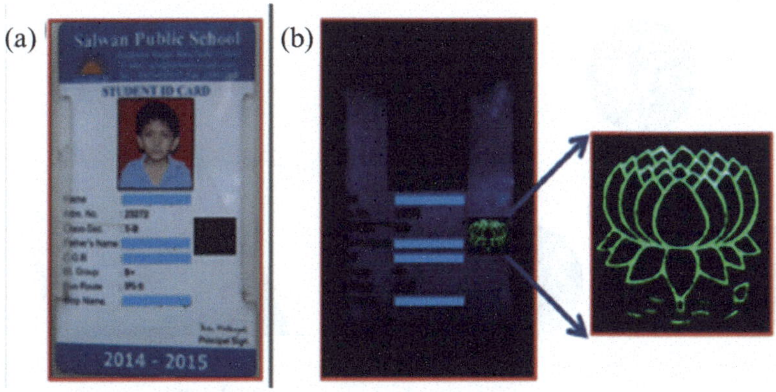

Fig. 3 The optical image of photo ID card **a** under room light and **b** after 5 s when the 365 nm wavelength UV excitation source was removed; it exhibits a phosphorescent lotus flower image emitting green color, which cannot be seen under room light. (*Source* Gupta et al. [72]. Reprinted with permission of Journal of Applied Physics)

has been applied to regular filter paper for the security application. The characters "SNU" and the Indian Rupee sign (TOC and TOF) were written. Green and yellow afterglow characteristics have been noticed after turning off the lamp.

3 Use of Persistent Luminescent Materials for Information Protection

Information exchanges can be made safe and efficient by using information encryption technology as a protective technique. The communication sector needs to improve the security level of information and provide enough characters and encoding techniques [76–79]. Information encryption is essential for information security, optical communication, and data storage given the rapid growth of digital information. For instance, increasing the information storage capacity and the anti-counterfeiting label by adding a second information encryption layer can be quite beneficial. Anti-counterfeiting technology frequently employs labels for optical security using photoluminescent or photochromic materials. Even while classic fluorescent anticounterfeiting technology has improved, there are still several issues, including a poor decoding technique, a weak anti-counterfeiting encryption standard, and a restricted capacity for information storage. For this reason, developing robust anti-counterfeit label technology with a high-security factor is essential to halting the spread of counterfeit goods [80]. The basic property of persistent luminescent materials for information protection/ storage is the largest trap depths. Greater trap depths guarantee higher storage at room temperature and allow stored energy to be released in response to light radiation.

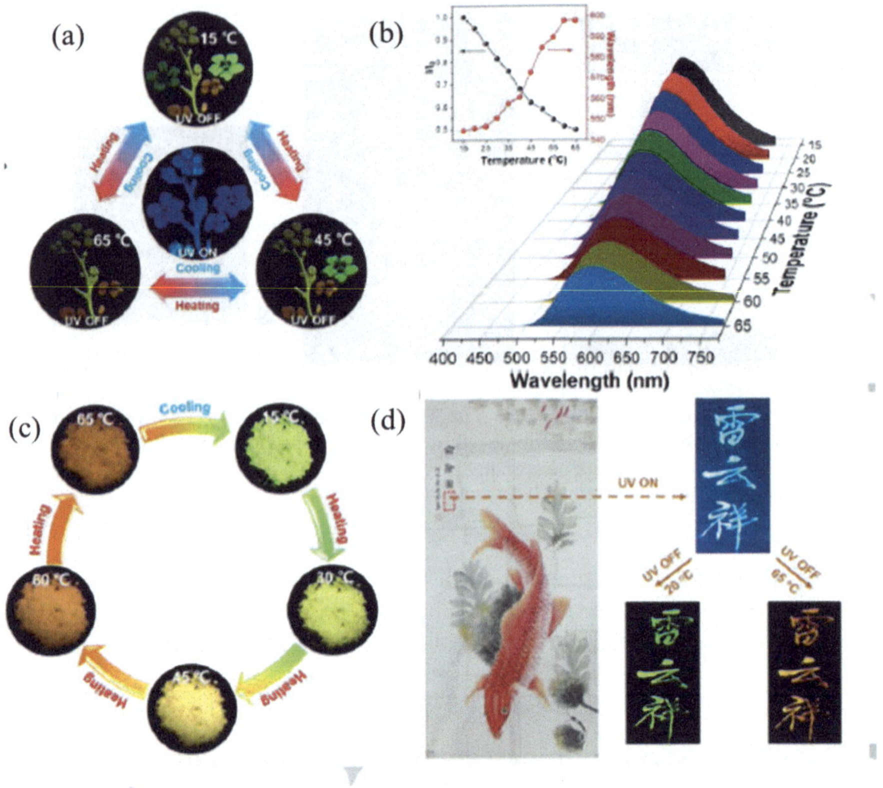

Fig. 4 **a** Photographs of the flower before (in the middle) and after removing the excitation source at different temperatures 15–65 °C. **b** Phosphorescence wavelength and intensity of DTA-DQD/TPP materials at different temperatures **c** Phosphorescence color of DTA-DQD/TPP host–guest material at different temperatures **d** Application of DTA-DQD/TPP in anti-counterfeiting of calligraphy and artworks. (*Source* Lei et al. [74]. Reprinted with permission of Angewandte Chemie International Edition)

To improve the luminous signal or increase the number of communication channels for highly secure information encryption, many researchers have taken advantage of the distinctive optical phenomena created by photonic structures. A bilayer, inverse opal PhC film with UCNPs incorporated was created by Meng et al. [81]. Based on synergistic upconversion increase generated by numerous physical effects, Zhou and colleagues showed angle-dependent anti-counterfeiting applications [82]. Physical unclonable function (PUF), another anti-counterfeiting method with high security, was offered. Wan and colleagues created the micro- and nano-structures found on the surfaces of real plants using a straightforward, eco-friendly manufacturing procedure [83]. The mentioned techniques can also be used in persistence luminescence-based improved information protection.

Liu et al. [84] reported Bi^{3+} doped persistence luminescence using $CaSnO_3$ host and Nd^{3+}/Gd^{3+} as codopants. Upon 510 nm excitation the luminescence obtained due to the transition $^2P_{3/2(1)} \rightarrow {}^2P_{1/2}$ peaking at 805 nm from Bi^{3+} ions existed for more than 24 h. The orthorhombic structure codoped $CaSnO_3$ phosphors have been produced using the solid-state reaction technique. The persistence emission has been well explained with the trapped state of developed phosphors. The phosphors have been used to write the Chinese characters and stick and numbers upon 365 nm UV irradiation (Fig. 5) [84].

Zhao et al. [85] developed strong blue emitting persistence glass ceramic hexagonal $Sr_{13}Al_{22}Si_{10}O_{66}$: Eu^{2+} crystals (named GC). Upon natural sunlight activation, the afterglow was found for more than 200 h. The phosphors were found suitable for outdoor optical information displays that does not need an electricity source. The phosphorus was dissolved in either water or ethanol (0.1 g/mL) to generate display ink. The prepared phosphor ink was examined by writing letters 'A', '123', 'EXIT' on a white paper and after 15 min of exposure to sunshine, degradation has been seen after 10 min, 30 min, 1 h, 3 h, 6 h, and 12 h (Fig. 6) [85].

Zhang et al. [86] reported 365 nm excited green emitting $Na_3(Y,Sc)Si_3O_9$: Eu^{2+} phosphors. Using the phosphor film, the 'Wuyi University icon' was written using a 365 nm laser, and the sample was then left in a dark environment for one hour. The

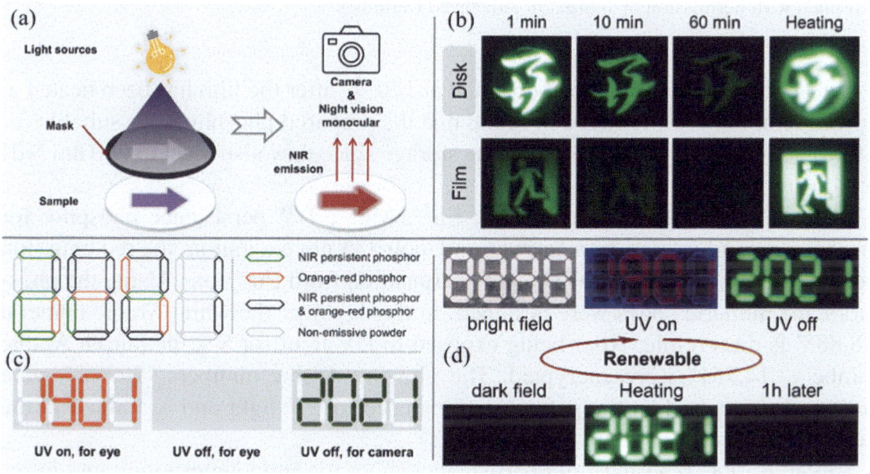

Fig. 5 Demonstration of optical data storage application (**a**). NIR afterglow images of the CNGSO: Bi ceramic disk and thin film after charging with 365 nm UV lamp (phosphor disk) and LED Flash (phosphor film) for 10 min, respectively (**b**). The imaging is recorded for 60 min. The heat-assisted NIR luminescence images are taken on a 150 °C hot plate. Imaging parameters: manual/ ISO 200/0.3 s (phosphor disk), manual/ISO 200/0.5 s (phosphor film). Schematic presentation of information encryption and decryption application by using CNGSO: Bi persistent phosphor and red fluorescent materials (**c**). Luminescence pictures of phosphor film at different light conditions (**d**). The luminescence pictures under bright ambient light and 365 nm UV light are taken by using a digital camera, while other pictures are taken by the digital camera combined with a night vision monocular. (*Source* Liu et al. [84]. Reprinted with permission of Journal of Luminescence)

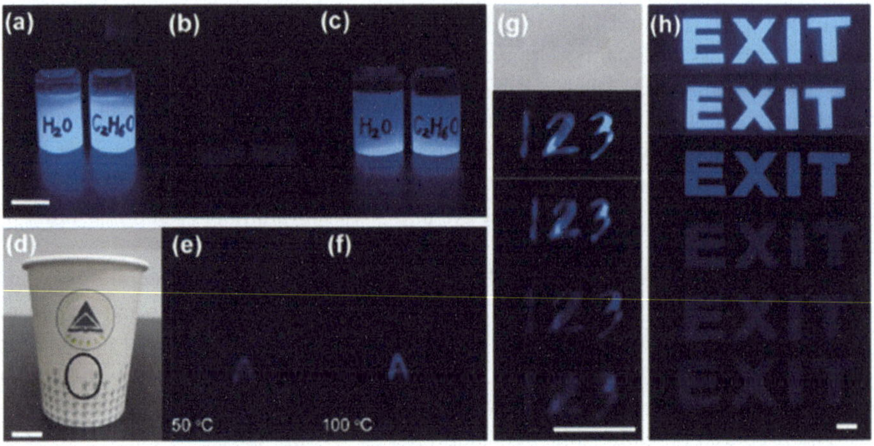

Fig. 6 PeL photographs of the GC: Eu powder contained water and ethanol suspension at 1 min (**a**), 2 h (**b**) (sunlight irradiation for 15 min), and then shaking (role as an external force) the glass bottle for 1 min (**c**). **d** Paper cup with a letter "A" written in a black circle at room temperature. **e**, **f** PeL photographs with pouring 50 and 100 °C water into the cup. **g** Photographs of the handwritten numbers "123" on an A4 paper and the corresponding PeL at 5 min, 30 min, 1 h, and 2 h. **h** PeL photographs of the printed "EXIT" pattern at different time, from top to bottom are 10 min, 30 min, 1 h, 3 h, 6 h, and 12 h, respectively. The scale bar represents 15 mm. (*Source* Zhao et al. [85]. Reprinted with permission of Journal of Advanced Ceramics)

brightest PersL pattern has been obtained at 120 °C after the film has been heated at various temperatures. It has been found that the prepared phosphors are suitable for information readout rate for optical data storage since they also react to 980 nm NIR photon stimulation [86].

Shen et al. [87] developed Y_2O_2S: Eu^{3+}, Mg^{2+}, Ti^{4+} persistence phosphor for smart textile and information encryption Upon 325 nm excitation, the red emission has been observed due to the $^5D_0 \rightarrow {}^7F_2$ transition from Eu^{3+} ions. Using the phosphors, the numeric codes were designed. In the daytime, the white Arabic numeral '88,888' is discernible. After being exposed to UV light for 8 s, the purple Arabic numbers "14,751" were encrypted. The glowing Arabic numbers "79,863" were further encrypted when the pattern was exposed to UV light and examined in the dark [87].

Liu et al. [88] reported $LuPO_4$:Bi^{3+} phosphors for anticounterfeiting and information encryption. The encryption patterns were created by screen-printing technique. The patterns have been excited with X-rays and the persistence luminescence obtained in UV region.

Han et al. [89] developed upconversion-based afterglow $KCaF_3$:Yb^{3+}, Mn^{2+} phosphors. The long-lived exchange coupled Yb^{3+}-Mn^{2+} dimers generated in $KCaF_3$: Yb^{3+}, Mn^{2+} is the source of the UC afterglow, which is unrelated to the traditional 'afterglow trap'. The Mn^{2+} emission at 570 nm and the Yb^{3+} at 971 nm NIR emission band can both be seen under excitation at 395 nm due to the transitions $^4T_{1(4G)} \rightarrow {}^6A_{1(6S)}$ of Mn^{2+} and $^2F_{5/2} \rightarrow {}^2F_{7/2}$ of Yb^{3+}, respectively. After being dynamically

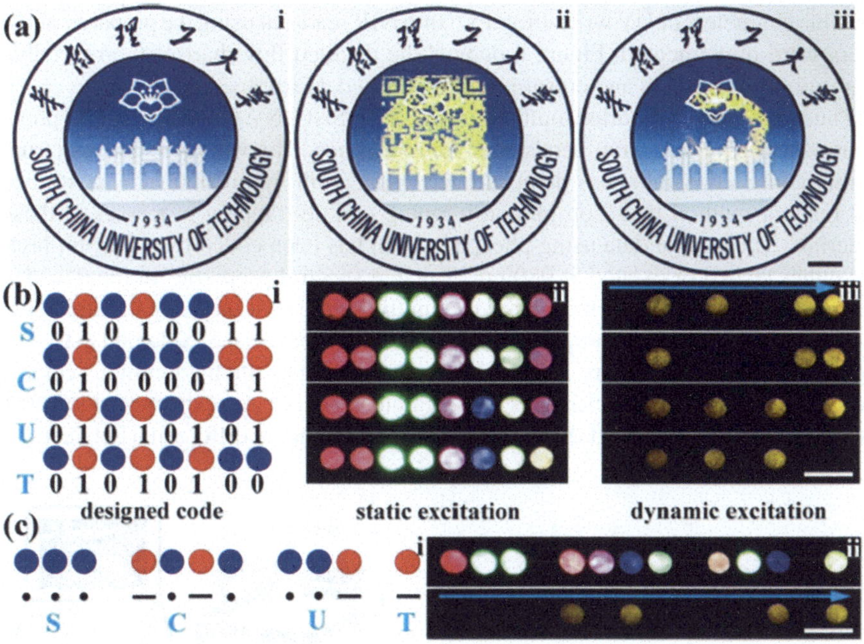

Fig. 7 **a** Photograph of the school badge (i) and the luminescent images of the printed two-dimensional code upon 980 nm LD excitation {both steady irradiation (ii) and dynamic irradiation (iii)}. Photographs of **b** binary coded and **c** Morse-coded dot matrices printed with the luminescent inks containing commercial UC phosphors as well as the as-prepared $KCaF_3:Yb^{3+}$, Mn^{2+} phosphor. The designed red and blue circles were printed with luminescent inks with and without the prepared $KCaF_3:Yb^{3+}$, and Mn^{2+} phosphor, respectively. (*Source* Han et al. [89]. Reprinted with permission of Journal of Materials Chemistry C)

exposed to a 980 nm laser, the upconverting phosphors were used on various bottle caps printed with $KCaF_3$: 5%Yb^{3+}, 10%Mn^{2+} luminous ink for information security. Both static and dynamic 980 nm LD irradiation have been applied on printed school badges and the two-dimensional codes (binary code and Morse code) as shown in Fig. 7 [89].

4 Persistent Luminescent Materials and Data Encryption

Information protection and data encryption techniques using the persistence phosphors have been reported by many researchers. Wang et al. [90] developed Bi^{3+} activated garnet phosphors by solid state reaction method. The afterglow from $LuY_2AlGa_4O_{12}$: Bi^{3+} phosphors has been found for more than 72 h. To utilize 8-bit ASCII binary codes for highly secure data encryption and decryption applications, invisible characters of UV write-in and 870 nm NIR read-out using the prepared phosphors were implemented. Figure 8 depicts the detailed flowchart of the reversible encryption, decryption, and erasure of microarray data storage [90].

Zhuang et al. [51] reported multicolor $(Sr_{1-x}Ba_x)Si_2O_2N_2$: Eu^{3+}/Yb^{3+}, Dy^{3+} phosphors which give deep trap persistence luminescence. The whole visible spectrum of multicolor emissions was demonstrated using a bit-by-bit optical data storage and readout method based on photon trapping and de-trapping processes in these materials. The optical data using phosphors film has been coded by a 405 nm laser and either heated or excited at 980 nm. Tunable spectral properties in the decoded information enable either emission-wavelength or emission-intensity multiplexing. In this work, optical data storage and readout strategies have been explained [51]. The high-energy photons that are incident during recording excite the luminous centers (i) and some of the excited electrons may migrate to nearby trap centers via the host's conduction band (CB). (ii) If the trap depth, that is the energy differential between the

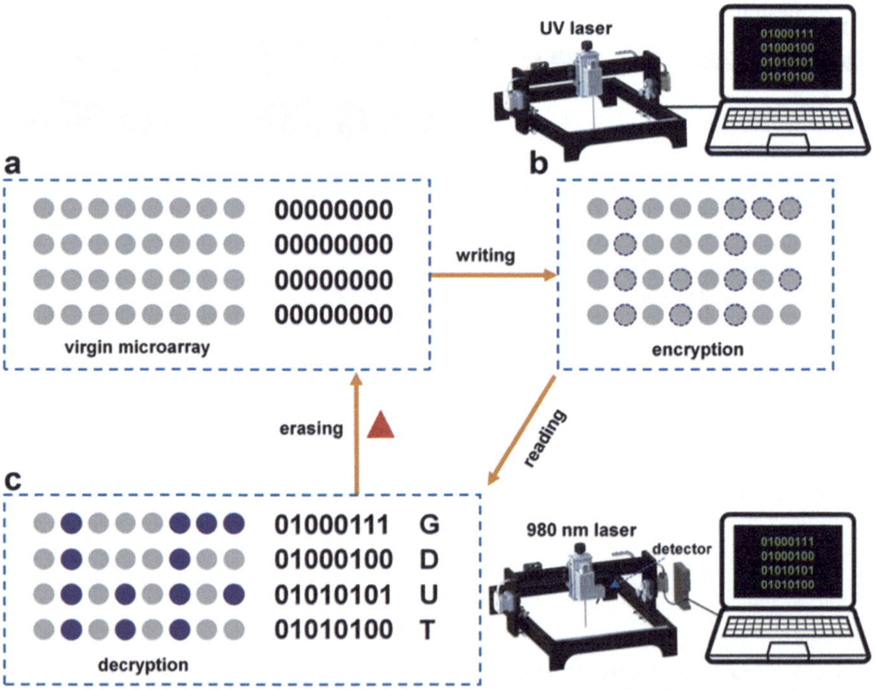

Fig. 8 Schematic illustration of the reversible encryption, decryption, and erasing of microarray information storage. **a** The uniformly dispersed micro-rods of the virgin microarray. Gray dots do not contain any information. **b** The process of the optical data encoding using a UV laser coupled with an optical fiber. Purple-dash dots containing encryption information. **c** The process of the optical data was decoded using a 980 nm laser combination with a fiber spectrometer. The micro-rods without (gray dots) and with (purple-dash dots) UVB emission signals are recognized as "0" and "1," respectively. (*Source* Wang et al. [90]. Reprinted with permission of Journal of Alloys and Compounds)

trap level and the CB bottom, is noticeably greater than the thermal activation energy at RT, the incident photon energy is frozen in the phosphors. (iii) The phosphor film is heated to a higher temperature during the readout procedure, which provides the trapped electrons with the activation energy they need to escape the trap levels. (iv) Emission of photons occurs upon recombining with the luminescent centers. Using the direct electron transition from the trap level to the CB, the photostimulation (980 nm, 1.3 eV) offers an additional effective de-trapping pathway. (v) To preserve recorded data (i.e., stored photon energy) at RT for an extended period, persistent luminescent materials must have a large trap depth within a limited distribution.

Another long persistence (M, Ca)$AlSiN_3$: Eu^{2+} (M = Sr, Mg) phosphors for display and optical storage has been reported by Zhu et al. [25]. Synthesis of the phosphors was carried out using the combustion method. The write-in using phosphors pattern has been achieved with 275 nm excitation and the read-out by 980 nm NIR light. The afterglow of $SrAlSiN_3$: 0.2 at% Eu^{2+} phosphor appears up to 200 min due to a suitable trap depth of (0.71 eV), which is larger for the $Mg_xCa1_{-x}AlSiN_3$: 0.2 at% Eu^{2+} phosphors (0.98 eV), thus assuring a greater rate of storage at ambient temperature. Gao et al. [91] reported multicolor emission in $ZnGaGeO_4$: Mn phosphors synthesized by solid-state reaction route. The $ZnGaGeO_4$: Mn phosphors release color-tunable luminescence with a decrease in the concentration of Ge^{4+} when exposed to a fixed 365 nm UV lamp. Zn_2GeO_4: Mn exhibits intensely bright green PL, PersL, PSL, and mechanoluminescence (ML), while $Zn_3Ga_2GaO_8$: Mn emits broad red luminescence and the mechanism has been explained with trap analysis (Fig. 9a). The process of creating luminous inks for printing that prevent counterfeiting was explained. Using a modified process, luminescence inks were created using persL phosphors. Typically, a mixture of 1.0 mL hydrochloric acid (0.2 mol/L) and 1.0 mL anhydrous C_2H_5OH was used to scatter the $ZnGaGeO_4$: 0.15% Mn PersL phosphors. The resultant suspension was then centrifuged and ultrasonically cleaned five times in alternating cycles with deionized water and C_2H_5OH. The finished products were then completely blended using a specific volume of PVA solution to create luminous encryption inks. The screen-printing technique was used to create the glowing patterns that were printed on the paper. 'Squirrel' pattern has been made with the prepared $ZnGa_2O_4$: Mn^{2+} phosphor inks and excited with varying the irradiation wavelengths of 254, 330, 335, 340, 365, 375, 488, 808 and 980 nm and thus, there is change in corresponding emission color from green to yellow, orange, pink, and red (Fig. 9b). Further, the information encryption patterns like '8888', '1111' and 'LOVE' also prepared and irradiated with 365 nm (Fig. 9c). Luminescent inks are typically distinguished by their high stability, low cost, availability, high-quality distribution of luminescent nanomaterials, strong luminescence intensity, suitability for printing on various surfaces, high surface tension, and viscous nature of the ink medium [92].

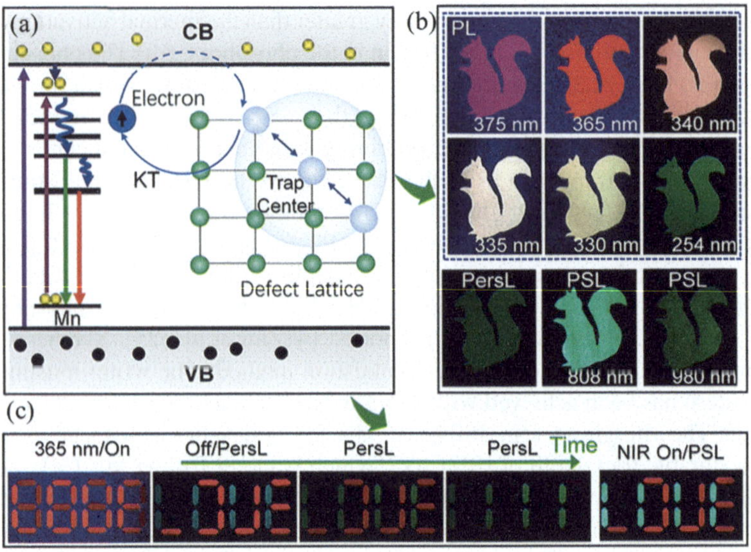

Fig. 9 **a** Mechanism of multimode luminescence. **b** Digital images of multicolor and multimode luminescence (PL, PersL, PSL) of squirrel pattern printed using $ZnGa_2O_4$: Mn phosphor inks. PL images were obtained under irradiation with varying wavelengths. PersL images were obtained after ceasing the irradiation of various wavelengths PSL images upon irradiation of an 808 nm or a 980 nm NIR laser (0.7 W). **c** Application of $Zn_3Ga_2GeO_8$: Cr (black part), $Zn_3Ga_2GeO_8$: Mn (blue part), and $ZnGa_2O_4$: Mn (red part) phosphors to information storage and anti-counterfeiting. PL and PersL images under and after the illumination using a UV lamp of 365 nm for 5 min; PSL images upon 980/808 nm NIR laser (at 0.7 W) irradiation after the disappearance of the PersL pattern. (*Source* Gao et al. [91]. Reprinted with permission of Materials Today Physics)

5 Implications of Persistent Luminescent Materials in Latent Fingerprinting

A total of 99.9% of men in the diverse human population share the same DNA, whereas 0.1% makes them biologically distinct from one another. Due to the aforementioned issue, no two persons in the world, not even identical twins, have fingerprints that are the same [93]. Experts in forensic science rely on certain equipment to record evidence at the site of the crime to identify the culprit. Fingerprints are utilized as physical evidence in forensic science to identify the suspects. The permanent nature of finger wrinkle designs causes them to leave stains on object handles when handled with bare hands. The majority of latent fingerprint-detecting materials displayed a two-tone difference (dark/clear) between the latent fingerprints and background surfaces, which makes it difficult for forensic experts to see the actual situation following crimes [94]. Latent fingerprints occupy a significant place in the fields of forensic research and criminal justice because of their singularity, universality, and permanence. Latent fingerprints are a crucial component of identification and crime investigation since they are frequently found at crime scenes.

One of the methods utilized by forensic experts is fingerprinting. The ability to see fingerprints is essential for the identification of criminals because they are sometimes either impossible to see or very dimly visible to detectives. The traditional techniques for detecting fingerprints include powder dusting, cyanoacrylate fuming, and ninhydrin spraying [95]. Again, the fact that the background of the fingermarks is a different color makes them exceptionally challenging. When exposed to UV/white light, long-lasting phosphors exhibit distinctive characteristics of visible light production and prolonged afterglow. Therefore, rather than using other traditional dusting agents or fluorescent powders, the powder of long-lasting phosphors can be employed for latent fingerprint detection due to the benefit of a dark background to high-contrast fingermarks. Recently, $SrAl_2O_4$: Eu^{2+}, Dy^{3+} long-lasting phosphor was used by Sharma et al. [96] to detect latent fingerprints on a variety of porous, semi-porous, and non-smooth surfaces.

To successfully use PersL phosphors in latent fingermark detection, a phosphor must have strong luminescence efficiency, constant particle morphology, and size, be dusty by nature, and be durable in humid environments. A small amount of gritty powder is typically dusted in the fingermarks area as part of a powder dusting technique. The powder is then applied to the area with a gentle brush. Using a gentle air flow, the extra powder is cleared from the surface after spreading. Some phosphorescent particles adhered to the fingerprints' ridges.

'Lip Print' is the second significant piece of evidence at a crime scene. Cheiloscopy is a forensic examination that focuses on identifying people based on their lip prints. Cheiloscopy is the branch of science that examines lip traces. In the region where the inner labial mucosa and the outside skin of the human lip meet to form the lip prints, there are common lines and fissures in the form of wrinkles and grooves. Another distinctive element of identification, they are arranged in a variety of carvings and forms that are particular to various people. Since lip prints are supposed to remain constant and distinctive, no two people have the same arrangement of lip furrows. Certain patterns can be identified as early as the sixth week of intrauterine life [98]. Like fingerprints, lip prints are distinctive, individual characteristics that can reside on surfaces (such as glass, coffee cups, cigarette butts, etc.) and are accepted as evidence in criminalistics. For this purpose, usually, the lip prints are created using synthesized nanophosphors that are triggered by UV light, and each of their distinguishing characteristics is marked with an afterglow.

Chen et al. [97] developed $NaLuF_4$: Mn^{2+} persistence phosphors by solvothermal synthesis method. The phosphors have been excited with X-rays for 5 min and a broad green emission band at 525 nm due to the $^4T_1 \rightarrow {}^6A_1$ transition of Mn^{2+} ions. The afterglow luminescence decayed for 12 h at room temperature. Suitable trap depths of 0.75 eV were found for the developed phosphors. The phosphor was tested for latent finger applications on different substrates, like aluminum foil and plastics and found universal for proposed time-gated latent finger print imaging method (Fig. 10).

Sharma et al. [94] developed the $CaAl_2O_4$: Eu^{2+}, Dy^{3+} phosphors for latent fingerprint application and lip mark detection. In their work, the phosphors were produced by combustion synthesis method and characterized with various techniques such as

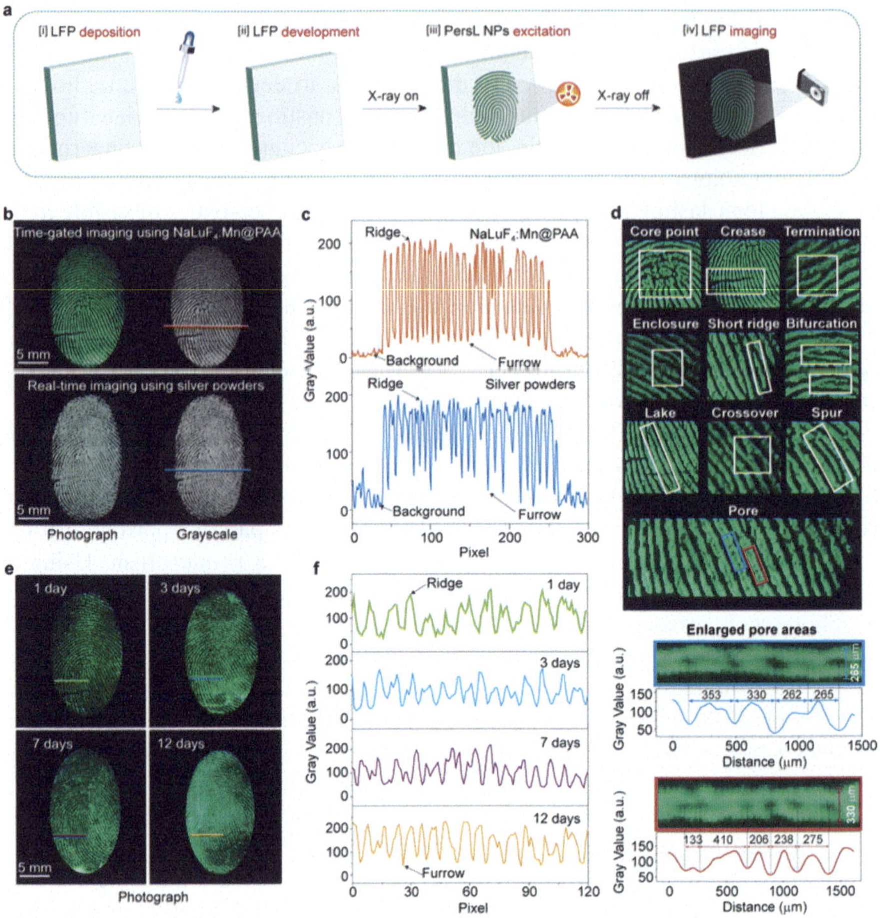

Fig. 10 Time-gated LFP imaging with Level 3 details. **a** Schematic diagram of the LFP imaging route using the $NaLuF_4$: Mn@PAA nanoparticles. **b** Photographs and grayscale images of LFPs deposited on glass. The top is time-gated imaging using $NaLuF_4$: Mn@PAA, and the bottom is real-time imaging using commercial silver powders. **c** Variation in grayscale values along the straight line highlighted in (**b**). **d** Enlarged features of LFP details, including Level 1, Level 2, and Level 3, in (**b**) developed by the $NaLuF_4$: Mn@PAA nanoparticles. The bottoms are enlarged pore areas and the corresponding grayscale value distribution. **e** Photographs and grayscale images of LFPs deposited on glass and aged for 1, 3, 7, and 12 days. **f** Variation in the grayscale value along the straight lines highlighted in (**e**). In the above cases, LFPs were developed with the $NaLuF_4$: Mn@PAA nanoparticles and charged with X-rays for 5 min. (*Source* Chen et al. [97]. Reprinted with permission of ACS Applied Materials and Interfaces)

XRD, HRTEM, EDS, UV-vis spectroscopy and photoluminescence study. The phosphors were found with monoclinic structure with average particles size of 33 nm. The photo luminescence spectra of $CaAl_2O_4$: Eu^{2+} phosphors have been optimized at 1.5 mol% recorded upon 245 nm excitation. The emission spectra show an intense peak at 614 nm with less intense peak at 590 nm. Further, the Dy^{3+} ions were optimized at 2 mol%, excited upon 326 nm. The emission from $CaAl_2O_4$: Eu^{2+}, Dy^{3+} phosphors lie in 447 nm, 577 nm, and 616 nm wavelength. The strong long phosphorescence from prepared phosphors have been used for finger printing and lip print detection purposes using powder dusting method. Figure 11 shows the "Lip print" formed with the designated identification qualities of the produced $CaAl2O_4$: Eu^{2+}, Dy^{3+} nanophosphor activated under UV light.

Liu et al. [98] developed Eu^{3+} doped $SrAl_2O_4$ long phosphorescence material for finger printing detection. The phosphors developed by hydrothermal method excited upon 365 nm wavelength. The green afterglow emission has been obtained at 526 nm due to Eu^{3+} ions. The finger print detection technique has been applied to porous, semi-porous or non-porous surfaces [98].

Gao [99] reported Zn_2GeO_4: Mn, Li, K long persistence phosphors synthesized by hydrothermal method. Rhombohedral structured phosphors show afterglow upon irradiation of 254 nm, 808 nm and 980 nm excitations. Upon 313 nm excitation, a broadband emission occurred peaking at 540 nm for Mn^{2+}, due to the $^4T_{1(4G)} \rightarrow$

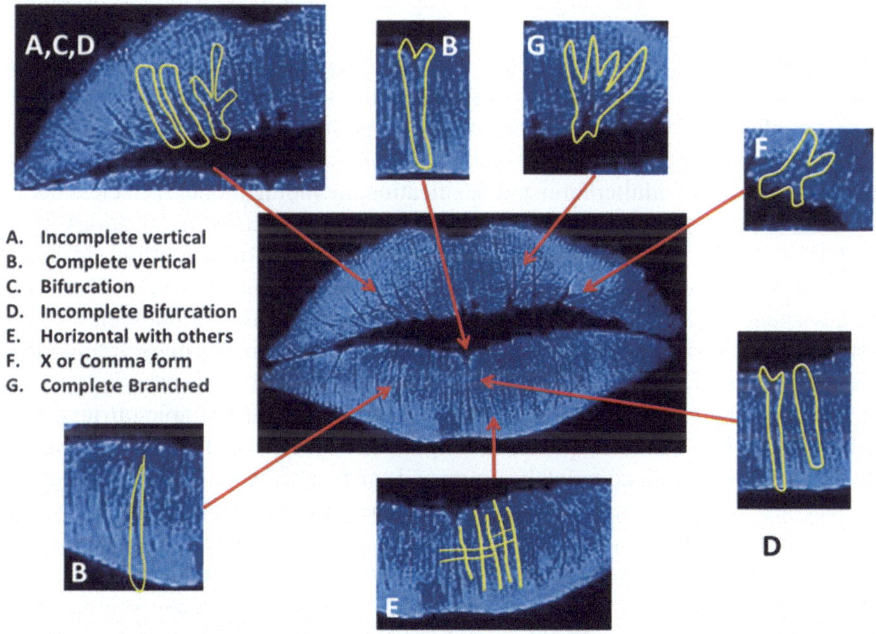

Fig. 11 Lip print developed using the synthesized nanophosphor excited under UV light, and their identification traits are labeled. (*Source* Sharma et al. [94]. Reprinted with permission of Physica B: Condensed Matter)

$^6A_{1(6S)}$ electron transition. The developed phosphors have been used for latent finger printing purposes. A 'sun logo' and 'Chinese paper money' made with finger prints on a note paper using this phosphor. An afterglow has been seen upon different excitations for 5 min. In this way, Mn^{2+} doped persistent phosphors has been used for finger printing application.

Again, Mn^{2+} doped silicate long afterglow phosphors has been reported by Liu et al. [100]. They developed Zn_2SiO_4: 0.6 mol % Mn^{2+}, 1.0 mol% Yb^{3+} phosphors. The average particle size found to be 200 nm. With the incorporation of Yb^{3+} to the Zn_2SiO_4: Mn^{2+} phosphors, the afterglow time increases by 23 times, i.e., from 5 to 114 min. The trap depths range from 0.696 to 1.112 eV. To achieve afterglow, the phosphors has been excited with 254 nm irradiation for 10 min. The emission peak found at 521 nm and 523 nm occurred from $^4T_{1g\,(4G)} \rightarrow {}^6A_{1g\,(6S)}$ transition of Mn^{2+} ions. The finger prints were white under daylight, but after excitation the green patterns appeared.

6 Implications of Persistent Luminescent Materials in Anti-Counterfeiting

With the development of technology and its implications in day today lives of human being the norms and values are downgrading. To prevent the situation of deception there is a need of establishing techniques for separating the authentic from the counterfeit. The prevalence of fake documents, including checks, money, certifications, and identity cards, is a major factor in global corruption and fraud. Making an unapproved replica of a product without receiving permission from the proprietor to benefit from an expensive brand is known as counterfeiting. The modern world is full of incidences of adulteration and falsification, including money, priceless documents, cosmetic brands, apparel, purses, electronics, passports, medications, food products and other items result in significant economic loss for any nation and pose a continuing threat to the health and safety of customers around the world. Therefore, there is a huge market for the development of anti-counterfeiting technologies. Quick response (QR) codes, holography, isotope tracking and luminescent printing radio frequency identification (RFID) and other conventional anti-counterfeiting systems have the drawbacks of high cost, complicated production, and simple infringement [101–107]. RFID-enabled chips can be integrated into a variety of products, however, the data they hold is not completely secure and can be altered. Hologram technology has recently been used in the marketplace to verify product authenticity. However, because of its simplicity and production ease, it is a very poor defense against forgers. Techniques based on isotope tracking are restricted to specific regions where radiation levels are within acceptable limits. One of the most effective strategies for preventing counterfeiting is the luminescence printing-based technology, which has low commercial production costs and is environmentally gentle and difficult to copy, among other advantages [108]. PersL phosphors have been used to develop several printing processes, including mask photolithography, inject printing, and aerosol jet printing. Gangwar et al. [108] prepared $NaYF_4{:}Eu^{3+}@SrAl_2O_4{:}Eu^{2+}/Dy^{3+}$ phosphor by hydrothermal synthesis technique. Upon single excitation of 254 nm wavelength, the dual red (611 nm) and green (532 nm) color emission occurred. The luminescent security ink has been developed by dispersing the phosphor pigment in a commercially available PVC gold medium for anti-counterfeiting applications. On non-fluorescent white bond paper, an image (of 'Tajmahal') was printed via a conventional screen printing procedure.

Zhao et al. [109] reported the upconversion-based afterglow phosphorescence for anticounterfeiting and encoding applications. In their work, platinum octaethylporphyrin (PtOEP) and 9,10-diphenylanthracene (DPA) were used to achieve triplet–triplet annihilation upconversion. Using dual-mode 365 and 532 nm excitations, triplet emission modes (fluorescence, upconversion, and afterglow) have been achieved. A diagram illustrating the preparation of a multi-dimensional luminous design using a direct-writing technique has been illustrated in Fig. 12. The printed patterns 76 and 49 were excited with both excitations. A hidden pattern '15' appeared with afterglow green emission when 532 nm excitation was turned off. An encoding system has been developed using binary codes. For this purpose, a part of printed dots were excited with UV irradiation and blue afterglow has been obtained. The other part excited with 532 nm irradiation for 10 s. '1' and 0' binary codes used for the ON/OFF state written with afterglow phosphors and then translated into the actual message "USTS". Again, using both irradiations on Morse code encoding system has been developed. The coded 'dash' and 'dot' give week green and strong green afterglow and then converted into message "USTS.".

Cai et al. [110] developed $ZnGaSnO_4$: Er^{3+}, Ni^{3+}, Cr^{3+} persistent luminescence phosphors active in near infrared windows. The phosphors synthesized with the solid-state reaction method have a cubic spinal structure. For the anticounterfeiting applications, the 'UTCBS' patterns are written in white paper using the developed $ZnGaSnO_4$: Er^{3+}, Ni^{3+} persistent luminescence phosphors. 'MPOE' patterns were written using $ZnGaSnO_4$: Er^{3+}, Ni^{3+}, and Cr^{3+} persistent luminescence phosphors. The patterns were excited with 365 nm irradiation for 5 min and then the emission was recorded for 15 min. The persistence luminescence found for ${}^3T_{2\,(3F)} \rightarrow {}^3A_{2\,(3F)}$ transition from Ni^{3+} ions.

Ge et al. [111] developed long-persistence upconversion material $Zn_3Ga_2SnO_8$: $1\%Cr^{3+}, 5\%Yb^{3+}, 0.5\%Er^{3+}$ by the sol–gel method. The phosphors have been excited with 274 nm excitation and the emission was found at 656 nm due to the ${}^2E \rightarrow {}^4A_2$ forbidden transitions from the Cr^{3+} ions. For the development of anti-counterfeiting ink, the phosphors were dipped with a toothpick and written the letters 'SD' on white paper and a stamp was also written on the rice paper and were excited by a 500 mW UV lamp for 10 min. Thus, the afterglow phosphors have been made applicable for anticounterfeiting purposes.

Chen et al. [30] reported organic persistence phosphors for high-level anti-counterfeiting applications. Successfully designed and synthesized organic compounds by one-step Suzuki coupling reaction having twisted electron donor–acceptor structures were then dispersed in methyl methacrylate (MMA), which is

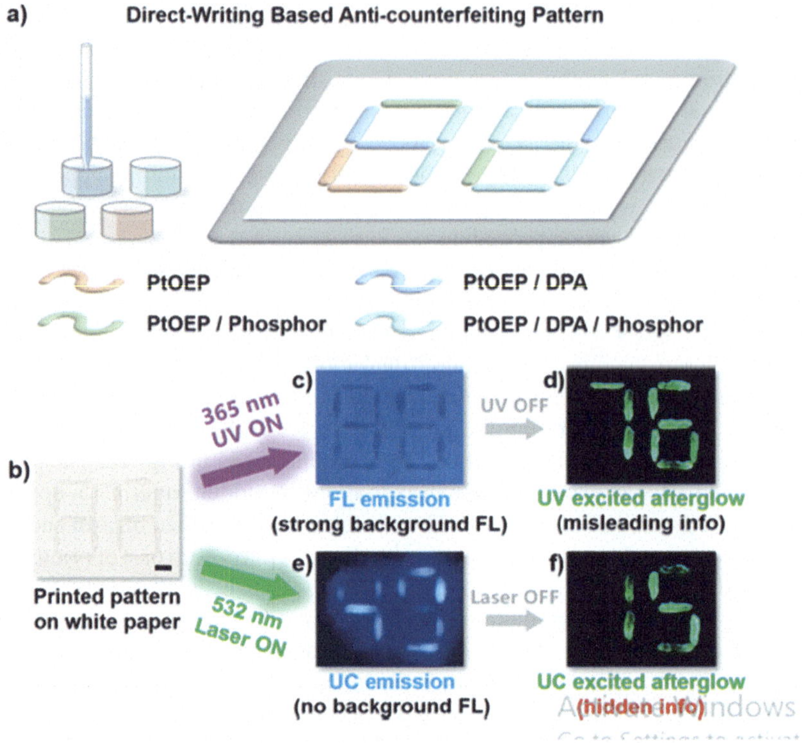

Fig. 12 a Schematic illustration of the multi-dimensional luminescent pattern preparation based on a direct-writing method. The luminescent property in each part of the pattern was elaborately designed and programmed by controlling the composition of the inks. **b** The printed pattern on a white paper (left, scale bar: 2 mm). **c** The printed pattern under the excitation of 365 nm UV light (FL mode). FL emission pattern exhibited strong background fluorescence from the paper substrate. **d** The printed pattern exhibited afterglow emission after UV irradiation, displaying a clear misleading message "76" (UV excited afterglow mode). **e** The printed pattern under the excitation of 532 nm CW laser area light source (UC mode). The UC emission image exhibiting a clear message "49" was recorded through a 520 nm short pass filter. **f** The printed pattern exhibited afterglow emission after 532 nm laser irradiation, displaying the real hidden message "15" (UC excited afterglow mode). (*Source* Zhao et al. [109]. Reprinted with permission of Journal of Materials Chemistry C)

followed by in situ polymerization to produce single-molecular organic compounds that are room-temperature phosphorescent (RTP) (P1–P4). All RTP materials display a long lifespan, but P2, in particular, displays an ultra-long lifetime of 1.51 s. Because of the simultaneous phosphorescence and delayed fluorescence emissions when the compounds were formed into single crystals, multicolored adjustable afterglow ranging from blue to yellow to purplish-red was created at various delay durations and shows suitable material for use in high-level anti-counterfeiting.

Das et al. [112] developed a Cr^{3+} doped $Zn_3Ga_2Ge_2O_{10}$ phosphor by solid-state reaction method. Upon 260 nm excitation, the NIR emission occurred at 700 nm. The emission can only be recorded with the night vision device; therefore, the phosphors have been found suitable for military applications viz., tagging, tracking, and locating concerns. Before capturing the image, the phosphors were excited with 254 nm irradiation. The process of fabricating tag, NIR paint, and NIR security ink using the prepared phosphors has been explained. To make the tag, 75 weight percent of the as-synthesized fine powder of the optimized $Zn_3Ga_{1.96}Cr_{0.04}Ge_2O_{10}$ long persistent phosphor was combined with transparent epoxy resin. Epoxy to binder proportion was 2:1. It was then poured into a disc-shaped mold. To obtain epoxy with a phosphor tag inserted, it was left in the air for one day. The tag was placed on a human garment, activated by exposing it to 254 nm UV light (8W) for 5 min, and then tracked and located by a night vision monocular.

For the preparation of NIR paint, 30-weight percent acrylic binder was used with powdered $Zn_3Ga_2Ge_2O_{10}$ long-lasting phosphor. The surfaces of various body materials were then coated with it. Different body surfaces were painted with NIR paint. It was coated on the cloth that was used for battle vehicles, weapons, and defense people. Depending on the situation, either ambient light or sunshine was used to activate the NIR paint. The picture was then captured in the dark using a camera linked to a night vision monocular. To identify friends or enemies during night combat, NIR paint is therefore very helpful. The $Zn_3Ga_2Ge_2O_{10}$ long persistent phosphor powder and a clear epoxy resin were combined to create the NIR security ink. The plain white paper was scribbled on with NIR security ink. A 254 nm UV laser was used to irradiate the NIR security ink, activating it.

The powder particles of the epoxy were joined together by a binder. Epoxy and binder were combined in a 2:1 ratio. For the development of NIR security ink, 30 weight percent mixture of the long-lasting $Zn_3Ga_2Ge_2O_{10}$ phosphor powder was added to the epoxy solutions. To create optically rewritable paper for security encryption and decryption, the same security ink was used. A sheet of plain white paper was cast with security ink, which was then gradually spread across the sheet using a glass rod and allowed to dry overnight. Since imaging instruments that detect NIR light are the only ones that can read the writing on the paper. The letter was written by optical radiation on the screen using a green screen-printing technique. Using a mask on the paper, the letters "IIT" were encrypted before being subjected to a 365 nm UV light. The Cr^{3+} doped $Zn_3Ga_2Ge_2O_{10}$ phosphor, which exhibits near infrared long persistence luminescence, was activated by the light. Night vision sensors that detect NIR light were used to decipher the texts. Even after a few days, the decryption procedure can be completed. The luminous paper was ready for reuse after the encrypted letter was decrypted by exposing it once more to 365 nm UV light for 10 s. In this way the prepared Cr^{3+} doped $Zn_3Ga_2Ge_2O_{10}$ afterglow phosphor made suitable for the applications of night vision surveillance and anti-counterfeiting.

Zhang et al. [113] prepared $Ca_3Al_2Ge_3O_{12}:Mn^{2+}$, Cr^{3+} phosphors by solid-state technique. The phosphors have been irradiated with 254 nm and 365 nm excitations. Upon 265 nm excitation the green emission found at 568 nm due to ${}^4T_{1(G)} \rightarrow {}^6A_{1(S)}$ Mn^{2+} ion. By varying the concentration of Cr^{3+} ions there was a change in emission color has been observed. With the variation of external temperature, the color emission varies from green to orange, to deep red. The afterglow emission shows that the phosphors can be utilized in industrial inspection and anti-counterfeiting measures.

The industrial inspection markings were created using metal 3D printing design security and screen printing. The polydimethylsiloxane (PDMS) and powdered sample (1.3: 1 by volume by weight) were blended by a metal 3D printing method for later application in a plastic mold. The powdered material was mixed with a transparent screen-printing ink (commercial PVC), stirred, and then subjected to an ultrasonic treatment to create the luminous ink for screen printing. Using the Huawei Mobile Mate40 Pro phone camera, the ink was uniformly applied to the tires of the Mercedes AMG-ON W08 EQ Power racing model and the chip plating of an LM.R70.S1 W motherboard. In order to take multicolor pictures while being exposed to radiation, a near-infrared camera has been used. Figure 13 shows the multimode luminescent image demonstration for information encryption and anti-counterfeiting using the prepared printing ink.

Wang et al. [114] reported $Sr_3Ga_4O_9:Tb^{3+}$ persistence phosphors developed by solid state reaction method for anti-counterfeiting application. The afterglow exhibits for 7.5 h, due to the $^5D_4 \rightarrow {}^7F_5$ transition from Tb^{3+} ions upon UV excitation. The afterglow has been tested on the letters 'HUNAU' under 254 nm and 365 nm excitations.

Pei et al. [115] developed $NaCa_2GeO_4F:Tb^{3+}$ multicolor phosphor synthesized by solid-state reaction method. The color tunability varies from blue to cyan and

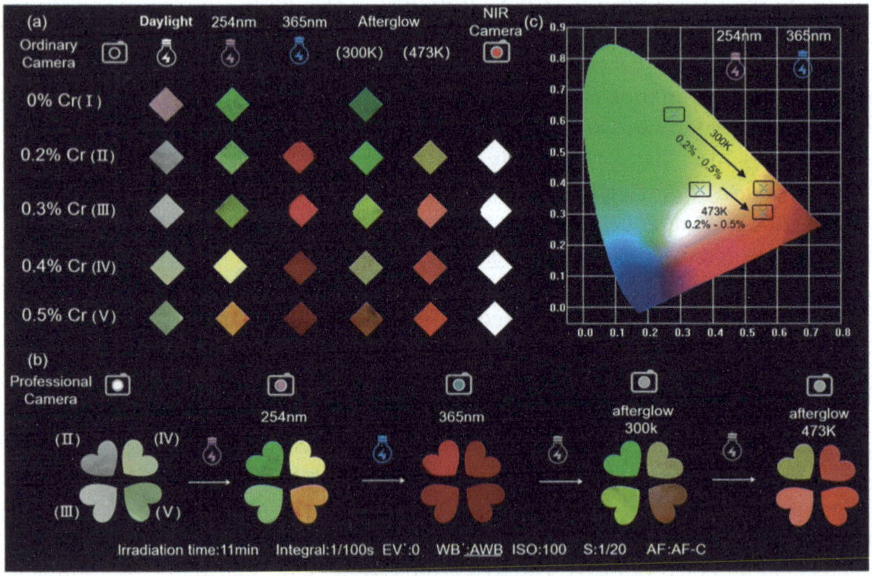

Fig. 13 **a** CAG: xMn^{2+}, yCr^{3+} (x = 2%, y = 0–0.5%) powder irradiated with a UV lamp, digital photographs at different excitation wavelengths and temperatures under an ordinary camera and near infrared shot with an infrared camera. **b** Image of a four-leaf clover composed of CAG:2%Mn^{2+}, yCr^{3+} (y = 0.2%-0.5%) phosphors taken with a professional camera. **c** CIE chromaticity coordinates of samples at different concentrations and temperatures. (*Source* Zhang et al. [113]. Reprinted with permission of Inorganic Chemistry Frontiers)

bright green by changing the concentration of Tb^{3+} ions. At different concentration of Tb^{3+} ions say 0.11%, 0.4% and 0.8% the afterglow exhibit for 5.62 h, 8.52 h and 7.14 h, respectively. The afterglow phenomenon is responsible due the suitable traps and cross-relaxation from Tb^{3+} ions. To create a viscous phosphor ink, polydimethylsiloxane (PDMS, a curing agent in a 5: 1 ratio), was uniformly mixed with $NaCa_2GeO_4F:Tb^{3+}$ phosphor in a weight ratio of 1: 1. For the testing of developed ink, typical anticounterfeiting design on banknotes '100' has been made and excited under 254 nm irradiation.

6.1 Cause of Dynamic Anti-Counterfeiting

The persistent luminescent materials are used for anti-counterfeiting application in both the static and dynamic ways [27, 116]. In the static anti-counterfeiting phosphors monochromatic afterglow is obtained. However, dynamic anti-counterfeiting phosphors can be achieved by adjusting the concentration of doping ions, types of codopants, trap depths etc. The concentration quenching assisted by cross relaxation with the increase in dopants concentration is the basic phenomenon behind the admirable change in emission color. Here, the use of persL phosphors for different dynamic anticounterfeiting methods has been discussed.

Time-dependent dynamic multicolor afterglow has been reported by Zhang et al. [117]. The developed $LiGa_5O_8$: Eu^{3+}/Tb^{3+} phosphors by solid-state reaction method has been found for anticounterfeiting and encryption with high security levels. The phosphors excited with 254 nm/365 nm wavelength show emission from blue to red/ green. Once they stopped excitation, the phosphors showed multicolor emissions. The available deep and shallow traps and electronic transitions from phosphors are responsible for the dynamic multicolor emissions. The long phosphorescence from the phosphors has been applied to flower patterns, number systems, QR codes, and cipher diagrams.

Yang et al. [118] developed multimode persistence luminescence phosphors for dynamic anticounterfeiting and encryption. The Mg_2GeO_4: Mn^{2+}, Yb^{3+} phosphors have been synthesized by solid-state reaction method. The emission has been tuned from red ~650 nm/NIR ~735 nm to pure NIR region by changing the quantity of Mn^{2+} ions. With the incorporation of Yb^{3+} ions, suitable traps are introduced. The phosphors have been excited with 254 nm for 5 min. The emission intensity has been improved by further exciting with 980 nm wavelength. The letters 'SCUT' have been written for dynamic anticounterfeiting and encryption applications. The white '8888' encrypted information turned "red" upon being excited by UV light (254 nm) and the actual information was not visible. The useful information '6006' can be read after the UV light has been switched off. Additionally, '6006' changed into the final piece of information 'LOVE' after 5 min, which could be seen with a camera or the human eye. These details also suggest that these phosphors can be applied to anti-counterfeiting or time-resolved information encryption.

Gao et al. [119] developed multi-mode, multi-color dynamic anti-counterfeiting $Zn_3Ga_2GeO_8$: Cr^{3+}, Yb^{3+}, Er^{3+}, and $Zn_{1.6}Li_{0.4}GeO_4$: Mn^{2+} persistent phosphors. The phosphors have been synthesized by the hydrothermal method and found with cubic and rhombohedral structures, respectively. The phosphors have been excited with three different irradiations 365 nm, 808 nm, and 980 nm, and luminescence in four modes {photoluminescence (PL), persistent luminescence (PersL), photostimulated luminescence (PSL), photostimulated persistent luminescence (PSPL)} have been obtained. Using these multi-mode persistent phosphors, a "dragonfly on-lotus" pattern has been printed that exhibits the dynamic growth of emission colors and brightness under various luminescence modes. For the development of anti-counterfeiting patterns and inks, a modified process has been used. The persL phosphor was typically combined with 0.2 M HCl solution and 1 mL of anhydrous ethanol to scatter it. After centrifuging, anhydrous ethanol and deionized water were used to wash the suspension. The finished materials were then combined with PVA aqueous solution to create anti-counterfeiting inks. Screen printing was used to imprint the 'dragonfly-on-lotus' anti-counterfeiting designs on the paper.

Dubey et al. [120] reported bi-luminescent security ink using $Eu(TTA)_3Phen$ and $CsPbBr_3$ nanocrystals, which provide multi-level anti-counterfeiting strategies. The dynamic anticounterfeiting has been achieved by varying the concentration of $Eu(TTA)_3Phen$ and $CsPbBr_3$ nanocrystals. The obtained dynamic anticounterfeiting is associated with defect traps. The material has been excited with 367 nm and 450 nm irradiation. Upon 367 nm excitation, emission was found at 610 nm from Eu^{3+} ions.

Again, Zhang et al. [121] explained the dynamic anticounterfeiting in $Ca_3Al_2Ge_3O_{12}$:Cr^{3+} phosphors. Upon 467 nm excitation a broad band peaking at 722 nm occurred from Cr^{3+} ions. The phosphors has been used to design multi-layer information encryption methods such as digital tube, Moss code, QR code, bar-code, school celebration pattern, and love letter information encryption.

Dynamic anti-counterfeiting has been reported by Tian et al. [27] in $LiTaO_3$: Pr^{3+} phosphors. The color-tunable has been achieved by varying the concentration of Pr^{3+} ions. The shallow traps (0.668–0.8 eV) and deep traps (0.916 eV) are available in the prepared phosphors, which are responsible in the process of charge carriers being trapped and released. QR codes have been generated with the phosphors excited with 254 nm, there is a change in color from dark orange-red to yellow and the afterglow exhibits for 600 s. For the development of quad-mode QR code technology a finely engraved substrate of QR-code was created using 3D printing. Then, the phosphor@PDMS was poured into the grooves of the mold. The final device is obtained after 4 h of curing as given in Fig. 14 [27]. Similar works have also been reported in the articles by Zhang et al. [122] and Wang et al. [123], in which a persistent multicolored luminous phosphor $Ca_3Al_2Ge_3O_{12}$: Re^{3+} ($Re^{3+} = Tb^{3+}$, Pr^{3+}, Dy^{3+}) and β-$NaYF_4$: RE^{3+} (RE = Sm, Tb, Dy, Pr) have been developed, respectively for dynamic anti-counterfeiting application.

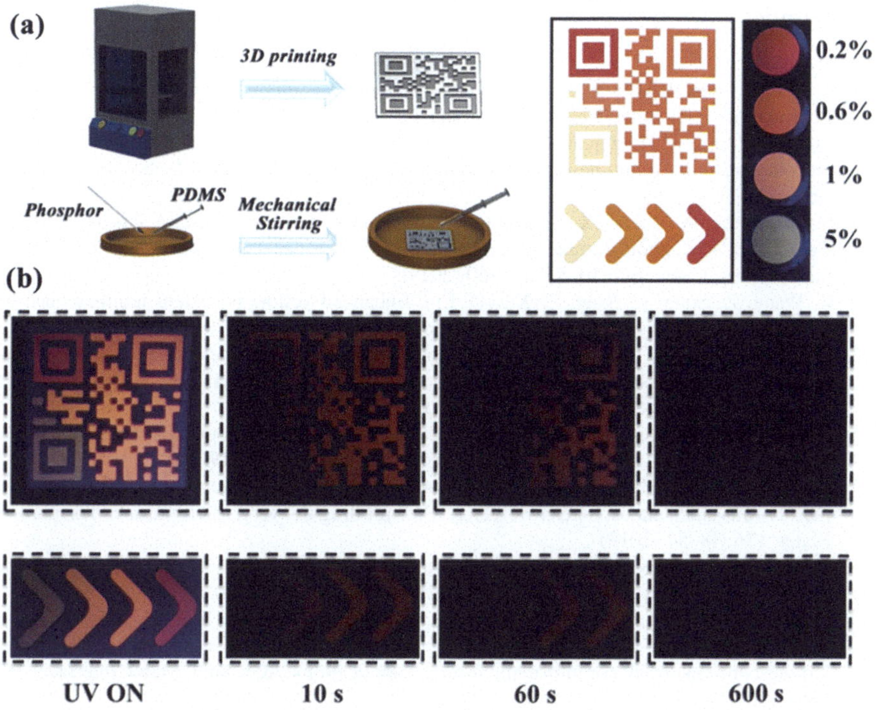

Fig. 14 **a** The preparation flow chart of the PersL dynamic anti-counterfeiting device. **b** The persistent quick response code (QR code) of $LiTaO_3$: Pr^{3+} images. (*Source* Tian et al. [27]. Reprinted with permission of Journal of Alloys and Compounds)

7 Conclusion and Future Perspective

The properties of persistence phosphors have been explained and the effect of exposure time, trap depths, decay time, etc. on the emission of persistence luminescence is seen in various works. The colorful persistence phosphorescence has been achieved with different lanthanides and transition metals doping, varying doping concentration, and irradiation wavelengths. The diverse afterglow organic and inorganic phosphors have been discussed in security, information protection, data encryption, latent fingerprinting, lip printing, and anti-counterfeiting applications. The advancement of security ink and the techniques of several securities and anti-counterfeiting patterns have been incorporated, which can be progressed with the manufacturing industries.

References

1. S. Wu, Z. Pan, R. Chen, X. Liu, *Long Afterglow Phosphorescent Materials* (Springer International Publishing, Cham, Switzerland, 2017)
2. Q. Ye, J. Zhang, Y. Hongquan, S. Jiashi, L. Xiangping, L. Cheng, C. Baojiu, Long persistent and photostimulated luminescence properties of $Sr_2Al_2SiO_7$: Eu^{2+}/Tm^{3+} phosphors. J. Rare Earths **34**, 1–6 (2016)
3. Y.Gao, R. Li, W. Zheng, X. Shang, J. Wei, M. Zhang, J. Xu, W. You, Z. Chen, X. Chen, Broadband NIR photostimulated luminescence nanoprobes based on CaS: Eu^{2+}, Sm^{3+} nanocrystals. Chem. Sci. **10**, 5452–5460 (2019)
4. Z. Wang, Z. Song, L. Ning, Z. Xia, Q. Liu, Enhanced yellow persistent luminescence in Sr_3SiO_5: Eu^{2+} through Ge incorporation. Inorg. Chem. **58**, 8694–8701 (2019)
5. T. Lyu, P. Dorenbos, Towards information storage by designing both electron and hole detrapping processes in bismuth and lanthanide-doped LiRE (Si, Ge)O_4 (RE = Y, Lu) with high charge carrier storage capacity. Chem. Eng. J. **400**, 124776 (2020)
6. A. Bessière, S. Jacquart, K. Priolkar, A. Lecointre, B. Viana, D. Gourier, $ZnGa_2O_4$: Cr^{3+}: a new red long-lasting phosphor with high brightness. Opt. Express **19**, 10131–10137 (2011)
7. Z.J. Li, Y.J. Zhang, H.W. Zhang, H.X. Fu, Long-lasting phosphorescence functionalization of mesoporous silica nanospheres by $CaTiO_3$: Pr^{3+} for drug delivery. Microporous Mesoporous Mater. **176**, 48–54 (2013)
8. S. Zhang, Z. Song, S. Wang, Z. Wang, F. Wang, Q. Liu, Red persistent and photostimulable phosphor $SrLiAl_3N_4$: Eu^{2+}. J. Mater. Chem. C **8**, 4956–4964 (2020)
9. Y. Zhuang, Y. Lv, L. Wang, W. Chen, T.L. Zhou, T. Takeda, N. Hirosaki, R.J. Xie, Trap depth engineering of $SrSi_2O_2N_2$: Ln^{2+}, Ln^{3+} (Ln^{2+} = Yb, Eu; Ln^{3+} = Dy, Ho, Er) persistent luminescence materials for information storage applications. ACS Appl. Mater. Interfaces **10**, 1854–1864 (2018)
10. Z. Zhou, Y. Li, M. Peng, Near-infrared persistent phosphors: synthesis, design, and applications. Chem. Eng. J. **399**, 125688 (2020)
11. M. Yamaga, Y. Ohsumi, T. Nakayama, T.P. Han, Persistent phosphorescence in Ce-doped Lu_2SiO_5. Opt. Mater. Express **2**, 413–419 (2012)
12. L. Wang, Z. Shang, M. Shi, P. Cao, B. Yang, J. Zou, Preparing and testing the reliability of long-afterglow $SrAl_2O_4$: Eu^{2+}, Dy^{3+} phosphor flexible films for temperature sensing. RSC Adv. **10**, 11418–11425 (2020)
13. J. Kong, W. Zheng, Y. Liu, R. Li, E. Ma, H. Zhu, X. Chen, Persistent luminescence from Eu^{3+} in SnO_2 nanoparticles. Nanoscale **7**, 11048–11054 (2015)
14. S. Lin, C. Xiong, D. Ma, H. Li, S. Long, B. Wang, Persistent luminescence found in Mg^{2+} and Pr^{3+} co-doped $LiNbO_3$ single crystal. J. Mater. Chem. C **6**, 10067–10072 (2018)
15. S. Tian, L. Zhao, W. Chen, Z. Liu, X. Fan, Q. Min, H. Yu, X. Yu, J. Qiu, X. Xu, Abnormal photo-stimulated luminescence in $Ba_2Ga_2GeO_7$: Tb^{3+}, Bi^{3+}. J. Lumin. **202**, 414–419 (2018)
16. Z. Gong, W. Zheng, Y. Gao, P. Huang, D. Tu, R. Li, J. Wei, W. Zhang, Y. Zhang, X. Chen, Full-spectrum persistent luminescence tuning using all-inorganic perovskite quantum dots. Angewandte Chemie **131**, 7017–7021 (2019)
17. D. Zhou, Z. Wang, Z. Song, F. Wang, S. Zhang, Q. Liu, Enhanced persistence properties through modifying the trap depth and density in $Y_3Al_2Ga_3O_{12}$: Ce^{3+}, Yb^{3+} phosphor by Co-doping B^{3+}. Inorg. Chem. **58**, 1684–1689 (2019)
18. H. Guo, Y. Wang, G. Li, P. Feng, D. Liu, Q. Ye, Activating persistent luminescence and thermal stability of $K_2Ba_7Si_{16}O_{40}$: Eu^{2+} via traps modulation. J. Alloy. Compd. **801**, 295–301 (2019)
19. V. Castaing, A.D. Sontakke, A.J. Fernández-Carrión, N. Touati, L. Binet, M. Allix, D. Gourier, B. Viana, Persistent luminescence of $ZnGa_2O_4$: Cr^{3+} transparent glass ceramics: effects of excitation wavelength and excitation power. Eur. J. Inorg. Chem. **2017**, 5114–5120 (2017)
20. H. Lin, G. Bai, T. Yu, M.K. Tsang, Q. Zhang, J. Hao, Site occupancy and near-infrared luminescence in $Ca_3Ga_2Ge_3O_{12}$: Cr^{3+} persistent phosphor. Adv. Opt. Mater. **5**, 1700227 (2017)

21. Y. Li, Y.Y. Li, K. Sharafudeen, G.P. Dong, S.F. Zhou, Z.J. Ma, M.Y. Peng, J.R. Qiu, A strategy for developing near infrared long-persistent phosphors: taking $MAlO_3$: Mn^{4+}, Ge^{4+}(M = La, Gd) as an example. J. Mater. Chem. C **2**, 2019–2027 (2014)
22. T. Hu, H. Lin, J. Xu, B. Wang, J. Wang, Y. Wang, Color-tunable persistent luminescence in oxyfluoride glass and glass ceramic containing Mn^{2+}: α-Zn_2SiO_4 nanocrystals. J. Mater. Chem. C **5**, 1479–1487 (2017)
23. Q. Ye, Y. Wang, H. Guo, X. Zhou, P. Feng, S. Ding, Designing a novel red to near-infrared persistent phosphor $CaMgGe_2O_6$: Mn^{2+}, Sm^{3+} based on a vacuum referred binding energy diagram. Dalton Trans. **48**, 11052–11062 (2019)
24. W. Wang, S. Yan, Y. Liang, D. Chen, F. Wang, J. Liu, Y. Zhang, K. Sun, D. Tang, A red-light-chargeable near infrared $MgGeO_3$: Mn^{2+}, Yb^{3+} persistent phosphor for bioimaging and optical information storage applications. Inorg. Chem. Front. **8**, 5149–5157 (2021)
25. K. Zhu, Z. Chen, Y. Wang, H. Liu, Y. Niu, X. Yi, Y. Wang, X. Yuan, G. Liu, (M, Ca) $AlSiN_3$: Eu^{2+} (M = Sr, Mg) long persistent phosphors prepared by combustion synthesis and applications in displays and optical information storage. J. Lumin. **252**, 119288 (2022)
26. Y. Lin, Z. Tang, Z. Zhang, X. Wang, J. Zhang, Preparation of a new long afterglow blue-emitting $Sr_2MgSi_2O_7$-based photoluminescent phosphor. J. Mater. Sci. Lett. **20**, 1505–1506 (2001)
27. S. Tian, P. Feng, S. Ding, Y. Wang, Y. Wang, A color-tunable persistent luminescence material $LiTaO_3$: Pr^{3+} for dynamic anti-counterfeiting. J. Alloy. Compd. **899**, 163325 (2022)
28. B. Qu, B. Zhang, L. Wang, R. Zhou, X.C. Zeng, Mechanistic study of the persistent luminescence of $CaAl_2O_4$: Eu, Nd. Chem. Mater. **27**, 2195–2202 (2015)
29. Z. Hong, P. Zhang, X. Fan, M. Wang, Eu^{3+} red long afterglow in Y_2O_2S: Ti, Eu phosphor through afterglow energy transfer. J. Lumin. **124**, 127–132 (2007)
30. R. Chen, Y. Guan, H. Wang, Y. Zhu, X. Tan, P. Wang, X. Wang, X. Fan, H.L. Xie, Organic persistent luminescent materials: ultralong room-temperature phosphorescence and multicolor-tunable afterglow. ACS Appl. Mater. Interfaces **13**, 41131–41139 (2021)
31. S. Wu, S. Zhang, Y. Liu, J. Yang, The organic ligands coordinated long afterglow phosphor. Mater. Lett. **61**, 3185–3188 (2007)
32. P. Wang, D. Zheng, S. Liu, M. Luo, J. Li, S. Shen, S. Li, L. Zhu, Z. Chen, Producing long afterglow by cellulose confinement effect: a wood-inspired design for sustainable phosphorescent materials. Carbon **171**, 946–952 (2021)
33. X. Chen, C. Xu, T. Wang, C. Zhou, J. Du, Z. Wang, H. Xu, T. Xie, G. Bi, J. Jiang, G. Zhang, Versatile room-temperature-phosphorescent materials prepared from N-substituted naphthalimides: emission enhancement and chemical conjugation. Angewandte Chemie Int. Edn. **55**, 9872–9876 (2016)
34. J. Tan, J. Zhang, W. Li, L. Zhang, D. Yue, Synthesis of amphiphilic carbon quantum dots with phosphorescence properties and their multifunctional applications. J. Mater. Chem. C **4**, 10146–10153 (2016)
35. X. Yang, D. Yan, Long-afterglow metal–organic frameworks: reversible guest-induced phosphorescence tunability. Chem. Sci. **7**, 4519–4526 (2016)
36. Y. Liu, B. Lei, C. Shi, Luminescent properties of a white afterglow phosphor $CdSiO_3$: Dy^{3+}. Chem. Mater. **17**, 2108–2113 (2005)
37. R.P. Wei, Z.H. Ju, J.X. Ma, D. Zhang, Z.P. Zang, W.S. Liu, A novel white afterglow phosphorescent phosphor $Ca_3SnSi_2O_9$: Dy^{3+}. J. Alloy. Compd. **486**, L17–L20 (2009)
38. J. Xu, S. Tanabe, Persistent luminescence instead of phosphorescence: history, mechanism, and perspective. J. Lumin. **205**, 581–620 (2019)
39. Y. Li, M. Gecevicius, J. Qiu, Long persistent phosphors—from fundamentals to applications. Chem. Soc. Rev. **45**, 2090–2136 (2016)
40. L. Yuan, Y. Jin, Y. Su, H. Wu, Y. Hu, S. Yang, Optically stimulated luminescence phosphors: principles, applications, and prospects. Laser Photonics Rev. **14**, 2000123 (2020)
41. Z. Wang, Z. Song, L. Ning, Q. Liu, Sunlight-activated yellow long persistent luminescence from Nb-doped Sr_3SiO_5: Eu^{2+} for warm-color mark applications. J. Mater. Chem. C **8**, 1143–1150 (2020)

42. K. Van den Eeckhout, A.J. Bos, D. Poelman, P.F. Smet, Revealing trap depth distributions in persistent phosphors. Phys. Rev. B **87**, 045126 (2013)
43. C. Richard, B. Viana, Persistent X-ray-activated phosphors: mechanisms and applications. Light: Sci. Appl. **11**, 123 (2022)
44. C. Zheng, Q. Liu, Luminescent properties of a new cyan long afterglow phosphor $CaSnO_3$: Lu^{3+}. RSC Adv. **9**, 33596–33601 (2019)
45. T. Jiang, H. Wang, M. Xing, Y. Fu, Y. Peng, X. Luo, Luminescence decay evaluation of long-afterglow phosphors. Physica B **450**, 94–98 (2014)
46. J. Wang, H. Zhang, B. Lei, H. Dong, H. Zhang, Y. Liu, N. Lai, Y. Fang, Z. Chen, Red persistent and photo-stimulated luminescence properties of $SrCaSi_5N_8$: Eu^{2+}, Tm^{3+} solid solution. Opt. Mater. **36**, 1855–1858 (2014)
47. J. Wang, H. Zhang, B. Lei, H. Dong, H. Zhang, Y. Liu, M. Zheng, Y. Xiao, Optical energy storage properties of $(Ca_{1-x}Sr_x)_2Si_5N_8$: Eu^{2+}, Tm^{3+} solid solutions. J. Am. Ceramic Soc. **98**, 1823–1828 (2015)
48. Y. Zhuang, L. Wang, Y. Lv, T.L. Zhou, R.J. Xie, Optical data storage and multicolor emission readout on flexible films using deep-trap persistent luminescence materials. Adv. Func. Mater. **28**, 1705769 (2018)
49. T. Matsuzawa, Y. Aoki, N. Takeuchi, Y. Murayama, A new long phosphorescent phosphor with high brightness, $SrAl_2O_4$: Eu^{2+}, Dy^{3+}. J. Electrochem. Soc. **143**, 2670 (1996)
50. H. Takasaki, S. Tanabe, T. Hanada, Long-lasting afterglow characteristics of Eu, Dy codoped SrO-Al_2O_3 phosphor. J. Ceramic Soc. Jpn. **104**, 322–326 (1996)
51. Y. Lin, Z. Tang, Z. Zhang, Preparation of long-afterglow $Sr_4Al_{14}O_{25}$-based luminescent material and its optical properties. Mater. Lett. **51**, 14–18 (2001)
52. Y. Lin, Z. Tang, Z. Zhang, C.W. Nan, Anomalous luminescence in $Sr_4Al_{14}O_{25}$: Eu, Dy phosphors. Appl. Phys. Lett. **81**, 996–998 (2002)
53. P. Zhang, M. Xu, Z. Zheng, B. Sun, Y. Zhang, Rapid formation of red long afterglow phosphor $Sr_3Al_2O_6$: Eu^{2+}, Dy^{3+} by microwave irradiation. Mater. Sci. Eng., B **136**, 159–164 (2007)
54. P. Zhang, L. Li, M. Xu, L. Liu, The new red luminescent $Sr_3Al_2O_6$: Eu^{2+} phosphor powders synthesized via sol–gel route by microwave-assisted. J. Alloy. Compd. **456**, 216–219 (2008)
55. Z. Ju, R. Wei, J. Zheng, X. Gao, S. Zhang, W. Liu, Synthesis and phosphorescence mechanism of a reddish orange emissive long afterglow phosphor Sm^{3+}-doped Ca_2SnO_4. Appl. Phys. Lett. **98**(12) (2011)
56. P. Boutinaud, L. Sarakha, E. Cavalli, M. Bettinelli, P. Dorenbos, R. Mahiou, About red afterglow in Pr^{3+} doped titanate perovskites. J. Phys. D Appl. Phys. **42**, 045106 (2009)
57. P.T. Diallo, P. Boutinaud, R. Mahiou, J.C. Cousseins, Red luminescence in Pr^{3+}-doped calcium titanates. Physica Status Solidi A **160**, 255–263 (1997)
58. B. Wang, H. Lin, J. Xu, H. Chen, Z. Lin, F. Huang, Y. Wang, Design, preparation, and characterization of a novel red long-persistent perovskite phosphor: $Ca_3Ti_2O_7$: Pr^{3+}. Inorg. Chem. **54**, 11299–11306 (2015)
59. M. Sójka, J.F. Ramalho, C.D. Brites, K. Fiaczyk, L.D. Carlos, E. Zych, Bandgap engineering and excitation energy alteration to manage luminescence thermometer performance. The case of Sr_2(Ge, Si)O_4: Pr^{3+}. Adv. Opt. Mater. **7**, 1901102 (2019)
60. H. Li, R. Pang, Y. Luo, H. Wu, S. Zhang, L. Jiang, D. Li, C. Li, J. Feng, H. Zhang, Commendable Pr^{3+}-activated $Ba_2Ga_2GeO_7$ phosphor with high-brightness white long-persistent luminescence. J. Mater. Chem. C **7**, 6698–6705 (2019)
61. F. Xue, Y. Hu, L. Chen, G. Ju, T. Wang, S. Zhang, M. He, Luminescence properties of a novel greenish-blue emission long persistent phosphor $Sr_3TaAl_3Si_2O_{14}$: Pr^{3+}. Ceram. Int. **42**, 11039–11044 (2016)
62. Y. Wang, P. Feng, S. Ding, S. Tian, Y. Wang, A promising route for developing yellow long persistent luminescence and mechanoluminescence in $CaGa_2O_4$: Pr^{3+}, Li^{+}. Inorg. Chem. Front. **8**, 3748–3759 (2021)
63. Y. Liang, F. Liu, Y. Chen, X. Wang, K. Sun, Z. Pan, Red/near-infrared/short-wave infrared multi-band persistent luminescence in Pr^{3+}-doped persistent phosphors. Dalton Trans. **46**, 11149–11153 (2017)

64. P. Xiong, M. Peng, K. Qin, F. Xu, X. Xu, Visible to near-infrared persistent luminescence and mechanoluminescence from Pr^{3+}-Doped $LiGa_5O_8$ for energy storage and bioimaging. Adv. Opt. Mater. **7**, 1901107 (2019)
65. L.E. Shea, R.K. Datta, J.J. Brown, Photoluminescence of Mn^{2+}-activated $ZnGa_2\ O_4$. J. Electrochem. Soc. **141**, 1950 (1994)
66. S. Parke, Effect of temperature on the fluore-scence of divalent manganese in glasses. J. Phys. Chem. Solids **32**, 669–675 (1971)
67. M. Prasad, V.K. Rai, Frequency upconversion in UCNPs containing transition metal ions. Upconverting nanoparticles: from fundamentals to applications, 141–169 (2022)
68. X.J. Wang, D. Jia, W.M. Yen, Mn^{2+} activated green, yellow, and red long persistent phosphors. J. Lumin. **102**, 34–37 (2003)
69. B. Struve, G. Huber, The effect of the crystal field strength on the optical spectra of Cr^{3+} in gallium garnet laser crystals. Appl. Phys. B **36**, 195–201 (1985)
70. L.S. Forster, The photophysics of chromium (III) complexes. Chem. Rev. **90**, 331–353 (1990)
71. X. Shen, Q. Hu, Y. Jin, C. Shi, M. Ge, Tunable full-color emission phosphors: enhanced security application via a patterned 3-dimensions code. Ceram. Int. **49**, 345–356 (2023)
72. B.K. Gupta, A. Kumar, P. Kumar, J. Dwivedi, G.N. Pandey, G. Kedawat, Probing on green long persistent Eu^{2+}/Dy^{3+} doped $Sr_3SiAl_4O_{11}$ emerging phosphor for security applications. J. Appl. Phys. **117**, 243104 (2015)
73. Y. Zhang, Z. Wang, Y. Su, Y. Zheng, W. Tang, C. Yang, H. Tang, L. Qu, Y. Li, Y. Zhao, Simple vanilla derivatives for long-lived room-temperature polymer phosphorescence as invisible security inks. Research (2021)
74. Y. Lei, W. Dai, J. Guan, S. Guo, F. Ren, Y. Zhou, J. Shi, B. Tong, Z. Cai, J. Zheng, Y. Dong, Wide-range color-tunable organic phosphorescence materials for printable and writable security inks. Angewandte Chemie Int. Edn. **59**, 16054–16060 (2020)
75. H. Bhatia, D. Ray, Use of dimeric excited states of the donors in d^4-a systems for accessing white light emission, afterglow, and invisible security ink. The J. Phys. Chem. C **123**, 22104–22113 (2019)
76. Y. Li, X. Zhou, Q. Yang, Y. Li, W. Li, H. Li, S. Chen, M. Li, Y. Song, Patterned photonic crystals for hiding information. J. Mater. Chem. C **5**, 4621–4628 (2017)
77. H. Wang, M. Li, Z. Yin, T. Zhang, X. Chen, D. Zhou, J. Zhu, W. Xu, H. Cui, H. Song, Remarkable enhancement of upconversion luminescence on cap-Ag/PMMA ordered platform and trademark anticounterfeiting. ACS Appl. Mater. interfaces **9**, 37128–37135 (2017)
78. X. Lai, Q. Ren, F. Vogelbacher, W. E. Sha, X. Hou, X. Yao, Y. Song, M. Li, Bioinspired quasi-3D multiplexed anti-counterfeit imaging via self-assembled and nanoimprinted photonic architectures. Adv. Mater. **34**, 2107243 (2022)
79. M. Prasad, V. Kesarwani, V.K. Rai, Strategies for designing low thermal quenching upconverting temperature sensors. RSC Adv. **13**, 15833–15842 (2023)
80. H. Chen, H. Hu, B. Sun, H. Zhao, Y. Qie, Z. Luo, Y. Pan, W. Chen, L. Lin, K. Yang, T. Guo, F. Li, Dynamic anti-counterfeiting labels with enhanced multi-level information encryption. ACS Appl. Mater. Interfaces **15**, 2104–2111 (2023)
81. Z. Meng, Y. Wu, J. Ren, X. Li, S. Zhang, S. Wu, Upconversion nanoparticle-integrated bilayer inverse opal photonic crystal film for the triple anticounterfeiting. ACS Appl. Mater. Interfaces **14**, 12562–12570 (2022)
82. D. Zhou, D. Liu, W. Xu, X. Chen, Z. Yin, X. Bai, B. Dong, L. Xu, H. Song, Synergistic upconversion enhancement induced by multiple physical effects and an angle-dependent anticounterfeit application. Chem. Mater. **29**, 6799–6809 (2017)
83. Y. Wan, P. Wang, F. Huang, J. Yuan, D. Li, K. Chen, J. Kang, Q. Li, T. Zhang, S. Sun, Z. Qiu, Y. Yao, Bionic optical physical unclonable functions for authentication and encryption. J. Mater. Chem. C **9**, 13200–13208 (2021)
84. J. Liu, Y. Liang, S. Yan, D. Chen, S. Miao, F. Xie, W. Wang, Low-energy visible light excitable Bi^{2+}-activated near-infrared persistent phosphors for information encryption application. J. Lumin. **251**, 119243 (2022)

85. J. Zhao, L. Lei, R. Ye, J. Zhang, X. Zhang, S. Xu, Sunlight activated ultra-stable long persistent luminescence glass ceramic for outdoor information display. J. Adv. Ceramics **11**, 974–983 (2022)
86. K. Zhang, B. Wang, T. Liu, H. An, S. Deng, Z. Hu, Y. Kong, Q. Zeng, Sc dopant induced tailoring of persistent luminescence in $Na_3YSi_3O_9$: Eu^{2+} for information recording. J. Mater. Chem. C **10**, 15967–15980 (2022)
87. X. Shen, Q. Hu, Y. Jin, M. Ge, Long-lived luminescence and photochromic cellulose acetate-based fiber: preparation, characterization, and potential applications. Cellulose **30**, 2181–2195 (2023)
88. L. Liu, S. Peng, Y. Guo, Y. Lin, X. Sun, L. Song, J. Shi, Y. Zhang, Designing X-ray-excited UVC persistent luminescent material via band gap engineering and its application to anti-counterfeiting and information encryption. ACS Appl. Mater. Interfaces **14**, 41215–41224 (2022)
89. X. Han, E. Song, Y. Zhou, T. Hu, Z. Xia, Q. Zhang, Photon upconversion afterglow materials toward visualized information coding/decoding. J. Mater. Chem. C **8**, 3678–3687 (2020)
90. C. Wang, Y. Jin, R. Zhang, L. Yuan, Z. Li, H. Wu, L. Chen, Y. Hu, Tunable ultraviolet-B full-spectrum delayed luminescence of bismuth-activated phosphors for high-secure data encryption and decryption. J. Alloys Compd **902**, 163776 (2022)
91. D. Gao, Q. Kuang, F. Gao, H. Xin, S. Yun, Y. Wang, Achieving opto-responsive multi-mode luminescence in $Zn_{1+x}Ga_{2-2x}Ge_xO_4$: Mn persistent phosphors for advanced anti-counterfeiting and information encryption. Mater. Today Phys. **27**, 100765 (2022)
92. M.M. Abdelhameed, Y. Attia, M. Abdelrahman, T.A. Khattab, Advances in the preparation of uncopyable photoluminescent security inks from long-persistent phosphors for secure documents. Egypt. J. Chem. **65**, 1199–1215 (2022)
93. J. Berry, A.D. Stoney, *History and Development of Fingerprinting Advances in Fingerprint Technology*, 2nd edn. (Boca Raton, FL, CRC Press, 2001)
94. V. Sharma, A. Das, V. Kumar, V. Kumar, K. Verma, H.C. Swart, Combustion synthesis and characterization of blue long lasting phosphor $CaAl_2O_4$: Eu^{2+}, Dy^{3+} and its novel application in latent fingerprint and lip mark detection. Physica B **535**, 149–156 (2018)
95. S. Oden, B.V. Hofsten, Detection of fingerprints by the ninhydrin reaction. Nature **173**, 449–450 (1954)
96. V. Sharma, A. Das, V. Kumar, Eu^{2+}, Dy^{3+} codoped $SrAl_2O_4$ nanocrystalline phosphor for latent fingerprint detection in forensic applications. Mater. Res. Express **3**, 015004 (2016)
97. W. Chen, Y. Song, W. Zhang, R. Deng, Y. Zhuang, R.J. Xie, Time-gated imaging of latent fingerprints with level 3 details achieved by persistent luminescent fluoride nanoparticles. ACS Appl. Mater. Interfaces **14**, 28230–28238 (2022)
98. L. Liu, Z. Zhang, L. Zhang, Y. Zhai, The effectiveness of strong afterglow phosphor powder in the detection of fingermarks. Forensic Sci. Int. **183**, 45–49 (2009)
99. D. Gao, F. Gao, Q. Kuang, X. Zhang, Z. Zhang, Y. Pan, R. Cahi, H. Jiao, Zinc Germanate nanophosphors with persistent luminescence for multi-mode imaging of latent fingerprints. ACS Appl. Nano Mater. **5**, 9929–9939 (2022)
100. L. Liu, F. Xie, D. Xu, C. Tao, H. Xu, S. Zhong, $Zn2SiO_4$: Mn^{2+}, Yb^{3+} long afterglow materials prepared employing Zn-based coordination polymer as precursor: properties, mechanism and application. J. Lumin. **255**, 119601 (2023)
101. A.K. Deisingh, Pharmaceutical counterfeiting. Analyst **130**, 271–279 (2005)
102. P. Kanika Kumar, S. Singh, B.K. Gupta, A novel approach to synthesise a dual-mode luminescent composite pigment for uncloneable high-security codes to combat counterfeiting. Chem.– A Eur. J. **23**, 17144–17151 (2017)
103. A. Mitrokotsa, M.R. Rieback, A.S. Tanenbaum, Classifying RFID attacks and defenses. Inf. Syst. Front. **12**, 491–505 (2010)
104. C. Lin, X. Shen, R. Tang, X. Zou, Multiple images encryption based on Fourier transform hologram. Opt. Commun. **285**, 1023–1028 (2012)
105. M. You, J. Zhong, Y. Hong, Z. Duan, M. Lin, F. Xu, Inkjet printing of upconversion nanoparticles for anti-counterfeit applications. Nanoscale **7**, 4423–4431 (2015)

106. J. Zhao, D. Jin, E.P. Schartner, Y. Lu, Y. Liu, A.V. Zvyagin, L. Zhang, J.M. Dawes, P. Xi, J.A. Piper, T.M. Monro, Single-nanocrystal sensitivity achieved by enhanced upconversion luminescence. Nat. Nanotechnol. **8**, 729–734 (2013)
107. K. Park, K. Jung, S.J. Kwon, H.S. Jang, D. Byun, I.K. Han, H. Ko, Plasmonic nanowire-enhanced upconversion luminescence for anticounterfeit devices. Adv. Func. Mater. **26**, 7836–7846 (2016)
108. A.K. Gangwar, G. Kedawat, G.S. Papanai, B.K. Gupta, Single excitable dual emissive novel luminescent pigment to generate advanced security features for anti-counterfeiting applications. J. Mater. Chem. C **7**, 13867–13877 (2019)
109. X. Zhao, S. Chen, C. Ye, L. Li, Y. Hu, X. Wang, Y. Song, Triplet–triplet annihilation upconversion combined with afterglow phosphors for multi-dimensional anti-counterfeiting and encoding. J. Mater. Chem. C **10**, 12853–12862 (2022)
110. G. Cai, T. Delgado, C. Richard, B. Viana, ZGSO Spinel nanoparticles with dual emission of nir persistent luminescence for anti-counterfeiting applications. Materials **16**, 1132 (2023)
111. P. Ge, S. Chen, Y. Tian, S. Liu, X. Yue, L. Wang, C. Xu, K. Sun, Upconverted persistent luminescent $Zn_3Ga_2SnO_8$: Cr^{3+}, Yb^{3+}, Er^{3+} phosphor for composite anti-counterfeiting ink. Appl. Opt. **61**, 5681–5685 (2022)
112. S. Das, S.K. Sharma, J. Manam, Near infrared emitting Cr^{3+} doped $Zn_3Ga_2Ge_2O_{10}$ long persistent phosphor for night vision surveillance and anti-counterfeit applications. Ceram. Int. **48**, 824–831 (2022)
113. J. Zhang, Z. Wang, X. Huo, Y. Wang, P. Li, Multimodal dynamic color-tunable persistent luminescent phosphor $Ca_3Al_2Ge_3O_{12}$: Mn^{2+}, Cr^{3+} for anti-counterfeiting and industrial inspection. Inorg. Chem. Front. **9**, 6517–6526 (2022)
114. J. Wang, W. Chen, L. Peng, T. Han, C. Liu, Z. Zhou, Q. Qiang, F. Shen, J. Wang, B. Liu, Long persistent luminescence property of green emitting $Sr_3Ga_4O_9$: Tb^{3+} phosphor for anti-counterfeiting application. J. Lumin. **250**, 119066 (2022)
115. P. Pei, P. DuanMu, B. Wang, X. Miao, C. Zhang, W. Liu, An advanced color tunable persistent luminescent $NaCa_2GeO_4F$: Tb^{3+} phosphor for multicolor anti-counterfeiting. Dalton Trans. **50**, 3193–3200 (2021)
116. W. Li, R. Hu, X. Wang, Y. Li, X. Yao, Color-tunable persistent luminescence in $LiTa_3O_8$ based phosphors for dynamic anticounterfeiting. J. Alloy. Compd. **936**, 168034 (2023)
117. P. Zhang, W. Xie, Z. Wang, Z. Lin, X. Huang, Z. Ju, W. Liu, Time-dependent dynamic multicolor afterglow of simple $LiGa_5O_8$: Eu^{3+}/Tb^{3+} particles for advanced anticounterfeiting and encryption. Inorg. Chem. Front. **9**, 4022–4029 (2022)
118. J. Yang, Y. Zhou, H. Ming, E. Song, Q. Zhang, Site-selective occupancy of Mn^{2+} enabling adjustable red/near-infrared multimode luminescence in olivine for dynamic anticounterfeiting and encryption. ACS Appl. Electron. Mater. **4**, 831–841 (2022)
119. D. Gao, J. Gao, F. Gao, Q. Kuang, Y. Pan, Y. Chen, Z. Pan, Quintuple-mode dynamic anti-counterfeiting using multi-mode persistent phosphors. J. Mater. Chem. C **9**, 16634–16644 (2021)
120. C. Dubey, S.K. Singh, A.K. Singh, Halide perovskite nanocrystals and lanthanide complex-based bi-luminescent security ink for multilevel static-dynamic anticounterfeiting. Mater. Res. Bull. **155**, 111977 (2022)
121. J. Zhang, Z. Wang, X. Huo, Y. Wang, Z. Yang, D. Wang, J. Zhao, P. Li, Garnet-structured persistent luminescence phosphor $Ca_3Al_2Ge_3O_{12}$: Cr^{3+} for dynamic anticounterfeiting applications. Dalton Trans. **51**, 12137–12146 (2022)
122. J. Zhang, Z. Wang, X. Huo, Y. Wang, P. Li, A multi-color persistent luminescent phosphor $Ca_3Al_2Ge_3O_{12}$: Re^{3+} ($Re^{3+} = Tb^{3+}$, Pr^{3+}, Dy^{3+}) for dynamic anti-counterfeiting. J. Lumin. **254**, 119518 (2023)
123. B. Wang, Z. Wang, P. Mao, Y. Wang, A multi-color persistent luminescent phosphor β-$NaYF_4$: RE^{3+} (RE = Sm, Tb, Dy, Pr) for dynamic anti-counterfeiting. RSC Adv. **12**, 11534–11542 (2022)

SPRINGER NATURE

GPSR Compliance

The European Union's (EU) General Product Safety Regulation (GPSR) is a set of rules that requires consumer products to be safe and our obligations to ensure this.

If you have any concerns about our products, you can contact us on ProductSafety@springernature.com

In case Publisher is established outside the EU, the EU authorized representative is:

Springer Nature Customer Service Center GmbH
Europaplatz 3
69115 Heidelberg, Germany

The manufacturer's authorised representative in the EU is Springer Nature Customer Service Centre GmbH, Europaplatz 3, 69115 Heidelberg, Germany. If you have any concerns regarding our products, please contact ProductSafety@springernature.com

Printed and bound by CPI Group (UK) Ltd, Croydon, CR0 4YY
01/10/2026
02225859-0013